Forest and Plant Bioproducts Division

AA003100

Core Programming Topic at the 2011 AIChE Annual Meeting

Minneapolis, Minnesota, USA
16 – 21 October 2011

ISBN: 978-1-61839-737-9

Printed from e-media with permission by:

Curran Associates, Inc.
57 Morehouse Lane
Red Hook, NY 12571

Some format issues inherent in the e-media version may also appear in this print version.

Printed by Curran Associates, Inc. (2012)

For permission requests, please contact AIChE
at the address below.

AIChE
3 Park Avenue
New York, NY 10016-5991

Phone: (203) 702-7660
Fax: (203) 775-5177

www.aiche.org

Additional copies of this publication are available from:

Curran Associates, Inc.
57 Morehouse Lane
Red Hook, NY 12571 USA
Phone: 845-758-0400
Fax: 845-758-2634
Email: curran@proceedings.com
Web: www.proceedings.com

TABLE OF CONTENTS

() From Risk Management to Sustainability: in Theory and in Practice

Subhas Sikdar, National Risk Management Research Lab, US EPA, Cincinnati, OH

Environmental risk management for years has allowed interventions, which depended on understanding and quantification of risk. Sustainability now is superseding various earlier risk management approaches such as waste treatment, waste minimization, pollution prevention, and design for environment. Industrial, and government entities at state, city, and municipal levels, have espoused the principles of sustainability in implementing development and growth agenda. Approaches vary but all use criteria, indicators and metrics to determine favorable sustainability outcomes. There is still much subjective judgment in these determinations, despite attempts to quantify such assessments. Of the three sustainability domains, industry struggles with quantifying societal impacts or benefits, especially at the product and process levels. Lack of transparency from one to another company is a significant barrier to researchers for building a uniform framework and acceptable set of metrics for quantifying the sustainability state of products and processes. At the level of the corporation, much public reporting is now available. But that sort of reporting, while it is important for public relations, has little research value. In practice, the technical actions to improve the environmental stewardship for products and processes have moved not far from traditional yet locally effective approaches of risk management. Sustainability being an inherently multidisciplinary idea, only integrative approaches that use life cycle thinking in technical enquiry and data gathering will move us to a better position to judge sustainability of products and processes.

See more of this Session: Plenary Session: 1st Annual World Congress On Sustainable Engineering
See more of this Group/Topical: 1st Annual World Congress on Sustainable Engineering

() Sustainable Forest Resource Biorefineries Vision for the Future

Shri Ramaswamy, Bioproducts and Biosystems Engineering, University of Minnesota, St Paul, MN

The forest products industry has been a globally competitive industry in the US and North America for much of the last century. Abundance of forest resources, high quality infrastructure for its collection and processing at large scale, availability of human and financial capital and accessibility to the markets are the key competitive advantages. This industry has been primarily focused on producing low cost cellulose fiber and fiber products including building materials. The industry's markets are mature and the products have become commodities and as a result, profit margins have dwindled. High investment costs, increasing costs of raw materials, and leveling of operational efficiency and the technology playing field have thrown the industry's future economic viability into question. The industry is responding by looking to expand the the product portfolio to increase revenue streams and profitability. A diverse portfolio of bioenergy and biomaterials products can transform conventional pulp and paper mills into full fledged biorefineries. With significant expertise in the understanding, logistics, handling and processing of lignocellulosic biomass and sustainable forestry practices, the forest industry has much to offer in contributing to the future development and engineering of sustainable biorefineries. In addition to high value cellulose fiber and fiber based products including nano-cellulose composites, the individual constituents of biomass namely hemi-cellulose, lignin, extractives and such can be sustainably extracted and converted to value-added products. In addition to second and third generation biofuels, bioenergy and other bio-based products offer a tremendous renewable products opportunity for the US and much of the world. Thermochemical or biochemical approaches or a combination of the two hold promise for the future. There are still unsolved scientific and engineering challenges in achieving future sustainable biorefineries including recalcitrance of biomass, density of biomass feedstock, low cost biomass pretreatment approaches, separation and purification of sugar and non-sugar streams, bioengineering of fermentative organisms, removal of fermentation inhibitors and purification of product streams, novel bioprocessing approaches suitable for mixture of biomass feedstocks and value added co-product development. The approaches and solutions need to include the local and regional natures of biomass species and sustainability considerations and they can be in combination with existing forest and other biomass based industries.

See more of this Session: Plenary Session: 1st Annual World Congress On Sustainable Engineering
See more of this Group/Topical: 1st Annual World Congress on Sustainable Engineering

() Sustainable Hydrogen, Gasoline, and Diesel Biofuels From Grasses and Crop Residues

Ram B. Gupta, Chemical Engineering, Auburn University, Auburn, AL

Today world is facing two critical problems: (1) high fuel prices, and (2) climatic changes. Both of these are linked to the overdependence on the fossil fuels: petroleum, natural gas, and coal. The transport is almost totally dependent on fossil particularly petroleum based fuels such as gasoline, diesel fuel, liquefied petroleum gas, and natural gas. Of the special concerns are the liquid fuels for use in the automobiles. Hence, there is a widespread recent interest in obtaining liquid fuels from non-fossil sources. The combination of rising oil prices, issues of security, climate instability and pollution, and deepening poverty in rural and agricultural areas, is propelling governments to enact powerful incentives for the use of biomass-based fuels, which is in turn sparking investment.

Production of grain-based ethanol and vegetable-oil-based bio-diesel are being practiced today with difficulties due to the competition with the food supply. Now the industry is looking toward using biomass which does not compete with the food or fiber supplies; however it is facing unsolved technological/scientific difficulties. Cellulosic ethanol involves the use of expensive cellulase enzymes. Biomass can be gasified to produce synthesis gas which can be converted into liquid fuels by Fischer-Tropsch technology. Another exciting technology is hydrothermal process in which biomass components (cellulose, hemi-celluloses, and lignin) are first depolymerized to their respective monomers, followed by hydrogenation to produce liquid fuels. The process is carried out in sub- or supercritical water, which can utilize wet biomass without any need for drying. Since the carbon dioxide released upon burning of the biomass-based fuels is captured-back in producing new biomass, there is no net emission of CO_2. The technology can significantly address both fuel and climatic challenges as today's world has a large surplus of biomass or biomass-producing capacity.

Biomass can also be used to produce hydrogen, which can be directly used as a clean fuel or combined with heavy crude in the upgradation step. The former use can produce petroleum which has part of the energy from biomass, suitable for near term sustainability.

See more of this Session: Plenary Session: 1st Annual World Congress On Sustainable Engineering
See more of this Group/Topical: 1st Annual World Congress on Sustainable Engineering

(90a) Sustainability of Biomass Crop Production

Shijie Liu, Paper and Bioprocess Engineering, SUNY College of Environmental Science and Forestry, Syracuse, NY

Woody biomass is renewable only if sustainable production is imposed. How to evaluate the benefit of different biomass crop species on a sustainability scale is a difficult task. Site dependence and regional diversity are the most important factors, while biomass productivity determines the viability and selection of a biorefinery facility. An optimum and sustainable biomass stand production rate is found to be one with the incremental growth rate at harvest equal to the average overall growth rate. Forest biomass is found to fevorable over algae and other higher cost crops.

See more of this Session: Feedstock, Supply Chain, and Logistics of Forest Biorefineries
See more of this Group/Topical: International Congress on Energy 2011

(90b) Determining and Addressing Feedstock Variability for Thermochemical Biomass Conversion

Jessica L. Olstad[1], Kim Magrini[1], Richard Boardman[2] and Garold L. Gresham[3], (1)National Renewable Energy Laboratory, Golden, CO, (2)Chemical Engineering, Idaho National Laboratory, Idaho Falls, ID, (3)Interfacial Chemistry, Idaho National Laboratory, Idaho Falls, ID

A common and significant problem in quantitative analysis is working with a small sample, taken from tons of material, for characterization and generating a variety of values for a single property. The question that arises from the situation is whether the variability comes from the analysis or the sample itself. In many instances, it is assumed that the analysis holds inherent errors and no thought is given to the inconsistency or variability of the given sample.

The quality of feedstock for any production process is extremely important for consistent performance, and the same can be said about biomass feedstock for green energy processes. Having a uniform biomass feedstock can ensure good quality of the resulting fuel and other chemical products. Biomass feedstocks are naturally different, starting with the fact that feedstocks cover a range of lignin based materials including corn stover, switch grass, wheat straw, as well as hard and soft woods; in addition, the material properties for each of these feedstock categories differ depending on the location of the feedstock, time of year for growth, water availability and soil properties, as well as how it was harvested, processed, and stored. The Idaho National Laboratory (INL) has been tasked with developing uniform feedstocks for use in biomass thermochemical processes. The National Renewable Energy Laboratory (NREL) is developing thermochemical biomass conversion processes to fuels and chemicals. Consequently, INL and NREL have formed a collaboration to address and minimize these issues in the future. INL and NREL will jointly characterize an assortment of biomass feedstocks in an effort to define sample heterogeneity and determine the most effective sampling proceeses that address these variabilites. Analytical techniques to be used include ultimate and proximate analysis, surface area, and particle size analyses. These results, to be discussed, will be tabulated in an online database for thermochemical research.

See more of this Session: Feedstock, Supply Chain, and Logistics of Forest Biorefineries
See more of this Group/Topical: International Congress on Energy 2011

(90c) Variability of Switchgrass, a Dedicated Bioenergy Crop

Paul B. Filson[1], Nicole Labb[1], Samuel W. Jackson[2], Elena Doukhanina[3], Parthiban Radhakrishnan[3] and Joon-Hyun Park[3], (1)Center for Renewable Carbon (Forestry Wildlife and Fisheries), University of Tennessee, Knoxville, TN, (2)Office of Bioenergy Programs, University of Tennessee, Knoxville, TN, (3)Ceres Inc., Thousand Oaks, CA

The utilization of switchgrass as a dedicated energy crop is well documented in the scientific community. However, there has not been any large-scale production of switchgrass as a feedstock for the production of biofuels. In this work, three different varieties of switchgrass were grown on 600 acres of land as a feedstock in the production of bioethanol in Tennessee. Samples of each variety were harvested from the various fields at an interval of eight weeks for three consecutive periods. The variability, quality and suitability of the three different varieties of switchgrass as a feedstock for the production of bioethanol were assessed using near infrared spectroscopy, standard chemical analyses, and other structural properties in a process analytical technology.

See more of this Session: Feedstock, Supply Chain, and Logistics of Forest Biorefineries
See more of this Group/Topical: International Congress on Energy 2011

(90d) Anaerobic Bacterial Treatment of Corn Stover to Prepare a Biorefining Feedstock

James MacLellan, Biosystems and Agricultural Engineering, Michigan State University, East Lansing, MI

Anaerobic Bacterial Treatment of Corn Stover to Prepare a Biorefining Feedstock

James MacLellan, Zhengbo Yue, Robert Kraemer, Wei Liao*

Department of Biosystems and Agricultural Engineering, Michigan State University

East Lansing, Michigan 48823

Abstract

The objective of this study is to apply an anaerobic bacterial consortium to uptake the hemicellulose in corn stover to prepare a relatively cellulose-enriched feedstock for biofuel production. Previous research demonstrated that submerged anaerobic bacterial culture is capable of converting hemicellulose in raw lignocelluloses to methane gas. The in-depth study is needed to optimize the culture, enhance the hemicellulose utilization efficiency, and improve the quality of treated cellulose as a biorefining feedstock. An anaerobic bacterial consortium from animal manure will be screened and enriched as the seed for this study. Two feedstocks of raw corn stover and steam-treated corn stover will be used to compare the effects of structure on the anaerobic culture of hemicellulose uptake. A completely Randomized Design (CRD) on seed-to-feedstock ratio, culture temperature, and culture time will be carried out. Fiber composition, methane production, and microbial consortia will also be analyzed. An optimal condition can then be established based on the quality and quantity of cellulose in the treated corn stover. The ethanol production on the treated corn stover will be fulfilled in order to elucidate the effects of anaerobic bacterial treatment on biofuels production.

*: Corresponding author

Dr. Wei Liao

Department of Biosystems & Agricultural Engineering

Michigan State University

202 A.W. Farrall Hall

East Lansing, MI 48824-1323

Tel: 517-432-7205

Fax: 517-432-2892

E-mail: liaow@msu.edu

See more of this Session: Feedstock, Supply Chain, and Logistics of Forest Biorefineries
See more of this Group/Topical: International Congress on Energy 2011

(90e) Pellets From Pretreated Biomass

M. Toufiq Reza[1], Charles J Coronella[2], Joan G Lynam[2] and Victor R. Vasquez[2], (1)Chemical and Materials Engineering, University of Nevada, Reno, Reno, NV, (2)Chemical Engineering, University of Nevada, Reno, Reno, NV

Biomass supply is complicated by diversity of feedstocks, seasonal availability, and widely distributed feedstock. The costs of biomass supply and logistics hinder commercialization for advanced fuel and power production. Hydrothermal carbonization (HTC, or wet torrefaction) is a pretreatment process for making a homogenized, carbon rich, and energy dense solid fuel, called HTC biochar, from lignocellulosic biomass. Compared to raw biomass, HTC biochar is both more hydrophobic for better storage and more friable for better processing. In this pretreatment method, the biomass is treated with hot compressed water in an inert environment in the temperature range of 200-260°C. The range of mass yield of biochar is 60-90% and its energy yield is 75-95%, indicating that the process causes an energy densification of 6-25%. Higher reaction temperature decreases mass yield and increases energy densification.

Kinetic data show that hemicelluloses and water extractives react very quickly compared to cellulose and that lignin is relatively inert, resulting in a biochar with increased lignin content. Lignin extracted from raw wood by Van-Soest method exhibits a glass transition temperature at about 135-165 ° C. Similarly, lignin extracted from HTC biochar exhibits a glass transition temperature in the same range. Lignin acts as a natural binder for pelletization of HTC biochar, if pelletizing is done in the temperature range of glass transition. A hydraulic press with a controllable heated die was used for producing uniform pellets. The materials used to make pellets were raw loblolly pine and pine pretreated at 200°C, 230°C, and 260°C. The sample was compressed into a 13 mm die to make cylindrical pellets 8-10 mm long. The pressure applied was 1000 MPa while a temperature of 140 ° C was maintained for 30 seconds to allow the lignin showing the glass transition behavior.

Pellets made from HTC biochar exhibit favorable properties resulting from high levels of lignin, including increased abrasion resistance, increased durability, increased modulus of elasticity, increased energy density, and increased mass density. On the other hand, ultimate breaking strength of pellets is decreased with increasing pretreatment temperature. Pellets produced from biomass pretreated at 260°C, have volumetric fuel value 70% greater than pellets produced from untreated pine and 142% greater than the raw biomass. Equilibrium moisture content (EMC) of pellets of HTC biochar pretreated at 260 ° C is about one fourth that of pellets produced from raw biomass. Pellets produced from HTC biochar maintain structural integrity significantly longer than do pellets produced from raw woody biomass.

See more of this Session: Feedstock, Supply Chain, and Logistics of Forest Biorefineries
See more of this Group/Topical: International Congress on Energy 2011

CHANGE OF STRUCTURE OF SWEET SORGHUM BAGASSE AFTER SOLID-STATE FERMENTATION

Ji-hong Li[1], Shi-Zhong Li[1], Zhi-pei Yan[1] , Qing Li[2], Liang-cai Peng[3]

[1]*Institute of New Energy Technology, Tsinghua University, Beijing, China*

[2]*Colledge of Science, Huazhong Agricultural University. Wuhan, China*

[3]*Colledge of Plant Science & Technology, Huazhong Agricultural University, Wuhan, China*

Abstract

Solid–state fermentation technology of sweet sorghum is an efficient, simple and low cost process to produce ethanol. In this study, the structure of sweet sorghum bagasse before and after solid-fermentation was determined by analysis of composition, IR, SEM and XRD. The results indicated that the sweet sorghum bagasse after solid-fermentation (SSBA) has lower crystallized cellulose (15.87%), higher amorphous cellulose (3.69%) and significantly higher xylan (23.32%) than the sweet sorghum bagasse before solid-fermentation (SSBB). And the corresponding composition of SSBB was 22.41%, 3.39% and 14.44% respectively. Moreover, the crysrallinity index and degree of polymerization (DP) of SSBA was also slightly lower than that of SSBB. The image obtain from SEM indicated the surface of SSBA was rougher than that of SSBB. This means yeast destruct the cell wall of sweet sorghum stem to some extent. It is interesting that the composition of lignin of sweet sorghum also changed significantly after solid-fermentation. Total lignin increased after solid-fermentation, while the ratio of S and H unit of lignin decreased. All results revealed that yeast not only utilized the soluble sugars in sweet sorghum stem, but also metabolized some composition of cell wall. So this sweet sorghum bagasse with relative higher xylan could be a potential feedstock for xylose or furfural. It is also a potential feedstock for lignocellulosic ethanol due to its relative feasibly accessible structure. b

(93a) Sustainable Business Operation and Environmental Preservation - the Evolving 21st Century Thinking On the Achievability of Both

Jacqueline Sibblies, Environmental Compliance Consulting, Somerset, NJ

The governmental implementation of environmental protection measures of the 1970's generated passionate debate among stakeholders on the results that these effected procedures could or would make. A large section of public and several public advocacy groups felt that requirements were strict enough on the industry to bring about the kind and degree of operational changes needed to appreciably mitigate the state of environmental deterioration. Members of the industrial sector viewed the environmental protection measures as very restrictive, undermining business profitability and possibly threatening to business survival. The regulatory agencies considered the requirements reasonable and adequate in addressing the then prevailing environmental concerns.

The arguments and diverse perspectives on environmental protection requirements have persisted over the decades, intensifying at pivotal junctures. As scientific research and human health evaluation data revealed increasing and/or new negative environmental effects from industrial operations, governmental agencies have sought to reduce these factors through further regulations, either by amendments of existing regulations or promulgation of new ones. Such moves have left industry with a heightened sense of being subjected to overly burdensome requirements that could not be complied with economically. Some thought that there were requirements that were simply not necessary and that industry was in the best position to devise procedures for controlling practices that negatively impact the environment, should be allowed to develop appropriate countermeasures.

Indeed by the 1980's, the chemical industry embarked on a set of self derived standards of operation known as Responsible Care, developed in Canada. Participation is voluntary and requires that members follow recommended practices of safe handling and disposal of chemicals, monitor these practices and keep records of specified metrics. Later, the European non-governmental organization (NGO), International Organization for Standardization (ISO), developed standards for an environmental management system which included requirements for regulatory compliance.

Despite these efforts, the 1990's brought even more stringent environmental regulations due to demands for a cleaner environment and growing concerns regarding climate change by the public, public advocacy groups and environmental groups. Internationally, consumers began to indicate their preferences for goods and services that incorporate environmental preservation. Members of the industrial sector started to look for ways in which they could fulfill consumer expectations and still maintain acceptable business returns.

During the first decade of the 21st century, various sectors of industry led by prominent leader in those sectors developed operational/business models that have succeeded in sustaining high environmental standards and economic performance. This paper will present discuss the development of the new thinking that generated this industrial paradigm shift and offer a projection into an era of fuller stake holder collaboration towards sustainability.

See more of this Session: Green Engineering and Sustainability In Collegiate Curriculum
See more of this Group/Topical: Environmental Division

(93b) Preparing Students for Careers In Which Sustainability Will Grow In Importance

Larry Erickson, Keith L. Hohn and John R. Schlup, Chemical Engineering, Kansas State University, Manhattan, KS

There is a growing interest in sustainability among students because of the growing importance of renewable energy, sustainable ecosystems, life cycle analysis, and biobased products. Recent experience with a Research Experiences for Undergraduates program on sustainable energy, a sustainability seminar program, new courses for a Biobased Products and Bioenergy Certificate Program, and other sustainability efforts will be described. Distance education processes enable universities to work cooperatively to offer more new courses. Recent progress of the Great Plains Interactive Distance Education Alliance will be presented.

See more of this Session: Green Engineering and Sustainability In Collegiate Curriculum
See more of this Group/Topical: Environmental Division

(93c) Sustainability Education In Progress At New York Institute of Technology

Nada Marie Anid, School of Engineering and Computing Sciences, New York Institute of Technology, Old Westbury, NY

Sustainability education in progress at NYIT.

See more of this Session: Green Engineering and Sustainability In Collegiate Curriculum
See more of this Group/Topical: Environmental Division

(93d) Teaching the Concepts of Life Cycle Assessment Through the Evaluation of Pharmaceutical Syntheses

Mariano J. Savelski, C. Stewart Slater and David Hitchcock, Chemical Engineering, Rowan University, Glassboro, NJ

The concepts of green engineering and life cycle assessment can be integrated into the curriculum using a drug manufacturing example. The pharmaceutical industry has one of the highest waste generation rates per pound of product produced and the highest amount of organic solvents used per pound of product produced for any commercial sector. The solvents result in wastes released into the environment through the drug manufacturing process and throughout the life cycle of the solvent. Students can easily evaluate chemical use and waste reduction that result from new drug synthesis routes, but few understand how these new processing routes impact the overall life cycle emissions from cradle to grave. This paper describes our efforts to educate students in life cycle thinking in making green design decisions in chemical syntheses.

A case study in drug manufacture is used to illustrate how a sustainability analysis is performed to select the design with the lower overall carbon footprint. A tutorial was developed to show how a life cycle assessment is performed in a process and the key issues such as defining the boundaries of the system and calculating the appropriate life cycle inventory data for all chemicals involved in the analysis. The case study can be used in a condensed version for lower level instruction or expanded for upper level process design classes. Students see how changes made in manufacturing lead to environmental impacts well beyond the production plant and define the true carbon footprint of a manufacturing operation.

This project is sponsored by the NSF Engineering Research Center on Structured Organic Particulate Systems.

See more of this Session: Green Engineering and Sustainability In Collegiate Curriculum
See more of this Group/Topical: Environmental Division

Sustainability Initiatives at Several U.S. Institutions of Higher Education

M.P. Sharma, Ph.D.
Chemical and Petroleum Engineering Department
University of Wyoming
Laramie, Wyoming

Matthew Winslett
Engineering Manager, Energy Management
Facilities Management Department
University of Alabama at Birmingham
1530 3rd Avenue South
Birmingham, Alabama 35294-4564

Robert W. Peters, Ph.D., P.E.
Department of Civil, Construction, and Environmental Engineering
University of Alabama at Birmingham
1075 13th Street South
Birmingham, Alabama 35294-4440

Abstract

Recently the American College and University Presidents' Climate Commitment (ACUPCC) released its 2009 Annual Report showing building momentum on comprehensive action towards clean energy and greenhouse gas pollution reductions in higher education. The following samples of observations made by some of the leaders of institutions of higher education point to a trend (perhaps and a new phenomenon) at campuses of institutions of higher education in the nation (and the world): "Colleges and universities must exercise leadership in their communities and throughout society by modeling ways to eliminate global warming emissions, and providing the knowledge and educated graduates to achieve climate neutrality campuses that address the climate challenge by eliminating global warming emissions..... will better serve their students and meet social mandate..." – The signatories of the ACUPCC.

"As the national debate on climate change moves in a positive direction after years of neglect, leaders of higher education are inspiring the rest of society to address the critical challenge of global climate disruption by reducing emissions on campus and by starting by starting to reorient the education of the nation's future professionals." -- Dr. Anthony D. Cortese, one of the ACUPCC's supporting organizations). "More than ever, universities must take leadership roles to address the grand challenges of the twenty-first century, and climate change is paramount amongst these". – Michael Crow, President, Arizona State University.

The authors of this paper have been involved with the Campus Sustainability Programs at their own universities in conducting and studying projects. This paper will present the results and discussions of design, implementation and outcome of campus sustainability projects initiated, completed and underway at several colleges and institutions of higher education including the University of Wyoming (UW) and the University of Alabama at Birmingham (UAB). The vision, the lessons learnt, and impacts obtained from the implementation of these campus sustainability projects will be discussed in the paper. A part of this paper is a work-in-progress. The goal is to compare, contrast, and draw conclusions to learn from different models of "campus sustainability initiatives and projects" that are emerging at different institutions of higher education.

The University of Wyoming (UW) is positioning itself to become more sustainable [1]. Over the last five years, it has made great advances. President of UW created the Campus Sustainability Committee (CSC) in 2005 "to serve in an advisory capacity to all campus departments, colleges and other entities in an effort to advance environmental and economic sustainability at the University of Wyoming." In 2007, the President signed the American College and University Presidents Climate Commitment (ACUPCC), committing UW to become carbon neutral by 2050. In 2009, UW submitted its Climate Action Plan (CAP), UW's guiding document to reducing carbon emissions and becoming more sustainable. This is a review of UW's advances in sustainability in 2010 and serves as a marker for where it currently stands on its path to more sustainable operations.

Since signing the ACUPCC, the University of Wyoming has made considerable progress in its goal. Each fiscal year since the signing of the ACUPCC, the University has conducted a greenhouse gas emissions inventory (GGEI) [2]. The GGEIs show that the University has remained relatively steady in the amount of greenhouse gases (GHG) it emits since the baseline year of 2005. As more efficient campus buildings begin to come online – including the Kendall House, the Berry Center, Visual Arts, and the College of Business – and as the development and use of alternative energy sources become more feasible at UW, emissions relative to the total square footage are expected to decrease more significantly. Creating sustainability awareness among the campus community and beyond, and encouraging behavior that leads to less resource consumption, is also critical to reducing carbon emissions. UW has made great strides to that end, as well. Now on to presenting and discussing selected accomplishments toward sustainability commitments and GHG reductions at UW [3].

Sustainability Initiatives Underway at UW

Accomplishments

The University of Wyoming has taken steps toward reducing institutional greenhouse gas emissions, particularly within the following key areas: energy, facilities, grounds, water, transportation, purchasing and procurement, recycling and waste management, curriculum and research, and behavioral change.

Energy

UW has begun to reduce its energy consumption through a phased approach. Phase I (2009 – 2015) calls for ramping up energy efficiency across campus. Phase II (2015 – 2030) will include the implementation and use of alternative energy. Phase III (2030 – 2050) will continue building on the first two phases and include purchase of carbon offsets where increased efficiency and alternative energy are inappropriate. The following list highlights progress made in 2010 toward energy efficiency.

- Student interns completed the third annual Greenhouse Gas Emissions Inventory in the spring 2010 semester.
- A team has been created as a sub-committee of the CSC to focus on energy efficiency in utilities and facilities.
- Interior lighting systems in several core campus buildings have been upgraded.
- Windows in five core campus buildings were replaced in 2009.
- One biomass test burn in the Central Energy Plant was conducted in 2009 and a second in 2010.
- Phase one of a steam trap upgrade project was completed in summer 2010.

- Phase I of a campus metering program is complete with installation of condensate, water and electric meters for core campus buildings.
- Interior lighting upgrades continue to replace T-12 lamps with T-8 lamps, add lighting controls where appropriate and replace incandescent bulbs with compact fluorescent bulbs.
- Twenty-five vending misers (mechanisms that turn vending machines off during non-peak hours or when outdoor beverage machines do not need to cool due to weather) have been installed with funding from Auxiliary Enterprises. These machines are in the lowest traffic areas on campus so the savings will be the greatest. Also, after analysis of the usage of vending machines, 17 were deemed underperforming and were removed. This program will be expanded to the cardswipe type of vending machines if a demonstration test of the vending miser is successful.
- All light bulbs/ tubes in Washakie Center have been switched from T-12 to energy saving T-8 bulbs as they have needed to be replaced.
- The Residence Hall Association and the Residence Life & Dining Services Sustainability GA piloted a program to reduce electrical consumption by 6,379 kilowatt hours against the previous three year average in the month of February.

Facilities

Through the Facilities Planning office, all new buildings on campus are now constructed to Leadership in Energy & Environmental Design (LEED) standards. More efficient buildings directly affect UW's greenhouse gas emissions and result in a more comfortable and healthy indoor environment for students, faculty and staff. The following highlights UW's commitment to building toward LEED principles.

- All new buildings are being constructed to at least LEED Silvers standards.
- Two buildings will be LEED certified. The College of Business is certified LEED Gold, and the Kendall House is expected to be certified LEED Silver by spring 2011.
- ACRES Farm has a new passive solar shed, which is expected to be LEED certified in 2011.

Grounds

University Grounds Services continues to operate and develop in ways that reduce impact on the campus and community ecosystem. Tree-planting is at the heart of this goal. Trees beautify campus and sequester carbon. In the last two years, the UW Physical Plant has added nearly 200 trees.

- Over 100 trees were planted in several key campus areas in 2009.
- An additional 60 trees were planted in several targeted campus locations in spring 2010.
- More efficient maintenance equipment is being acquired as replacement purchases are made.
- Environmentally friendlier fertilizer and pesticides are being used to maintain campus grounds.

Water

Water conservation involves educating the campus community, reducing water use, and upgrading plumbing fixtures. The first step in water conservation at UW is to better understand use so that education programs can be developed and upgrades can be made in key areas. A metering program is now in place and will guide conservation measures in the coming years.

- Installation of water meters for each major facility on core campus will provide necessary data to track water usage, identify problem areas and recognize usage improvements.
- An irrigation system upgrade is being planned to replace individual controls with a centrally controlled automated system.

- Use of ground water and rain water is being evaluated for feasible applications.

Transportation

Transportation practices and services impact UW's institutional carbon emissions significantly. Facilities Planning, TransPark and Transit Services aim to help reduce UW's emissions through expanded shuttle services, bike and pedestrian routes, and adding hybrid and other alternative fuel vehicles to the fleet.

- The campus shuttle system has been expanded to include express shuttles from a remote south lot and long-term east lot.
- As part of the Long Range Development Plan (LRDP), defined transportation routes for shuttles, vehicles, bicycles and pedestrians have been developed.
- The number of electric fleet vehicles was expanded in 2009 from three to eight.
- A number of fleet vehicles can run on flex fuel including a 20% bio fuel mix.
- Six efficient diesel buses are on order and should be received by mid-2011.

Purchasing and Procurement

From paper and cleaning products to food and food service containers, UW is working to purchase products with post-consumer recycled content and products that are compostable.

- A contract was negotiated and executed to purchase 100% recycled content paper products for restrooms in campus facilities.
- Through the Procurement Office, most departments contract for 30% recycled content paper.
- Dining Services uses green cleaning products for every application where there is a certified green product on the market including: Geo Fusion 9000 Machine Detergent, Geo Fusion Silver Fusion Presoak, Geo Mag Fusion (Pot and Pan detergent), Geo Crystal Fusion Rinse Additive.
- Wherever possible, dining locations use compostable boxes, flatware, cups, plates, bowls, to-go boxes, sushi containers, salad and sandwich containers, and other packaging. The number of compostable/biodegradable products issued in February 2010 was 8,038,800, which approximates an average month.
- Dining Services sourced a supplier and product to be able to serve only cage-free liquid egg mix.
- Dining Services has worked to increase procurement of locally-grown foods when possible. Sources within 200 miles have been found for beef patties, eggs, lamb, and lettuce when in-season. A source for apples has also been found in Western Colorado when in-season. Milk is purchased through a local dairy in Denver, CO.
- Dining Services continues to work with ACRES Farm by purchasing locally grown produce when available.
- Dining Services serves organically-grown, fair-trade coffee in Washakie Dining Center, University Catering, Book and Bean, Elements and Ross Hall.

Recycling and Waste Management

Recycling efforts at UW have advanced significantly over the last year. The improvements can be generally categorized as campus-wide collection efforts and residence hall specific efforts. The following list details the landfill diversion achievements UW has enjoyed in 2010.

- UW participated in Recyclemania in 2009 and 2010 – coordinated by UW Recycling, the Campus Sustainability class and support by the Wyoming Union. In 2010, UW placed 51/267 in the Grand Champion category; 120/346 in the Per Capita Classic category; and 62/199 in the Waste Minimization category.

- The campus recycling operation has been expanded to include additional pickup locations, one additional truck, one additional employee and a drive-through Recycling Drop-Off Center.
- Recycling pickups have been increased from bi-monthly to weekly resulting in increased tonnage collected.
- Twenty cardboard-specific dumpsters have been purchased and placed resulting in significantly increased cardboard collections.
- Recycling efforts in the Athletic venues have been expanded by adding collection points, coordinating pick-ups after events and expanding the Recycling Cyclers Program in the tailgating lots.
- Recycling and composting efforts have been expanded in Washakie Dining Hall.
- Through the collection efforts in the residence halls, a total of 11.5 tons of material collected over the academic year was redirected away from the landfill.
- Three tons of cardboard were collected from the Residence Halls on move-in weekend.
- CSC and Students for Sustainability purchased two pedal pickups (modified three-wheel bicycles) for recycling collection at UW events.
- In May 2010, 2.383 tons of reusable items were collected and donated to local charities during the "When You Move Out, Don't Throw It Out" campaign.
- One project of the ENR 4600/5600 Campus Sustainability Class (spring 2010, 20 students) was a Furniture Round-Up to gather, store, and redistribute unwanted furniture after a semester ends and before a semester starts.
- Summit '10 Welcome Picnic on Prexy's Pasture (August) was a zero waste event, where all waste from the event was either recycled or composted.
- Washakie Dining Center conducted surveys/research on going tray-less within the dining hall and held two tray-less days, one each semester. These tray-less days netted a savings in pulp waste of approximately 300 pounds each day in addition to raising awareness about food waste.
- Dining Services removed drink straws from the Washakie Dining Center.
- Insulated drink containers are given to all students living in the residence halls, in an effort to promote the use of reusable, rather than disposable containers.
- Approximately 70-105 gallons of compostable waste are sent to ACRES farm everyday from Washakie Dining Center, University Catering, and the Wyoming Union Food Court.
- Dining Services worked with the Laramie landfill to test the composting of food waste from the Washakie Dining Center.
- Dining Services worked with the College of Engineering to test a machine that would convert used fryer oil to Bio-Diesel.
- Students from the Center for Volunteer Service assisted in the going tray-less project at Washakie Dining Center.
- The Good Mule Project Conference used compostable dining utensils and compostable materials.
- Summit View Apartments deconstruction project estimates 85% of materials from deconstruction are being recycled (approximately 15,000 yards of materials).

Behavioral Change

Increasing awareness and encouraging changes in behavior among the students, faculty and staff can result in a significant decrease in GHG emissions. The following list highlights some of the lifestyle and educational programming sponsored by the CSC and others last year.

Behavioral Change – lifestyle

- The University of Wyoming ranked among the nation's 286 most environmentally responsible colleges, according to the Green Guide, published by the Princeton Review and the U.S. Green Building Council.
- University of Wyoming received a B on the Sustainable Endowment Institute's green report card. This is an improvement from the 2009 score of B-.
- A Green Cleaning program was implemented in 2010 with a first year reduction of 25% in routine cleaning chemicals.
- Three students were hired as interns to work on CSC efforts - two on the FY 2010 GHG emissions inventory and one on planning for the Wyoming Sustainability Summit.
- The CSC wrote a successful proposal for an AmeriCorps intern position for 2010-2011. A UW student was hired into the position to create and conduct online surveys for students, faculty and staff in 2011.
- Over 100 students helped weatherize homes around Laramie for the 2010 Days of Dialogue Service Day.
- Alternative Spring Break trips were led to the Grand Canyon and to the Everglades where students spent spring break helping preserve the native ecology and wildlife.
- Insulated drink containers were given to all students living in the residence halls, in an effort to promote the use of reusable, rather than disposable containers.
- RHA and the GA of Sustainability awarded the winning residents of the Recycling and Electrical Consumption Challenge with recycled grocery bags.
- RHA created a new executive position, the Sustainability and Community Service Officer, committed to sustainability issues within Residence Life and Dining Services.
- The RHA menu committee worked with Dining Services on a tray-less campaign in Washakie. A trayless day was held November 2 in the Washakie Dining Hall.

Behavioral Change - education and outreach

- A team has been created as a sub-committee of the CSC, which focuses on education and outreach to the campus community.
- A website has been created, and is continually maintained for the CSC, at http://www.uwyo.edu.sustainability/ to make the sustainability vision, goals, reports and information readily available.
- A Campus Sustainability group has been created on WyoWeb, the UW intranet portal, to share files, collaborate on action steps and provide work areas for subcommittees and teams.
- A media relations program was implemented in cooperation with UW Media to publish sustainability-related articles twice per month.
- Development of a CSC logo was unveiled at Earth Day 2010 events. The logo was chosen through a logo contest held in the spring of 2010. The winner was a UW student, chosen from 58 entries and awarded a $300 prize.
- The first Wyoming Sustainability Summit was held April 12-13, 2010 at the UW Conference Center in Laramie. Over 100 people attended the two-day event. The Summit was co-sponsored by the Campus Sustainability Committee, the Haub School of Environment and

Natural Resources, and Residence Life and Dining Services with support from the Argosy Foundation.

- UW's collegiate chapter of the American Marketing Association and Alpha Kappa Psi co-sponsored the 3rd annual Sustainable Business Practices forum in October 2010.
- The Environment and Natural Resources program organized a "Green for Green" 5k on April 17th, 2010. Volunteers held signs along the race route to inform runners of various sustainability tips such as eating locally, properly insulating homes, and using CFL bulbs.
- Academic Affairs and the Office of Research and Economic Development sponsored a public lecture by world-renowned ecologist, E.O. Wilson in September 2010.
- Over 10 members of the CSC attended the Association for the Advancement of Sustainability in Higher Education (AASHE) annual conference in Denver in October. Several committee members gave presentations and participated in panel discussions on topics ranging from implementing a CAP with a campus-wide committee to institutional recycling to teaching a project-oriented sustainability class.
- The CSC participated in the UW Homecoming Parade in October. The CSC entry featured the Physical Plant's electric truck and the Recycling Cylers pedal pickups. Recyclables were collected along the parade route.
- The CSC hosted an informational booth in the Union Breezeway on Campus Sustainability Day (October 20).
- University of Wyoming was presented with the first Channel 5 Sustainability Award by Channel 5 News in Cheyenne. A 30-minute program featured some of UW's achievements in green building and recycling.
- RHA hosted information booths on campus on Campus Sustainability Day (October 20) and Earth Day (April 22nd).
- The GA of Sustainability, with support from RHA, hosted many educational events throughout the school year that educate residents on various topics such as drinking water, electrical usage and real benefits of recycling.

Sustainability Initiatives Underway at UAB

The University of Alabama (UAB) Energy Management Department's function is to ensure the most energy efficient operation of UAB's facilities. UAB's Energy Management Department is responsible for managing all of the energy usage, billing, metering, and various conservation projects for the more than 150 UAB campus and hospital buildings. We are constantly striving to save energy and promote sustainability without sacrificing a comfortable environment throughout the campus. Some of the things we do include customer support regarding utilities, building space surveys and analysis, lighting surveys and analysis, utility billing, usage and rate analysis, monthly utility billing, utility sub-metering, project reviews, track usage and rebated funds from sewer credits, maintain all of UAB's utility records with the use of Energy-CAP database, provide customer requested utility usage reports, and energy conservation and sustainability projects.

<u>Examples of Sustainability Projects at UAB</u>

With the increasing cost of utilities, energy conservation and sustainability projects have been instrumental in achieving ongoing costs reductions campus-wide. Some of the recent projects UAB has accomplished include building equipment setbacks (during unoccupied times), reduced air exchanges of research areas (during unoccupied times), carbon dioxide (CO_2) monitoring of occupied space to control introduced outside air quantities, efficient upgrades of air handling equipment – controllable by building automation systems, installation of variable frequency drives, replacement of water cooled medical/laboratory gas pumping systems, ice makers, and freezers, demand controlled domestic water pumping systems, T-12 to T-8/T-5/LED fluorescent lighting upgrades, installation of occupancy sensed lighting controls, steam trap surveys and system upgrades, utility rate renegotiations, fin water condensation recovery reclaimed to cooling tower, evaporative cooling tower upgrades, vegetative (green) roof installations, hot water temperature setbacks based on outside air temperatures, and boiler combustion control upgrades.

<u>Equipment Setbacks and Reduced Air Exchanges</u>

Unoccupied space conditioning equipment setbacks and reduced air exchanges have by far been the greatest return on investment. Starting in 2007 many of UAB's building were surveyed at intervals around the clock to determine occupancy frequency habits of students and employees. Occupancy data was gathered and analyzed with help from students from the UAB Civil, Construction, and Environmental Engineering Department and resulting analysis determined that most buildings are unoccupied nightly from hours between 10:00 p.m. and 6:00 a.m. as well as most weekends and holidays. Each building's automation system was programmed to shut-off the air handling motors and decentralized chillers during unoccupied time saving btu's and electrical horsepower. If an internal space temperature reaches above 85F in the summer or below 55F in the winter the respective system starts up insuring safe operations. At present, 41 UAB buildings are being setback during unoccupied times generally from 10:00 p.m. until 6:00 a.m. nightly with savings realized at over $1,000,000 per year.

Other unoccupied setbacks include reduced air exchanges of research laboratories. Normally used in lighting control applications, ceiling mounted occupancy sensors have been utilized to signal the building's automation system (BAS) unit to decrease air flows. Normal air exchanges for laboratories are 8–12 per hour depending on the building during occupied times. After 30 minutes of vacancy, the occupancy sensor directs the BAS to lower the number of air exchanges to three per hour. A group effort between UAB's Departments including School of Medicine, Campus Maintenance, Occupational Health and Safety, and Energy Management has set a cost saving precedence with this endeavor saving utility costs in all areas of utilities including electricity, steam, natural gas, and chilled water. At present, two research buildings are utilizing this control strategy with four others planned for the future.

<u>Domestic Water Conservation and Recovery Projects</u>

Domestic water conservation and recovery projects have also been very beneficial in reducing costs at UAB. Compressed air and vacuum pumping systems are found throughout the UAB campus laboratories and hospital settings. These traditionally inefficient pumps require water for compression as well as a cooling source. With escalating water and sewer costs this method of cooling was deemed unfeasible. These projects replace liquid-ring pumps with ones that do not require water for cooling including claw-type vacuum systems and scroll-type compressed air systems. In most applications less electricity is required to do the same work due to increased pump efficiencies. Depending on building type, which determines electrical rates ($0.08 – $0.10/kW-hr), for every horsepower that is saved by a more efficient pump, electrical costs are reduced approximately $400 - 500 per year for continuous duty operations. A constant single-pass through water flow for cooling is especially expensive if sewer costs are also required. The local cost for each constant year-round one gallon per minute cooling water source is almost $7,500 per year including sewer charges. Therefore system paybacks are very short – usually within one to two years. See Table 1 below for details and Figures 1 and 2 for an example typical installation.

Table 1. UAB Air/Vacuum Pump Replacement Savings.

Pumping System Building Location	System Type	Required GPM Cooling Water	Annual Savings Gallons	Annual Savings Water/Sewer	Horse Power Reduction	Annual Savings Electrical
Tinsley-Harrison Research	Vacuum	4.5	2,366,820	$33,750	0	$0
School of Dentistry	Medical Air	4.5	2,366,820	$33,750	50	$20,000
Bevill Biomedical Sciences	Lab Air	3.5	1,840,860	$26,250	5	$2,000
McCallum Health Science	Vacuum	2.5	1,314,900	$18,750	0	$0
Spain-Wallace Hospital	Medical Air	3.5	1,840,860	$26,250	15	$6,000
Kaul Human Genetics	Vacuum	2.5	1,314,900	$18,750	15	$6,000
Biomedical Research II	Vacuum	2.5	1,314,900	$18,750	15	$6,000
TOTALS		23.5	12,360,060	$176,250	100	$40,000

Figure 1. Spain-Wallace Hospital Medical Vacuum Pumping System (before).

Figure 2. Spain-Wallace Hospital Medical Vacuum Pumping System (after).

UAB has been a leader in recovering and utilizing condensate water from air handler coils. Many of UAB buildings incorporate 100% outside air into conditioned space. The harsh and humid Alabama summer weather causes huge amounts of condensation to form from the air hander chilled water coils. Collection and recovery systems were designed and installed to utilize this source of water. As condensate water is gravity drained from air handler chilled water coils, it is intercepted and collected in small basins positioned locally at each air handler. Each local basin incorporates a small level-controlled sump pump sized to transfer collected water to a larger (500 gallon) collection tank. The 500 gallon collection tank employs level controls that pump high-pressure, filtered, 55°F condensate back into the centralized chilled-water return line before it leaves the building. This rise in pressure is sensed blocks away at the central plant and water there is locally relieved from the chilled water return line into the evaporative cooling tower. This process reduces tower makeup water required during the evaporative cooling process. If the small basin sump pump, high pressure return pump, or Central Plant tower relief valve fails, the water safely exits down the local drain found in the building as it has for many years. While this process may have been accomplished before we may indeed be the first to incorporate this manner of utilizing pressure relieved water in our cooling towers from long distances. We presently have two sites collecting fin water condensate since May 2011 including McCallum/Kaul and Zeigler/Tinsley-Harrison Tower (see Table 2 below) with future sites underway including Bevill Biomedical Research, Biomedical Research II, Research Support Building, North Pavilion Hospital, and Shelby Biomedical Research Building yielding a hopeful 3,000,000 gallons per month during the cooling season.

Table 2. UAB Fin Water Recovery Savings.

Fin Water Recovery Location	Month	Gallons	GPM	Value
McCallum/Kaul Genetics	MAY	166,200	3.72	$699.81
	JUN	430,300	9.96	$1,811.85
	JUL	615,604	13.79	$2,592.10
Zeigler/Tinsley-Harrison	MAY	183,700	4.12	$773.50
	JUN	285,100	6.60	$1,200.46
	JUL	410,497	9.20	$1,728.47
TOTALS		2,091,401		$8,806.19

Figure 3. Zeigler/Tinsley-Harrison Fin Water Collection System.

UAB utility usage and costs are on the decline as a result from these many projects. Operating on an October to September fiscal year (FY), FY0910 realized close to a $3 million reduction over the previous fiscal year. It is also worth noting that during this time period Alabama experienced the harshest winter and summer weather on record – all within the same FY. FY1011 looks to be an exceptional year of savings as well. At the half-way mark of the FY1011 fiscal year (October through March) savings compared to the previous year are down $5 million.

Pilot Vegetated Roof on Top of Hulsey Center

UAB implemented a pilot vegetated roof system installed on top of Hulsey Center in July 2008. The roof occupies approximately 14,940 ft^2 (i.e., occupying ~81.2% of the roofing area, or approximately 0.34 acres). Figure 4 shows a photograph of the pilot green roof, and Figure 5 shows several of the sedum plants on the roof. [Prior to installing the pilot green roof, the civil engineering senior design class investigated the loading of a wet vegetative roof on top of Hulsey Center, and found that the current building infrastructure could withstand the loading associated with the vegetative roof. Hulsey Center was originally designed to have two additional floors in the building]. This pilot green roof contains more than 20,000 sedum plants. In the construction of this vegetated roof, the existing roof was removed down to the structural concrete deck. A waterproof roofing membrane was installed directly to the concrete deck. A layer of 3-inches of extruded polystyrene roofing insulation was then applied, to which an 8-oz non-woven geo-textile scrim sheet was applied. Then, 2-ft x 2-ft x 2-in prestressed pavers or brick pavers were installed for design, decoration, and access to the vegetative roof. Then, 3½-inches of light weight engineered soil were applied. Sedum plants were planted at the rate of 150 plants per 100-ft^2. The cost of retrofitting the roof and installing a pilot green roof on top of Hulsey Center was ~$150,000. The usage of natural gas, water, and electricity are lower than that compared to the utility usage prior to installation of the green roof on top of Hulsey Center. Implementation of the vegetated roof system has reduced the utility bills of Hulsey Center by 20–25% compared with the utility bills prior to implementation of the vegetated roof system.

Figure 4. Vegetative roof on top of Hulsey Center (June 2009).

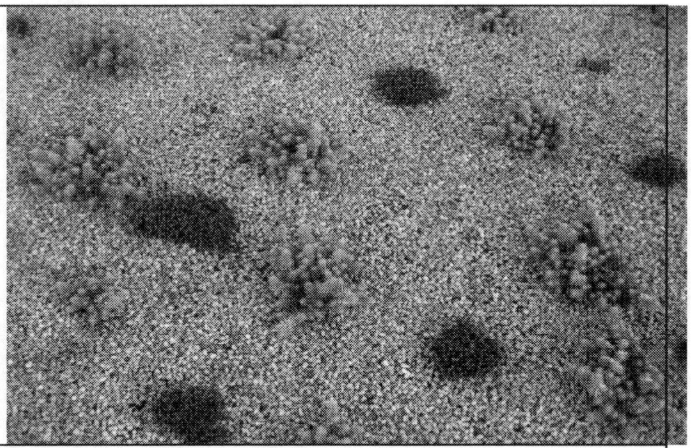

Figure 5. Close-up of vegetative roof on top of Hulsey Center (June 2009).

The UAB Energy Management team is passionate and dedicated to reducing utility usage and costs for the University. Consisting of four engineers and one clerical personnel, everyone is focused on achieving our common goal – *Think Green*, *Save Gold*!

Summary and Conclusions

Over the past eighteen months, the University of Wyoming and the University of Alabama at Birmingham have been successful in advancing campus sustainability in several major areas. The University of Wyoming's particular focus in the areas of energy, facilities, grounds, water, transportation, purchasing and procurement, recycling and waste management, policy, and behavioral change suggest that the UW campus is on its way to achieving its goal of significantly reducing its carbon emissions. In fact, even in light of tremendous growth in square footage (about 400,000 SF since 2006), campus stationary and purchased electricity have remained fairly steady. This is a testament to ramping up energy efficiency in existing buildings and adding more efficient buildings through capital construction projects.

In addition to the University of Wyoming's progress in reducing GHG emissions, the education and outreach efforts show a commitment to encouraging behavior change among the campus community. Approaches that involve students directly and provide them with hands-on experience, such as the greenhouse gas emissions inventory internships and the sustainability GA in Residence Life and Dining Services, conserve staff resources and provide valuable learning opportunities for students.

As illustrated by the activities and initiatives underway at the University of Wyoming and the University of Alabama at Birmingham, campuses are becoming more proactive in the area of sustainability to make the campuses more energy efficient and more environmentally friendly.

Acknowledgements

Sections in this document correspond to different sections in the three references cited below in the List of References. Reference [1] is the main source for use of direct citations of the sections. The authors also acknowledge the technical assistance provided by UAB's Facilities Management Department.

Reference Citations

[1] Campus Sustainability Committee (CSC) Progress Report, University of Wyoming, February 2011.

[2] www.uwyo.edu/sustainability.

[3] Climate Action Plan, University of Wyoming, September 15, 2009.

(93f) Teaching Life Cycle Assessment and Related Concepts

Arunprakash T. Karunanithi, Department of Civil Engineering, University of Colorado Denver, Denver, CO

This presentation shares the authors experience in teaching a graduate course on Life Cycle Assessment based concepts. The presentation will cover various concepts discussed in the course including Life Cycle Assessment, Exergy analysis, Emergy analysis, and Ecological foot printing. Impact assessment tools used for teaching will be discussed. Effective case studies developed for this course will be shared.

See more of this Session: Green Engineering and Sustainability In Collegiate Curriculum
See more of this Group/Topical: Environmental Division

(124a) GREENSCOPE: A Method for Modeling Chemical Process Sustainability

Raymond L. Smith, Gerardo J. Ruiz-Mercado and Michael A. Gonzalez, US EPA, Cincinnati, OH

Current work within the U.S. Environmental Protection Agency's National Risk Management Research Laboratory is focused on the development of a method for modeling chemical process sustainability. The GREENSCOPE methodology, defined for the four bases of Environment, Economics, Efficiency, and Energy, can evaluate processes with over a hundred different indicators. These indicators provide a means for realizing the principles of green chemistry and green engineering in the context of sustainability.

Development of the methodology has centered around three focal points. One is a taxonomy of impacts that describe the indicators and provide absolute scales for their evaluation. The setting of best and worst limits for the indicators allows the user to know the status of the process under study in relation to understood values. Thus, existing or imagined processes can be evaluated according to their relative indicator scores, and process modifications can strive towards realizable targets.

A second area of focus is in advancing definitions of data needs for the many indicators of the taxonomy. Each of the indicators has specific data that is necessary for their calculation. Values needed and data sources have been identified. These needs can be mapped according to the information source (e.g., input stream, output stream, external data, etc.) for each of the bases. The user can visualize data-indicator relationships on the way to choosing selected ones for evaluation.

Finally, the methodology is applied to an example, considering the production of biodiesel with various process parameters. Evaluations and process improvements are shown using the taxonomy and data developments described above. Analyses show where biodiesel processes can be improved and made more sustainable in terms of environmental, economic, efficiency, and energy measures.

See more of this Session: Sustainability Metrics At the Process and Product Level
See more of this Group/Topical: Environmental Division

(124b) Life Cycle Comparison of Lithium Ion Battery Manufacturing Methods

Qiang Dai and Christian M. Lastoskie, Civil & Environmental Engineering, University of Michigan, Ann Arbor, MI

Abstract

Although lithium ion batteries(LIB) are considered to be a promising power storage option for batteryelectric vehicles (BEVs), little work has been done to improve and optimize themanufacturing of LIBs from a life-cycle perspective. In this study, theenvironmental impacts of five commercially available LIBs based on differentcathode chemistries, including $LiCoO_2$, $LiMn_2O_4$,$LiCoMnO_4$, $LiNiMnO_4$ and $LiFePO_4$, are examined toaddress the differences in material flows in the production phase of LIBs. In addition, solid state and sol-gelsynthesis pathways are investigated for each cathode materials to account forthe differences between wet chemistry and dry chemistry, with a particularinterest in process heat input and organic solvent use. The results indicatethat using solid state pathway, $LiMn_2O_4$appears to be the most environment-benign cathode material, followed by $LiFePO_4$. For all of the cathode materials, the environmental burdens of sol-gel pathwayoutweigh that of the solid state synthesis method,with organic solvent as the largest contributor. In both cases, startingmaterials containing nickel and cobalt are responsible for a considerable shareof the environmental impacts pertinent to the manufacturing of LIBs. Acomparison of the environmental impacts of anode active materials and theoperational phase impacts of BEVs relative to internal combustion enginevehicles will also be presented in this talk.

See more of this Session: Sustainability Metrics At the Process and Product Level
See more of this Group/Topical: Environmental Division

(124c) Use of Life Cycle Assessment to Evaluate the Sustainable Manufacture of the Active Pharmaceutical Ingredient Pregabalin

Mariano J. Savelski[1], **C. Stewart Slater**[1], Peter J. Dunn[2], Donald J. Knoechel[3] and Christine M. Visnic[4], (1)Department of Chemical Engineering, Rowan University, Glassboro, NJ, (2)Global Research and Development, Pfizer, Inc., Sandwich, United Kingdom, (3)Global Supply, Pfizer, Inc., Kalamazoo, MI, (4)Global Environment, Health and Safety, Pfizer, Inc., Peapack, NJ

Three different routes to synthesize pregabalin, the active ingredient in the pain relieving medicine Lyrica®, have been analyzed. A life cycle assessment has been conducted to evaluate emissions generated from cradle to gate for the three syntheses. The first process studied was a classical chemical synthesis resolution route. The second manufacturing path involves the use of an enzyme to perform a biocatalytic resolution. The third production path, also a biocatalytic route, integrates the recycle of a wrong enantiomer. The analysis shows how transitioning from an entirely chemical synthesis to an enzyme-based biosynthesis greatly reduces organic solvent use and also the usage of the sponge Nickel catalyst. A life cycle analysis associated with the raw materials used, energy consumed, and waste generated was performed. Life cycle emissions resulting from the raw materials manufacture represent the majority of the carbon footprint for the three processes. Upon using the biocatalytic process, the resulting amount of raw materials, waste and energy, and their respective carbon footprint are significantly reduced. These changes are quantified and their contribution to the life cycle evaluated. For example, the sponge nickel catalyst, is one of the largest contributors on a per mass basis and three times that of the lipolase enzyme, but small when the LCA is performed. Organic solvents are a major contributor to the environmental footprint for the classical route. Since the biocatalytic route is performed mainly in water, the majority of organic solvents are eliminated and their associated life cycle emissions are significantly reduced.

See more of this Session: Sustainability Metrics At the Process and Product Level
See more of this Group/Topical: Environmental Division

(124d) Sustainable Supply Chain Design by the P-Graph Framework

Istvan Heckl, Department Computer Science and Systems Technology, University of Pannonia, Veszprem, Hungary, Heriberto Cabezas, Office of Research and Development, U.S. Environmental Protection Agency, Cincinnati, OH, Ferenc Friedler, Department of Computer Science and Systems Technology, University of Pannonia, Veszprm, Hungary and L. T. Fan, Department of Chemical Engineering, Kansas State University, Manhattan, KS

The present work proposes a computer-aided methodology for designing sustainable supply chains in terms of sustainability metrics by resorting to the P-graph framework. The methodology is an outcome of the collaboration between the Office of Research and Development (ORD) of the U.S. EPA and the research group led by the founders of the P-graph framework. U.S. EPA/ORD has substantial creditable experience with the development of indicators and metrics for sustainability, while the P-graph framework is deemed to be highly effective for algorithmic design of process networks including supply chains. The integration of supply chain design and sustainability is the main focus of the collaboration.

The P-graph framework provides a mathematically rigorous procedure for synthesizing optimal and alternative suboptimal networks subject to multiple objectives and constraints, which include profitability and sustainability in the proposed methodology. Specifically, to evaluate the sustainability of a given process under construction including its supply chain, sustainability metrics are incorporated into the design procedure.

The proposed methodology is demonstrated with the optimal design of a supply chain for the infrastructure providing heat and electric power to an agricultural region with relatively limited area where agricultural wastes can potentially be recovered as renewable resources. The objective functions for optimization comprise the profit, ecological footprint, exergy dissipation, green net regional product, and ratio between renewable and total emergies.

See more of this Session: Sustainability Metrics At the Process and Product Level
See more of this Group/Topical: Environmental Division

(124e) Risk Assessment Tools Use Through Process/Product Life-Cycle: Start with Questions, Not Data !

Joao Machado, Tiago Robalo and **Jose Menezes**, Technology Services & Products, 4Tune Engineering Ltd, Lisbon, Portugal

Risk Assessment is done today in the Pharma and Bio-pharma industries with a wide range of tools (FMEA, Ishikawa diagrams, Process Layouts, etc.) that are scattered across the organization and addressed in different meetings and sometimes by different teams. This arises to many problems on the workflow because the communication between each tool is almost non-existing and in order to keep consistency, all tools must be revisited each time a change is made in one of them. Also, most of these tools are not suitable for brainstorm meetings but only to aggregate the key-points. All of these problems lead to redundant workflow and interruptions in the middle of very productive brainstorm sessions on complex and multidisciplinary subjects.

Therefore, it is crucial to tackle this problem with a solution that holds in itself all the necessary tools for risk analysis and, most important, all data is connected between the tools so there is no wasted time revisiting the project.

Moreover, Risk Assessment is many times done in a brainstorming and intense environment where the ability to transport all the ideas in a quick but, at the same time, accurate and professional way is crucial.

For this, an interactive and user-friendly platform where anyone can contribute is crucial not only to hasten the Risk Assessment process, but actually improve it. The use of such tool will be demonstrated with the implementation of a case study.

See more of this Session: Sustainability Metrics At the Process and Product Level
See more of this Group/Topical: Environmental Division

(124f) Sustainable Energy Bio-Sources, An Environmental Performance Approach: Life Cycle Assessment From WWTP to BESs

Gabriel Blejman[1], Roberta Pacciani[2], Sonia Guri[2] and **Lourdes F. Vega**[2], (1)MATGAS Research Center, Barcelona, Spain, (2)Carburos Metlicos, Air Products Group, Barcelona, Spain

Sustainableenergy bio-sources, an environmental performance approach: life cycleassessment from WWTP to BESs

G. Blejman[1], R.Pacciani[1,2], S. Guri[1,2], and L.F. Vega[1,2]

[1] *MATGAS Research Center, Campus UAB, 08193 Bellaterra,Barcelona, Spain*

[2] *Carburos Metlicos,Air Products Group, C/ Aragn300, 08009 Barcelona, Spain*

There is a huge impetus for developing newtechnologies for the production of both energy and bioproductsin a sustainable way. For example, Bio-Electrochemical Systems (BESs), such as MicrobialFuel Cells (MFCs) and Microbial Electrolysis Cells (MECs) are noveltechnologies which promise to replace energy intensive wastewater treatmentprocesses (WWTPs) by converting the organic waste fraction present inwastewater into hydrogen or electricity. Their potential as sustainabletechnologies for combined energy and biofuelproduction is very high, considering the huge amount of sludge generated(average 30 kg dry matter/inhabitant-year). Moreover,defining the characteristics and the environmental burdens of the wastewater cominginto a WWTP plays a key role in understanding the possible applications of thesludge for generating energy and hydrogen with MFCs and MECs, respectively.

Inthis work we analyse the environmental impact of (i)a Wastewater Treatment Plant and (ii) the production of Hydrogen and biogas fromthe sludge generated in the WWTP itself.

Firstlywe assess the environmental impact of 1 m3 of water entering the Waste Water Treatment (WWTP) process using Life Cycle Assessment methodology. We then identify theenvironmental hot spots in the process chain, focusing on the most important environmentalburdens of a WWTP: energy use, main pollutants from personal care products,heavy metals and final utilization of sludge as a bioproduct.

Theenvironmental impact is assessed with Simapro,Version 7.2, with CML2000 method. The impact categories considered are: AP(acidification potential, global, kg SO2 eq.), GWP100a (global warmingpotential, kgCO2 eq.), EP (eutrophication potential,global, kg PO_4 eq.), PHO (photochemical oxidation, kg formed ozone),DAR (depletion of abiotic resources, kg antimonyeq.), ODP (ozone depletion potential, kg CFC-11 eq.), ETP (ecotoxicitypotential, kg 1,4-DCB eq.) as well as CED (Cumulative EnergyDemand). Bearing in mind that water reuse and reclamation is one of the mainissues of water sustainability, we add to the standard CML impact categories theFreshwater Ecosystem Indicator, (direct effect on biodiversity of freshwaterecosystem) and the Freshwater Depletion (FD), to properly estimate the impactof the process in the water cycle.

This project is partially financedby Carburos Metlicos andthe Spanish Government, CDTI, in the CENIT BIOSOS, belonging to the Program Ingenio 2010.

See more of this Session: Sustainability Metrics At the Process and Product Level
See more of this Group/Topical: Environmental Division

(192a) Flocculation Kinetics of Precipitated Calcium Carbonate and Its Response to Addition of Various Polymers and Salts

Bhavin Bhayani, Empire State Paper Research Institute, Syracuse, NY and Bandaru V. Ramarao, Paper and Bioprocess Engineering, Empire State Paper Research Institute, Syracuse, NY

Precipitated calcium carbonate has proved to be economical filler in paper industry due to its technical compatibility, high availability and low price. An optimum amount of filler is important as an excess amount of fillers weaken paper strength by decreasing the fiber–fiber bonded area. In this study, the kinetics of aggregation of industrially applied precipitated calcium carbonate (PCC) were studied by measuring the particle size distribution, zeta potential, and their response to various flocculants, dispersants and salts. Aggregate diameters (as measured by dynamic light scattering) increased with time but anionic dispersants suppressed agglomeration. Simple electrolytes suppress agglomeration, while cationic polyelectrolytes caused flocculation. A model based on population balances and incorporating orthokinetic and perikinetic flocculation and also the effect of charge neutralization according to the DLVO theory was developed. The calculations from the model are in good agreement with observations of agglomerate size evolution. The model further enabled predictions of turbulence levels necessary to keep the PCC particles in a dispersed state under different conditions.

See more of this Session: Poster Session: Sustainable Forest Bioresources Engineering
See more of this Group/Topical: International Congress on Energy 2011

Correlating Blending of Pulps of different qualities

T. J. G. Dattatreya, N. K. Bhardwaj, A. K. Ray

Department of Paper Technology, IIT Roorkee, Saharanpur 247 001, India

Abstract

In this present study the physical and mechanical properties of paper handsheets made from mixture of different qualities of pulp stocks were evaluated. The properties studied were : Canadian Standard Freeness, CSF, water retention value, WRV, apparent density/bulk, grammage, thickness, tensile index, burst index and tear index as well as double folds . The pulp mix were obtained by blending different proportions of short and long fibre pulps. The values were then compared with those predicted from the models developed statistically. The pulp fibres were collected from a nearby integrated pulp and paper mill located in Northern India. The short fibre collected was obtained from a mixed hardwood chemical pulp (Eucalyptus Tereticornis and Poplar), and the long fibre pulp was a softwood chemical pulp (Pinus Roxburghii). The initial CSF value of the hardwood pulp was 460 ml, and those of the softwood pulps 380 ml and 270 ml, respectively . The hardwood pulp was blended with two types of softwood pulps varying in the CSF value. Several blends of the two fibre types were performed by adding different proportions of long fibre pulps to the short fibre pulp. The pulp and paper properties of these blends were determined with TAPPI standard procedures. It is well known that the property development of long fibre pulp of any CSF value when mixed with short fibre pulps of different CSF values (even for the same CSF value) cannot be predicted easily. Therefore, this extensive investigation was undertaken to examine the hypothesis whether the pulp blends follow the super position principles of additivity or not. In order to fulfil the aim statistical regression equations were developed with least square techniques. Both linear and polynomial parabolic/quadratic type models were tried. The correlations obtained prove satisfactory with reasonable accuracy permitted in engineering estimates.

1. Introduction

Indian pulp and paper industry, either integrated or nonintegrated are suffering from crisis of raw materials. As there is acute shortage of softwood pulps (long fibre), industry uses various kind of good quality hardwoods of fast growing species which is also in short supply. Eucalyptus and Poplar are only two hardwood species upon which Indian Industry depends.This scarcity of plant fibres forces the Indian paper industry to use blend of hardwood pulps along with small quantity of softwood pulp. Mixing with Softwood fibres is required as it increases the mechanical strength properties of paper products. This is important as per the requirement of quality which can compete in the international market . In order to achieve the goal Industry must rely on different kind of pulps to achieve the required strength properties in the paper. With more availability of recycled paper, limited hard wood resources and abundant non wood resources the required strength properties of the paper cannot be achieved unless and until a stronger and longer pulp is used along with the indigenous materials. The raw materials in India are mainly short fibred but they still compensate for the shortage of raw materials in India. But qualities being a major concern, these short fibres are blended with long fibred stock so as to attain the desired properties.

In fact both type of pulps, short fibre and long fibre have their unique and distinct properties. The furnish values are also different. The moment they are mixed there will be different type of characteristic developments, radically distinguishable from the individual ones. On the other hand the mixture of the two fibres is inevitable in the present day context of Indian Pulp and Paper Industry.

Indian Pulp and Paper industry relies on both the type of raw materials i.e., soft wood and hard wood because of the following reasons. The indigenous raw material i.e., the hard wood fibre has good bulk and more fines which make the paper bulkier, smoother and gives greater yield. While on the other hand the soft wood fibre has greater fibre length such that more surface area is available for bonding. Thus the paper formed has excellent strength characteristics. The Soft wood pulp is mixed with the hard wood pulp to enhance the strength of the paper while getting good yield.

Blending models were developed in microscopic level to calculate apparent density of the sheet as a function of blend composition and fibre properties. The interactive multi planar model [1] developed by Görres, Amiri, Wood & Karnis assumes that (a) paper is a super imposition of fibre layers, (b) the layer weight is a function of fibre width and coarseness and is given in a point when 1% of the network area is covered by fibre crossings, (c) layer structure is described adequately by network statistics, (d) translayer bonding occurs when fibres are deflected by wet pressing, and (e) average fibre properties are sufficient to describe the sheet structure. To compute the mean fibre properties of a blend, the fibre properties of each pulp are weighted by the fraction of total fibre length that the individual pulps contribute to the sheet. Very little work was done in the area of effect of blending on freeness and prediction of various properties of pulp and paper in macroscopic level. The properties may be additive or counteracting based on the interaction and orientation of individual fibres in a whole mass. The perfect macroscopic blending model can be used in industry to give better productivity.

Srivatsava and Sharma[2] tried to model the effect of blending on optical and surface properties of pulps made from softwood and annual nonwood fibres. They have developed a regression equation for the interaction of blend percentage, amount of filler and amount of size added in the blend furnish on the properties such as burst strength,

cobb-60, scattering coefficient, and air permeability. The regression equation developed was a three parameter non linear regression equation [2] of the form
$Y=B_0+B_1.X_1+B_2.X_2+B_3.X_3+B_{12}.X_1.X_2+B_{13}.X_1.X_3+B_{23}.X_2.X_3+B_{123}.X_1.X_2.X_3.$
Similar attempts have been made by Rajesh and Krishanan Ravi[3] and also by Brecht[4]. All the above investigation tried to give an idea of the properties obtained by the blending of two furnishes of different compositions. The experimental procedure was first designed using statistical experimental design methodology followed by performing factorial experiments . The experimental data were subjected to regression techniques for different strength properties of the paper formed from different blends.The models developed were compared with the experimental data.

2. Experimental

2.1. Pulp samples

The collected pulps both short fiber as well as long fiber were first conditioned and kept in the laboratory carefully. The long fibre pulp at 10% consistency was treated in PFI mill to attain a freeness values of 380 and 270 CSF. The refined short fibre mix hardwood pulp of 460 CSF freeness was taken for various experiments.The pulp mix was kept in conditioned laboratory for further investigations.

2.2. Evaluation of pulp and pulp blending

Individual pulps were blended in different proportions in percentages (95-5, 90-10, 85-15, 80-20, and 75-25). Slurries of each blend were prepared carefully. The slurries were well stirred for uniform distribution of the fibres of the two pulps before measuring CSF values. Various readings for the freeness values of the blends were taken and parameters influencing CSF were adjusted properly.

2.3. Hand sheet making and paper testing

Standard hand sheets (of approximate 60 gsm) were made on British sheet forming machine as per Tappi standard method T-205 cm-80. The hand sheets made from samples of pulp obtained after blending were subsequently tested for physical strength properties according to various Tappi test methods after conditioning at 27 ± 2 ^{0}C and 65 ± 2 % relative humidity following conditions of Bureau of Indian Standards(BIS).

3 Results and discussion

The data obtained from laboratory experiments are analysed by plotting graphs. The results include the quality of pulps used in the study, papermaking properties of their blends and regression equations developed through least square technique showing the relationship between various pulp and paper properties. The experimentally obtained data and the regression model predicted data are compared in the graphs.

3.1. Influence of blending on Pulp Properties

Canadian Standard Freeness values ,CSF and water retention values, WRV as a function of the percentage of softwood in the blend are drawn in figs. 1-4.respectively.

3.1.1. Effect of blending on Freeness

Figures 1& 2 show the effects of blending on experimentally determined CSF values. The data when subjected to statistical analysis lead to a single variable regression models. The fitness of the models are tested in a number of ways such as, the values of R^2, estimating percentage of error and residual analysis. The models predicted data are also compared in figures 1& 2.
It is evident from the from the graphs that with the increase of the percentage of softwood pulp in the blend, the CSF values decrease. This is equally true for both the pulps having different initial CSF values of the softwood component of the pulp. On comparing the effect of initial freeness (as depicted in figures 1 & 2) it reveals that the slopes of the two linear regression lines are different. The slope of the pulps blends having lower initial CSF values is steeper. The expected behaviour of the property development is additive [3] due to the linearity of the fit. To confirm this remarks further model predicted data are also critically analysed. The model predicted data closely agrees with the experimental data. Therefore the model developed is found to be very accurate with R^2 value of the order of 0.99.The percentage error with experimental data as a basis is in the ranges +0.7 and -1.2. Further residual plot also clearly shows that the values are very accurate. The percentages of error obtained are permissible within the required accuracy of the engineering estimates.

3.1.2. Effect of blending on Water Retention Value.

Figures 3 & 4 reveal the effects of blends on experimentally determined WRV values. Figures 3 & 4 are the comparisons of predicted models with actual experimental values. It is very clear that with the increase in the percentage of softwood pulp in the blend, the WRV values increases for both the pulps having different initial WRV values of the softwood component of the pulp. From the figures it is evident that the relationships are distinctly linear.The property may therefore be considered additive. The percentage error with experimental data as a basis is in the ranges +1.7 and -0.7. When statistically interpreted ,the model developed is also found to be very accurate with R^2 value of the order of 0.99.

3.2. Influence of blending on Paper Properties

The properties of handsheets made from blended pulps were determined in the laboratory using Tappi testing methods. The properties included Burst index, Tensile index, Tear index and Folding endurance . These properties are shown as a function of percentage of softwood in the blend .

3.2.1 Effect of blending on Bursting Index and Tensile Index.

Figures 5 & 6 are drawn with the experimental data to explain the effects of blending on Burst Index values. The graphs reveal that with the increase of the percentage of softwood pulp in the blend and decrease in the freeness of the blend, both the Burst Index and Tensile Index values increase.This are expected behavior as more is the softwood pulp content in the blend , more will be the burst and tensile strength of the paper.

3.2.2. Effect of blending on Tear Index.

The effects of blending on Tear Index values are drawn in figures 7 & 8. As Tear strength is a function of both fibre strength and fibre bonding, the addition of soft wood pulp in the blend has caused an increase of both of the parameters, resulting in increase of tear index values of pulp blend. Further the graphs clearly show that the values of the tear index are found to be small for low freeness value pulp than than that of the higher one. It is clear that with the increase of the percentage of softwood pulp in the blend and increase in the freeness of the blend, there is an increasing trend of Tear Index values reflected.

3.2.3. Effect of blending on Folding Endurance

The effects of blending of pulp on Folding Endurance values experimentally determined are depicted in figures 9&1. It is evident that with the increase the percentage of softwood pulp in the blend and decrease in the freeness of the blend, the folding Endurance values increase .It is an expected behavior.

3.2.4. Effect of blending on Bulk and Drainage time.

The influences of blending on the Bulk and Drainage time values of hand sheets from the hardwood-softwood pulp blends are shown in figures 11 and 12.
The Bulk and Drainage time of the sheets of various pulp blends are measured experimentally using Tappi procedures. Figure 11 indicates that the bulk decreases with increase in the percentage of soft wood. The brightness is not much changed because of the pulps of pure samples have similar brightness. Fig 12 indicates the increase in the Drainage time with increase of the soft wood in the blend and decrease in the freeness of the blend. Though the former is known , the later can not be interpreted quantitatively.

3.3. Statistical Interpretation of Data

Based on the experimental data obtained from the laboratory are analysed from graphs and then subjected to statistical regression analysis. Least square techniques have been used. For simplicity linear regressions and 2nd degree polynomial regressions have been tried according to the applicability of the data. Some are found to be distinctly linear whereas some possess nonlinear characteristics. Predicted data from regression models are plotted against experimental values in various graphs and percentage errors have been determined.

Percentage error is defined as [(the experimental value – model predicted value)/ experimental value] * 100.
These regression models along with the maximum and minimum percentage error are depicted in following Tables 5&6.
It is evident that all the regression models developed are very accurate as the regression coefficients, R^2 are close to unity (on an average above 0.98) and the percentage error does not exceed in any case ± 7%. The figures 13 & 14 show the plot between model predicted data versus experimental data and residual analysis respectively which show an excellent agreement between them. The values are shown clearly in table 7. Therefore statistical predictions are claimed to be very good and reproducible with a reasonable degree of accuracy permitted in engineering estimates. Therefore the statistical regression can be used reliably for analysis purposes.

4. Conclusions

The linearity in the relations of the long-fibred softwood pulp proportion in the blended pulps versus freeness CSF and water retention values WRV shows the additivity of freeness and water retention value in the blends. With increasing proportion of long-fibre pulp, the burst strength and the tensile strength of the paper increase because of larger surface area available for bonding. The tear strength and the folding endurance of the paper also increase with increasing proportion of the long-fibre pulp. When the proportion of softwood pulp increases, drainage time and bulk decrease. The developed regression equations agree excellently with the experimental data having a regression coefficient close to unity.

Acknowledgements
The authors thankfully acknowledge Dr. Sanjay Tyagi, Scientist B ,Dr. Y. V. Sood, Retired Scientist-F & In charge SPPMC Division, Central Pulp and Paper Research Institute (CPPRI) for their guidance and providing us the necessary required facilities in order to complete the experimental work. The authors also thankfully acknowledge STAR Paper mill for providing the pulp.

References

1 Görres, J., Amiri, R., Wood, J.R., Karnis, A. 1996, “The Apparent Density of Sheets. Made from Pulp Blends”, TAPPI Journal, Vol 79, no. 1, pp. 179-187.

2 Srivatsava, S., Sharama, A. (1994-1995)The optical and surface properties of paper based on blending non wood fibre with long fibre pulp. B.Tech project report, Department of Paper Technology, IIT Roorkee, India

3 Rajesh D. and Krishnan Ravi. A project report on " Role of blending on paper properties-A statistical evaluation".(1995-96) ,Department of Paper Technology,IIT Roorkee.
4 Brecht, W., "Das Mischen verschiedenartiger Halbstoffe in seinem Einfluss auf diephysikalischen Eigenschaften der Papiere". Svensk Papperstid. 66(1963)5, 159-167.

Table 1: *Various Notations used.*

S.No	Notation	Definition
1	S.W.3	Soft wood pulp with CSF value of 270ml
2	S.W.4	Soft wood pulp with CSF value of 380ml
3	H.W.	Hardwood pulp
4	S.W.	Softwood pulp
5	CSF	Canadian Standard freeness
6	WRV	Water Retention Value

Table 2: *Compositions of various blends.*

S.NO	Compositions of blend of H.W to S.W.3	Compositions of blend of H.W to S.W.4
1	95:05	95:05
2	90:10	90:10
3	85:15	85:15
4	80:20	80:20
5	75:25	75:25

Table 3*. Various pulp and paper properties of the blends of H.W & S.W.3.*

% of S.W.3	CSF, ml	Bursting Index, $kPa.m^2/g$	Tensile Index, Nm/g	Tear Index, mNm^2/g	Folding Endurance	WRV %
0	460	2.29	20.32	1.63	1	138.2
5	448.33	3.18	22.45	1.72	1.23	151.4
10	441.67	3.39	24.44	1.84	1.462	157.6

15	430	3.72	25.17	1.96	1.832	168.2
20	415	3.99	26.47	2.39	1.806	177.8
25	405	4.47	27.86	2.51	1.944	189.3

***Table 4**: Various pulp and paper properties of the blends of H.W & S.W.4.*

% of S.W.4	CSF, ml	Bursting Index, $kPa.m^2/g$	Tensile Index, Nm/g	Tear Index, mNm^2/g	Folding Endurance	WRV %
0	460	2.29	20.32	1.63	1	138.2
5	453.33	3.16	21.87	1.77	1.176	146.2
10	446.67	3.38	23.09	1.9	1.544	151.8
15	443.33	3.74	24.36	2.14	1.579	160.8
20	438.33	3.84	25.69	2.33	1.785	167.9
25	433.33	4	26.45	2.51	1.88	176.9

***Table 5:** Various Regression equations and Regression coefficients, R^2 as a function of different parameters for blends of H.W & S.W.3.*

S.No	Y	X	Regression type	Regression equation	R^2
1	Freeness	% of S.W.3	linear	$y = -1.8881x + 457.2$	$R^2 = 0.9973$
2	Bursting Index	% of S.W.3	polynomial	$y = -0.0005x^2 + 0.0914x + 2.4826$	$R^2 = 0.9888$
3	Tensile Index	% of S.W.3	polynomial	$y = -0.0013x^2 + 0.3221x + 20.716$	$R^2 = 0.9975$
4	Tear ndex	% of S.W.3	polynomial	$y = -0.0002x^2 + 0.0406x + 1.5369$	$R^2 = 0.9868$
5	Folding Endurance	% of S.W.3	polynomial	$y = -0.0003x^2 + 0.0462x + 1.0304$	$R^2 = 0.9845$
6	Water retention value	% of S.W.3	linear	$y = 1.9946x + 138.84$	$R^2 = 0.9996$
7	Bursting Index	CSF	polynomial	$y = 0.0001x^2 - 0.1503x + 41.754$	$R^2 = 0.9817$
8	Tensile Index	CSF	polynomial	$y = -0.0008x^2 + 0.5803x - 74.576$	$R^2 = 0.955$
9	Tear Index	CSF	polynomial	$y = 1E-04x^2 - 0.1007x + 27.624$	$R^2 = 0.9788$
10	Folding Endurance	CSF	polynomial	$y = -0.0005x^2 + 0.413x -$	$R^2 = 0.9418$

				83.117	
11	Water retention value	CSF	linear	y = -0.8421x + 529.27	$R^2 = 0.9931$

***Table 6:** Various Regression equations and Regression coefficients, R^2 as a function of different parameters for blends of H.W & S.W.4.*

S.No	**Y**	**X**	**Regression type**	**Regression equation**	**R^2**
1	Freeness	% of S.W.3	Linear	$y = -0.7702x + 455.68$	$R^2 = 0.9908$
2	Bursting Index	% of S.W.3	Polynomial	$y = -0.0004x^2 + 0.0732x + 2.5712$	$R^2 = 0.9776$
3	Tensile Index	% of S.W.3	Polynomial	$y = -0.0011x^2 + 0.2754x + 20.432$	$R^2 = 0.9995$
4	Tear *ndex*	% of S.W.3	Polynomial	$y = -0.0001x^2 + 0.0392x + 1.5873$	$R^2 = 0.9984$
5	Folding Endurance	% of S.W.3	Polynomial	$y = -0.0002x^2 + 0.0415x + 1.0249$	$R^2 = 0.9915$
6	Water retention value	% of S.W.3	Linear	$y = 1.417x + 139.16$	$R^2 = 0.9993$
7	Bursting Index	CSF	Polynomial	$y = -0.0007x^2 + 0.5597x - 110.73$	$R^2 = 0.9538$
8	Tensile Index	CSF	Polynomial	$y = -0.0007x^2 + 0.4159x - 14.491$	$R^2 = 0.9838$
9	Tear Index	CSF	Polynomial	$y = 0.0004x^2 - 0.4151x + 102.73$	$R^2 = 0.9768$
10	Folding Endurance	CSF	Polynomial	$y = -0.0009x^2 + 0.7991x - 167.82$	$R^2 = 0.9853$
11	Water retention value	CSF	Linear	$y = -1.5819x + 861.48$	$R^2 = 0.9765$

***Table7:** The experimental value and model predicted values for pulp blends of H.W-S.W.3 of regression eqn* $y = -1.8881x + 457.2$, $R^2 = 0.9973$

% s.w.3 in blend *H.W-S.W.3* (x)	CSF(y), ml	Predicted Data($\hat{Y}$)	% Error	Residuals (e_{ij})
0	460	457.2	0.608695652	2.8
5	448.33	447.7595	0.127250017	0.5705
10	441.67	438.319	0.758711255	3.351
15	430	428.8785	0.260813953	1.1215
20	415	419.438	-1.06939759	-4.438
25	405	409.9975	-1.233950617	-4.9975
100	270	268.39	0.596296296	1.61

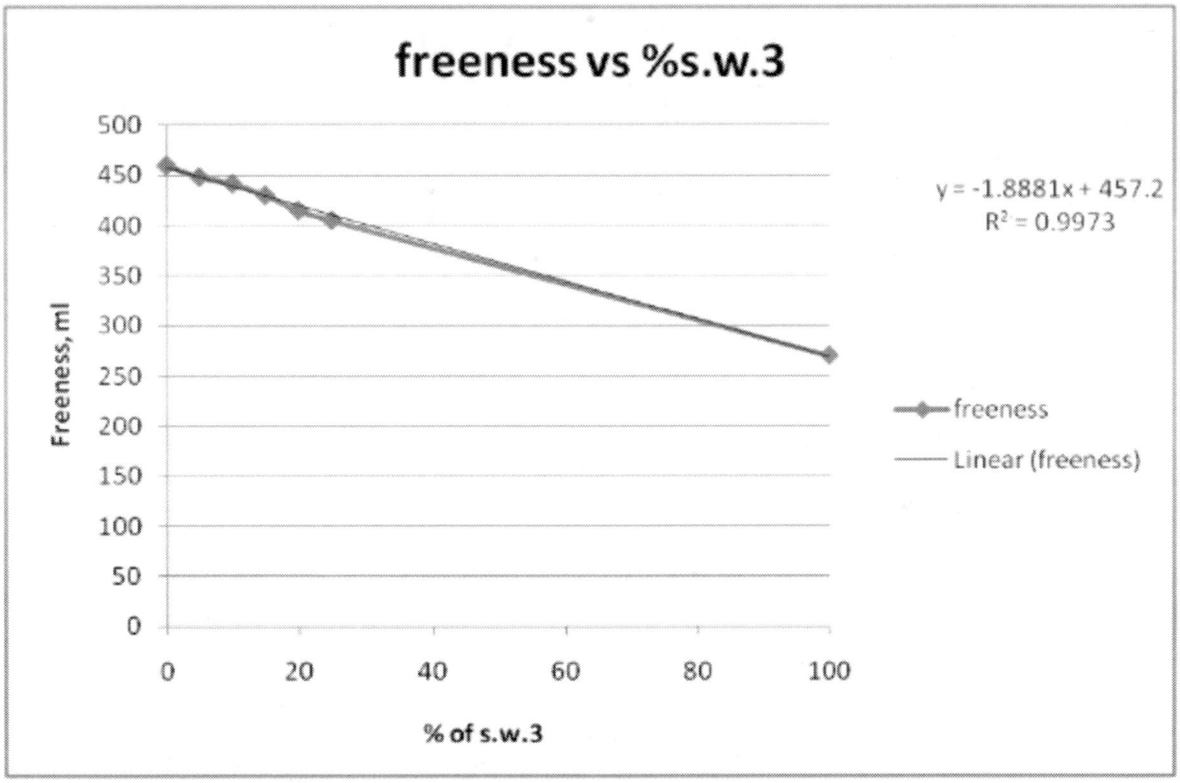

Fig1: *Effect of % of S.W.3 on the freeness in the blends of H.W-S.W.3.*

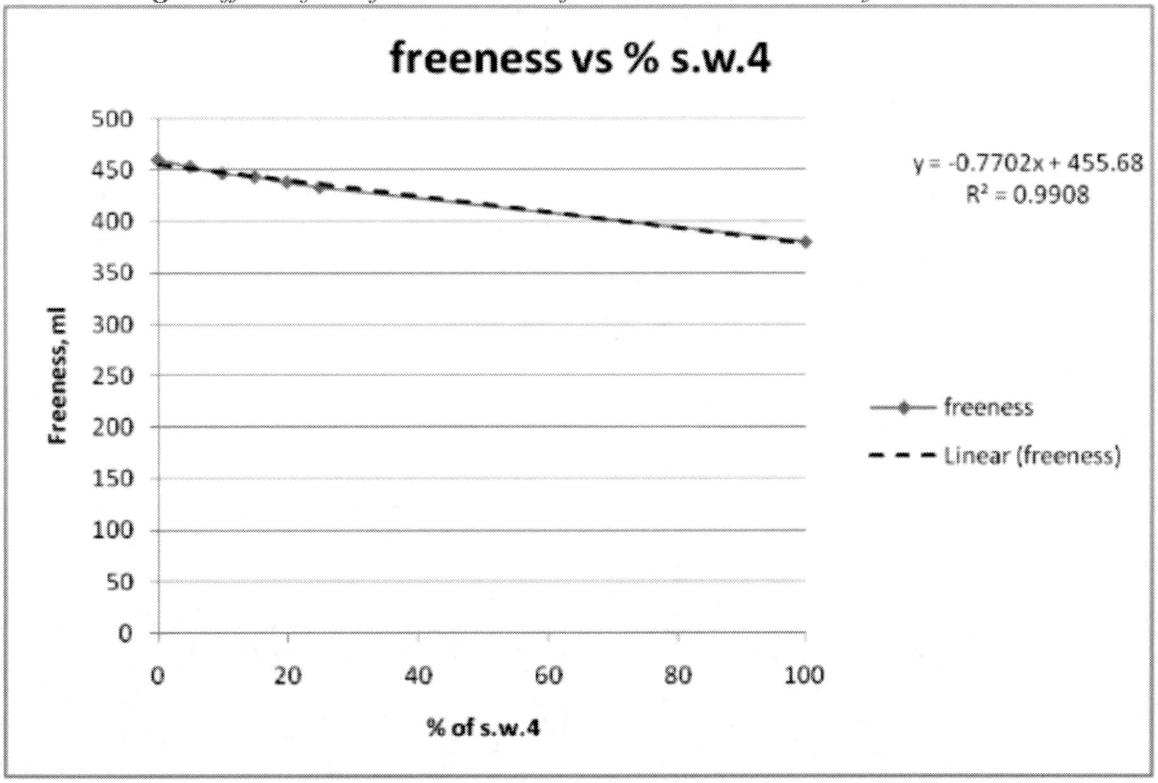

Fig 2: *Effect of % of S.W.4on the freeness in the blends of H.W-S.W.4.*

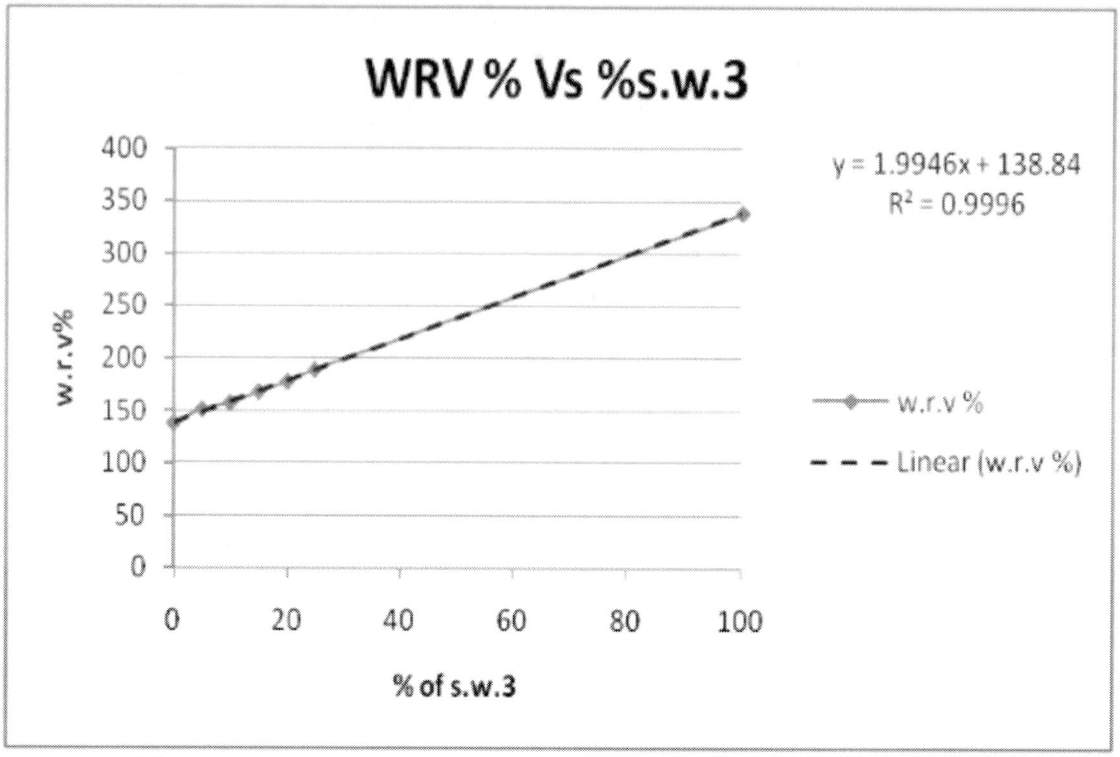

Fig 3: *Effect of S.W.3% on the W.R.V in the blends of H.W-S.W.3.*

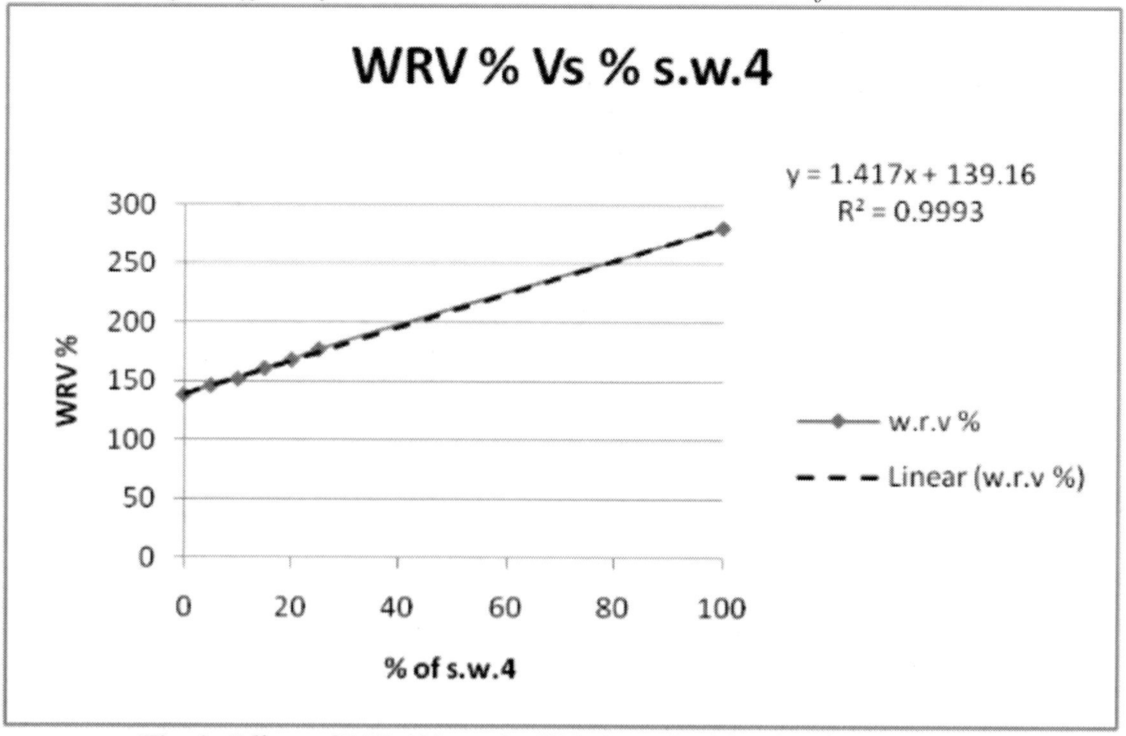

Fig 4: *Effect of S.W.4% on the W.R.V in the blends of H.W-S.W.4.*

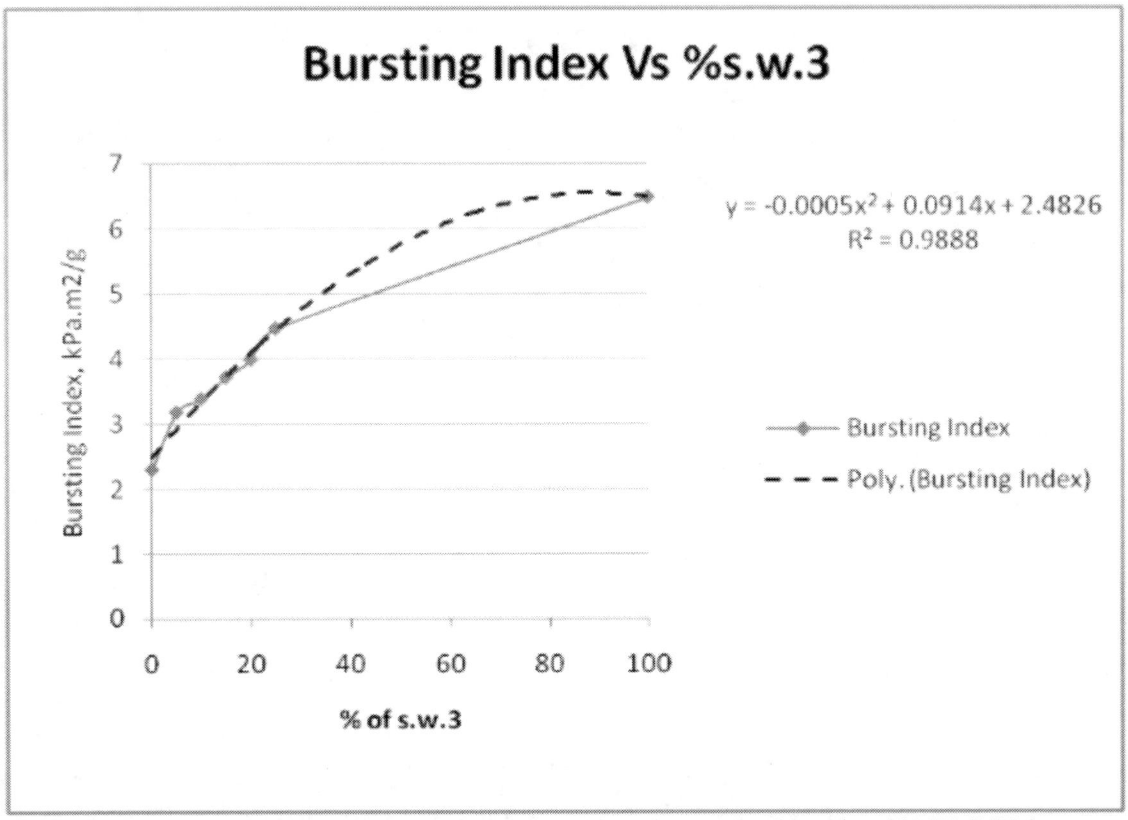

Fig 5: Effect of S.W.3% on the Bursting Index in the blends of H.W-S.W.3.

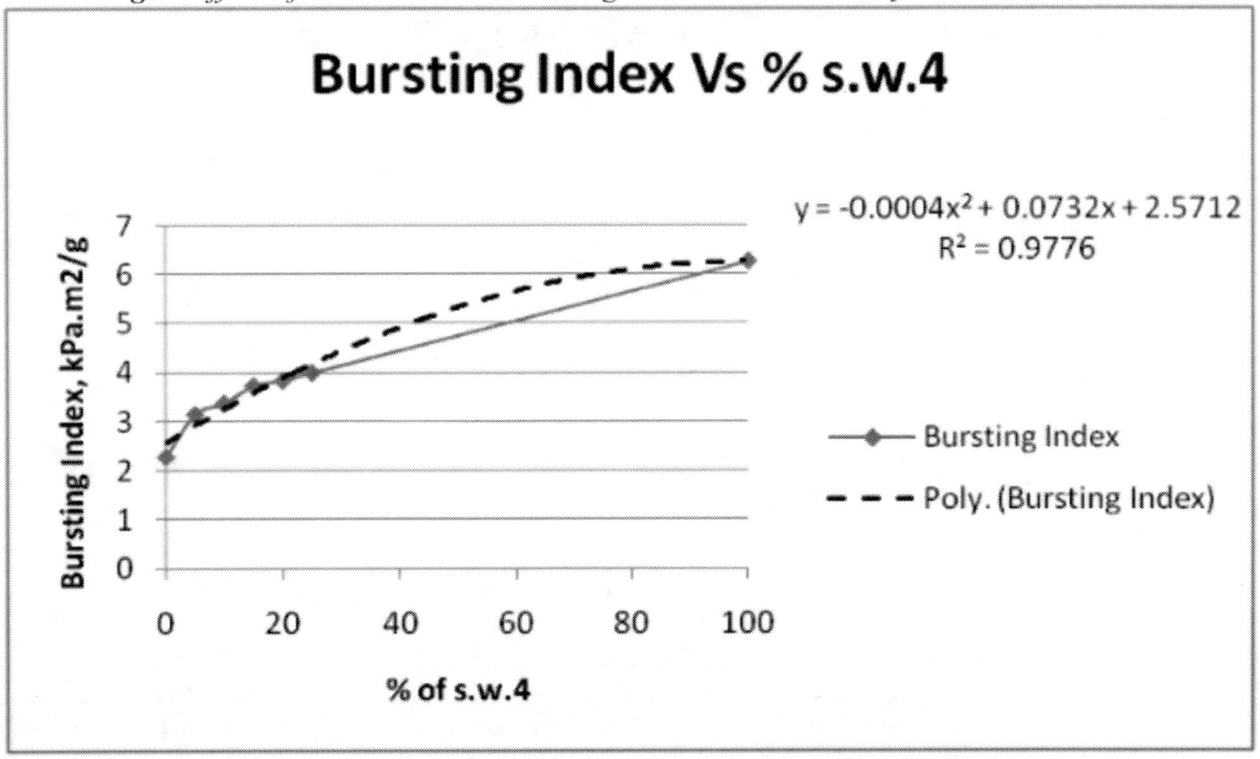

Fig 6: Effect of S.W.4% on the Bursting Index in the blends of H.W-S.W.4.

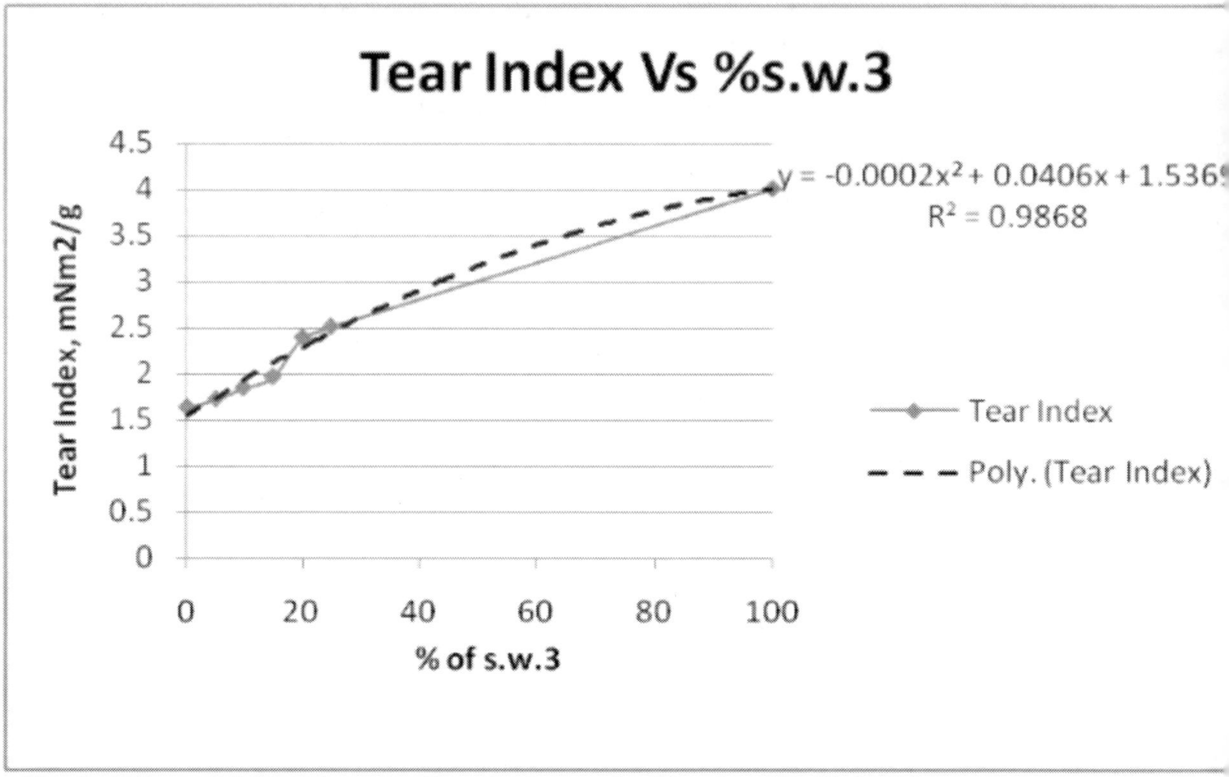

***Fig** 7: Effect of S.W.3% on the Tear Index in the blends of H.W-S.W.3.*

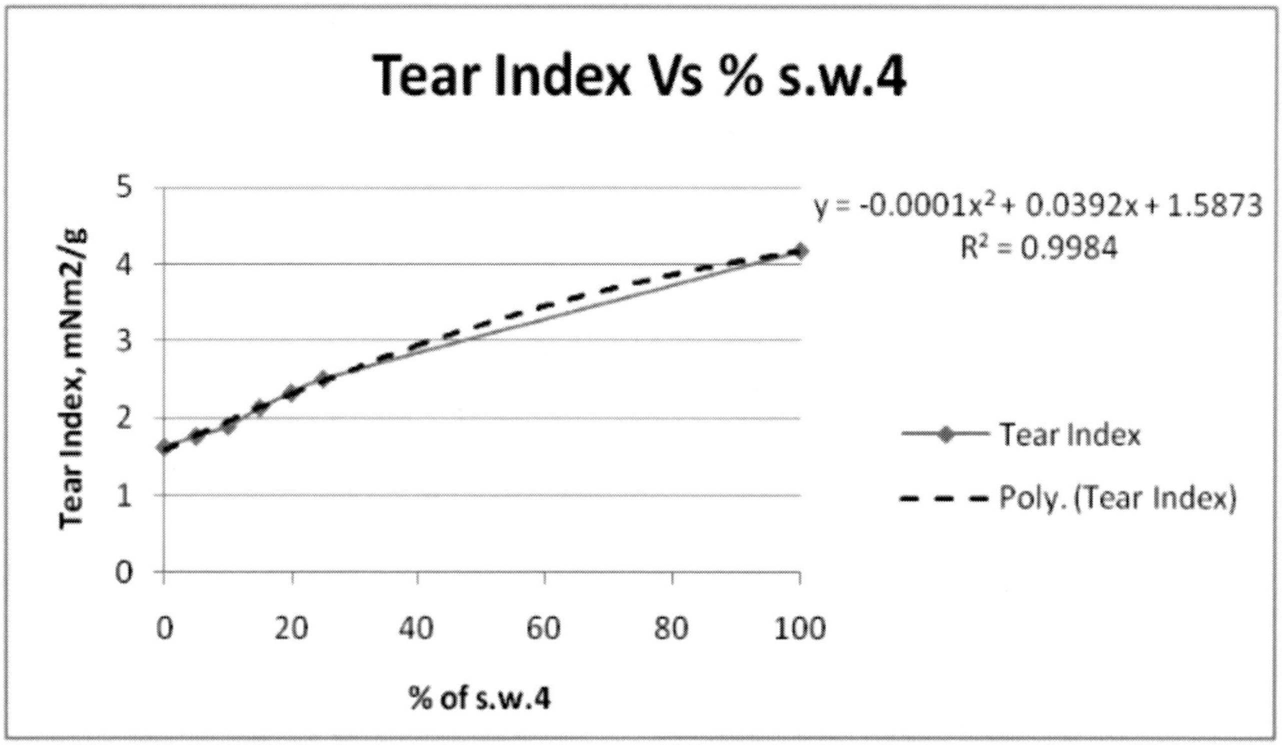

***Fig 8:** Effect of S.W.4% on the Tear Index in the blends of H.W-S.W.4.*

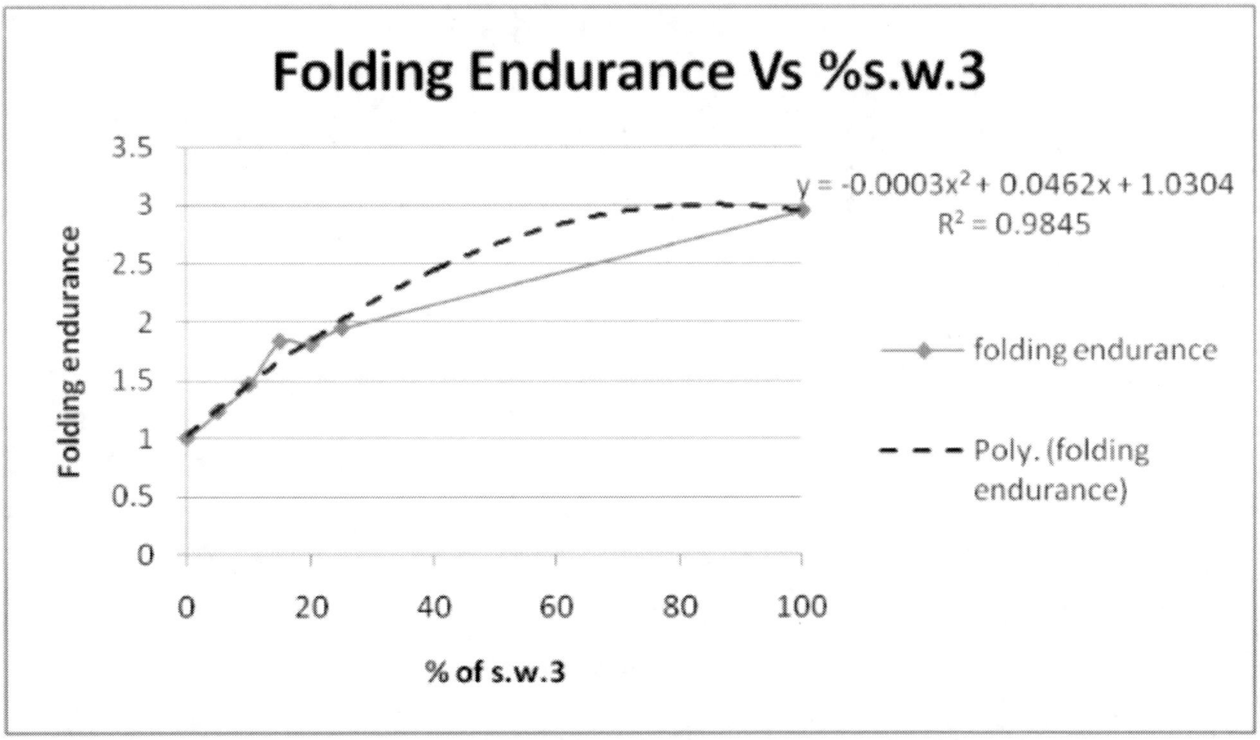

Fig 9: Effect of S.W.3% on the Folding Endurance in the blends of H.W-S.W.3.

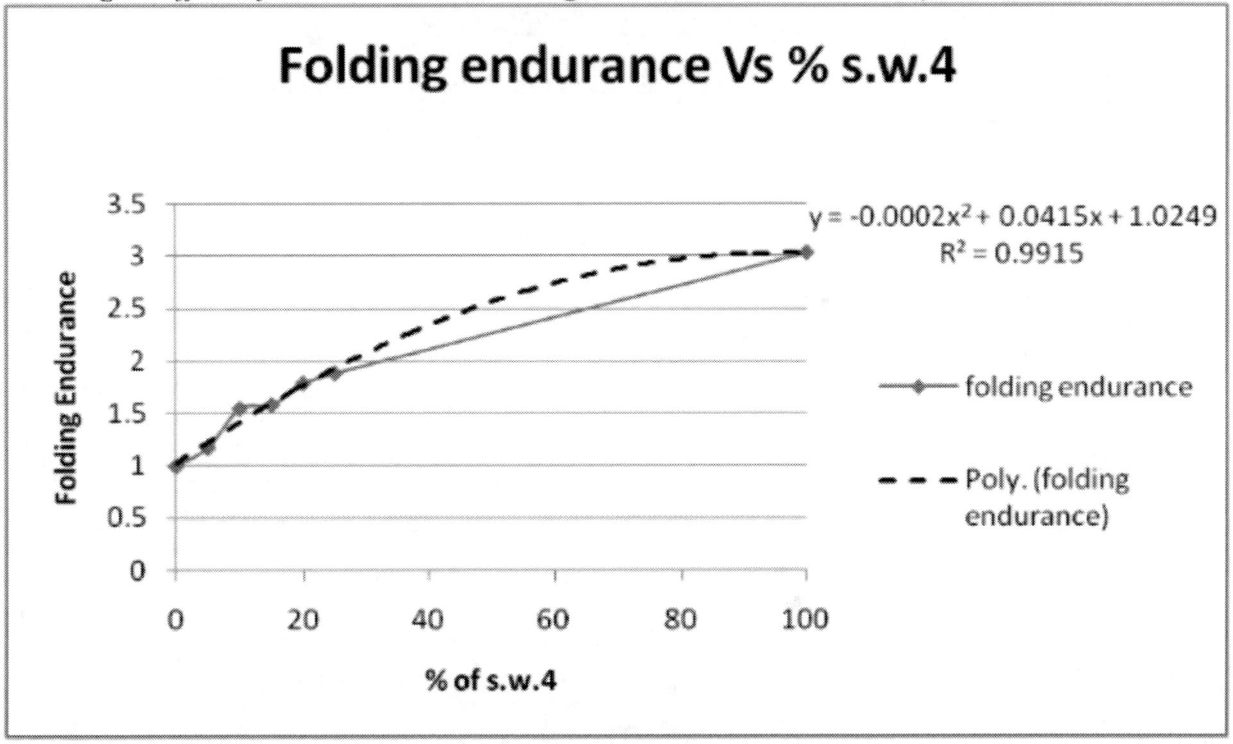

Fig 10: Effect of S.W.4% on the Folding Endurance in the blends of H.W-S.W.4.

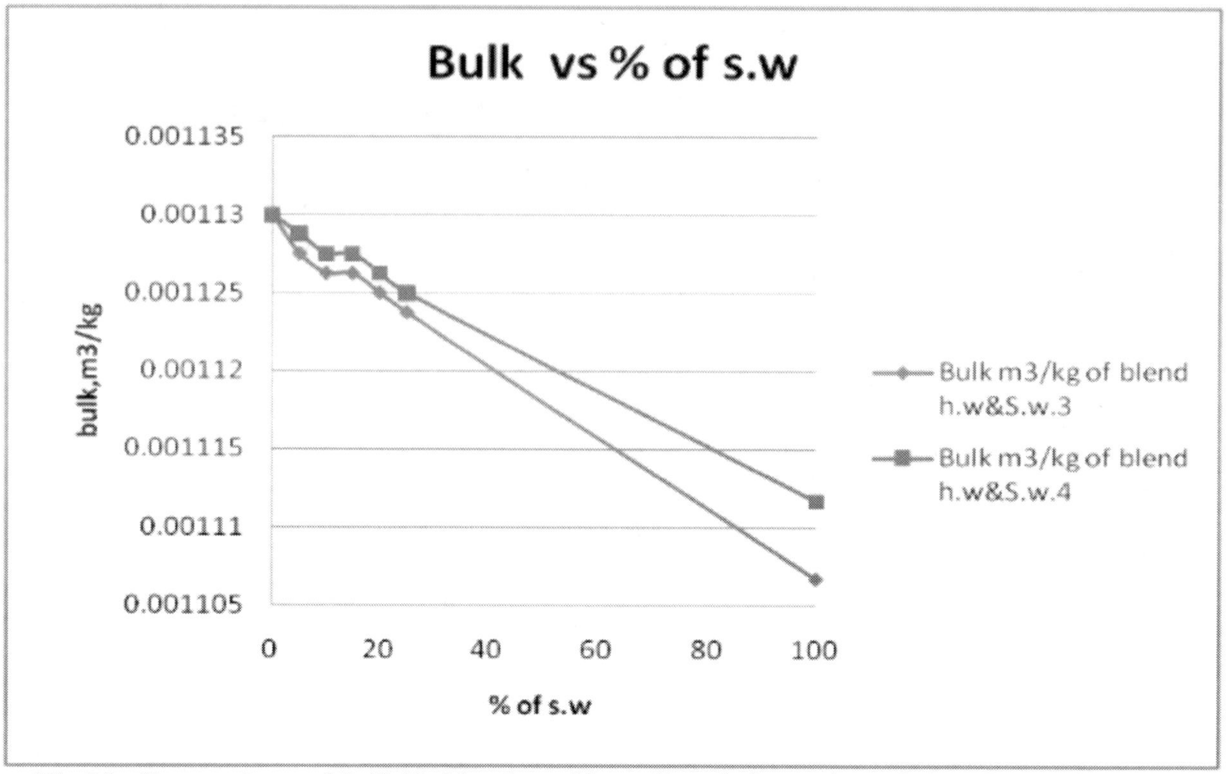

Fig 11: *Comparison of Bulk Vs blend profiles of hand sheets from S.W pulps with two different initial Bulks.*

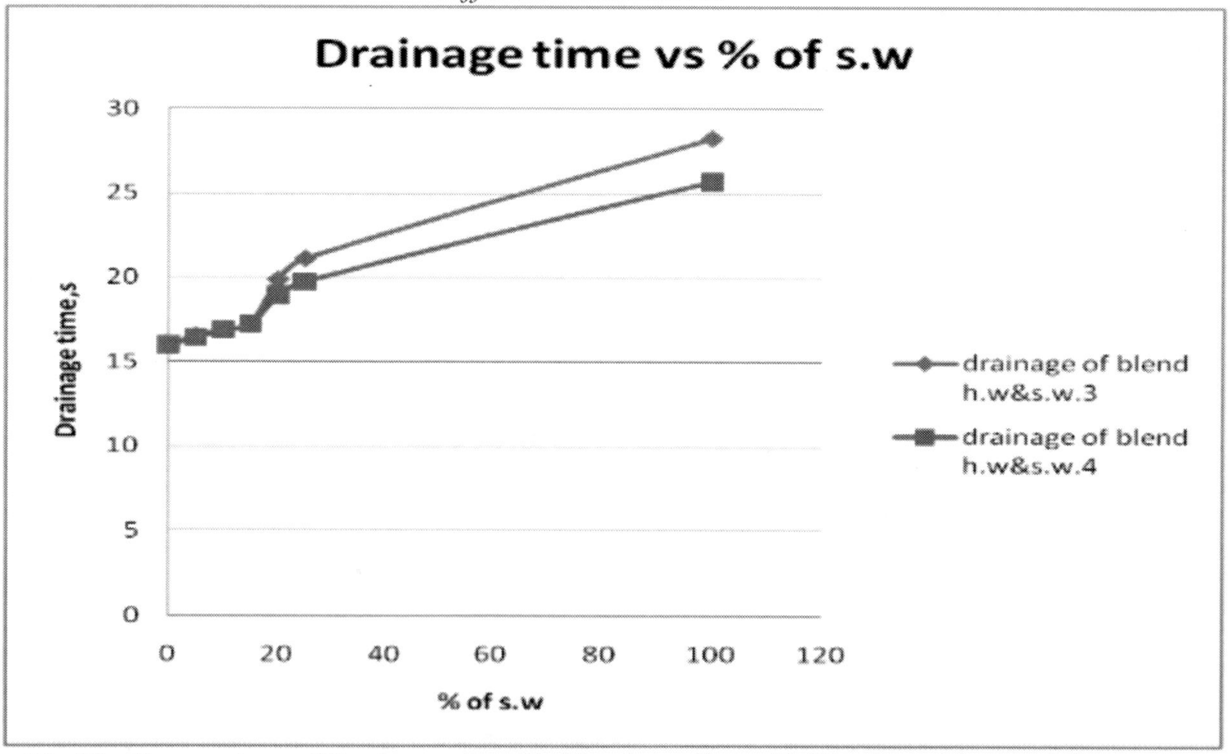

Fig 12: *Comparison of Drainage time Vs blend profiles of S.W pulps with two different initial Drainage time.*

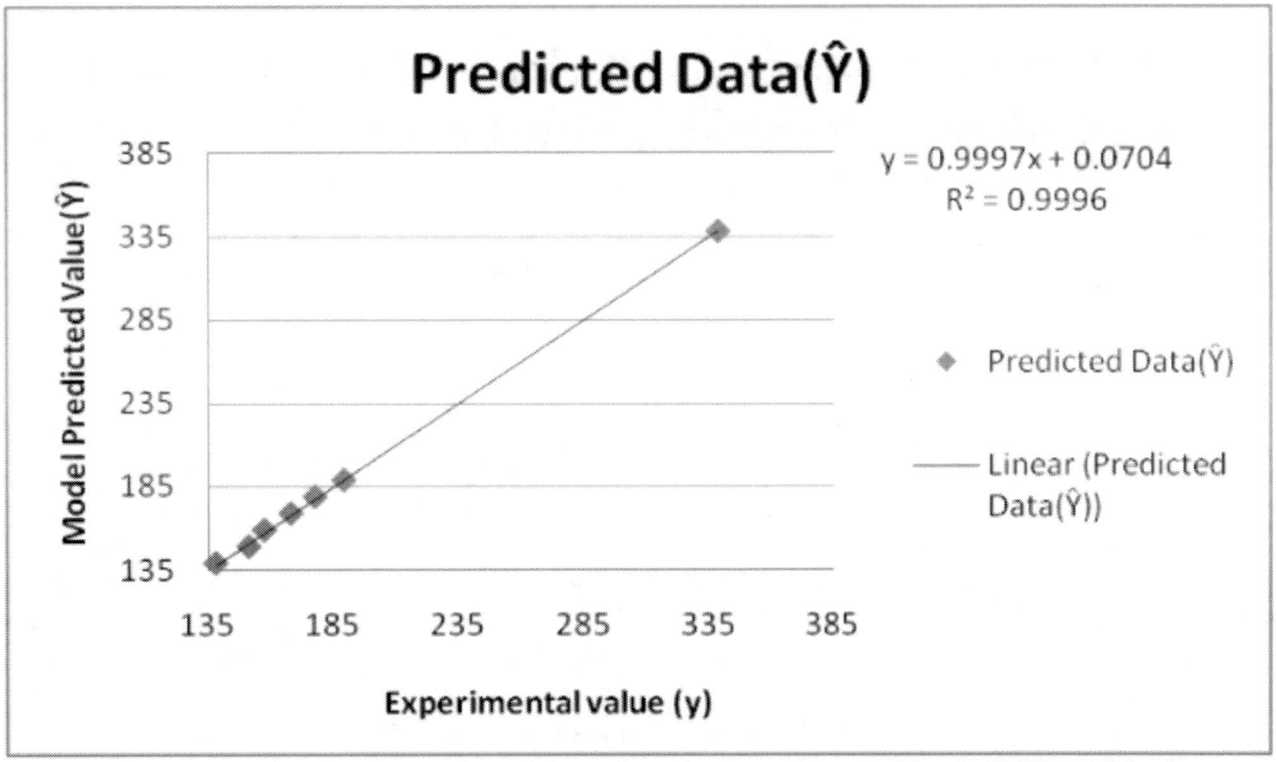

***Fig13:** The experimental value and model predicted values for pulp blends H.W- S.W.3 of regression eqn* $y = -1.8881x + 457.2$, $R^2 = 0.9973$

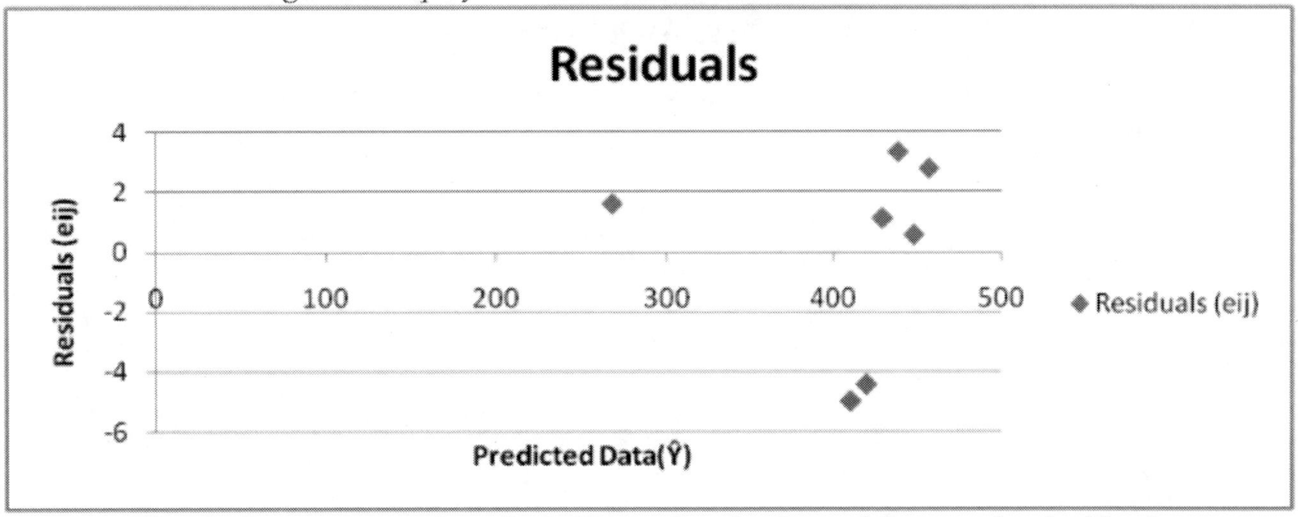

***Fig14:** Comparision of residuals Vs predicted values of freeness for the blends H.W- S.W.3.*

(192c) Identification and Quantification of Sugar and Oil From Arkansas-Native Algae

Ching-Shuan Lau, Blake Johnson, Heather Sandefur, Marty Matlock, Tom Costello and Danielle Julie Carrier, Department of Biological and Agricultural Engineering, University of Arkansas, Fayetteville, AR

Green algae grow commonly at phosphate removal raceway placed at Fayetteville waste water treatment plants. Samples from May to September 2010 were harvested from the Noland waste water treatment facility. The algae were analyzed for its oil and sugar content. Dilute acid pretreatment with 1% sulfuric acid, 160 °C for 60 minutes combined with an enzymatic hydrolysis step were used to release the carbohydrates from the algae. Separately, hexane was used to extract oil from the algae sample. The results indicated that the oil and sugar content of algae did vary according to months: 5.4% glucose and 1.8% xylose for May, 3.6% glucose and 4.0% xylose for June, 3.1% glucose and 3.4% xylose for July, 3.5% glucose and 3.6% xylose for September. All the algae samples had glucose and xylose in less than 10% total of their dry mass, and the oil content in algae is less than 3% mass of the overall algae mass, which is lower than algae used for biofuel that typically contains 30% oil.

See more of this Session: Poster Session: Sustainable Forest Bioresources Engineering
See more of this Group/Topical: International Congress on Energy 2011

(192d) Mechanistic Insights Into Cellulose III Formation and Its Depolymerization

Shishir PS Chundawat[1], Leonardo da Costa Sousa[1], Giovanni Bellesia[2], Nirmal Uppugundla[1], Dahai Gao[1], Paul Langan[3], Venkatesh Balan[1], Sandrasegaram Gnanakaran[4] and Bruce Dale[1], (1)Chemical Engineering and Material Science, Michigan State University, Lansing, MI, (2)T6 and CNLS - Los Alamos National Laboratory, Los Alamos, NM, (3)Biosciences Division - Los Alamos National Laboratory, Los Alamos, NM, (4)T6 - Los Alamos National Laboratory, Los Alamos, NM

Enzymatic conversion of cellulose can be improved via either reducing cellulose crystallinity by thermochemical treatments or enhancing cellulase activity by protein engineering. A third approach is via reorganizing the hydrogen bonding network within crystalline cellulose to produce a less recalcitrant allomorph. Cellulose III is one such allomorph, however, there is little known about its formation and depolymerization by glycosyl hydrolases. Detailed studies were carried out to determine the effect of ammonia-water-acetone loading, total residence time and reaction temperature on the conversion of cellulose I to III. Cellulose crystallinity and extent of conversion between I and III allomorphic states was determined using X-ray diffraction (spectral deconvolution approach). The substrates were then subjected to hydrolysis by various combinations of purified *Trichoderma reesei* cellulases. Based on protein structural analysis and activity assay data, it appears that a cellulase such as Cel7B with a more open and unrestricted active site cleft may further enhance the degradation of cellulose III. Molecular dynamics simulations also revealed that cellulose III fibrils are readily hydrated, explaining its reduced hydrophobically-driven binding to cellulases and that their surface has structural and dynamical features that significantly influences binding of cellulases to it. Glucan chains within cellulose III are more readily accessible by cellulases due to reduced intra-chain and intra-sheet hydrogen bonding compared to native cellulose.

See more of this Session: Poster Session: Sustainable Forest Bioresources Engineering
See more of this Group/Topical: International Congress on Energy 2011

(192e) Analysis of Heat Transfer Fouling Characteristics During Evaporation of Lignocellulosic Biomass Process Streams

Raghu N. Gurram, **Patrick C. Gilcrease** and **Todd Menkhaus**, Chemical and Biological Engineering, South Dakota School of Mines and Technology, Rapid City, SD

Lignocellulosic feedstocks such as pine wood waste are becoming viable biomass sources for liquid fuel production. Within the ethanol production process, concentrating the sugar-rich hydrolysate stream by evaporation prior to fermentation can enhance the fermentation productivity and reduce downstream distillation costs. Similarly, evaporation can be used to concentrate the lignocellulosic thin stillage during solids recovery/water treatment. However, fouling of the evaporator surface by soluble and suspended solids within the process steams can become problematic. This can reduce the heat transfer rate and ultimately increase energy consumption of the process. In this study we analyzed fouling characteristics of pine wood hydrolysate and thin stillage using an annular fouling probe containing a heated surface to mimic a continuous evaporator operation. The heated region of the probe caused solids in the hydrolysate to adhere to the surface, thus fouling the probe over time. 3.5 L batches of process liquor were circulated through the annulus of the fouling probe while temperatures and power supplied to the probe were measured over time and used to calculate fouling resistance, rate of fouling and induction period. Hydrolysate samples were collected at regular intervals to monitor the sugar and inhibitory compound concentrations. Liquor pH, Reynolds number (laminar region 440 and 880 Re) and feed composition were analyzed for their effects on fouling rate, induction period and fouling deposit ash concentration. Two analytical methods: Inductively Coupled Plasma (ICP) and Scanning Electron Microscopy-Energy Dispersive Spectroscopy (SEM-EDS) were used to evaluate and compare the mineral composition analysis of fouled material, while ash was determined by the dry ash method. Due to an elevated concentration of inhibitory compounds in solution following evaporation, an adsorption/ polyelectrolyte removal operation was performed prior to fermentation analysis. Optimizing operating conditions and evaporator materials of construction will lead to enhanced biorefinery process efficiencies.

See more of this Session: Poster Session: Sustainable Forest Bioresources Engineering
See more of this Group/Topical: International Congress on Energy 2011

(192f) Characterization, Evaluation, and Fouling Analysis of Reverse Osmosis Membranes for Concentrating Sugars In a Lignocellulosic Biomass Hydrolysate

Amitkumar Gautam, Todd Menkhaus, Patrick C. Gilcrease and Jason C. Hower, Chemical and Biological Engineering, South Dakota School of Mines and Technology, Rapid City, SD

In today's era, with increasing fuel prices and greater dependence on petroleum resources, lignocellulosic biomass offers tremendous potential as an alternative renewable feedstock for fuels and chemicals. An example of this is wood chips from Ponderosa Pine, which after pretreatment and enzymatic hydrolysis consists of a slurry containing fermentable sugars. Unfortunately, due in large part to the limited solids loading capacity of many pretreatment and hydrolysis operations, the concentration of sugars is relatively low. Thus, concentrating the sugars prior to fermentation can lead to more efficient production and purification of renewable fuels and chemicals by allowing for higher titers in the fermentation broth. In this study, wood chips from Ponderosa Pine, after pretreatment and enzymatic hydrolysis, contained approximately 20 g/L of glucose. This hydrolysate was clarified with a 0.2 micron microfiltration and then fed to a reverse osmosis (RO) system for concentrating the sugar to over 100 g/L. During the RO operation, severe fouling was observed on the membrane. Thus, in order to develop improved membranes for processing the complex lignocellulosic hydrolysate, an understanding of the fouling behavior was needed. Here we investigated the membrane fouling phenomena including operational variables (mixing versus no mixing), identification of components in the complex hydrolysate responsible for fouling, and properties of the membrane that contributed to greater observed fouling. Different RO membranes were tested to measure the final concentration of the sugars (as well as other soluble and insoluble species) and the fouling characteristics of each of the membranes were studied using SEM, AFM, contact angle, BET and FTIR spectroscopy. In addition, the molecular modeling software Charmm was used to theoretically evaluate the interactions between different membrane chemistries and the components present in a lignocellulosic enzymatic hydrosylate. Molecular modeling was also used to guide design of membrane modifications that could be used to potentially improve performance.

See more of this Session: Poster Session: Sustainable Forest Bioresources Engineering
See more of this Group/Topical: International Congress on Energy 2011

(192g) Evaluation and Optimization of Biofuel Production Fromsunflower Hulls Using Central Composite Design Analysis

Srinivas Reddy Kamireddy, Christopher Schaefer and Yun Ji, Chemical Engineering, University of North Dakota, Grand Forks, ND

Theannual production of sunflower seeds in USA is around 1.8 million metric tons.Sunflower seeds are used mostly in the production of edible oil. This resultsin large amount of sunflower hulls and stalks during the industrial processing.Sunflower hulls have less commercial value and these hulls decompose veryslowly. Sunflower hulls mainly consist of cellulose, hemicellulose and lignin andthe presence of these compounds is a major driving force for biofuel research. Thesunflower hull is rich in 5-carbon (C-5) and c-carbon (C-6) sugars that areideal for the production of advanced biofuels. The objective of this project isto evaluateand optimize the biofuel production from sunflower hulls as a non-food resourceandthus improvethe local economy and reduce the dependence of our nation onforeign sources of energy.

Thisproject is divided into two sections (pretreatmentand enzymatic hydrolysis) to evaluate thesunflower hull as a potential biofuel feedstock. Sunflower hulls arepulverized to an average particle size between 0.1 mm to 0.5 mm using Wileymill. Extractives in the sunflower hulls will be removed by both waterextraction and ethanol extraction using Soxhlet apparatus. Raw sunflower hullscompositional analysis will be performed by using NREL LAP procedures. The sunflower hull pretreatment uses dilutesulfuric acid at 160°C, 0.5% sulfuric acid and 10 minutes residencetime. Structural carbohydrates of all slurry samples will be analyzed usingAgilent 1200 High Performance Liquid Chromatography (HPLC)and acid soluble lignin will be evaluated by Thermo Scientific Evolution 600UV-VIS Spectroscopy. The enzymatic hydrolysis will be conducted using central composite design method. The experiments will be performed at different enzyme loadings (5, 10, and 20 mg/g biomass),different reaction temperatures (40, 50 and 60°C) and different solid contents(1%, 5%, and 10%). The enzymatic digestibility ofthe pretreated solids will be measured to evaluatethe enzyme accessibility and yield. A comparision study of hulls will be doneusing GC 220 and Accellerase 1500 enzymes supplied by Genencor.

See more of this Session: Poster Session: Sustainable Forest Bioresources Engineering
See more of this Group/Topical: International Congress on Energy 2011

(192h) Fermentablity of Organosolv Prehydrolysate From Loblolly Pine and Sweetgum

Mi Li[1], Maobing Tu[2] and Dongxu Cao[1], (1)School of forestry and wildlife science, Auburn University, Auburn, AL, (2)Auburn University, Auburn, AL

Forestry lignocellulosic biomass is a great potential source of sustainable materials for cellulosic ethanol bioconversion process, which requires pretreatment, enzymatic hydrolysis and fermentation. In order to improve accessibility of lignocellulose to enzymes for hydrolysis, high temperature pretreatment with acid or base is typically needed to break down the recalcitrant structure of cellulose, and solubilize hemicellulose. However, during the pretreatments, considerable amount of inhibitors will be generated from degradation of cellulose, hemicellulose and lignin; many of these inhibitors will significantly reduce the microbial growth and ethanol productivity during the subsequent fermentation step. This inhibitory effect has become one of the major barriers to develop an economic viable process for cellulosic ethanol production. Identification of these fermentation inhibitors is widely recognized as one of the roadblocks for developing any effective detoxification approach and improving stress-tolerance yeast strain. The objective of this study is to examine the fermentability of pre-hydrolysate after organosolv pretreatment of loblolly pine (*Pinus taeda*) and sweetgum (*Liquidambar*). New approach will be utilized to identify microbial fermentation inhibitors in the prehydrolysate of loblolly pine and sweetgum during the bioconversion of forestry biomass to ethanol. Various detoxification methods will be explored for improving the fermentation yield of lignocellulosic pre-hydrolysate.

See more of this Session: Poster Session: Sustainable Forest Bioresources Engineering
See more of this Group/Topical: International Congress on Energy 2011

(192i) Development of Ionic Liquid-Based Consolidated Bioprocessing (i-CBP) for Bioethanol Production

Nobuiro Ishida, Biotechnology Laboratory, Toyota Central R&D Labs., Inc., Nagakute, Aichi, Japan, Satoshi Katahira, Toyota Central R&D Labs., Inc., Nagakute, Aichi, Japan, Wataru Tokuhara, Organic Material Engineering Div, Toyota Motor Co., toyota, Aichi, Japan, Yoshiyuki Noritake, Toyota Motor Co., toyota, Aichi, Japan, Noriho Kamiya, Center for Future Chemistry, Kyushu University, Fukuoka, Japan, Kazunori Nakashima, Graduate School of Engineering, Tohoku University, Sendai, Japan, Chiaki Ogino, Graduate School of Engineering, Kobe University, Kobe, Japan and Akihiko Kondo, Department of Chemical Science and Engineering, Graduate School of Engineering, Kobe University, Kobe, Japan

Lignocellulose that is the primary polysaccharide of plant cell wall has been received considerable attention as a main feedstock for bio-refinery process such as bio-fuel production. The enzymatic hydrolysis of lignocellulose to soluble sugars is considered to be one of the environmental friendly processes for bio-ethanol production. But the spontaneous crystallization of cellulose due to the chemical uniformity of glucose and high degree of hydrogen bonding can form densely packed micro-fibrils which are inaccessible to cellulolytic enzymes. Therefore, efficient and cost-effective methods for the degradation and fermentation of lignocellulosic biomass to ethanol are required.

In this study, for deconstruction of biomass, the ionic liquid was used as a pre-treatment medium, and the effective degradation and assimilation procedure was investigated. This effective process has been named as 'Ionic liquid-based Consolidated Bio-Processing (i-CBP)'. In this process, pretreated biomass by ionic liquids would be easily hydrolyzed to glucose and directly converted to ethanol by functional transgenic yeasts that were simultaneously displayed four kinds celluloses, endoglucase (EG), cellobiohydrolase (CBH I & II), and beta-glucosidase (BGL) on the cell surface. Research findings from these studies will be presented.

See more of this Session: Poster Session: Sustainable Forest Bioresources Engineering
See more of this Group/Topical: International Congress on Energy 2011

Lipid Production From Lignocellulosic Materials Via Filamentous Fungi

Jianguo Zhang and Bo Hu, Bioproducts and Biosystems Engineering, University of Minnesota, Saint Paul, MN 55108

Introduction

Converting lignocellulosic materials to microbial lipids is a promising approach to produce "drop-in" biodiesel fuel because the raw materials are widely available and are not competing with human food [1]. Lignocellulose consists of three major components: cellulose, hemicellulose and lignin, as well as small amounts of other materials such as ash, proteins and pectin in different degrees based on the sources. The efficient conversion process is necessary to reach cost effective biodiesel production.

Due to the complicated structure lignocellulosic materials have, several pretreatment methods were reported to disrupt lignin crystallization and/or degrade hemicellulose in order to make cellulose more accessible for subsequent enzymatic hydrolysis to generate fermentable sugars [2]. These pre-treatment methods include acid or alkaline pre-treatment, wet oxidation, steam pre-treatment, ammonia fiber explosion, etc. Dilute acid pretreatment was found to be promising due to its simplicity, effectiveness and low cost [3]. Dilute acid pretreatment involves the solubilisation of hemicellulose and disruption of lignin by dissolving acid-soluble lignin [4] and mainly yields xylose as a major sugar with minor amount of glucose in the pre-treated liquid fraction. The residual solid (pretreated biomass) would be readily degrable by cellulase due to the improved enzyme accessibility[5].

In this research, oleaginous filamentous fungi were used as oil producer to accumulate lipid from treated lignocellulosic materials. Oleaginous filamentous fungi have the advantages on lipid production from lignocellulosic materials, such as: all sugar utilization, inhibitor tolerance, high lipid content, fast growth rate, high cell density, and production of hydrolytic enzymes that may decrease the need for external enzymes. Many filamentous fungi may form pellets during their cultivation, which significantly facilitate the cell harvest [6]. *Mortierella isabellina* (*M. isabellina*) was used as the model strain to accumulate lipid through several kinds of conversion processes, such as submerge cultivation, simultaneous saccharification and fermentation, solid state fermentation. *M. isabellina* can utilize some pure carbohydrates, such as xylose, glycerol, glucose [7] and pectin, starch, lactose to synthesize lipids [8]. Soybean hull and corn stover were used as lignocellulosic materials in this work. Soybean hull constitutes about 8-10 % of the soybean seed which is the 2rd largest cultivated plant in worldwide. Soybeans hulls are low valued byproducts from soybean processing. They are sometimes considered as waste only for combustion due to the low and anti-nutritional protein value even as the animal feed supplements. Some researchers tried to produce alcohol from soybean hull to obtain high protein value treated soybean hull [9]. The major components of soybean hull include 40–45% cellulose, 30–35% xylan, 9-12% protein and only 3-4% lignin. From the view of low lignin content, Soybean hull is a good resource for lipid production because lignin was the hardest part to be degraded [10, 11]. Another lignocellulosic material tested in this research is corn stover which is the largest source of agricultural materials, contributing 75% of all agricultural residual[12].

Materials and methods

M. isabellina IFO 7884 was purchased from National Biological Resource Center (NBRC), Chiba, Japan. *M. isabellina* spores were harvested from PDA agar medium (Potato dextrose 24g/L, Agar 15g/L) with 10 ml distilled and sterilized water. *M. isabellina* spores were stored in 25 % glycerol at -70 °C for usage.

Soybean hull was obtained from Minnesota Soybean Processor, Brewster, MN, USA. Corn stover was obtained from University of Minnesota Southern Research and Outreach Center (waseca, MN). Lignocellulosic materials were grinded into particle (Φ 0.625 mm) by grinder (475A Arthur H. Thomas Co. Philadelphia, USA). The lignocellulosic materials particles were dried in oven (Model 28 Precision Scientific, Chicago, IL, USA) at 105 °C for 24h.

Diluted acid pretreatment of corn stover

Corn stover was immersed into 2% sulfuric acid for 16 h after well-mixed at 5 % loading ratio. And then the corn stover was heated for 60 min at 121 °C. The liquid portion was neutralized by sodium hydroxide to pH 5.3 after filtration. The solid portion t was washed until pH at around 7, and then put in refrigeration for use.

Batch cultures of M. isabellina

M. isabellina was cultivated in 250 ml flask containing 100 ml medium (KH_2PO_4 2g/L, $MgSO_4 \cdot 7H_2O$ 1.5g/L, $CaCl_2 \cdot H_2O$ 0.1mg/L, $CoCl_2 \cdot H_2O$ 0.1mg/L, $MnSO_4 \cdot 5H_2O$ 0.1mg/L, Glucose 45g/L). *M. isabellina* spores were inoculated into the flask at concentration 1.05×10E5/L.*M. isabellina* cells were harvested after 6 days cultivation at 27 °C and 180 rpm. Different C/N ratio medium was obtained by changing nitrogen concentration and keeping initial glucose concentration constant.

Submerge fermentation

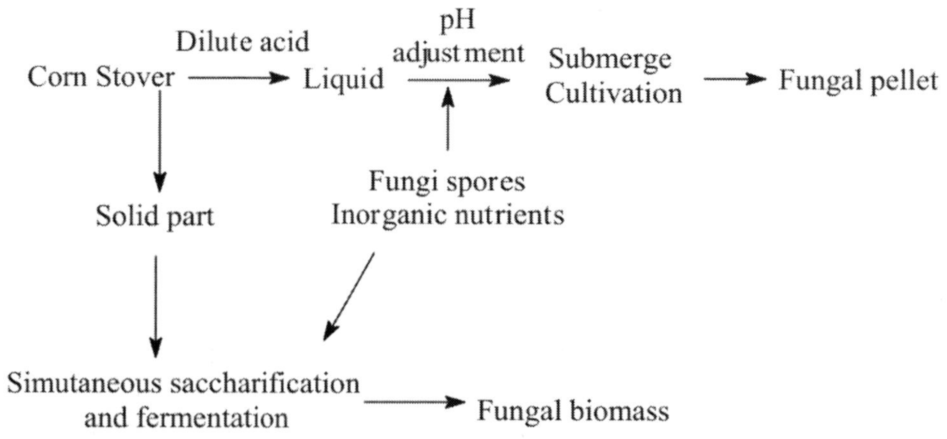

Fig.1 Submerge conversion processes of corn stover after diluted acid pretreatment.

The liquid part from diluted acid pretreatment of corn stover was used as carbon source for *M. isabellina* growth after different ratio of dilution. Other nutrients were added with the same concentrations mentioned in the batch cultures. *M. isabellina* cells were harvested after 6 days of cultivation at 27 °C and 180 rpm.

Simultaneous saccharification and fermentation

Soybean hull was used for this experiment and four enzymes (cellulase, β-glucosidase, pectinase, hemicellulase) were added at the beginning of the flask cultures to hydrolyze the soybean hull to generate sugars for lipid production with *M. isabellina*. The 250ml flasks containing 5 gram soybean hull and 33 ml water were cultivated in the same conditions as batch flask cultures described previously. All flasks were shaken at 100 rpm in room temperature (about 27 °C) for 12 days. Data from triplicate experiments of every test were analyzed using the Excel software, Microsoft (Redmond, WA, USA). Corn stover was also tested as the raw material for the simultaneous saccharification and fermentation, and cellulase and galactase were added for hydrolysis . The culture condition is same as Simultaneous saccharification and fermentation of soybean hull.

Solid state fermentation

Soybean hull (30g) in 250 ml flask was sterilized at 121 °C for 20 min after mixed with water. *M. isabellina* spores were inoculated into solid state fermentation system after cooling down. The solid state fermentation was carried out at 27 °C after addition of sterilized water to meet the initial moisture contents. Samples were taken every week after mixed manually. Fresh samples were weighted as well as dry samples after 24 h store in over at 105 °C. All experiments were triple for standard division analysis. Four factors were considered in this work. The first one was moisture content. The second one was inoculums size. The third was spores age. The last factor was nutrients addition to avoid the salts defiance.

Analysis

Dry biomass was measured by baking the solid portion of fermentation broth after centrifugation at 7,000 rpm for 7 min in oven (105 °C) to constant weight.

DNS method was used to determine residual sugar concentration in the fermentation broth. HPLC method was carried on to analyze sugar released from lignocelluloses (NREL LAP Determination of Structural Carbohydrates and Lignin in Biomass).

Elements (carbon, hydrogen and nitrogen) of soybean hull were analyzed by CE-440 element analyzer (Exeter Analytical, Inc. Chelmsford, MA, USA). Ash was weighted after soybean hull had been fired in the muffle oven (59344 Lindberg Hevi-Duty, Wisc, USA) at 550 °C over night.

The moisture content of soybean hull sample was determined as percentage of water in fresh soybean hull according to the equation 1. Weigh loss was calculated according to equation 2:

$$M_i\ (\%) = \frac{FW_i - DW_i}{FW_i} \times 100$$

Equation 1

$$W_i = 30 - (1 - M_i/100) \times TW_i$$

Equation 2

FW_i: fresh sample weight at week i,

DW_i: fresh sample weight at week i,

M_i: Moisture content at week i, i=1, 2, 3, 4, 5, 6, 7, 8,

TW_i: total weight of soybean hull at week i,

Glucosamine contents of soybean hull samples were determined through the method of glucosamine [13].

For total lipid analysis, dry sample (0.1-0.3 g) was mixed with 10 ml mixture (Chloroform: Methanol=2:1), and was shaken for 16 h on a shaker at 180 rpm. Down layer was harvested by Φ 0.45 um filter after 2.5 ml water added, vortex for 1 min and centrifuged for 7 min at 7,000 rpm. The total lipid was weighted after solvent was evaporated by compressed nitrogen feeding.

Lipid profile analysis was performed by two consecutive steps including preparation of fatty acid methyl ester (FAME), which was known as transesterification, and GC analysis.

Results

M. isabellina growth at different carbon sources

M. isabellina grow well on Petri dishes with single carbon source (Table 1). *M. isabellina* grows on different concentration of glucose or xylose. The results showed that *M. isabellina* grew well even glucose reached to 180g/L . As for xylose, *M. isabelina* grow well at 45g/L. The inhibition of growth occurred if xylose reached to 90g/L (Fig. 2). High sugar concentration leads to low lipid content, especially xylose (Fig.3). Therefore, *M. isabellina* can endure higher concentration of glucose compared to xylose.

Table 1 *M.isabellina* grow on different carbon sources.
(++: grow well, + grow not well-: no grow)

	glucose	fructose	galactose	Lactose	mannose	xylose
48h	++	++	++	++	++	++

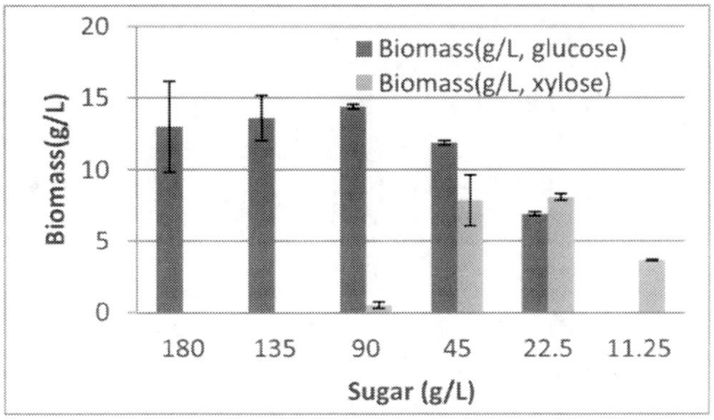

Fig.2 Effects of sugar concentration on *M. isabellina* biomass

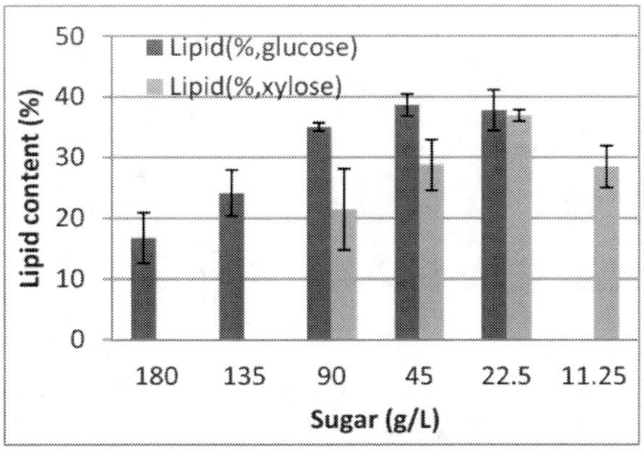

Fig.3 Effects of sugar concentration on lipid contents of *M. isabellina*

M. isabellina growth at different inhibitors at different concentration

Several inhibitors produced during the diluted acid pretreatment. Furfural, Hydroxymethylfurfural (HMF), acetate, and 4-Hydroxybenozic acid were reported as the major chemical compounds. These chemicals inhibit microorganism growth when certain concentration was reached. Table 2 showed the biomass and lipid content at different concentrations of inhibitors (Table 2). It showed that furfural is the leading inhibitor of all organic chemicals generated from the diluted acid pretreatment and the fungal strains have lower threshold of tolerance [5]. *M. isabellina* can endure higher concentration of furfural, compared with other fungal species which showed strong inhibition from furfural at 244.22 mg/L [14].

Table 2 Effect of inhibitor concentrations on *M. isabellina* biomass and lipid content

	control	Furfural (g/L)			HMF (g/L)				Acetate (g/L)			4-Hydroxybenozic acid (g/L)			
		0.1	0.3	0.5	0.3	0.8	1.2	2	2	6	10	0.2	0.4	0.6	1
Biomass (g/L)	10.62	10.65	6.54	-	12.59	12.76	12.84	1.16	10.45	12.79	7.93	10.38	8.66	6.94	6.35
STDEV	0.94	0.16	1.65	-	0.10	0.32	0.45	0.41	1.59	1.49	1.38	0.09	0.91	1.67	2.07
Lipid (%)	33.36	34.06	25.61	-	39.93	36.48	34.86	-	24.27	31.81	26.64	29.57	29.27	27.29	19.60
STDEV	2.52	2.30	5.29	--	1.59	1.84	0.34	-	4.18	1.05	1.73	2.00	2.39	3.29	7.66

Effect of C/N ratio on the biomass and lipid content of M. isabellina

M. isabellina biomass increased as C/N ratio increased from 5.47 to 21.90, and then the biomass decreased from 9.28g/L to 3.50g/L as C/N ratio continuous increasing to 218.96. Lipid content of *M. isabellina* increased as C/N ratio increased from 5.45 to 21.90 and kept stable when C/N ratio is higher than 21.90. This revealed that the stress condition to induce oil accumulation for *M. isabellina* is significant lower than C/N ratio of others [15]. Our result was also confirmed by Suzuki [16] which showed optimized C/N ratio is 40 at the 180 g/L glucose of high cell density cultivation *M. isabellina*. The lipid profile analyzed by GC, showed that lipid profile at different C/N ratio medium is constant, as well as different cultivation time (data not shown). The elements of this soybean hull were determined as shown in Table 3 with C/N at 21.38. Therefore, C/N ratio of soybean hull fitted for *M. isabellina* to accumulate lipid.

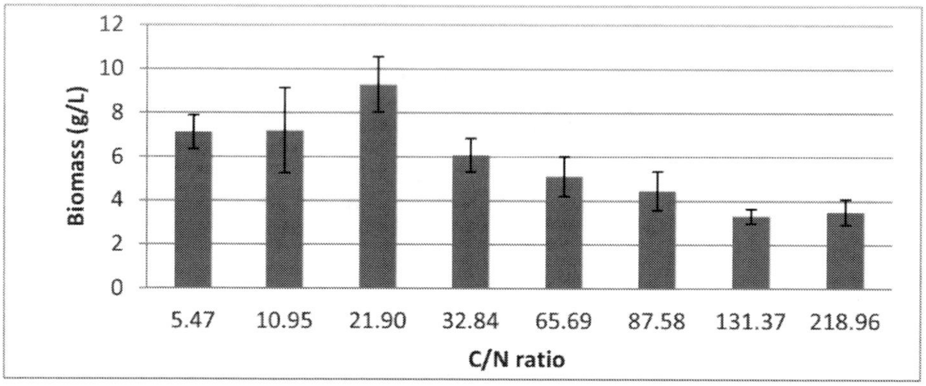

Fig.4 Effect of C/N ratio on biomass of *M. isabellina*

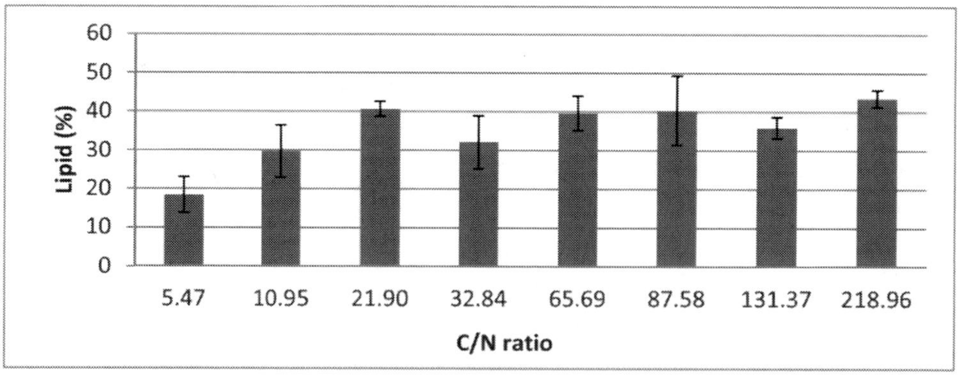

Fig.5 Effect of C/N ratio on lipid content of *M. isabellina*

Lignocellulosic materials and its pretreatments

As for diluted acid pretreatment of corn stover, 40% sugar was released by DNS method. Within the solid part, cellulose contributed 65.92%according to acid hydrolysis.

Table 3. Components of soybean hull used in this work

Element	Percentage (%)	STDEV
C	42.47	0.39
H	6.10	0.17
N	2.32	0.07

Ash	5.25	0.11

Submerge cultivation of M. isabellina in liquid part after diluted acid pretreatment

M. isabellina grew on the liquid portion of corn stover after acid pretreatment after different ratio of dilution with the highest biomass obtain at 2 times of dilution (Fig.6). The lipid yields from sugar were constant among the different dilution (1 to 4 dilution ratio). Fungal pellets were formed after 4 days cultivation and cells were harvested after 6 days cultivation. Lipid content was not affected by dilution at this condition. Fungal pellets were formed well during the cultivation which is easy to harvest. Therefore, Dilution (2 to 4 dilution ratio) is a possible method for lipid accumulation and *M. isabellina* growth.

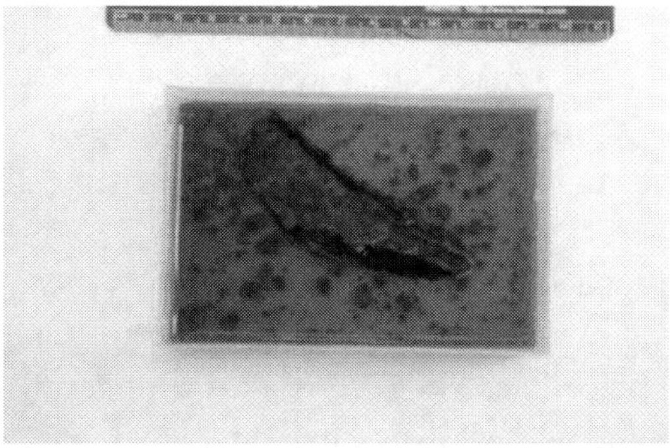

Fig. 6 Fungal pellets in submerge cultivation of *M.isabellina* in liquid part

Enzymatic hydrolysis of solid part

Enzymatic hydrolysis of pretreated corn stover was investigated at different cellulase loading ratio (Fig.7), and different solid loading (Fig. 8). Cellulase activities waere enough at 30U/g solid for sugar release at this condition, which has 55% sugar released after 6 days of hydrolysis. Sugar release ratio didn't change significantly when solid loading ratio increased from 1% to 5%. Sugar released ratio decreased when loading ratio was higher than 5%.Therefore, 5% loading ratio will be used for further research.

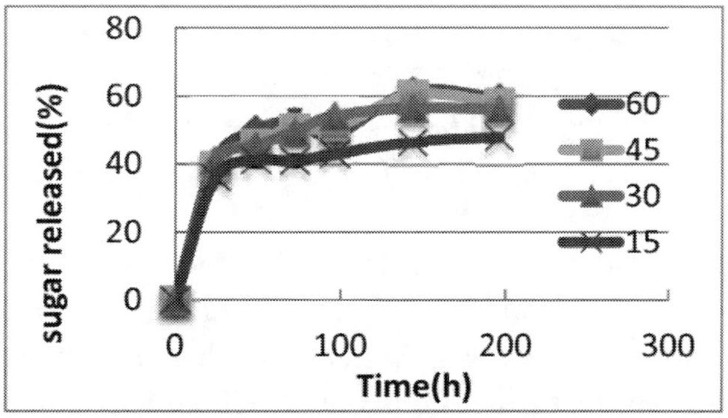

Fig.7 Effect of cellulase loading ratio on sugar release from pretreated corn stover after dilute acid pretreatment

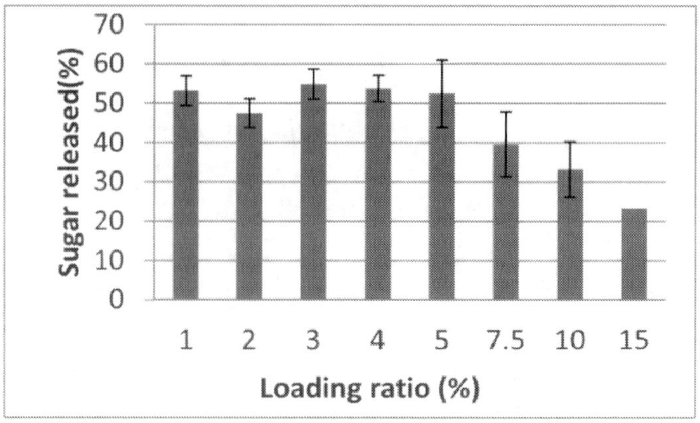

Fig.8 Effect of solid loading ratio on sugar released from solid part of corn stover after dilute acid pretreatment

Simultaneous saccharification and fermentation

Weight loss of soybean hull reached above 30% when cellulase was added, which was significant compared to that of control (4.4%). Weight loss and fatty acid increased as cellulase loaded increased until 12 U/g soybean hull. And there were a little decrease at 24 U/g soybean hull although total lipid increased continuesly. The total lipid of sample with cellulase loading is a little higher than that of control with or without inocululation. In was showed that fatty acid part of total lipid increased significantly when cellulase loaded. Glucosamine content determination indicated that *M. isabellina* biomass increased as cellulase activity increasing.The curve of weight loss was a bell shape as glucosidase loaded increased to 81.6U/g soybean hull. The highest weight loss reached 47.80% when 54.4 U/g soybean hull glucosidase loaded. Generally, the trends of total lipid and fatty acid content were decreasing and increasing respectively with some fluctuations as β-glucosidase loaded increased from 0 to 81.6 U/g boybean hull. At the same time, the glucosamine content increased from 44.11mg to 91.92mg/g soybean hull, which indicated that glucosidase was helpful to *M. isabellina*'s growth.

Weight loss increased from 37.40% to 47.40 %, and then decreased to 37.20 as pectinase loaded increasing to 6835.64 U/g soybean hull. Glucosamine content increased from to 59.53 mg to 70.63 mg. And the total lipid and fatty acid content were increased from 33.62 mg and 24.76 to 43.61 mg and 33.84 respectively at the same time.

As hemicellulase loaded increasing from 15U/g to 120 U/g soybean hull, weight loss increased to 47.40% and then decreased to 43 %. Glucosamine contents decreased to 57.58 mg when 30 u/g soybean hull hemicellulase was loaded, which was not significant to weight loss of other hemicellulase activities loaded. Total lipid fluctuated from 34.08 mg to 35.97 mg at different activities of hemicellulase loaded. Fatty acid decreased when activities of hemicellulase loaded increased.

Lipid profile analysis showed that about 70 % of total lipid was fatty acid (C13-C25). There were similar fatty acid profiles in different kinds of hydrolytic enzymes and different hydrolytic enzymes activities. The total percentage of C18:1 and C16:1 was 39.58 %. The lipid produced fitted for biodiesel production because the ideal biodiesel fatty acids were C18:1 and C16:1 [17] [18]. The total fatty acid increased when cellulase or peinctinase activity increased, and decrease when hemicellulase activity was higher than 15 u/g soybean hull. B-glucosidase hadn't significant effects on the fatty acid accumulation in *M. isabellina*. This was consistent to the chemical composites of soybean hull as showed in snyder's research (Snyder, 1987). The insoluble carbohydrate fraction of soybean hull consists of 30 % pectin, 50 % hemicellulose and 20 % cellulose. Similar lipid profile was got when lipid profile was used to compared to the lipid profile of submerge in suzuki's report [16].

Solid state fermentation

Initial moisture content of soybean hulls significantly affected the microbial cell growth and lipid accumulation although the general trends of these parameter changes were similar. Moisture levels of different initial moisture content increased at the beginning six weeks, and kept stable in the last two weeks The moisture value increased to 76.22%, 78.53%, 83.63%, and 88.12% from 50%, 66.70%, 75%, and 78.57% respectively after eight weeks of cultivation. The largest moisture increase, 26.22%, was seen at samples with 50 % of initial moisture level. We also tracked the weight loss of these samples and the results showed that all samples had relatively stable rate of weight loss over the eight weeks. More weight loss was obtained at samples with lower moisture content within our four initial moisture levels experiments. The total lipid produced after eight weeks cultivation increased as moisture increasing although the total lipid value at 75 % moisture content is relatively lower. The highest lipid production (38.90 mg per gram soybean hull) was obtained at the initial 78.57 % moisture.

The moisture values of four inoculums sizes increased significantly after three weeks cultivation. And the moisture contents increased slowly after six weeks cultivation, especially of 10×E4 and 10×E7 inoculums. At week eight, the highest weight loss was obtained as 11.32% at inoculums 10×E7. Lipid production decreased according to the increasing inoculums size from 10×E4 to 10×E7. The highest total lipid value (47.92 mg per gram soybean hull) was obtained at 10×E4 inoculums.

The moisture values of soybean hull with 10 days spores and 14 days spores didn't have significant difference (P=0.13). The moisture content of soybean hull with 7 days spores was lower than that of other two spore ages. The total lipid values increased from 29.18 mg to 44.27 mg when *M. isabellina* spores age became elder from 7 days to 14 days).

As for lipid production, total lipid increased as 20 % inorganic salts addition to 45.95 mg per gram soybean hull, and decreased when 100% nutrients were added. Our results showed that total lipid increased at the beginning four weeks, and kept stable till the end. Morphology of *M. isabellina* was checked through microscope as yeast-like. We counted the spores number using microscope. Cell numbers increased at the beginning four week, and then, cell numbers kept stable till week eight. Fatty acid consisted 30.05% of total lipid. Most parts of fatty acid are the C16 (including C16:0 and C16:1) and C18 (including C18:0, C18:1, C18:2).

It was indicated that carbohydrates of soybean hull were converted into lipid through *M. isabellina* cultivation because total lipid was increased. *M.isabellina* cell utilized soybean hull as substrate to reproduce itself and accumulate lipid. Among the four factors, moisture content is the only factor that has the significant effect on weight loss which was confirmed by Sharma et al [19]. Moisture contents increased in all of the experiments which were consistent to Oriole's report [20]. At low moisture level (50%), more soybean hull was oxidized to produce carbon dioxide and water which was confirmed by the largest the moisture contents increasing. Low diffusion of nutrients and metabolites took place in lower moisture conditions [21].

Conclusion

M. isabellina is a promising oil producer based on its high lipid content, inhibitor tolerance, and good growth at high sugar concentration. To convert lignocellulosic materials to lipid efficiently, several conversion processes were investigated to accumulate lipid frower in solid statm corn stover and soybean hull. *M. isabellina* can form the pellets well in liquid part of corn stover after diluted acid pretreatment by 2 fold dilution. High solid loading ratio of soybean hull during simultaneous saccharification and fermentation process provided high lipid production. Compared to simultaneous saccharification and fermentation process, the conversion ratio is loe fermentation process.

Reference

1. Bajpai, R., et al., *Microbial lipids from renewable resources: production and characterization.* Journal Of Industrial Microbiology & Biotechnology, 2010. **37**(12): p. 1271-1287.
2. Mosier, N., et al., *Features of promising technologies for pretreatment of lignocellulosic biomass.* Bioresource Technology, 2005. **96**(6): p. 673-686.
3. Hendriks, A.T.W.M. and G. Zeeman, *Pretreatments to enhance the digestibility of lignocellulosic biomass.* Bioresource Technology, 2009. **100**(1): p. 10-18.
4. Wolfrum, E.J. and A.D. Sluiter, *Improved multivariate calibration models for corn stover feedstock and dilute-acid pretreated corn stover.* Cellulose, 2009. **16**(4): p. 567-576.
5. Mathew, A.K., et al., *Dilute acid pre-treatment of oilseed rape straw for bioethanol production.* Renewable Energy, 2011. **36**(9): p. 2424-2432.
6. Hu, B., et al., *A new cultivation method for microbial oil production: cell pelletization and lipid accumulation by Mucor circinelloides.* Biotechnology For Biofuels, 2011. **4**.
7. Fakas, S., et al., *Evaluating renewable carbon sources as substrates for single cell oil production by Cunninghamella echinulata and Mortierella isabellina.* Biomass Bioenerg, 2009. **33**(4): p. 573-580.
8. Papanikolaou, S., et al., *Lipid production by oleaginous Mucorales cultivated on renewable carbon sources.* European Journal Of Lipid Science And Technology, 2007. **109**(11): p. 1060-1070.
9. Schirmer-Michel, A.C., et al., *Production of ethanol from soybean hull hydrolysate by osmotolerant Candida guilliermondii NRRL Y-2075.* Bioresource Technol, 2008. **99**(8): p. 2898-2904.
10. Himmel, M.E., et al., *Biomass recalcitrance: Engineering plants and enzymes for biofuels production.* Science, 2007. **315**(5813): p. 804-807.

11. Sanchez, C., *Lignocellulosic residues: Biodegradation and bioconversion by fungi.* Biotechnol Adv, 2009. **27**(2): p. 185-194.

12. Matthew T. Carr, J.R.H., *Sustainable Production of Cellulosic Feedstock for Biorefineries in the USA*, in *Biofuels*, E.J.V. Wim Soetaert, Editor. 2009. p. 9-38.

13. Roopesh, K., et al., *Comparison of phytase production on wheat bran and oilcakes in solid-state fermentation by Mucor racemosus.* Bioresource Technology, 2006. **97**(3): p. 506-511.

14. Jian Zhang, Z.Z., Xiaofeng Wang, Nan Wang, Wei Wang, Jie Bao, *Biodetoxification of toxins generated from lignocellulose pretreatment using a newly isolated fungus, Amorphotheca resinae ZN1, and the consequent ethanol fermentation.* Biotechnology for Biofuels, 2010. **3**: p. 26.

15. Bhatia, I.S., Raheja, R. K., Sukhija, P.S., *fungal lipid IIa.effect of varing concentration of nigrogen in the medium and incubatioin period on the chemical composition of lipids of Pythium irregulare buisman.* J SciFd Agric, 1973. **24**: p. 779-787.

16. Suzuki, O. *Recent trends of oleochemicals by biotechnology.* in *World conference on oleochemicals*. 1990. Kuala lumpur.

17. Knothe, G., *Improving biodiesel fuel properties by modifying fatty ester composition.* Energy & Environmental Science, 2009. **2**(7): p. 759-766.

18. S. Pinzi, I.L.G., F. J. Lopez-Gimenez, M. D. Luque de Castro, G. Dorado,| andM. P. Dorado, *The Ideal Vegetable Oil-based Biodiesel Composition: A Review of Social, Economical and Technical Implications.* Energy & Fuels, 2009. **23**: p. 2325-2341.

19. Sharma, R.K. and D.S. Arora, *Production of lignocellulolytic enzymes and enhancement of in vitro digestibility during solid state fermentation of wheat straw by Phlebia floridensis.* Bioresource Technol, 2010. **101**(23): p. 9248-9253.

20. Oriol, E., et al., *Water and Water Activity in the Solid-State Fermentation of Cassava Starch by Aspergillus-Niger.* Appl Microbiol Biotechnol, 1988. **27**(5-6): p. 498-503.

21. Gervais, P. and P. Molin, *The role of water in solid-state fermentation.* Biochem Eng J, 2003. **13**(2-3): p. 85-101.

(192k) Characterization of a Forest Products Waste Effluent for Production of Fermentable Sugars and Co-Products During Hydrolysis

Jifei Liu[1], David Shonnard[1], Stephanie Groves[2] and Susan T. Bagley[3], (1)Chemical Engineering, Michigan Technological University, Houghton, MI, (2)Department of Biological Sciences, Michigan Techonological, Houghton, (3)Biological Sciences, Michigan Technological University, Houghton, MI

A forest product waste effluent with low solids percentage (≤10%) containing cellulose, hemicelluloses, lignin and other complex carbohydrates was treated with dilute acid to generate fermentable sugars. In order to understand the effects of the acid hydrolysis, Scanning Electron Microscopy (SEM) was employed to study the structure of the solids suspended in the waste effluent, and the solids remaining after the hydrolysis. Fourier Transform Infrared Spectroscopy (FTIR) and Nuclear Magnetic Resonance (NMR) were also used to evaluate the change in chemical composition due to the hydrolysis. In addition to the organic content, the mineral content of the solids was analyzed by Atomic Absorption (AA) Spectroscopy. Together with a sugar content analysis using a High Performance Liquid Chromatography (HPLC), characterization pre and post treatment provides a clear image of the acid hydrolysis process and compositional changes of the effluent solids during hydrolysis.

See more of this Session: Poster Session: Sustainable Forest Bioresources Engineering
See more of this Group/Topical: International Congress on Energy 2011

(192l) Enzymatic Hydrolysis of Spin Coated Regenerated Cellulose Thin Films Studied with a Quartz Crystal Microbalance

Halley Bernet[1], **Gary Brodeur**[2], John Telotte[3], John Collier[1] and Subramanian Ramakrishnan[4], (1) Chemical and Biomedical Engineering, Florida State University, Tallahassee, FL, (2)Chemical & Biomedical Engineering, Florida State University, Tallahassee, FL, (3)Department of Chemical and Biomedical Engineering, Florida State University, Tallahassee, FL, (4)Department of Chemical and Biomedical Engineering, Florida A&M University - Florida State University, Tallahassee, FL

During the last decade considerable effort has focused on understanding the mechanism of action of the cellulase enzymes, optimization of the enzyme mixture to result in maximum sugar yield and development of a comprehensive kinetics model to describe the hydrolysis process. However, a main difficulty (due to lack of experimental techniques) in studying enzymatic hydrolysis lies in directly measuring the amount of enzyme adsorbed onto the cellulose and the resultant hydrolysis rates. Recent developments in manufacture of thin cellulose films and in experimental techniques have helped overcome these difficulties. The aim of this thesis is to use a novel experimental technique - Quartz Crystal Microbalance with Dissipation monitoring (QCM-D) to study cellulose hydrolysis by different cellulases. Thin films of cellulose were successfully fabricated on gold crystals for hydrolysis study. Hydrolysis was carried out by three different cellulases: Accellerase™ 1000, Accellerase™ 1500 and Spezyme® CP; at different activities. The amount of enzyme adsorbed, the viscoelastic nature of the cellulose film and the subsequent hydrolysis rates were directly measured under different enzyme activities. It was found that the three enzymes had the same affinity for the cellulose film and that the cellulose thin films were hydrolyzed at a constant rate which was directly proportional to the amount of enzyme adsorbed. The measured specific rates for Accellerase™ 1000 and Spezyme® CP were similar but Accellerase™ 1500 was found to have a lower specific rate. This suggests that Accellerase™ 1000 and Spezyme® CP had a similar composition of endo and exo glucanases but the concentration of these enzymes in Accellerase™ 1500 composition was lower.

See more of this Session: Poster Session: Sustainable Forest Bioresources Engineering
See more of this Group/Topical: International Congress on Energy 2011

(192m) Separation and Purification of Lignocellulosic Wood Hydrolyzates by Polymer Flocculation and Scale up Studies

Rakesh Yasarla, Paper & Bioprocess Engineering, State University of New York, Syracuse, NY and Bandaru V. Ramarao, Paper and Bioprocess Engineering, Empire State Paper Research Institute, Syracuse, NY

Lignocellulosic hydrolyzates produced by hydrothermal treatment of wood chips contain hemicellulose sugars, acetic acid and significant quantities of colloidal material. The purification of the hydrolyzate stream and the separation of fermentable sugars from it thus constitute an important step in biorefinery processes. Raw lignocellulosic hydrolyzates produced by acid or water treatment contain hemicelluloses (primarily xylooligomers, xylose and xylan), acetic acid and significant quantities of colloidal material. The particulates are composed mostly of lignin and its derivatives which were ranging in between the nanometers to micrometers particle sizes and are anionic in nature.

Here, we report the separation of hot water based hydrolyzates of sugar maple by polymer induced flocculation. In the polymer induced flocculation, we investigated the dynamics of flocculation of wood extracts with particular emphasis on the action of four different flocculating agents – Alum, PEI, pDADMAC and PEO. The variation in zeta potential measurements showed that the colloidal particles in wood extracts are charge neutralized initially and particles were found to be increased in size after charge neutralization. The rate of sedimentation of aggregated particles was measured by turbidity of supernatants of the dispersed solution as a function of time. The optimal concentrations of flocculating agents to flocculate the colloidal particles were measured by both charge neutralization and rate of sedimentation. The pH sensitivity of flocculating agents was tested by varying the pH of wood extract between 2 and 8.0. The optimal concentrations of flocculating agents for sedimentation were found for different pH conditions. Flocculation and clarification with PEO were demonstrated on large pilot scale batches of 1000 kg of extracts. Further the flocculated extract was filtered and the filtrate showed a 99.5% reduction in turbidity, from 12000 NTUs in the raw extract to 50 NTUs in the filtrate.

See more of this Session: Poster Session: Sustainable Forest Bioresources Engineering
See more of this Group/Topical: International Congress on Energy 2011

STATISTICAL MODELING OF COATING COLOUR FORMULATION WITH A LOW COST BLEND OF STARCH-STYRENE BUTADIENE LATEX BINDER

NITIN MOON, N.C.MISHRA, AND A.K.RAY
DEPARTMENT OF PAPER TECHNOLOGY, IIT ROORKEE, INDIA

ABSTRACT

Styrene butadiene latex as binder and starch as co-binder are presently predominantly being used in Indian paper industry in coating color formulation along with pigments such as calcium carbonate and clay. Not only the cost of SB latex is much higher than that of starch, it is not environmentally friendly. However the use of starch and latex proportions in the binder are governed by the desired properties of paper .The effective cost of blends can be optimized in relation to optimizing coated paper properties.

The study planned in this present investigation mainly focuses on the variation of properties of coated paper with varying composition of coating color with clay as a pigment in which only binders and co-binders are changed. The study deals with various composition of starch and SB latex balance to get the targeted properties for all set of experiments. As the SB latex is used in almost all coating industries and more or less the properties remain constant, the starch has varying properties depending upon the source and the modification method used. The objective of the study is to find out the balance between costly SB latex and cheaper starch samples supplied by a well known Indian industry. The coating experiments were performed in "DT Paper science Finland make" laboratory coater.

Styrene Butadiene shows an increase in IGT, gloss, opacity, smoothness, brightness and whiteness with the increase in amount of SB latex in the binder, whereas porosity is found to decrease as the very small sized plastic particles of SB latex increase in the coating formulation. Starch samples show a similar nature of effect on properties of coated sheets. As the amount of starch in the coating color increases, the IGT value and the porosity are found to increase. But the opacity, gloss, smoothness, brightness and whiteness properties decrease with the increase in amount of starch in binder level in coating color formulation.

Based on the data experimentally evaluated in a pilot coater, various statistical regression analyses have been performed. A computer program was developed to optimize various options. A sample calculation has been shown to get the minimum cost of blend of latex and starch

Introduction:

Styrene butadiene latex and starch are being used in the paper industry as binders, either in combination or alone. The cost of the main binder SB latex is however higher than that of the co-binder, starch. However the use of starch and latex proportions in the binder are governed by the desired properties of paper. Large bodies of information are available regarding coating color preparation and application for preparation of pigment coated paper[1-14]. Experiments have been conducted to select a proper kind of starch .The effective cost of binder can be optimized by the detailed study on the mixing proportions of SB latex and starch samples in relation to coated paper properties. The reduction of overall binder cost in formulation studied here is not as a result of compromise of properties. The study deals with various composition of starch and SB latex balance to get the targeted properties for all set of experiments. As the SB latex is used in almost all coating industries and more or less the properties remain constant, the starch has varying properties depending upon the source and the modification method used.

OBJECTIVES:

The objectives of the study is to find out the balance between costly SB latex and cheaper starch samples supplied by Hindustan Lever Ltd. based on properties of coated paper which can give the lowest cost of total binder. The stepwise investigation is as follows:

1. to find out the quality of base paper supplied by an Indian mill for coating purpose.
2. to study the effect of latex on the various properties of coated paper
3. to formulate the various coating color with varying percentages of binder (latex) and cobinder (starch A, B, C) and to find out the characteristics
4. to coat the base paper with above coating formulations
5. to determine the optical properties (such as brightness, whiteness, opacity and gloss), porosity and surface strength in coated sheet

2.0 Experimental:

2.1 Materials: The binders used are Styrene Butadiene latex (SB latex) and specially prepared three starch samples manufactured by Hindustan Lever Ltd. The rest of the coating formulation is kept constant. The base paper has been supplied by Central Pulp and Paper Research Institute. The properties of the base paper measured in the laboratory are as follows:

Base Paper properties:

Property	Specifications
GSM	60 g/m^2
Brightness (ISO)	86.77 %
Whiteness (CIE)	90.99 %
Gloss	5 %
Opacity	81.14 %
Bendsten Roughness	120 ml/min.
Porosity	900 ml/min.

2.2 Methodology of investigation: The binder system in coating color can be written as follows:

Set 1: **pigment + SB latex**
Set 2: **pigment + starch A**
Set 3: **pigment + starch B**
Set 4: **pigment + starch C**

The specifications of the pigments, dispersing agent and optical brightening agent along with starch are shown in **Table 1.** The amount of $CaCO_3$, clay, dispersing agent and optical brightening agent taken in the experiment is shown in **Table 2.** The viscosities of the Starch A, B and C are evaluated through Brookfield synchrolectric viscometer available at M/S Hindustan Lever Ltd. Research Laboratory, Biopolymer Division, Pondicherry. The results are also shown inTable1.

Table 1: Coating component specifications

S.N.	Name	Specifications
1	$CaCO_3$	Brightness -93.5 % Particle size 90%<2μm.
2	Clay	Brightness – 88% Particle size 99%<2μm.
3	Styrene Butadiene latex	50% solids solution Particle size 99%<0.08-3 μm[3]
4	Starch A	Brightness – 86% Particle size 3μm avg. Viscosity – 52 cPS at 50^0C
5	Starch B	Brightness – 86% Particle size 3μm avg. Viscosity – 76 cPS at 50^0C
6	Starch C	Brightness – 86% Particle size 3μm avg. Viscosity – 80 cPS at 50^0C
7	Dispersing agent	Specifications unknown
8	Optical brightening agent	Specifications unknown

Table 2 Composition of coating color formulation without binder:

Coating ingredient	Parts based on 100 parts dry pigment by weight
$CaCO_3$	90 pph
Clay	10 pph
Dispersing agent	3.5 pph
Optical brightening agent	2 pph

In the set 1 the pigment and additives composition of coating formulation as shown in **table 2** is kept constant. The variation is made in the parts of SB latex in the coating formulation. The latex parts are varied from 10 -15 parts with the increment of 1 part in each subsequent experiment. The utmost care is taken to keep the coat weight in the range of 12-14 g/m^2 in all the experiments including replica experiments.

In set 2, the pigment and additive composition of coating formulation as shown in **table 4** is kept constant however the variation of binder is based on 14 parts equivalent strength of latex binder in the coating formulation when used alone. The approximate relation between latex binding power and starch binding power, obtained from literature[9], shows that the latex is about 1.5 times stronger than starch. So the starch A parts used in coating formulation in this set 2 can be given as

Starch A parts = (14 – latex parts) * 1.5
Or
Latex parts = 14 – (starch A parts/1.5)

The properties of coated paper and coating formulation are studied for this set of experiments for constant coat weight (12-14 g/m^2) and coating color solids (50%).

In set 3 and 4, the experimental composition of coating color is similar to the composition of set 2. The only difference is that the starch A is replaced by another variety of starch i.e. starch B and C respectively.

2.3 Equipment Description:
The coater used for the experimental coating trials was the DT Paper science Finland Laboratory coater, available at Central Pulp and Paper Research Institute, Saharanpur, India.

2.3.1 Main Features Laboratory Coater: Main features of this laboratory coater can be given as follows.

- Suitable for blade coating and film press coating studies
- Options for sheet and web blade coating
- Maximum speed 80 m/min
- Equipped with rod and blade coating facilities
- Equipped with infra-red drying with paper surface temperature from 66-85°C
- Equipped with 2-6 kW air dryer with maximum temperature of 180^{0}C
- Can handle web of any GSM in 25-30 cm width

2.4 Experimental Procedure: The Composition of the binder system and the viscosities of the coating color for four sets can be shown in the following tables.

Table 3: Composition for binder system for coating formulation of set 1:

Latex	Starch A	Starch B	Starch C	Viscosity in cps
Pph	pph	Pph	Pph	
10	0	0	0	184
11	0	0	0	176
12	0	0	0	189
13	0	0	0	203
14	0	0	0	215
15	0	0	0	231

Table 4: Composition for binder system for coating formulation of set 2:

Latex	Starch	Starch	Starch	Viscosity

	A	B	C	in cps
Pph	pph	Pph	Pph	
12	3	0	0	204
11	4.5	0	0	227
10	6	0	0	241
9	7.5	0	0	270
8	9	0	0	318

Table 5: Composition for binder system for coating formulation of set 3:

Latex	Starch A	Starch B	Starch C	Viscosity in cps
Pph	pph	Pph	Pph	
12	0	3	0	221
11	0	4.5	0	237
10	0	6	0	259
9	0	7.5	0	284
8	0	9	0	334

Table 6: Composition for binder system for coating formulation of set 4:

Latex	Starch A	Starch B	Starch C	Viscosity in cps
pph	Pph	Pph	Pph	
12	0	0	3	231
11	0	0	4.5	242
10	0	0	6	268
9	0	0	7.5	292
8	0	0	9	359

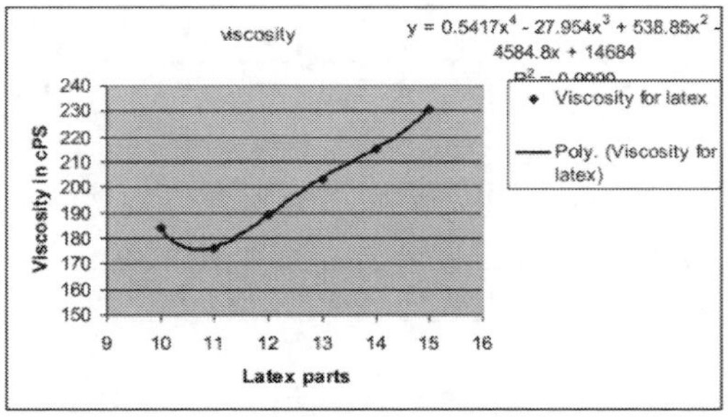

Fig 1: Effect of Latex Amount (parts) on viscosity of coating color.

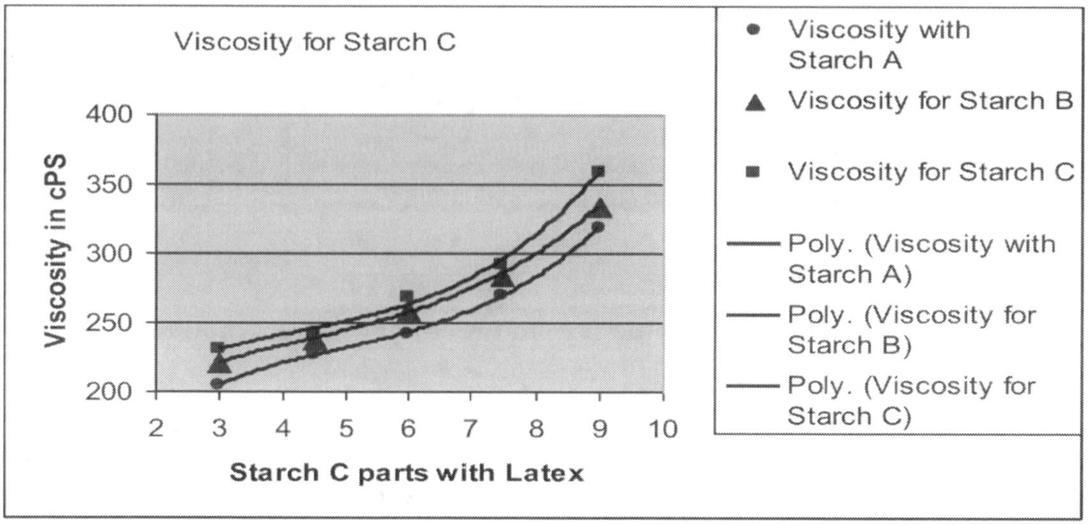

Fig. 2 Effect of Starch (A/B/C) with latex on viscosity for set 2 set 3 and set 4.

3.0 RESULTS AND DISCUSSION

The results obtained based on the various coating color formulations using latex and starch A, B, C are shown in table 7-10. The results are plotted in various graphs. The parametric influences are interpreted as follows.

The properties measured are: Brightness, Whiteness, Gloss, Roughness, Opacity and IGT Printability for each set of experiments. TAPPI standard methods are used to estimate the properties in Paper laboratory of Department of paper Technology, IIT Roorkee. The IGT printability test is performed at an accelerating speed at a constant force of 625 N.

Table 7: Properties of coated sheets using only latex as Binder

IGT	brightness	whiteness	Gloss	Roughness	Opacity	porosity
(m/s)	%	(CIE)%	%	ml/min	%	ml/min
0.73	86.19	89.94	46.37	30	86.59	95
0.77	86.43	90.37	49.33	30	87.34	85
0.85	87.21	90.88	50.41	20	89.41	85
0.85	87.88	90.85	52.57	15	91.44	80
0.88	88.3	91.25	54.51	15	92.84	70
0.91	88.91	91.31	55.16	15	92.79	70

Table 8: Properties of coated sheets using Latex with Starch A as binder

IGT	brightness	whiteness	Gloss	Roughness	Opacity	porosity
(m/s)	%	(CIE)%	%	Ml/min	%	ml/min
0.87	87.92	91.81	52.35	15	91.89	70
0.88	87.98	91.78	50.73	17	90.23	75
0.88	87.81	91.73	49.4	20	89.81	75
0.89	87.78	91.56	47.8	30	87.56	90
0.89	87.11	90.89	45.31	40	87.3	90

Table 9: Properties of coated sheets using Latex +Starch B as binder

IGT	brightness	whiteness	Gloss	Roughness	Opacity	porosity
(m/s)	%	(CIE)%	%	ml/min	%	ml/min
0.87	87.92	91.63	51.57	17	90.14	75
0.87	87.95	91.34	51.21	20	90.43	80
0.875	87.84	91.14	48.91	20	89.82	95
0.88	87.63	89.88	46.21	25	87.48	100
0.88	87.44	89.91	43.44	35	86.56	100

Table 10: Properties of coated sheets using Latex +Starch C as binder

IGT	brightness	whiteness	Gloss	Roughness	Opacity	porosity
(m/s)	%	(CIE)%	%	ml/min	%	ml/min
0.86	87.2	90.76	51.39	20	90.13	95
0.865	87.26	90.72	51.24	25	88.29	100
0.87	86.69	90.52	48.36	35	86.28	100
0.87	86.49	90.28	46.08	40	83.19	100
0.875	86.33	90.25	43.27	40	83.08	105

Effect of Latex concentration on porosity:

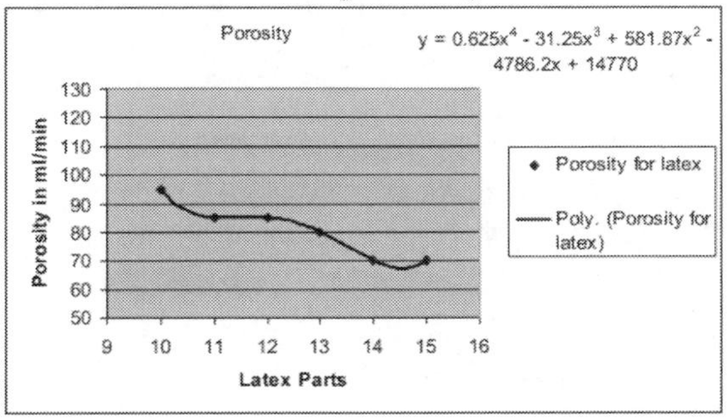

Fig3: Effect of Latex concentration on porosity

Fig 3 indicates that the porosity of the coated paper goes on decreasing with the increase in styrene butadiene latex. This decrease in porosity with the increase in latex is usual as it fills up the voids between pigment to pigment, pigment to fiber and fiber- fiber. The increase is much pronounced because of plastic nature of particles of latex.

Effect of Latex concentration on opacity:

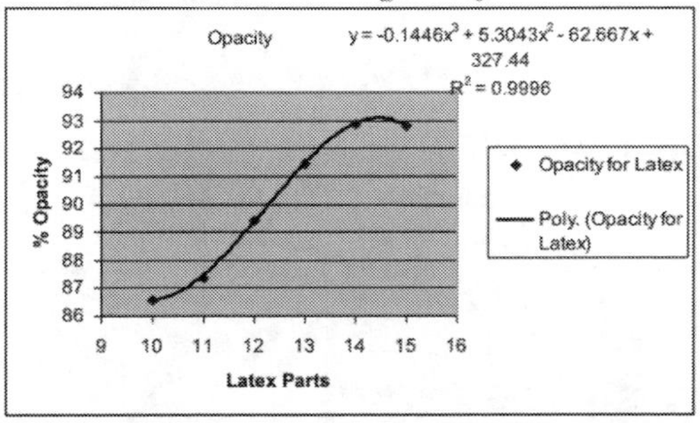

Fig4: Effect of Latex concentration on opacity.

The graph shows that the opacity of the coated sheet goes on increasing with the increase in the latex parts in the coating formulation. As already stated the particle size of SB latex is very small, these particles fill the voids between the coating pigments and the fiber structure in a more effective manner as the concentration of latex particles increases, thus making the coating structure more opaque.

Effect of Latex concentration on Bendtsen Roughness:

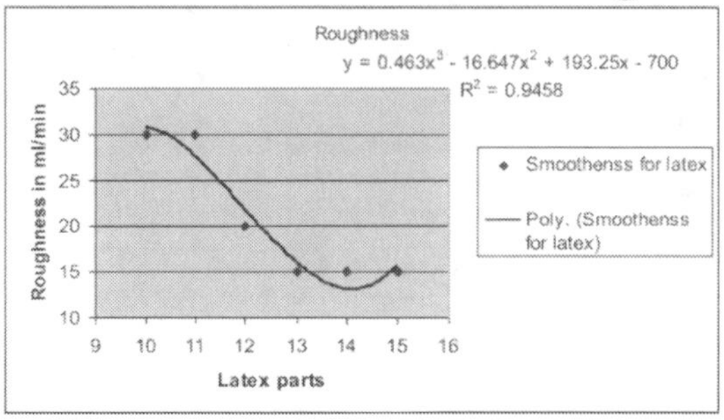

Fig 5: Effect of Latex concentration on Bendtsen Roughness.

Fig. 5 indicates decrease in Bendtsen Roughness (reverse in case of smoothness) with the increase in latex parts in the coating formulation. This is because of the decrease in pigment volume concentration (defined as Volume of pigment*100/(volume of pigment +volume of binder) which means that the voids between the pigments are being filled to a greater extent and also between the pigments on the surface, because of smaller size of latex particles than inorganic pigments.

Effect of Latex concentration on CIE Whiteness

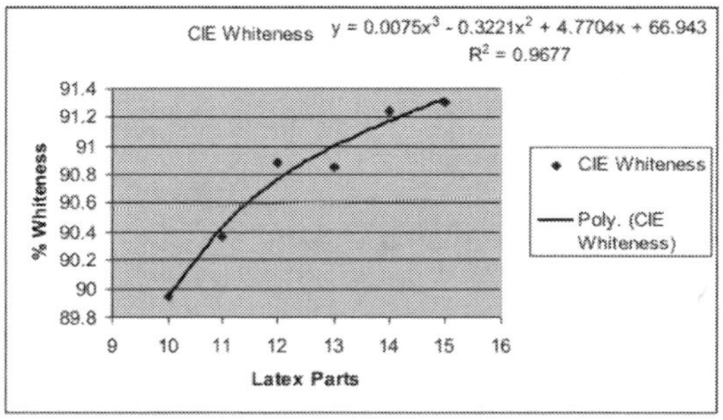

Fig6: Effect of Latex concentration on CIE Whiteness.

Fig.6 shows increase in the whiteness of paper with increase in latex parts. This result can be attributed to the increase in the optical brightening agent attaching sites with the increase in latex concentration.

Effect of Latex concentration on Brightness:

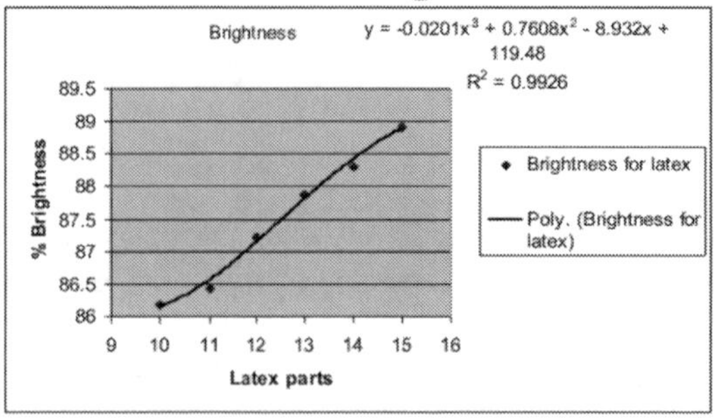

Fig 7: Effect of Latex concentration on Brightness

Fig 7 shows confirmation with fig 14 that increase in latex concentration increases the sites for the optical brightening agent, thus causing the brightness increase.

Effect of Latex concentration on IGT Printability:

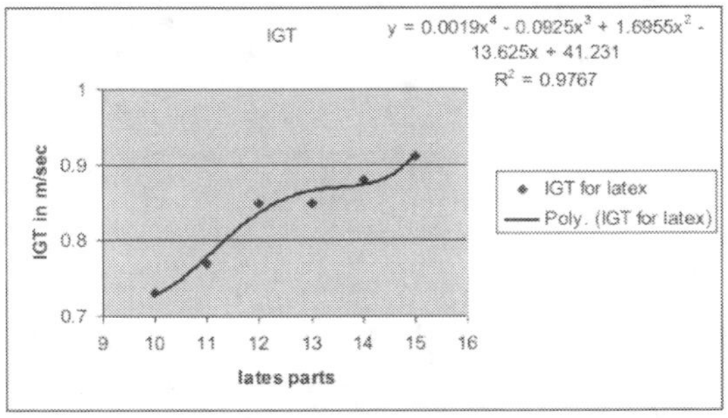

Fig 8: Effect of Latex concentration on IGT Printability

The increase in IGT value of coated paper with increase in latex binder confirms that increased binder level holds pigment particles to one another and to the paper surfaces. The increase binder level thus increases the surface strength of the coated paper.

Effect of Latex concentration on Gloss:

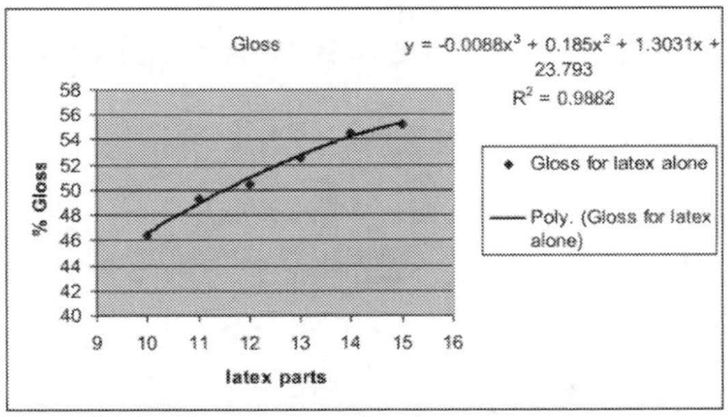

Fig 9: Effect of Latex concentration on Gloss

Fig 9 shows that the increase in latex particles in coating color enhances the gloss of paper. The reason governing this effect is once again the increase in concentration of plastic particles in the latex which respond better to calendering and produce a smoother and glossier surface.

Effect of Starch (A/B/C) and Latex concentration on Porosity:

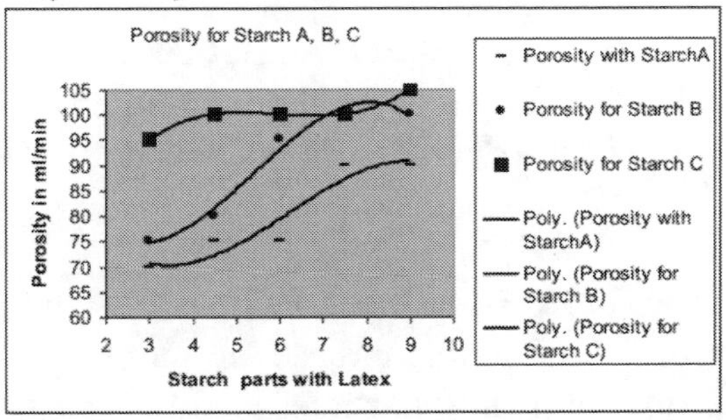

Fig 10: Effect of Starch (A/B/C) and Latex concentration on Porosity

Fig 10 shows the variation in porosity against Starch A, starch B and Starch C parts in coating color shows that the porosity of paper increases with increased Starch parts. The increase in porosity is due to two causes. First porosity increase is due to the reduction of latex parts as starch parts are increased (for equivalent binder being kept constant at 14 parts per 100 parts dry pigment by weight. Equivalent strength of latex is assumed to be 1.5 times better than starch). This decrease in latex parts in coating color with increased Starch A parts causes the concentration of plastic particles to go down. Thus filling up of the pores between pigment particles by deformation is reduced. The second cause for the increase in porosity is because of the difference in pigment and starch particle size. The difference causes more pores.

Effect of Starch (A/B/C) and Latex concentration on Opacity:

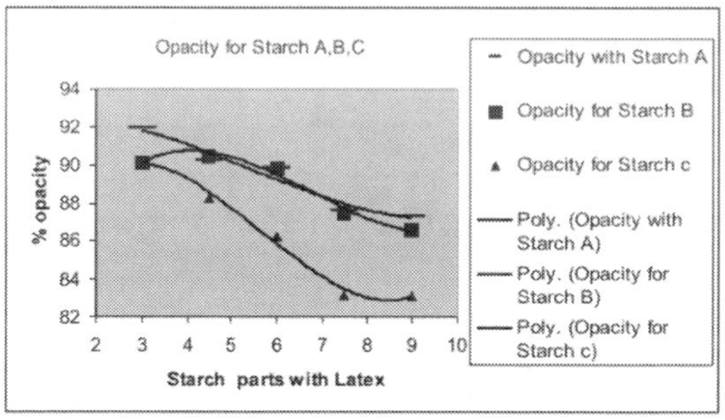

Fig 11: Effect of Starch (A/B/C) and Latex concentration on Opacity

The figure 11 shows that as the percentage and pph of starch A, B, C in coating binder goes on increasing with the corresponding decrease of latex particles; the opacity of the paper goes on decreasing. This is because the very small sized particles of SB latex are being replaced by comparatively large sized particles of Starch.

Effect of Starch (A/B/C) and Latex concentration on Roughness:

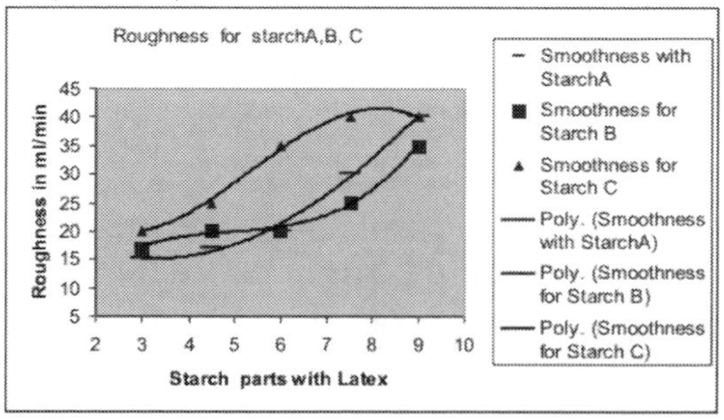

Fig 12: Effect of Starch (A/B/C) and Latex concentration on Roughness

Fig 12 shows that the increase in percentage of Starch A, B, C in binder causes the roughness of paper to increase. The reason behind this effect is the combined result of increase in non deforming particles of starch A, B, C and decrease in plastic deforming particles of latex in binder.

Effect of Starch (A/B/C) and Latex concentration on Gloss: As the fig 13 shows, there is a continuous decrease in the gloss of paper with increase in starch A, B, C. Again this is the combined result of decrease in plastic, deforming particles share of latex in the binder

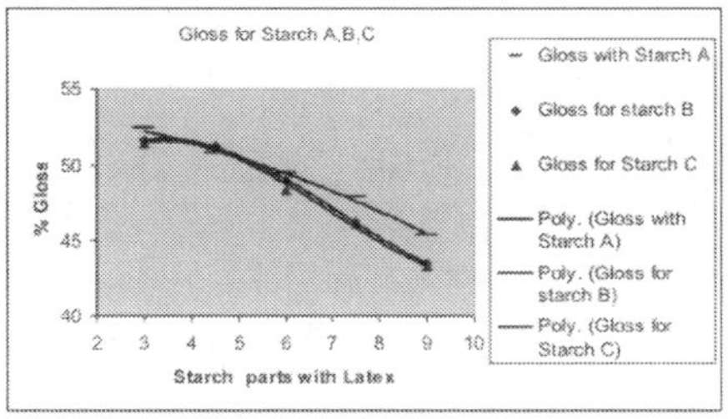

Fig 13: Effect of Starch (A/B/C) and Latex concentration on Gloss

Effect of Starch (A/B/C) and Latex concentration on Whiteness:

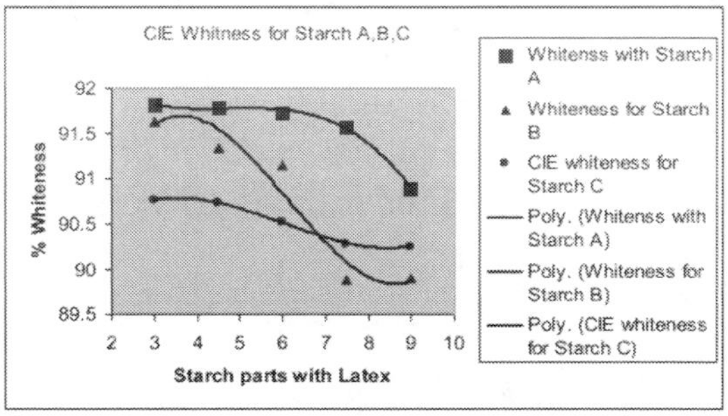

Fig 14: Effect of Starch (A/B/C) and Latex concentration on Whiteness

It is evident from figure 14 that the decrease in whiteness of paper with increase in starch A, B, and C parts in the binder level in each individual set of experiment for starch A, starch B and starch C. The decrease observed with Starch A, B, C parts increase is the result of brighter inorganic pigments being masked by the less bright Starch particles which give a turbid solution on cooking.

Effect of Starch (A/B/C) and Latex concentration on Brightness:

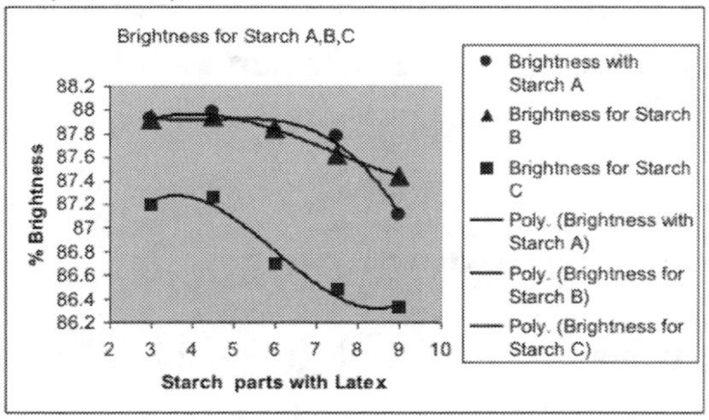

Fig 15: Effect of Starch (A/B/C) and Latex concentration on Brightness

Fig 15 confirms the masking effect of less brighter particles over the brighter inorganic pigments in accordance with fig. 14.

Effect of Starch (A/B/C) and Latex concentration on IGT value: As the binder level is increased with increase in starch parts in coating color, the network between pigment particles, binder particles and base paper gets stronger. This effect is clearly visualized by fig 16. It shows almost steady increase in IGT value of paper with increase in starch percentage in coating color. For same parts of starch A, IGT values are higher than starch B and starch C.

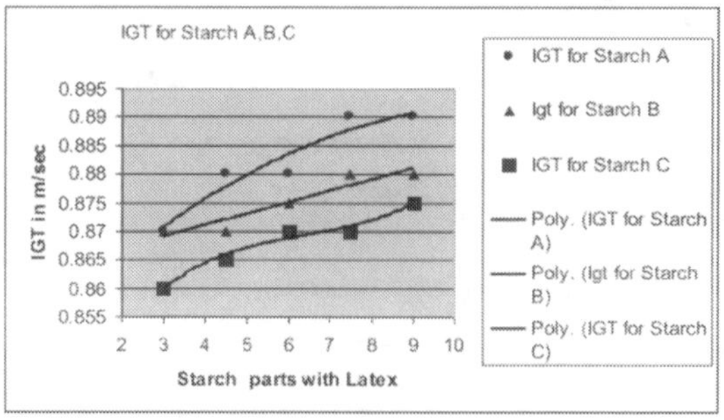

Fig 16: Effect of Starch (A/B/C) and Latex concentration on IGT value

From the experimental evaluation of properties of starch A, B and C, as a coating binder, it is found that starch A sample is better than that of starch B and starch C. Almost all properties show a higher value for starch A than for starch B and C. Also, for the same amount of Starch A, B and Starch C, the IGT value for experiments of starch A is found to be higher than starch B and starch C, indicating that starch A is a better binder than starch B and starch C.

STATISTICAL REGRESSION ANALYSIS

From the results obtained, graphs were plotted to obtain the relation between property and the binder composition.

To find out the optimum composition of the coating color based on the variables of binder composition, parameters are correlated through regression analysis for estimating the low cost binder through computer program in C^{++} language. From the graphs, a functional equation relating property and binder composition can be obtained (with minimum possible error) through least square technique. These equations obtained for each property for each set of experiments in the individual range of properties can be used to develop a computer program to predict the property of coated paper and coating formulation. The computer program can also be developed with these equations for each set to find out the composition of the binder or binder system which corresponds to the targeted values of properties namely IGT value, gloss, brightness and opacity, as required by the user.

The polynomial form of statistical regression equations are obtained from the results by using Microsoft Excel software. The equations are derived from the results plotted in respective figures.

It is evident that the plot is made using one independent variable and a dependent variable. The polynomial form is shown below.

$$\text{Property} = f(\text{Binder composition})$$

$$y = f(x)$$

The polynomials which best fit to the data obtained from the experiments are generated by computer with R^2 nearly equal to one. Sometimes residual plots and error analysis are used to further confirm the accuracy of the regression constant.

Table 1: Regression Equations for Latex

Property	Regression Equation	R^2

Porosity	y = 0.625x4 - 31.25x3 + 581.87x2 - 4786.2x + 14770	1
Opacity	y = -0.1446x3 + 5.3043x2 - 62.667x + 327.44	0.9996
Roughness	y = 0.463x3 – 16.647x2 + 193.25x – 700	0.9458
Whiteness	y = 0.0075x3 – 0.3221x2 + 4.7704x + 66.943	0.9677
Viscosity	y = 0.5417x4 – 27.954x3 + 538.85x2 – 4584.8x + 14684	0.9999
Brightness	y = -0.0201x3 + 0.7608x2 – 8.932x + 119.48	0.9926
IGT Printability	y = 0.0019x4 – 0.0925x3 + 1.6955x2 – 13.625x + 41.231	0.9767
Gloss	y = -0.0088x3 + 0.185x2 + 1.3031x + 23.793	0.9882

Table 2: Regression Equations for Starch A with Latex

Property	Regression Equation	R^2
Porosity	y = -0.2469x3 + 4.6032x2 - 23.016x + 105	0.898
Opacity	y = 0.0185x3 - 0.3025x2 + 0.6988x + 91.918	0.9543
Roughness	y = -0.0247x3 + 1.1746x2 - 7.0397x + 26.4	0.9945
Whiteness	y = -0.0119x3 + 0.1689x2 - 0.7933x + 92.994	0.9981
Viscosity	y = 0.6914x3 - 10.381x2 + 62.683x + 91	0.9993
Brightness	y = -0.0101x3 + 0.1403x2 - 0.6344x + 88.85	0.9616
IGT Printability	y = -0.0003x2 + 0.0071x + 0.852	0.9184
Gloss	y = -0.0638x2 - 0.3683x + 53.912	0.9953

Table 3: Regression Equations for Starch B with Latex

Property	**Regression Equation**	$\mathbf{R^2}$
Porosity	y = -0.3704x3 + 6.0317x2 - 24.881x + 105	0.9838
Opacity	y = 0.0573x3 - 1.1629x2 + 6.6554x + 79.036	0.9815
Roughness	y = 0.1975x3 - 2.9524x2 + 15.317x - 7.6	0.9955
Whiteness	y = 0.0296x3 - 0.5467x2 + 2.8067x + 87.28	0.936
Viscosity	y = 0.4691x3 - 6.1905x2 + 38.23x + 149	0.9989
Brightness	y = 0.004x3 - 0.0883x2 + 0.5168x + 87.056	0.9995
IGT Printability	y = 0.002x + 0.863	0.9
Gloss	y = 0.0462x3 - 0.9968x2 + 5.2047x + 43.698	0.9996

Table 4: Regression Equations for Starch C with Latex

Property	Regression Equation	R^2
Porosity	y = 0.2469x3 - 4.4444x2 + 26.111x + 50	1
Opacity	y = 0.0778x3 - 1.3244x2 + 5.6183x + 83.024	0.9909
Roughness	y = -0.2469x3 + 3.9683x2 - 15.397x + 37	0.9957
Whiteness	y = 0.0091x3 - 0.1651x2 + 0.8271x + 89.516	0.9989
Viscosity	y = 0.6914x3 - 8.9524x2 + 47.873x + 148.4	0.9947
Brightness	y = 0.0165x3 - 0.3x2 + 1.5194x + 84.914	0.9639
IGT Printability	y = 0.0001x3 - 0.0024x2 + 0.0166x + 0.828	0.9753
Gloss	y = 0.0543x3 - 1.1276x2 + 5.8225x + 42.668	0.9941

3.0 Principle of Selecting Binder:

Using these equations for predicting the properties in the corresponding binder range, the computer program can be formulated to give the binder composition according to the desired properties. The input properties by the user are chosen to be, IGT value, Brightness, Opacity and Gloss of the paper as desired values. Computer program, by relating above-mentioned equations of polynomials, for corresponding property, calculates the binder composition for the each set of data; the property lies in the range of the experiments. It also calculates the total cost of binder required for the targeted properties (based on individual cost of Latex: Rs.120/kg., Starch A: Rs.40/kg, Starch B: Rs.38/kg., Starch A: Rs.36/kg.

By comparing the cost of binder composition for each possible set of combinations for desired properties, program gives the minimum cost formulation as the optimum one. It also displays the other properties such as porosity, whiteness, smoothness, and viscosity of coating color along with desired input properties. An executive logic diagram with step by step calculations is shown in fig 1. Sample run of the computer program developed in C^{++} language is also shown.

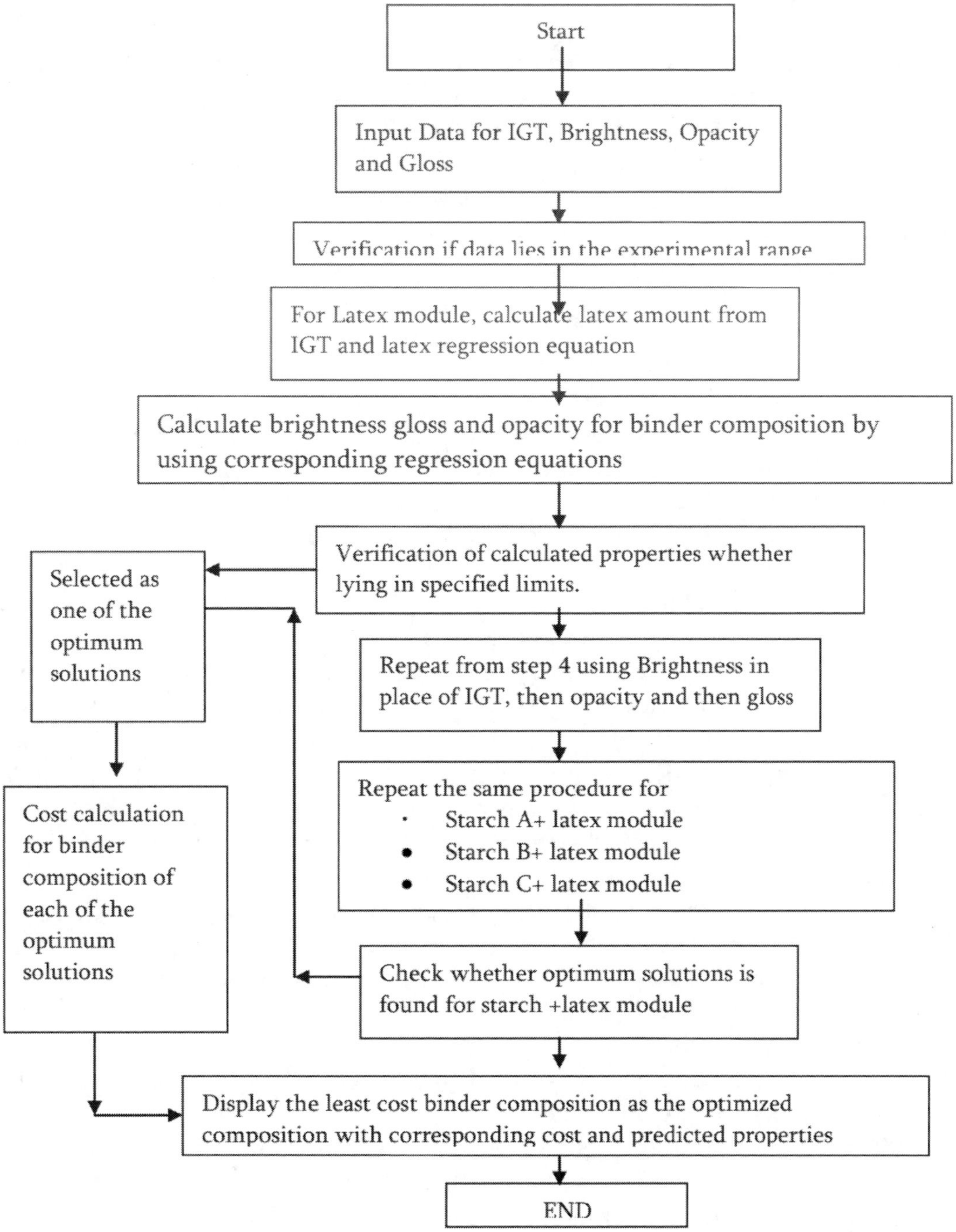

Fig 1: Logic Diagram for the Computer Program

4.0 CONCLUSION:

The present investigation deals with the variation of various properties of coated paper like, IGT, Brightness, whiteness, opacity, gloss and also the viscosity of the coating color, are studied with respect to various binder systems containing latex and Starch samples supplied by Hindustan Lever Ltd. Coating experiments have been conducted in a pilot plant coater "DT Paperscience Finland Laboratory coater". The coating on the base paper is made and the properties of the coated paper have been evaluated as per TAPPI specifications. The base paper characteristics have been kept constant with constant gsm. From the detailed properties of coated paper, the following conclusions can be drawn

- The porosity of the coated paper goes on decreasing with the increase in styrene butadiene latex. Plastic particles in latex deform on calendering and filling in the pores between the pigments.
- Also the opacity and smoothness of the sheet has been found to increase with increase in latex amount in binder. The increase in the concentration of these plastic particles of very small particle size (50-300nm) causing the possibility of filling more voids resulting into decreased porosity, increased smoothness and opacity.
- Increase in brightness and whiteness of the coated sheet is also observed with increase in latex amount in coating color.
- The IGT printability value and gloss value also shows increase with increase with increase in latex amount increases as the plastic particles of latex increased in color.
- Increase in Starch A, starch B and Starch C amounts in coating color shows that the porosity and roughness of paper increases with increased Starch parts.
- The IGT value of paper showed increase with increase in all starch grades amount but the opacity of the paper goes on decreasing.
- Continuous decrease in the gloss of paper with increase in starch A, B, C is observed. Again this is the combined result of decrease in plastic, deforming particles share of latex in the binder and increase in non deforming particles share of starch A, B, C in binder in each individual set of experiment for starch A, starch B and starch C.
- There is decrease in whiteness and brightness of paper with increase in starch A, B, and C parts in the binder level.
- Starch grades (Starch A, Starch B, Starch C) show a similar variation for paper properties where as SB latex shows different influence.
- Based on the data experimentally evaluated in a pilot coater, various statistical regression analyses have been performed. For all the four set of experiments, in most of the cases, the R^2 have been obtained above 94% except for porosity for starch A +latex and IGT printability for starch A and B. The principle of selecting binder has been discussed based on the market cost of latex as well as kind of starch supplied by Hindustan Lever Ltd.
- In order to compute through computer, a logic diagram has been proposed. For detailed evaluation, C^{++} program has been developed.

- A sample calculation has been shown to get the minimum cost of blend of latex and starch. The program can be used for optimizing various options.

REFERENCES:

1. Lehtinen, E., "Introduction to Pigment coating of paper", Pigment coating and surface sizing, Papermaking science and technology, Book 11,TAPPI Press, p 19.
2. Lehtinen, E., "Introduction to Pigment coating of paper", Pigment coating and surface sizing, Papermaking science and technology, Book 11,TAPPI Press, p 14.
3. Lee, D. I., "Latex", Pigment coating and surface sizing, Papermaking science and technology, Book 11,TAPPI Press, p 204.
4. Hagemeyer, R. W., "Pigment Coating Chemistry and Chemical Technology", IV, Wiley Interscience, P2023-p2055.
5. Lehtinen, E., "coating binders general", Pigment coating and surface sizing, Papermaking science and technology, Book 11, TAPPI Press, p 189-p195.
6. Bruub, S. E., "Starch", Pigment coating and surface sizing, Papermaking science and technology, Book 11,TAPPI Press, p 189-p241-p250.
7. Lee, D. I., "Latex", Pigment coating and surface sizing, Papermaking science and technology, Book 11,TAPPI Press, p 198.
8. Lee, D. I., "Latex", Pigment coating and surface sizing, Papermaking science and technology, Book 11,TAPPI Press, p 203.
9. Lehtinen, E., "coating binders general", Pigment coating and surface sizing, Papermaking science and technology, Book 11, TAPPI Press, p 189-p195.
10. Groves, R., Penson, J.E., Ruggles, C., "Styrene-Butadiene latex binders and coating structure",Tappi;1993 Coating conference Proceedings, Tappi Press, Atlanta, P187.
11. Schempp,W., Fiesen,W., Schurz,J., Tappi: 58(1); 116(1975).
12. Alince, B. and Lepoutre P., Tappi; 64(5); 67(198).
13. Aschan, P.,-J:,Tappi J. 69(12):25 1986.
14. Renvall, S.V., Rautiainen,P.J., Rossitto,J.C., Tappi J 75(5):143(1990).

Reactive extraction (in situ transesterification) of *Jatropha curcas L.* for methyl ester production

Newcastle University

Farizul Kasim, Adam Harvey

Overview

Methyl Ester

- METHYL ESTER (BIODIESEL)
  - Renewable fuel for diesel engines derived from natural oils, and which meet the specification (ASTM D 6751, EN 14214)
  - Compose of fatty acids alkyl ester
  - Can be transported and sold using today's infrastructure

Overview

Methyl Ester

- 4 factors determining biodiesel competitiveness as fuel
  - Feedstock price
  - Biodiesel production cost
  - Policy (tax, incentive) on biodiesel
  - Crude oil prices

Overview

Methyl Ester

- 4 factors determining biodiesel competitiveness as fuel
  - Feedstock price
  - Biodiesel production cost
  - Policy (tax, incentive) on biodiesel
  - Crude oil prices

Overview
Methyl Ester

- 4 factors determining biodiesel competitiveness as fuel
  - Feedstock price
  - Biodiesel production cost
  - Policy (tax, incentive) on biodiesel
  - Crude oil prices

Overview

Methyl Ester

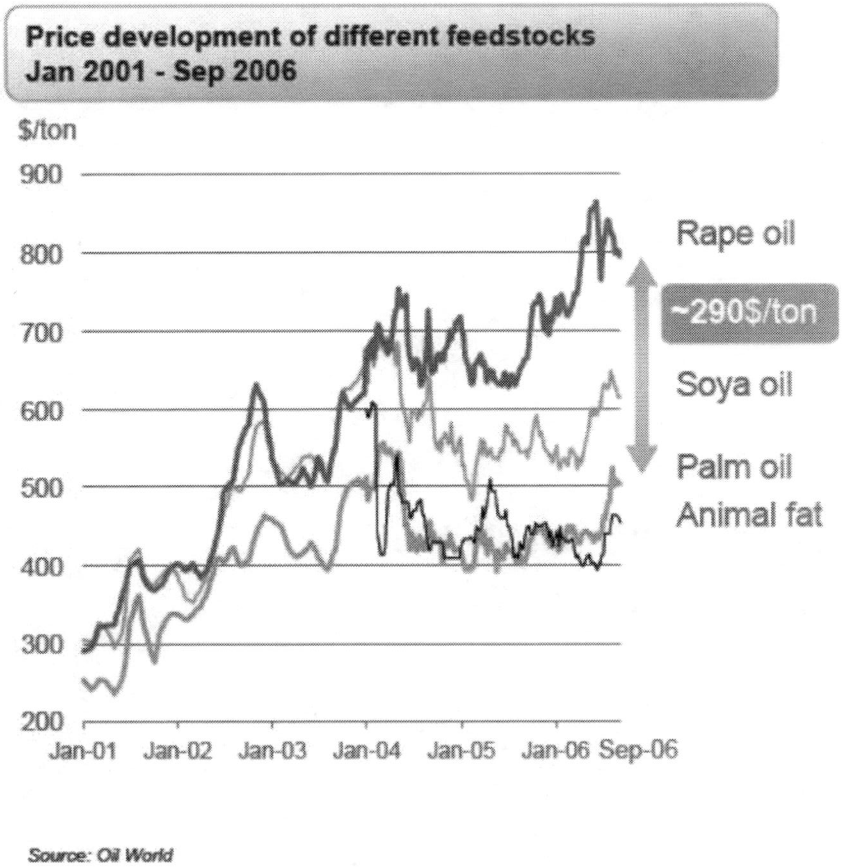

Source: Kimmo Rahkamo, Neste Oil presentation material

Overview

In situ transesterification

- Conventional transesterification of oil-bearing materials.

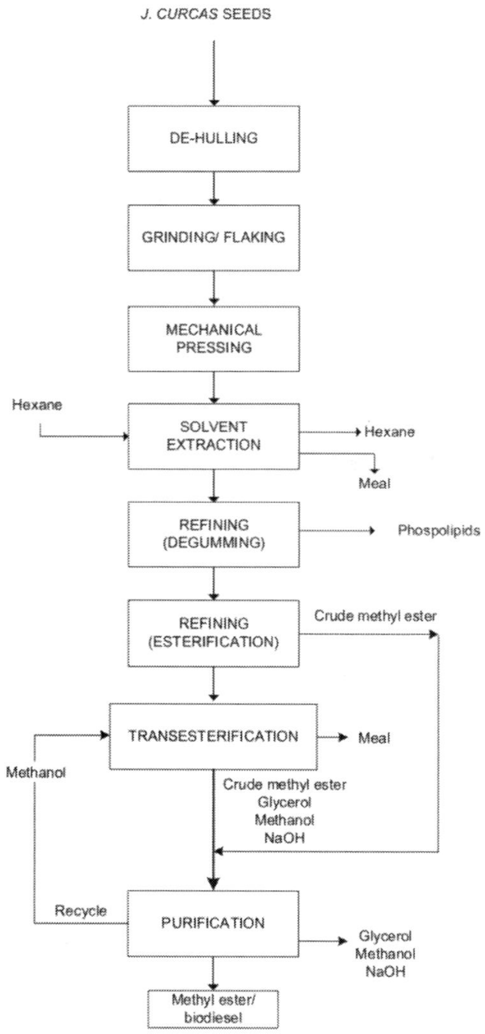

Conventional biodesel process from *J.curcas* seeds

Overview

In situ transesterification

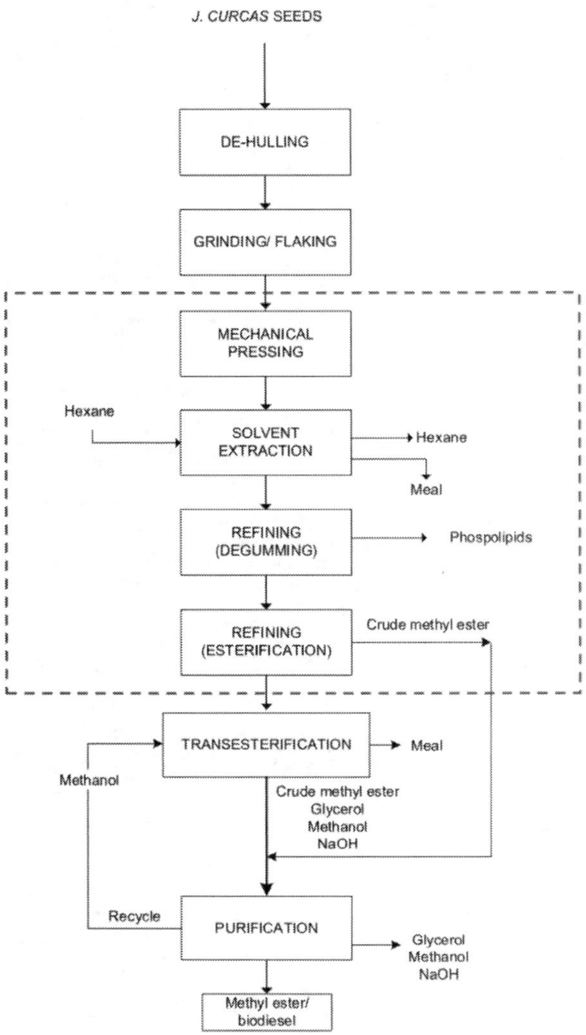

Conventional biodesel process from *J.curcas* seeds

- In situ transesterification is the direct transesterification of ground oil-bearing materials.

Overview

In situ transesterification

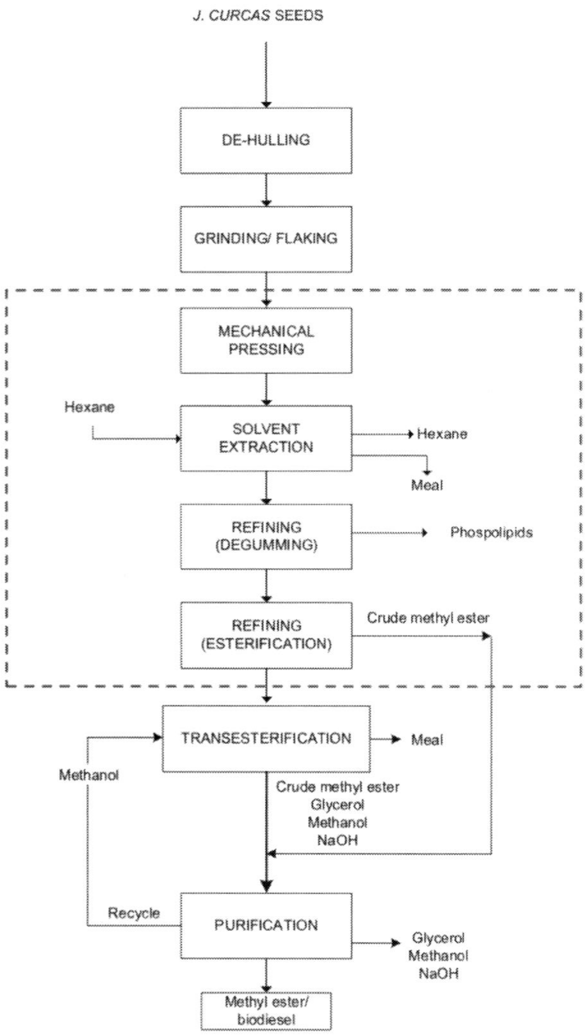

Conventional biodesel process from *J.curcas* seeds

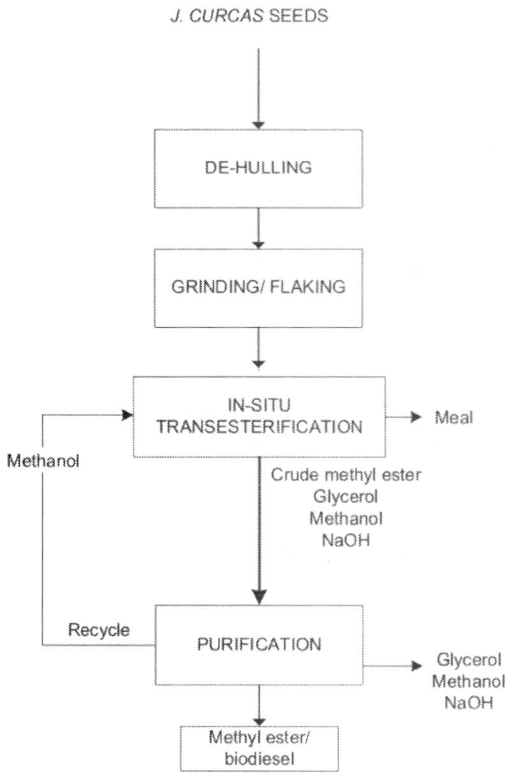

In-situ transesterification/ reactive extraction biodesel process from *J.curcas* seeds

Overview
Jatropha curcas

Overview

Jatropha curcas

High ecological adaptability

Able to grow in various types of soil

Widely cultivated in the tropics (Central/ South America, Africa, India subcontinent, SEA)

Grows well in area with rainfall range between 250 – 3000 mm/ yr

BUT: so far, there is no report on commercial, large-scale production of biodiesel from *J.curcas*

Low nutritional requirement

Seed production 7 months after plantation of 1 yr old sapling.

Overview

Jatropha curcas

- The kernel contains mostly fat and protein while the shell dominates by fiber
- Oil content varies very much, but generally between 30 – 60%

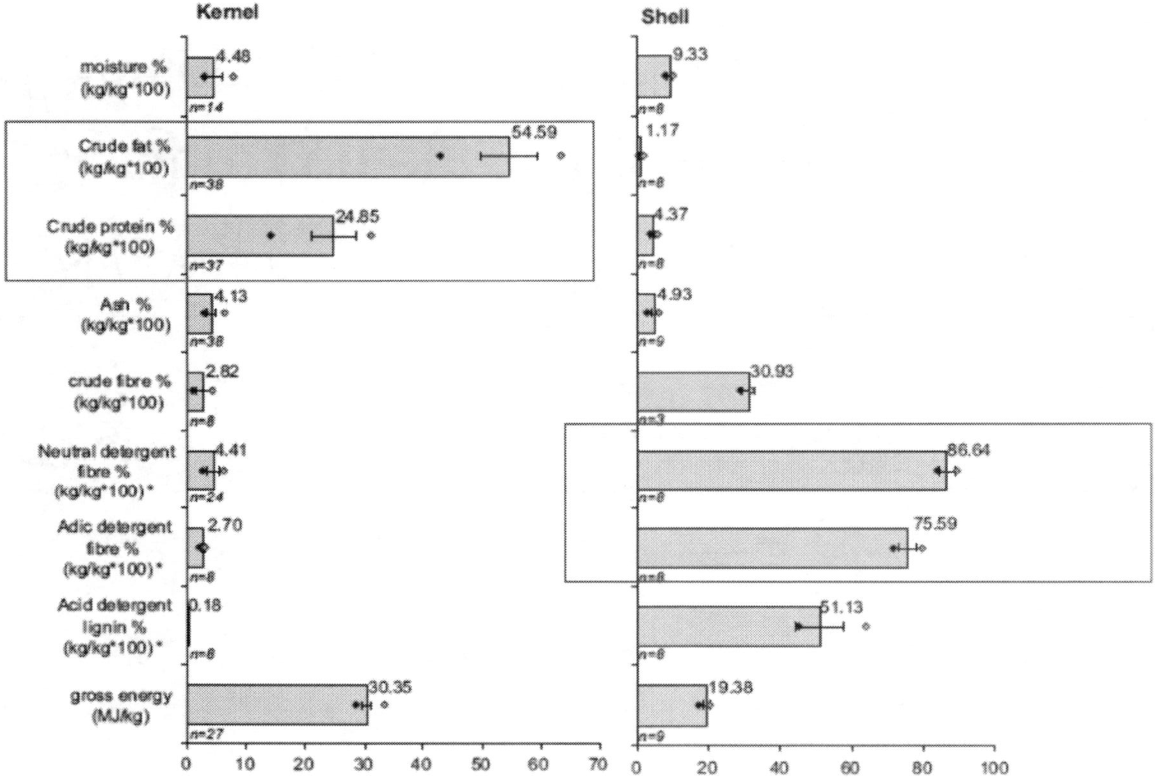

Overview

Jatropha curcas

Test	Assay method	Unit	
Oil content	BS 659: 2008	%	36
Moisture content	BS 665:2000	%	7.4
FFA content	BS 660:2009	%	9.2
Acid value	BS 660:2009	mg/g KOH	18
Fatty acids content	BS 684:2001 + BS 15304	%	C16:0 (13.6), C18:0 (6.0), C18:1 (42.3), C18:2 (38.1)

Overview

Jatropha curcas

- Problems (1):

  – Via conventional TE, when alkali catalyst is employed, high FFA contents in the JC oil promotes saponification rather than TE reaction and produce soap rather than methyl ester

$$RCOOH + NaOH \leftrightarrow RCOONa + H_2O$$

FFA — Soap

  – Consume the catalyst and halt the TE process. Effect on yield as well as on downstream processing

Overview

Jatropha curcas

- Problems (1):

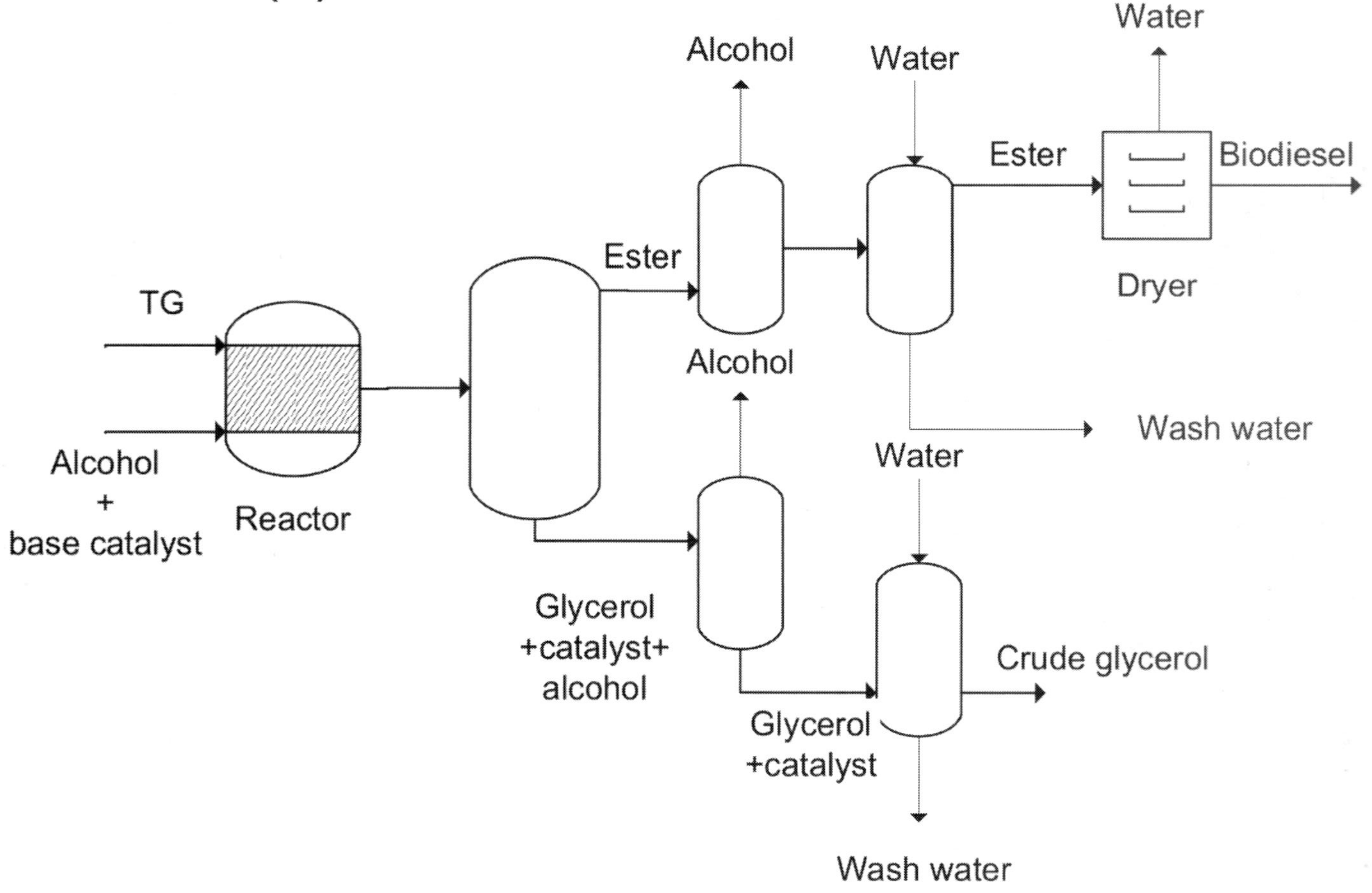

Conventional T.E

Overview
Jatropha curcas

- Problems (1):

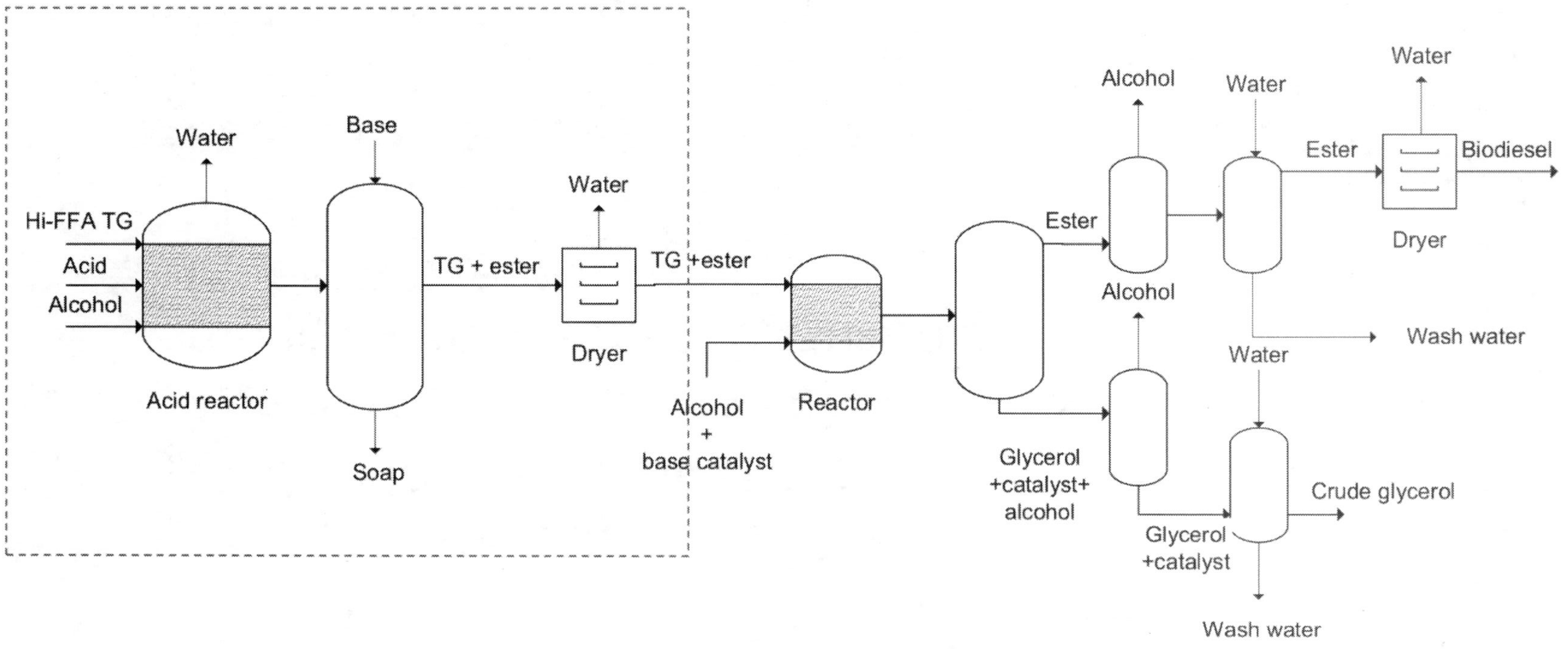

Conventional TE with high FFA raw material

Overview

Jatropha curcas

- Problems (2):
  - Phorbol esters in JC are toxic, even in low concentration
  - The phorbol esters end up in the meal, make the meal unsuitable for animal feed
  - There is a report that phorbol esters decompose in access methanol, but there is no report on the fate of phorbol esters in in situ transesterification of JC

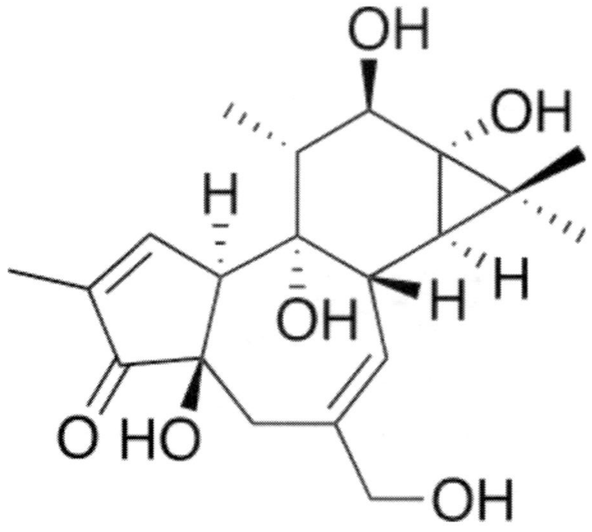

Methods and Materials

In situ transesterification

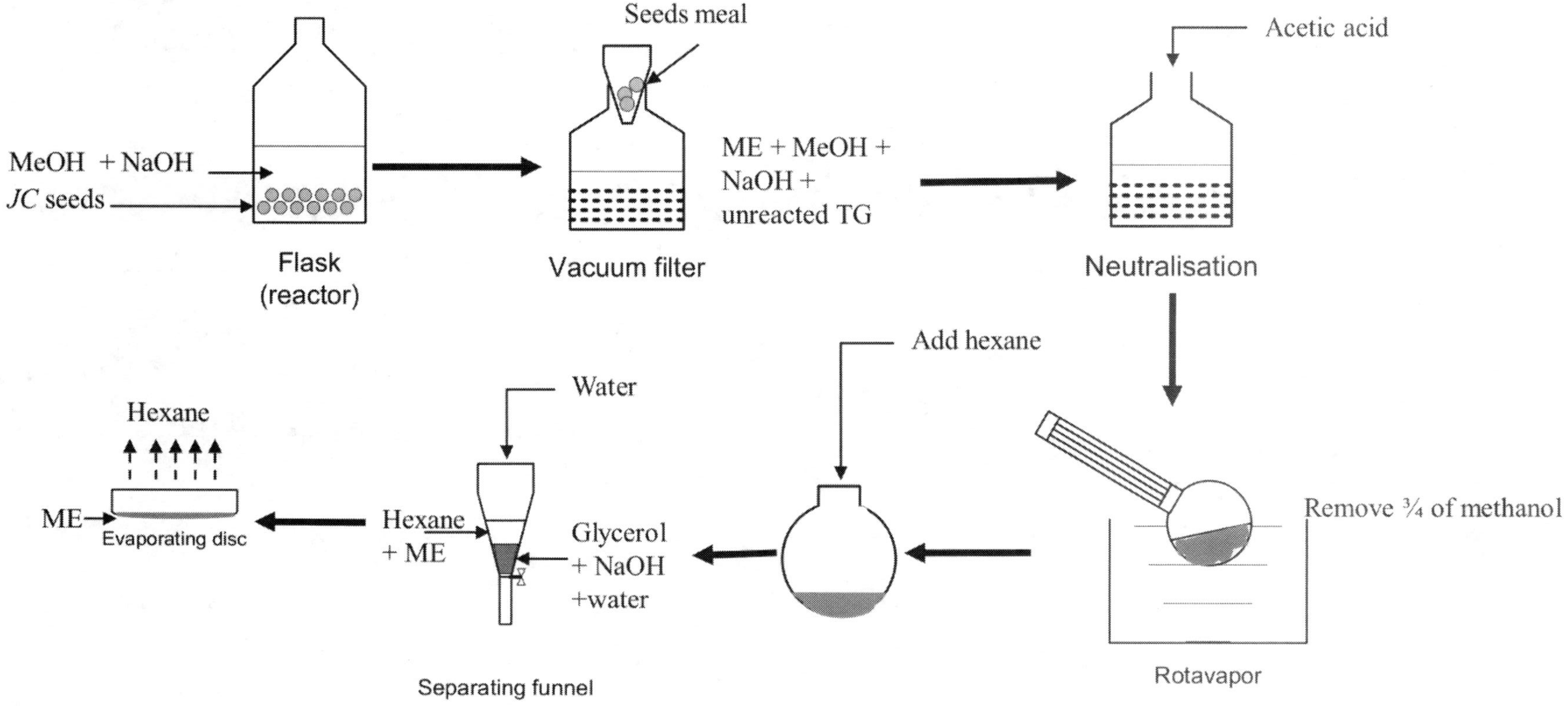

Flow diagram of IT experiments conducted in the lab

Methods and Materials

Full factorial design

Factor	Unit	Code	Level	
			-1	+1
Molar ratio of methanol to oil	-	A	100	400
NaOH concentration	N	B	0.1	0.2
Reaction time	Min	C	10	60
Reaction temperature	^{o}C	D	30	60
Mixing speed	rpm	E	100	400

Two level factorial design

Results

Full factorial design

- Executed on Design Expert software
- Model used was a 2^5 full factorial design
- Model based on first order model
- 5 central points were added to the design to assess the experimental error and also the curvature
- Overall there are 37 experiments

Results

Full factorial design

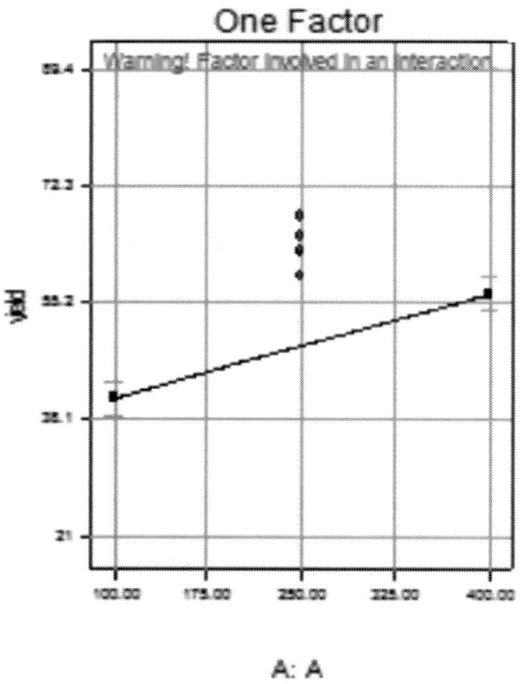

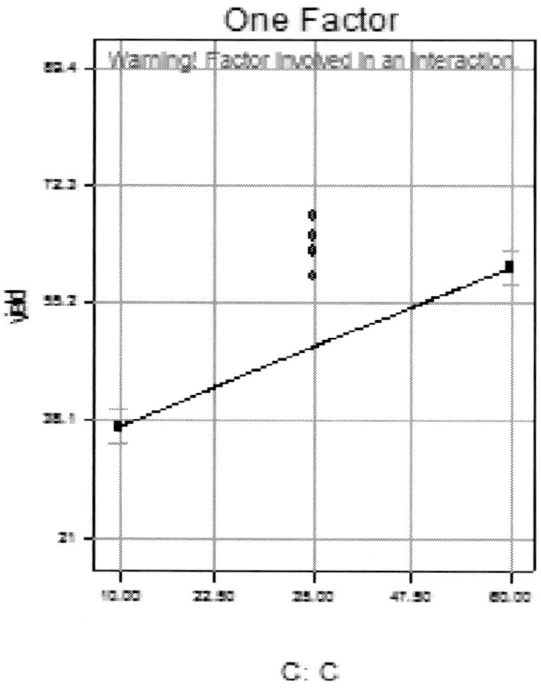

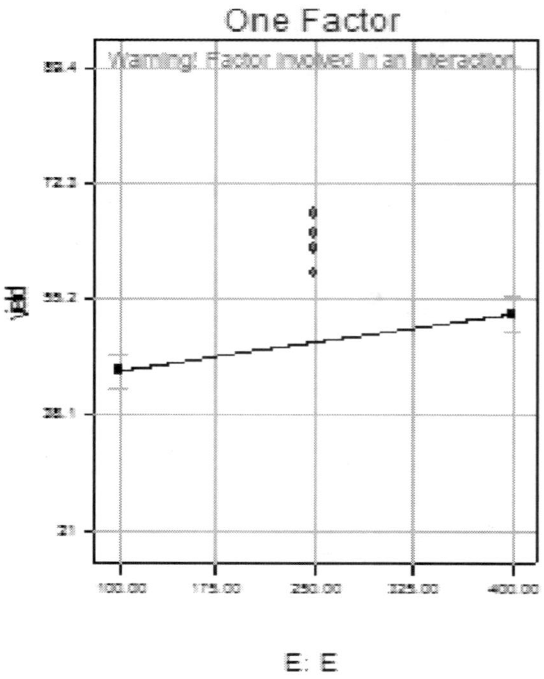

B= [NaOH]

D=temperature

A=molar ratio

C= reaction time

E=mixing speed

Results

Full factorial design

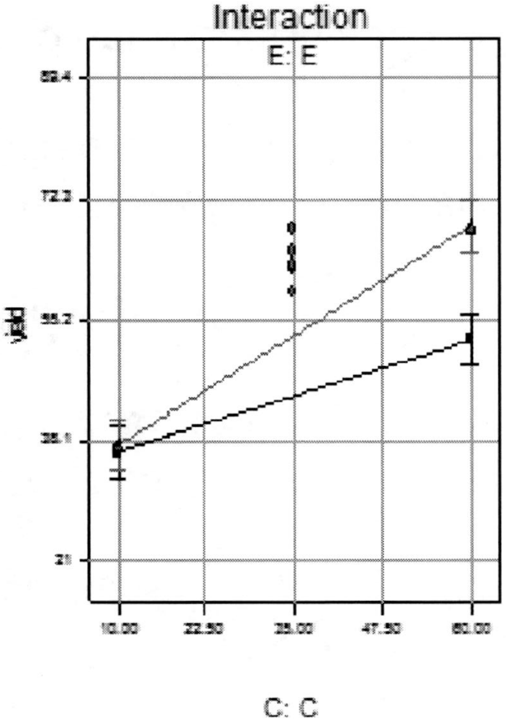

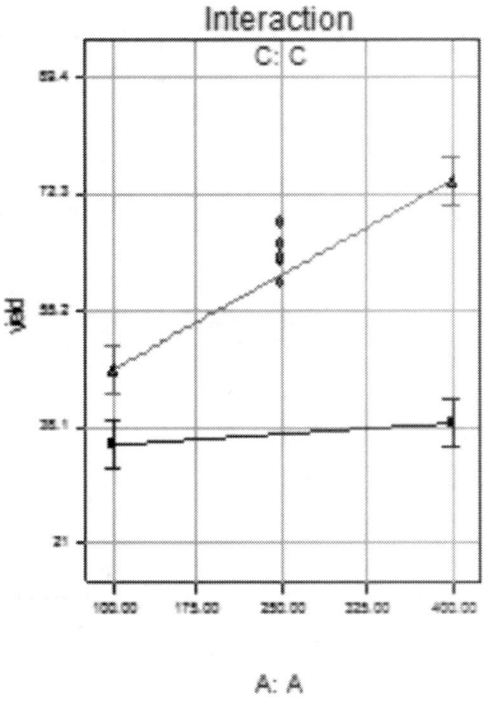

Results

Full factorial design

- A (molar ratio), C (reaction time) and E (mixing speed) are significant to the biodiesel yield
- B (catalyst concentration) and D (reaction temperature) are not significant to the biodiesel yield
- Interaction between A/C and C/E also significant to the yield
- The presence of curvature also significant, indicating that first order model is not suitable to predict the model
- Higher order model is needed to predict the model more accurately

Methods and Materials

Response surface methodology (RSM) design

Factor	Unit	Code	Level	
			$-\alpha$	$+\alpha$
Molar ratio of methanol to oil	-	A	25.70	474.30
NaOH concentration	N	B	0.08	0.22
Reaction time	Min	C	0	72.38
Reaction temperature	°C	D	22.57	67.43
Mixing speed	rpm	E	25.70	474.30

RSM design

Results

Response Surface Methodology

- To fit the data in second order model

- 10 additional experiment needed to test the yield at ±α

Standard Run	A	B	C	D	E	Y
1	25.70	0.15	35	45	250	13.2
2	474.30	0.15	35	45	250	82.1
3	250	0.08	35	45	250	51.2
4	250	0.22	35	45	250	55.4
5	250	0.15	0	45	250	0
6	250	0.15	72.38	45	250	57.3
7	250	0.15	35	22.57	250	47.8
8	250	0.15	35	67.43	250	58.7
9	250	0.15	35	45	25.70	13.5
10	250	0.15	35	45	474.30	56.9

- A quadratic model was selected to develop a mathematical model which includes all the five factors involved

- Can be represented by a 3D surface plot, predicting the yield in the range of parameters studied

Results

Response Surface Methodology

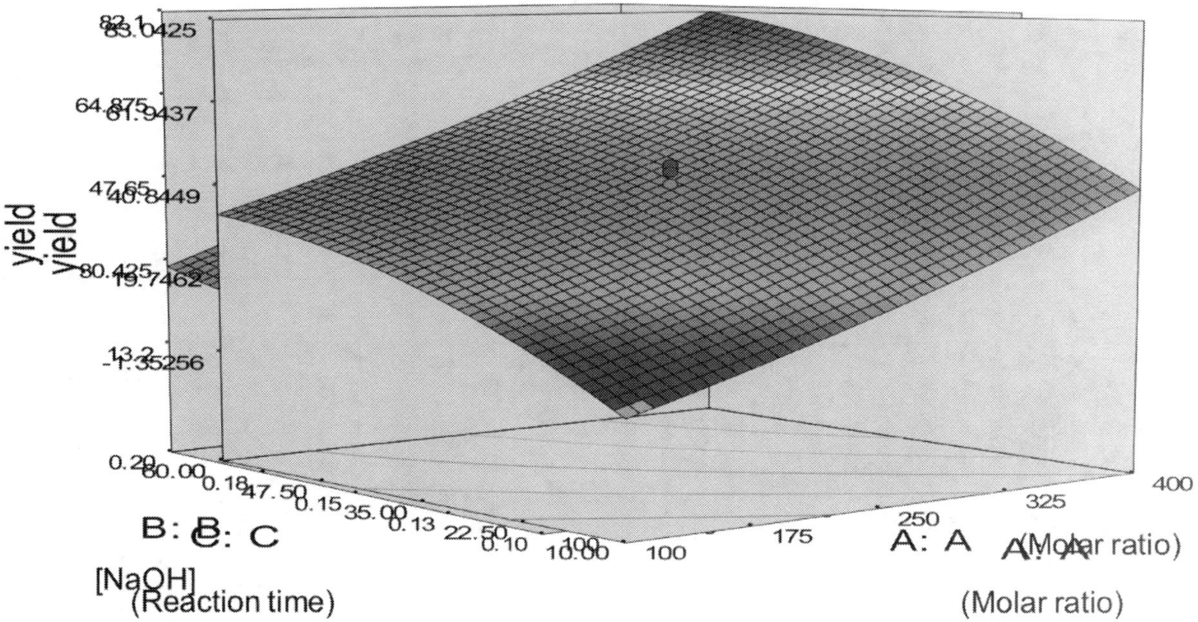

The 3D surface plot of interaction of A with B

The 3D surface plot of interaction of A with C

Results

Response Surface Methodology

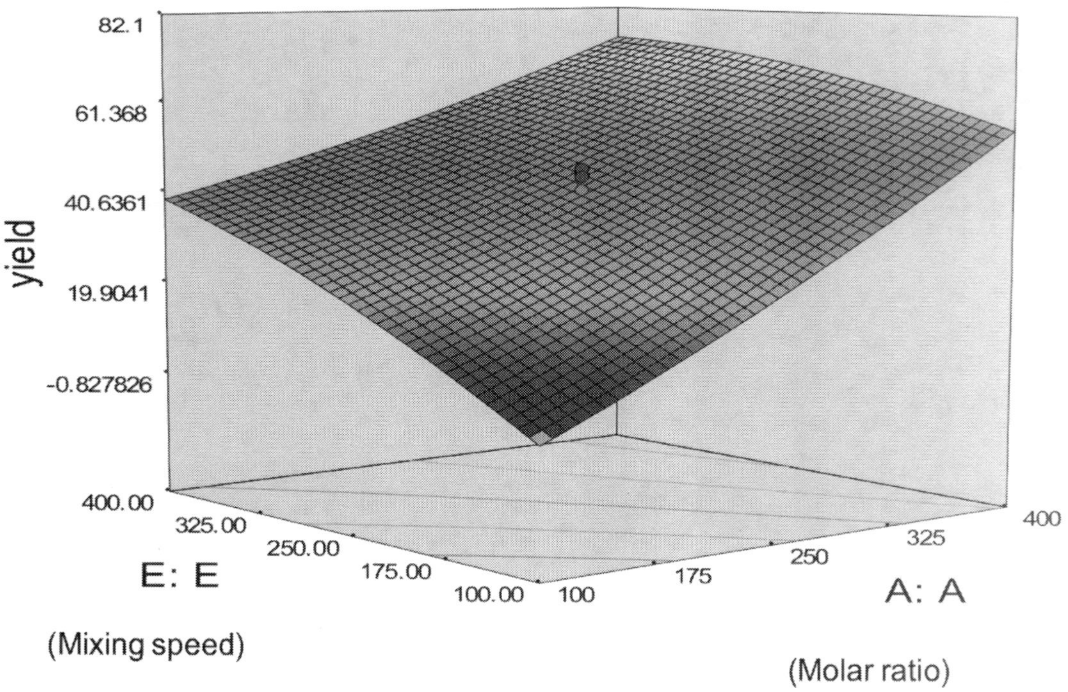

The 3D surface plot of interaction of A with E

Results

Response Surface Methodology

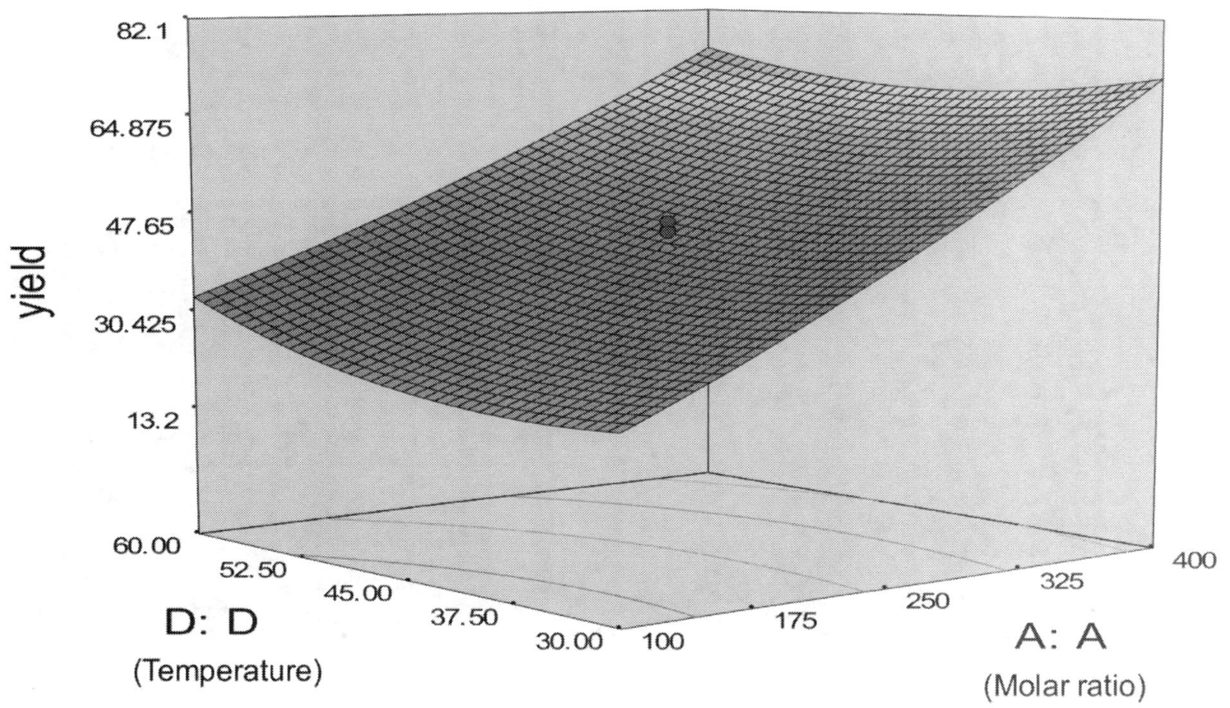

The 3D surface plot of interaction of A with D

Results

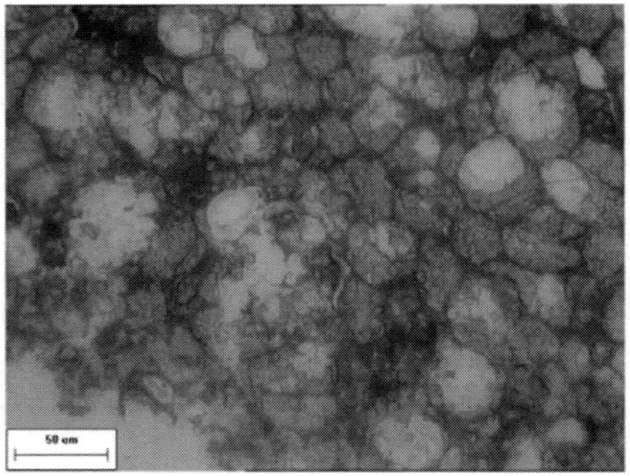

Before reaction

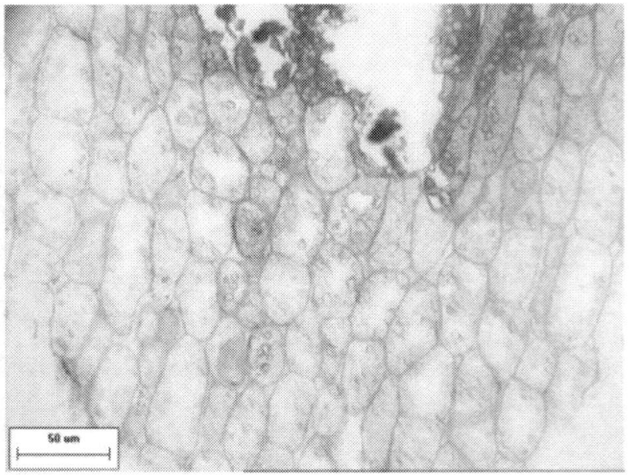

100 molar

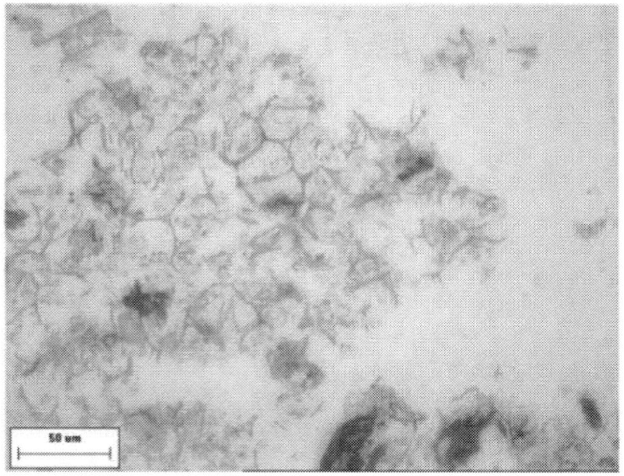

300 molar

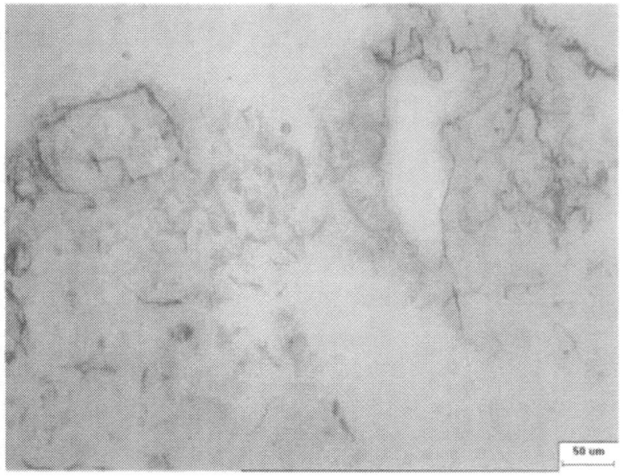

400 molar

Results

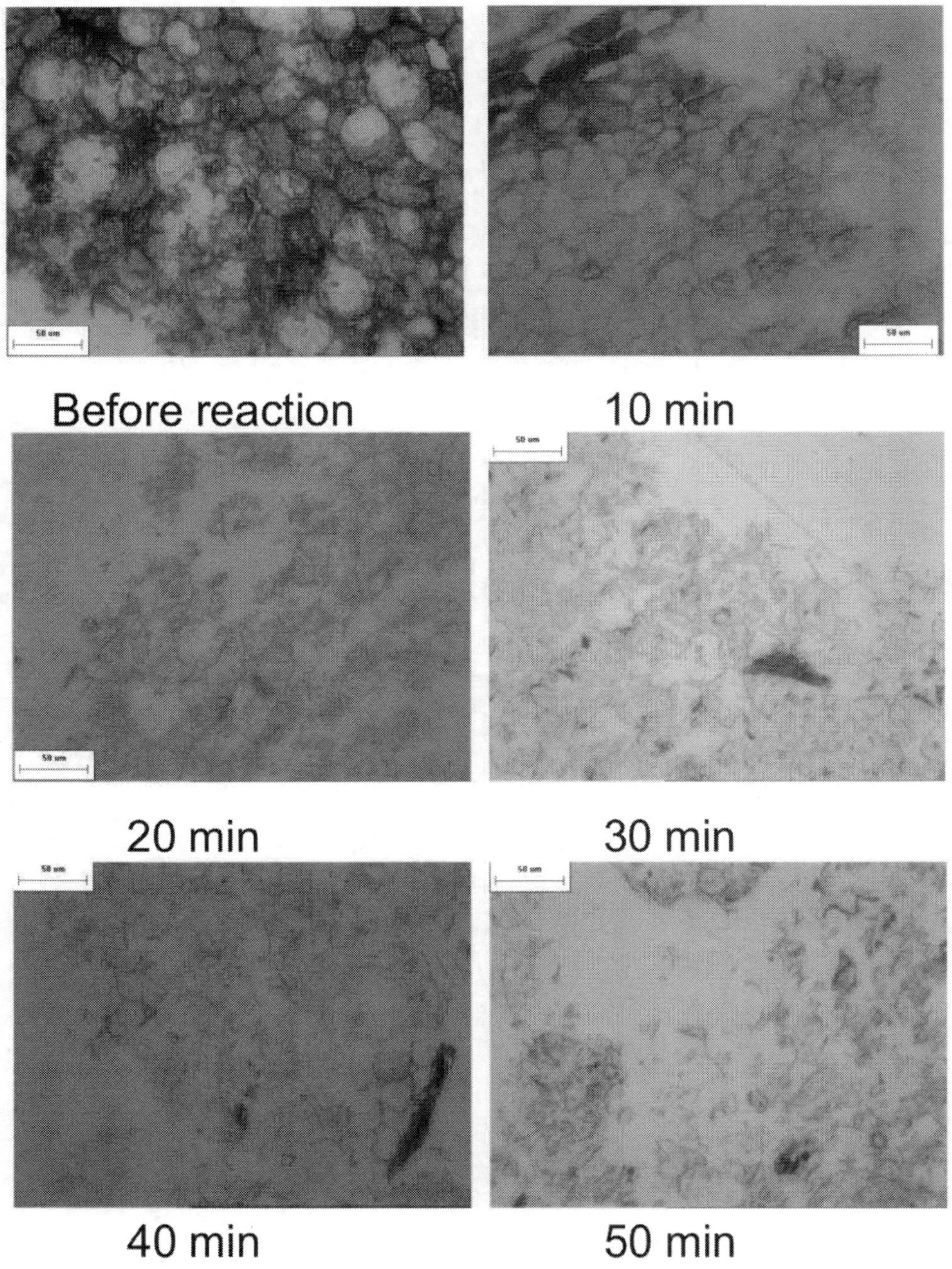

Conclusions

- Of the five factors investigated, three affect the response significantly:
  - molar ratio (methanol to oil)
  - reaction time
  - mixing speed
- The first-order model did not fit the process, but the second-order give a good fit
- NaOH concentration and reaction temperature did not have a significant effect to the response
- The cell walls break and release the oil to bulk environment at high molar ratio, longer reaction and high mixing speed

Q & A

HV | Mag | Spot | WD | Det
10 kV | 500 x | 2 | 10 mm | SE
50 µm

HV | Mag | Spot | WD | Det
10 kV | 500 x | 2 | 11.9 mm | SE
50 µm

STEAM EXPLOSION PULPING OF INDIAN NONWOOD FIBRES

By

B.V. KOKTA*, J.J. GARCEAU*, F. CARRASCO*& A.K.RAY **

*** University of Quebec, Three Rivers, Quebec, Canada**

****Department of Paper Technology,IITRoorkee, Saharanpur,India**

ABSTRACT

There is an increasing trend in Indian paper based biorefinery industries for efficient utilization of its abundantly available nonwood and especially annual renewable resources, mainly for two purposes, for developing a balanced ecosystem and at the same time to meet its paper and paper products using eco-friendly technology. Interests is also being explored to produce value added chemicals and low cost biofuels. Steam explosion technology is one of the innovative processes to achieve the above objectives. An attempt has been made to produce Steam Explosion Pulp from pre-impregnated Indian sugarcane bagasse(CO variety), rice straw and bamboo(Bambusa Dendrocalamus Strictus) cooked for 2 minutes at 463K (1.26 M Pa saturated steam pressure) using reactor designed by Stake Technology, Canada with varying sodium sulphite chemical charge as primary impregnating chemical, with or without the presence of secondary chemicals and some additional swelling agents. For the sake of comparison all the three species are impregnated at the same conditions of time and chemical charge and also cooked at the same conditions and then explosively decompressed and discharged. Explosion pulp properties were evaluated for all the species obtained from a variety of conditions. As many variables are involved in the above investigation a planned statistical experimental design is used to identify selective number of experiments

More detailed experimental investigations for bamboo, bagasse and straw were also carried out for finding out the optimum operational conditions. However for yield comparison purposes, change of pH and chemicals were studied. As bagasse and rice straw are radically different in terms of their structural openness compared to bamboo(even though all are non-wood species), the optimum conditions as expected are found to be quite different, bamboo required much more drastic impregnation and protracted time with severe chemical treatment and for the improved pulp quality the most effective pulping agents were either purely caustic soda at higher level of charge or an optimum mixture of sodium sulphite and alkali. Loss of pulp yield has been found in all species with treatment of NaOH. Total ionic content of bagasse and bamboo are comparable, being higher than straw.

The optical properties of unbleached pulp are acceptable atleast for manufacture of newsprint grade pulp. Combination of sodium sulphite with low percentage of soda may be a compromise between loss of pulp yield and gain in strength properties for bamboo. Drainage characteristics of bamboo pulp are found to be much more superior to those of bagasse or straw pulp.

From the detailed study, it is found that the exploded pulp is susceptible to changes in pH, the treatment of chemicals, the nature of chemicals and the concentration.

The waste liquor obtained after steam explosion were analyzed and evaluated for its pollution potential.

1.INTRODUCTION:

Due to acute shortage of fibrous raw materials, Indian pulp and paper industries are in quest for new processes for efficient utilization of its abundantly available nonwood and annual renewable agricultural resources to meet its demand for paper products. Indian large mills at present use generally bamboo and hardwood as a pulp mill feed stock (1.9 million tonnes) and some quantity of bagasse. On the contrary medium and small mills use principally bagasse wheat and rice straws. Sugar cane bagasse is being increasingly used as a major raw material in some Indian pulp and paper mills. But overwhelmingly majority of the mills use bagasse for production of chemical grade pulp, mainly used for offset papersThe pulp production from nonwood fibres is approximately 39 % of the total production of paper now days. Availability of 0.6 million tonnes of straw and nearly 1.8 million tonnes of bagasse can be of great value if suitable technology is developed. Steam explosion pulping at medium pressure range (SEP) **(1-20)**working on vapour phase short time exposure steam cooking can meet the desired objective of ever increasing demand of paper and at the dame time to reduce pollution load.

Steam explosion process (SEP) offers a rapid process, which can meet the above objectives by optimizing balance between properties and yield. This process uses the sequential stages of impregnation/pretreatment with appropriate chemicals, extracting the liquor, presteaming, steam cooking at medium pressure range (1.1-1.9 M Pa) at 443-483 K in vapour phase for 1-7 min, and finally by rapid pressure release.

The explosion process was regarded as an adiabatic expansion process because the release of steam was within a very short period of time. The explosion power could be determined according to enthalpy change of media. Mathematically the explosion power can be represented as:

$N = M\Delta^{H}/t$, kW/ kg of O.D. fiber .....(1)

The work of explosion, in kJ per kg of steam, E,

$E = \varphi\ a\ V\ \rho\Delta H$ .....(2)

$C = (2\Delta H)^{0.5} = [2(H_1\text{-}H_2)]^{0.5}$ .....(3)

$\Delta\ (\Delta H) = (\Delta H)\text{-}\ (\Delta H)_{0.1}$ .....(4)

Where N, the explosion power, kW/kg fiber (OD); E, the explosion work, kJ/kg steam, ρ the density of steam before explosion, kg/m^3; ΔH, enthalph change of steam, kJ/kg steam; V, the volume of fibrous material, m^3, t, the time of release, second; M, the mass of media, kg of media / kg of OD fiber; a, portion of steam in fibrous media; φ, the coefficient (<1) and c= speed (m/s) at the end of the nozzle of the explosion vessel; H_1= media enthalpy in explosion tank, J/kg (media) and H_2 = media enthalpy outside the explosion tank, J/kg (media).

The φ a V ρ can be considered the mass of steam**,** M in equation 2.

The calculated values indicate that the relative enthalpy ΔH reaches the highest value at the pressure of about 2.0 Mpa and the output of explosion work comes to its climax. Δ (ΔH) varies a little between 0.1 Mpa. Considering the explosion power output, operating at low pressure, such as 1.0 ~ 2.0 Mpa, is reasonable. Therefore, more disparity of enthalpy can result in a faster speed, which would cause a very strong shearing field in the nozzle, so that the fibers could be liberated.

The combined effects of pulping temperature and time can be quantified by expressing them in a single variable for the optimization of steam treatment, the severity parameter as defined below:.

The severity parameter, R_0-the reaction ordinate

R_0 [min.] = $\int_o^t \exp(T\text{-}100)/14.75)\,dt$, for constant T, $R_0 = t^*\exp^{[T-100)/14.75]}$
Where T, the temperature, ^{0}C and t, the time, min.
The severity factor increases exponentially with increase in reaction temperature and linearly with increase in reaction time. The equation will indicate the temperature- time sequence for not only re-pulping but also de-inking operation. The principles explained above can easily be manipulated to find out optimum conditions of temperature and time.

Enormous data are available on explosion pulping of hardwoods and softwoods (**1-4**); the same for nonwood fibre are extremely limited. No attempt has been made before 1991 to produce steam exploded pulp from Indian bagasse though some experiments on explosion pulping had been carried out with Australian and Hawaiian bagasse (14). Experiments have also been carried out recently on Bamboo, Rice straw & Bagasse of Indian origin by the present investigators (14-20).

Explosion pulping has been basically originated by Mason (14-20) using steam as the working fluid at 6.9 M Pa and 558 K on wood without any pretreatment of chemicals.
Subsequently a number of investigators attempted to produce either biomass or a paper grade pulp or particleboard from wood and nonwoods by steam treatment on the feed stock (with or without chemical treatments) at different temperature and pressures. These are extensively reviewed by Ray et al. (14-20). However most of the investigations on explosion pulping were due to two groups of investigators. Mamers et al. (3-13) of CSIRO Australia conducted experiments in SIRO pulper at 3.4-13.8 M Pa pressure in presence of Nitrogen gas (saturation steam pressure 0.54 - 1.55 M Pa) on both woods and nonwoods with the impregnating chemicals like sodium sulphite, kraft mixtures, caustic soda, ammonium hydroxide, ammonium bisulphate and urea to manufacture low yield chemical grad pulp. Kokta et al. (1-3) of UQTR, Canada also attempted to produce newsprint grade high yield/ ultra high yield pulp from impregnated woods and nonwoods in Stake Tech-make reactor at a milder pressure conditions (1.1 -1.9 M Pa) but at the same range of temperature (170^{0}C-210^{0}C) and lower time (1-7 min.) with or without the presence of Nitrogen gas pressure. The impregnating chemicals were mainly sodium sulphite with or without additional chemicals like caustic soda and other chemicals.
Ray et al. (14-20) conducted extensive experimentation on wood (Aspen) and nonwood fibres like bamboo , depithed bagasse and rice straw of Indian origin with various chemical treatments, chemicals to raw material ratio, and impregnation time to confirm the conclusions drawn by Mamers et al. and to examine the responses of these raw materials to sulphite impregnation charges.

In this present investigation, an attempt has been made to compare the properties of pulps from all the three species , namely bamboo, bagasse and rice straw with different principal chemical of sulphite or soda with the presence of other secondary (swilling agent) chemicals at a fixed condition of impregnation and pulping times. Comparison is also made on similar processes expected to produce similar quality pulp. Previous experiments show that though extensive sulphonation leads to increase in flexibility, conformability, sheet density and strength properties of paper in steam explosion cooking, is not really true in case of nonwood fibres, at least in case of bagasse and straw The data will then be compared with those from exploded soda pulp due to Mamers et al. produced in SIRO pulper and other investigations

3. EXPERIMENTAL

Experimental Design:

The pulping experimental conditions were performed in such a way that the resulting pulp can be conformed to ultra high yield, semi-chemical and chemical pulp by treatment of different percentages of impregnating chemicals, nature of chemicals, two level of temperatures with three level impregnation times.

Experimental Procedure:

3.1 Raw Material preparation:

All the raw materials, namely , chips of Dendrocalamus Strictus collected from chipyard of M/S Star paper mill (as such), bagasse (Saccharum Officinarum, CO variety) collected from M/S L.K. Sugar mills in mill wet conditions and rice straw from Western U.P of India. The bamboo chips were repeatedly washed; bagasse was depithed manually, first by dry depithing method, screening out the pith through 30-mesh screen followed by wet depithing in hydrapulper maintaining proper consistency.The rice straw were cut in 3-4 cm size and the same was wet cleaned. All materials individually were again thoroughly cleaned under high pressure by spraying water and dried. The pith content of the final bagasse was assessed by microscopic examination. Some of the proximate and chemical analyses of the raw materials are shown in **Table 1.** The raw materials were transported to University of Quebec,Trois Rivieres of the Quebec province of Canada and stored in the laboratory CRPP, University of Quebec,Trois Rivieres of the Quebec province of Canada under required humidity and temperature conditions.

3.1.1 Screening of bamboo chips

The bamboo chips were screened through a stack of laboratory gyratory screens for different sizes, 28.6 mm, 15.9mm, 9.5mm, 6.9mm and -6.9mm. The size of collected chips from different fraction were randomly measured and statistically averaged length (L), width (W) and thickness (Th) were calculated. The mean value of dimensions of chips were: length: 23.5mm; width: 13.3mm; thickness: 4mm whereas 95% of the chips have the values: L: 25 -27mm; W: 9-16mm and Th: 4-5mm. These were again kept in a conditioned storage room.

3.2 Impregnation and explosion pulping

or further treatment at bath ratio 1:6. Chip thickness for all categories were subjected to sodium Meta bi sulphite (pH=4.5) treatment at 333K for their impact on yield variation whereas mixed bamboo chips was impregnated at the conditions shown in **Table 2.** At the end of the impregnation period the residual liquor was extracted manually.

All the species were impregnated with chemicals (bath ratio= 1: 6) at different chemical charges for steeping time at 333 K. For bamboo three different sizes (15.9mm, 9.5mm, 6.9mm) and mixture of all bamboo chips (mixed) were processed. At the end of impregnation time, the raw materials were squeezed to drain out the residual liquor. The impregnated material was charged in to the batch reactor(Stake Technology make) . The design and operational parameters of the reactor is given in the following table.

Design and operational parameters of batch reactot of Stake Technology make

Design Parameters	Values	Operational Parameters	Values
Effective Volume,ml	300	Degassing time,min.	1
Int.dia Feed Spout,mm	53	Pressure,kg/cm^2(g)	12-13.8
Tot.length of reactor tube,mm Dia., exit nozzle,mm	685 38	Temperature,K	Varying
L/D of receiving tank, Dia.mm	2 450	Cooking Time to temperature,s Time at temp., min.	 15-20 2-8

Sealing of the reactor, presteaming, steam cooking and quick explosive discharge of the cooked material in collecting tank were followed sequentially. The time, temperature and pressure of the reactor were varied. The exploded pulp was screened out through 200-mesh screen and collected. This is then followed by washing with water thoroughly to remove any adhered chemicals. At the optimum conditions of impregnating chemicals and time, varying doses of secondary chemicals (Na_2CO_3, $NaHCO_3$ and NaOH) are added indicating the effect of nature of chemicals at different initial pH levels. Further experimental design is also attempted to examine the effect of different charges of caustic soda application (without the addition of Na_2SO_3).

3.3 Evaluation of Pulp Yield:

The washed exploded pulps produced under different conditions was refined for 90 seconds at 2-3% consistency in osterizer blender for desired freeness **and pulp yield was evaluated accordingly (1-2).** For severely chemical treated pulp (conforming to semi-chemical nature) did not demand for refining and after washing in the same way as the sulphite treated pulp yield was evaluated. Random samples from all explosion pulps were kept separately for detailed study regarding the physical shape of the disintegrated chips or fibres, the liberated fibres and the extent of deliberation by explosion.

3.4 Sufficient number of hand sheets were prepared in British Sheet former, dried and tested according to CPPA standard. Routine instruments were used to determine the strength, optical and surface properties for all pulps of bamboo and for high yield pulps of annual fibres). The pulp properties are compared with already reported data on explosion pulping of bagasse and straw at high pressure and low-pressure conditions. The ionic groups, namely sulphonate, and carboxylate were determined using conductometric titration followed by linear regression methods (1-2). Some of the typical data statistically averaged are shown in **Table 3.**
Drainage characteristics of all pulps from all kind of raw materials and ash from only bamboo pulp were evaluated as per CPPA standards.

McNett-Bauer classification and drainage time:

To compare with chemical and high pressure exploded pulp already reported **(3-13)** Bauer-McNett fractionation was performed with exploded pulp as such and after slight refining (for target CSF). Drainage times of the exploded unrefined pulp and those of refined ones are also compared.

Liquor Characteristics:
The characteristics of original liquors and the effluents from the impregnation and explosion stage were evaluated in terms of pH, % SO_2, residual alkali and the electrical conductivity (by CDM 83 conductivity meter).

.

4. RESULTS AND DISCUSSIONS:
The experimental data on explosion pulping of bamboo, bagasse and rice straw were compared. These are shown in previous publications.In majority of cases, common basis has been used. Extreme cases are also considered to show the yield variations by impact of severely treated chemicals (for all species) with protracted time of impregnation (bamboo). For open structured fibres (straw and bagasse) the effects of lower chemical charge on pulp properties were also shown.These are depicted in brief. In this present paper the effects of other additional impregnation chemicals ($NaOH$,Na_2CO_3, $NaHCO_3$) on yield, strength ,moptical properties,drainage characteristics and liquor properties are explained.

In previous publications (14-16), the following were discussed in detail and corresponding concluding remarks were drawn.

1. Effect of pH of original liquor on explosion pulp yield
2. Size distribution of industrial bamboo chips
3. Variation of pulp yield as a function of chip size
4. Effect of secondary chemicals on pulp yield and ionic group formation
5. Effect of cooking time, temperature on pulp yield and ionic group formation
6. Effect of chemical charge and impregnation time on ionic group formation
7. Effect of steam cooking time on ionic group formation for explosion pulp with only secondary chemical charge
8. Comparison of ionic group formations in pulp matrix
9. Comparison of breaking length-tear index and scattering coefficient inter-relationship
10. Comparison of porosity, roughness and zero span tensile breaking length
11. Effect of ash content of bamboo pulp on strength properties
12. Comparison of drainage characteristics
13. Comparison of optical properties
14. Comparison of spent Liquor characteristics
15. Characterization of liberated fibres by explosion

Sodium sulfite alone cannot produce bamboo pulp even with high charge and longer time of impregnation. 2.5%-5% NaOH addition to 40% sodium sulfite charge during impregnation gives acceptable pulp properties. Yield as high as 77% is possible at this condition. Higher addition leads to significant yield loss. Purely soda pulping with 25% charge provides paper of much

higher strength with consequent yield loss. Best conditions for sulfite pulping of rice straw and bagasse in this present investigation are found to be 40% sulfite impregnation charge with 12h impregnation time. The strength properties of annual fibres are far better than those of bamboo sulphite pulp even after protracted time of impregnation for the latter. Scattering coefficients, unbleached pulp brightness, and opacity are also better. Use of secondary chemical, like NaOH in necessary for bamboo to get paper of desired quality. On sulfite treatment to annual fibres pulps conforming to newsprint quality can be made.
At the optimum condition exploded pulp from all kind of fibres are highly difibrarted; Cells of annual fibres are vigorously ruptured. This causes the extremely poor drainage characteristics for at least rice straw pulp. The Bauer Mc Nettt fibre fractionation shows that even at lower initial unbeaten pulp freeness the results are comparable with the reported data by Makers et al. (3-13) for soda chemical and high pressure exploded bagasse pulp. But rice straw data could not be compared, as no results of fibre classification are available. However it gives much higher fines compared to bagasse and bamboo.

Sodium meta bisulphate solution (pH=4.5 though unsuitable for pulping of bamboo and straw, can be used for straw to improve brightness

Higher time of impregnation, higher temperature and longer time for steam cooking produce pulps of higher strength with concomitant yield loss. This is true for bamboo. They have detrimental effects on yield and properties of open structured fibers.

Influence of ionic group on strength properties is apparent. However it decreases with increase of cooking temperature and time. NaOH treatment increases the carboxyl ate content. Increase of time or increase of temperature decrease the total ionic content.

Ash content of bamboo pulp and change of ionic conductivity or change in % SO2 give much more reliable prediction of pulp quality than a change of pH or ionic content in the present investigation for nonwood fibres.

Further experimental data were analyzed for bagasse as well as rice straw in terms of the following parameters

1. Bauer-McNett classification:
2. Comparison of fibre fractionation data with other investigations:
3. Effect of Chemicals on Drainage properties:
4. Effect of Chemical addition on Optical Properties:
5. Effect of Pulping Parameters on Strength Properties
6. Liquor Characteristics

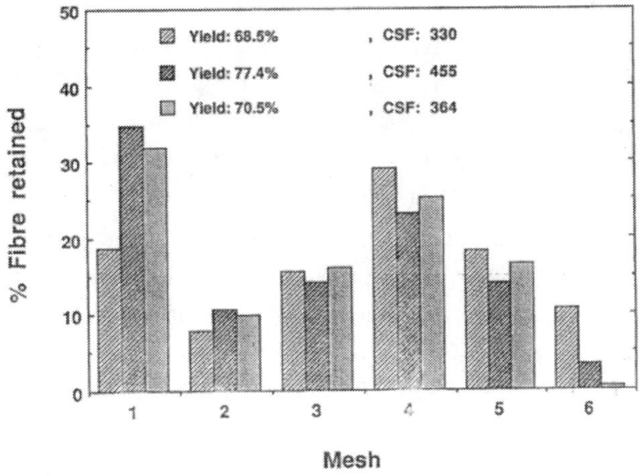

Fig.1 McNett Bauer classification of exploded bagasse pulp

Bauer-McNett classification:

Figure 1 shows the relative magnitude of fibres retained in different meshes of Bauer McNett classification unit as a function of explosion pulp yield during various impregnation charges with application of various nature of chemicals at initial freeness (**see, caption of figure**). As anticipated, fine fractions (-200 mesh) are highest in NaOH treated pulp, followed by Na-bicarbonate and then for Na-carbonate while for oversized fractions (+ 14 mesh) highest fraction is obtained followed by carbonate and the lowest was for NaOH as expected. Sum of fibres retained in +28, +48 and +100, the figures remain more or less invariant (48-52%; +200 fractions again as expected. NaOH overpowers other two and between carbonate and bicarbonate, the former preceeds the latter. Including +200 mesh the NaOH treated pulp supersedes others (70.6%), followed by carbonate (67.7%) and the bicarbonate pulp has the lowest value (61.9%). Addition of NaOH to sodium bicarbonate in equal magnitude (5%) to 40% sodium sulfite impregnation charge decreases the yield ad increases the usable fraction. An optimization is very much essential at this stage, keeping in view of the end objective.

Comparison of fibre fractionation data with other investigations:

Figure 2 indicates the comparison of the present fibre fractionation data with Mamers et al. **(5,6)** which is performed at high pressure,high temperature explosion process. On inspection of data it is revealed that any of the present pulps obtained by Stake technology make I reactor at 1.25 M Pa and 463 K has lesser initial freeness, lesser fines (-200 mesh) with higher yield but higher oversize (+14mesh). Good fibres (-30+100) and also moderate quality fractions (-100+200) are almost comparable even with chemical pulp. Evaluation of paper properties will give more clear picture to arrive at a definite unambiguous conclusion. This result clearly indicates that high pressure (3.4 M Pa, 13.8 M Pa) as used by Mamers et al. **(3-13) in** SIRO pulper of complicated tortuous flow channel is not necessary at all. Comparing the pressure effect on fibre size distribution from beyond 3.4 M Pa (saturated steam pressure 0.8 - 1.55 M Pa) is inadvertent on paper properties and fibre size distribution. Latter **Mamers et al. (6)** reported that quality and -

yield were relatively insensitive between temperatures 443K-473K for bagasse and straw and there is no significant advantages of operating digester above pressure 3.4 M Pa. The striking similarity of results between present investigation and those of **Mamers et al. (5,6)** may be explained by the fact that in the present study, chemical impregnating charge is comparatively high while unlike SIRO pulper, SEP uses straight path in Stake Tech rector (little tensile force compared to those due to **Mamers et al.)** under comparatively lower pressure (1.25 vs 1.55 M Pa). These factors counterbalance each other.

Comparing the present data with those reported by **Lai et al. (14-20),** it is found that much more lesser oversize (18-35% vs 88-89%) but probably higher fines fraction are obtained in this present investigation. On comparing the fluff pulp data on bagasse reported by **Saavedra et al. (14, Cuba-9)** the present pulp gives lower **+28** fractions (26-45% vs 68.4%), higher +48 fractions (14-16% vs 0.51%), + 100(23-29% vs 0.24%) and lesser p100 mesh (17-28% vs 30.2%). The above results indicate the potential of the present process.

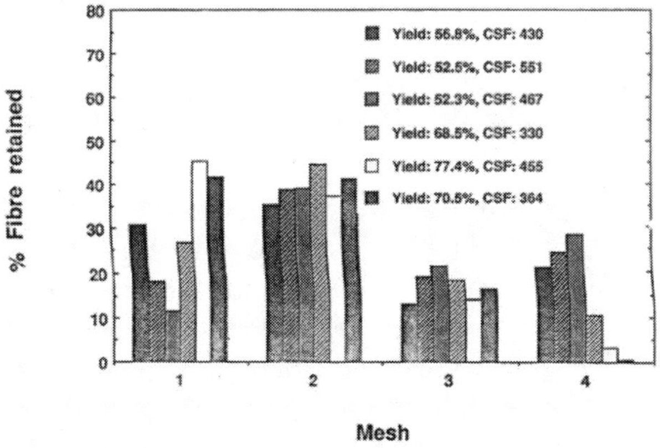

Fig. 2 Comparison of fiber classification between present investigation and those due to **Mamers etal.(3-13).**

Effect of Chemicals on Drainage properties:

Figures 3 & 4 indicate the drainage characteristics of bagasse pulp under various treatment of chemicals ,namely carbonate, bicarbonate and hydroxide and also as a function of yield values. A comparison is also made with sole treatment of the primary chemicals (sodium sulphite). It is evident that hydroxide pulp needs much more drainage time compared to carbonate and bicarbonate in case of bagasse. Sodium sulphite treatment gives better drainage characteristics. As expected 40% sodium sulfite has superior drainage characteristics. All other pulps produced by either extra addition of 5% secondary chemicals merge each other and negligible difference is noticed between them. Yield value prefers sodium bisulfite addition. Strength properties will dictate the superiority of secondary chemical addition.

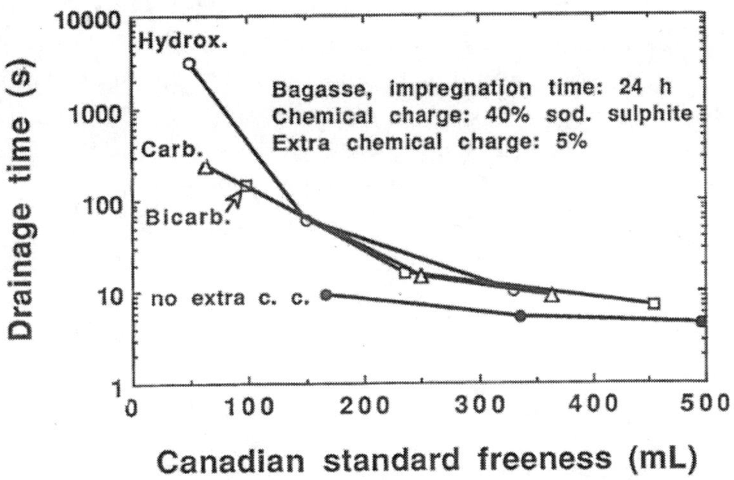

Fig.3 Drainage time as a function of chemicals and CSF for bagasse pulp

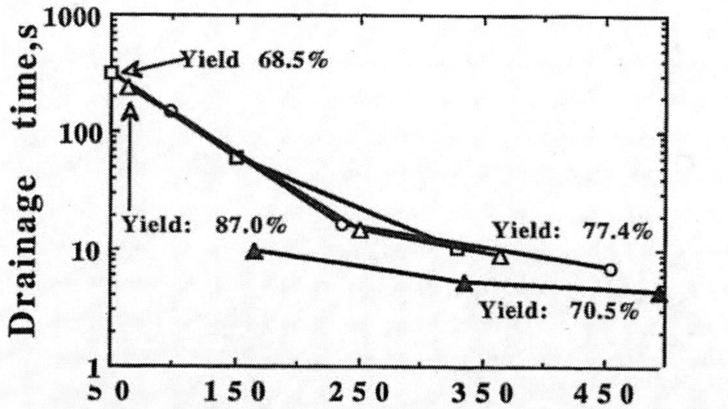

Fig. 4 Effect of CSF on drainability of pulp with yield as a parameter for exploded bagasse pulp

Effect of Chemical addition on Optical Properties:

Figures 5 and 6 give the inter-relationship between Canadian Standard Freeness with light scattering coefficient for bagasse explosion pulp. It is interesting to mention that without application of any swelling agents like carbonate, bicarbonate and hydroxide it gives highest scattering coefficient with highest yield while the carbonate application as secondary chemicals gives intermediate value between no extra Chemical, carbonate and hydroxide for bagasse. For newsprint grade pulp therefore purely sodium sulphite addition will exhibit better optical properties than those from combinations of primary and secondary chemical treatments. **Figure 7** indicates the brightness values of bagasse explosion pulp. This figure indicates that with the addition of secondary chemicals the rightness is reduced. This is an expected behaviour. However, the behaviour of hydroxide, carbonate and bicarbonate reflect slightly different for these two kinds of raw materials although the bicarbonate gives the lowest values. From printing paper point of view no chemical addition to sodium sulphite treatment will favour the quality.

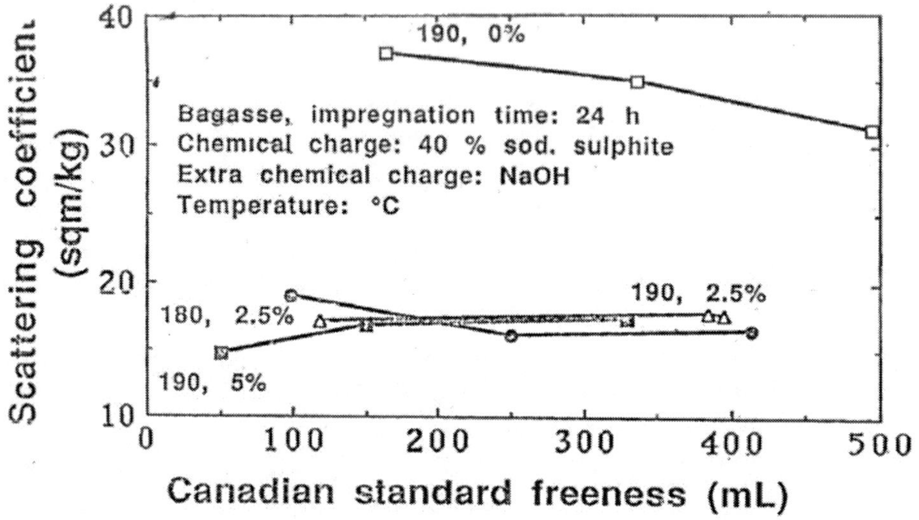

Fig. 5 Scattering Coefficient -CSF –Chemical dose relationship for exploded bagasse pulp

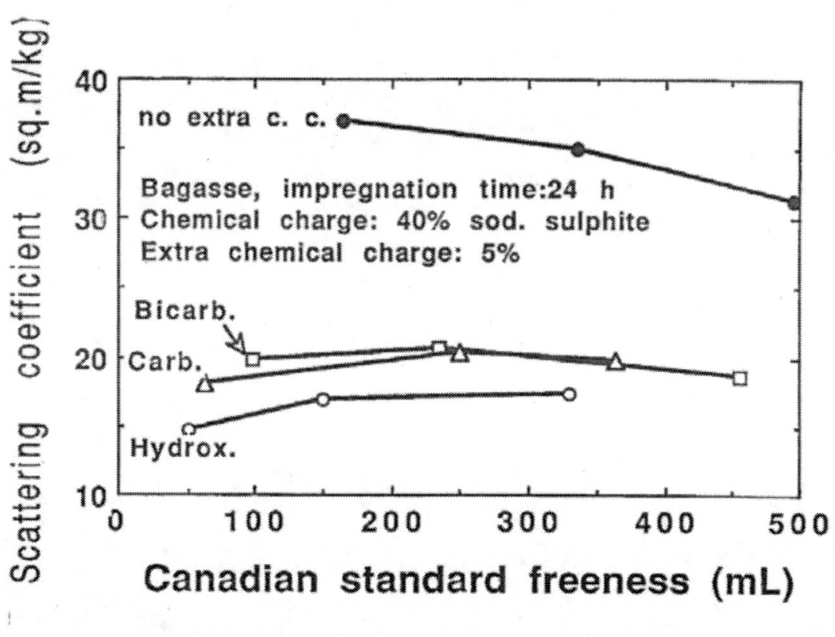

Fig.6 Scattering coefficients as a function of CSF and nature of chemicals for bagasse pulp

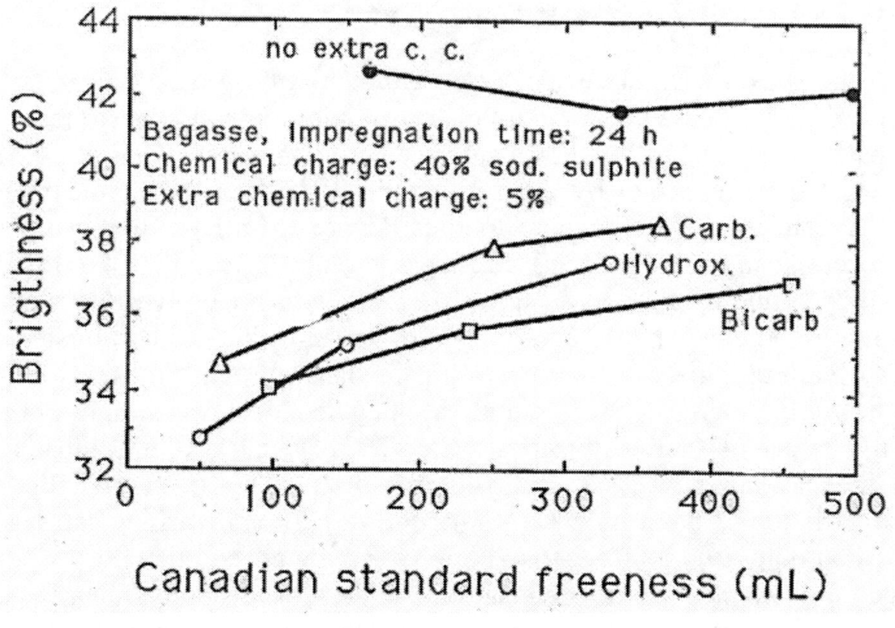

Fig.7 Brightness-Chemical dose –CSF inter-relationship for exploded bagasse pulp

Fig.8 indicates opacity as function of Canadian Standard Freeness with various chemical charges as parameter. It is evident that more opaque paper is produced if no extra chemicals are added.

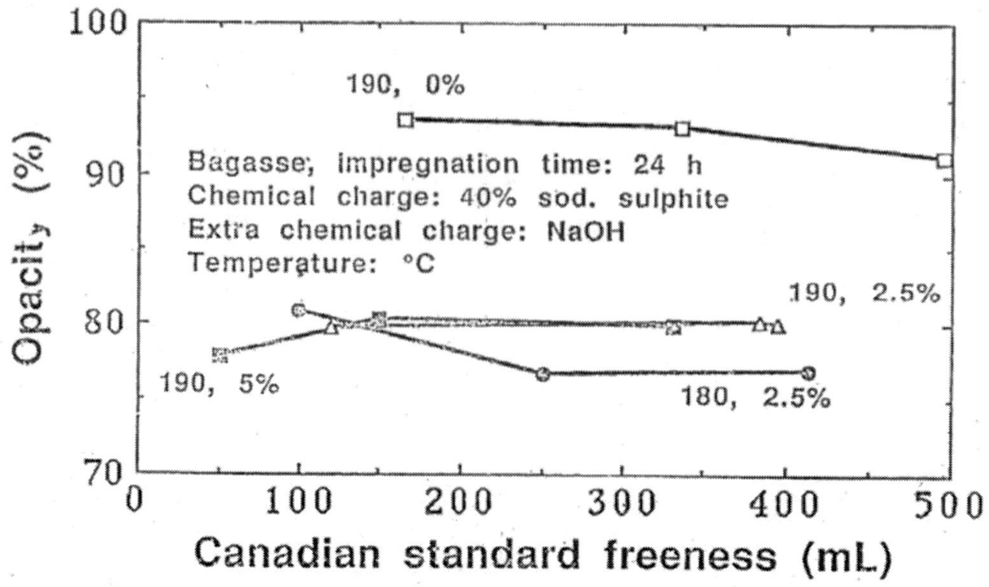

Fig. 8 Opacity-CSF –Chemical dose relationship for exploded bagasse pulp

Effect of Pulping Parameters on Strength Properties:

Fig.9 shows the stiffness (nM.m) as a function of CSF by treatments of various chemicals for bagasse. Surprisingly highest is found in case of no extra chemical addition and lowest incase of hydroxide addition. Carbonate, and bicarbonate shows intermediate behaviour. This is also an expected trend as paper becomes more soften by the application of hydroxide chemicals and become stiffer and stiffer as you go from carbonate to bicarbonate to only sodium sulphite treatment. **Fig.10** shows the zero span breaking length for bagasse pulp. Influence of hydroxide, carbonate, bicarbonate and only sodium sulphite are shown therein. Hydroxide gives the highest breaking length whereas purely sodium sulphite addition gives lower value. Bi-carbonate and carbonate gibes intermediate values. This is very interesting in the sense that these swelling chemicals will decrease the yield; increase the strength at the cost of lowering of optical properties. **Fig.11** shows the effect of CSF and various chemicals on energy of rupture. As expected hydroxide gives highest value whereas sole sodium sulphite treatment give lowest value. **Figure-12** shows the variation of burst index as a function of CSF and various natures of chemicals for bagasse explosion pulp. It is found that sodium hydroxide gives much higher burst index at a given CSF compared to when no extra chemical is applied. Carbonate and bicarbonate gives in between value (marginally equal). Therefore, it is evident that all the strength properties are increasing in proportion with higher pH chemical addition. This is scientifically true. **Figures13 & 14** give the bulk and tear index for all the species. Bulk become highest at only sodium sulphite treatment and lowest in case of hydroxide treatment. Addition of hydroxide therefore enhances the bulk values giving higher denser pulp. However, tear index values become lowest in case of hydroxide almost at par with the no chemical addition other than sodium sulphite but highest value with carbonate and bicarbonate (marginally equal). This is also very much expected in practice.

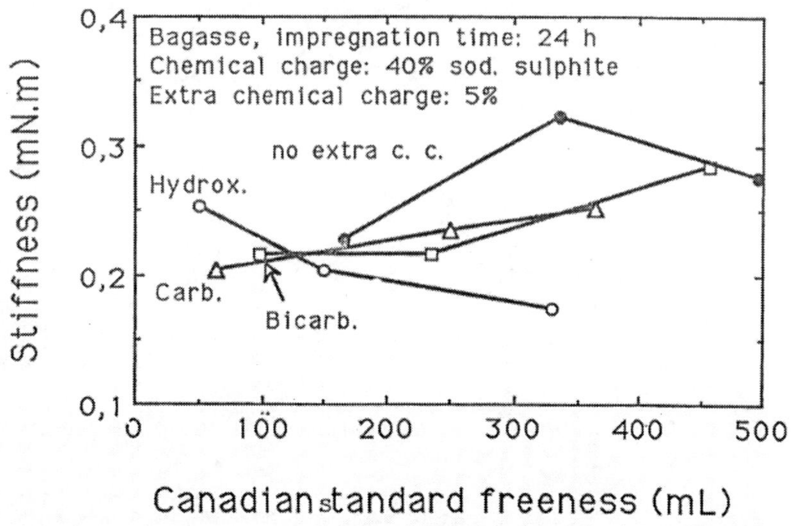

Fig.9 Influence of chemicals and CSF on Stiffness for bagasse pulp.

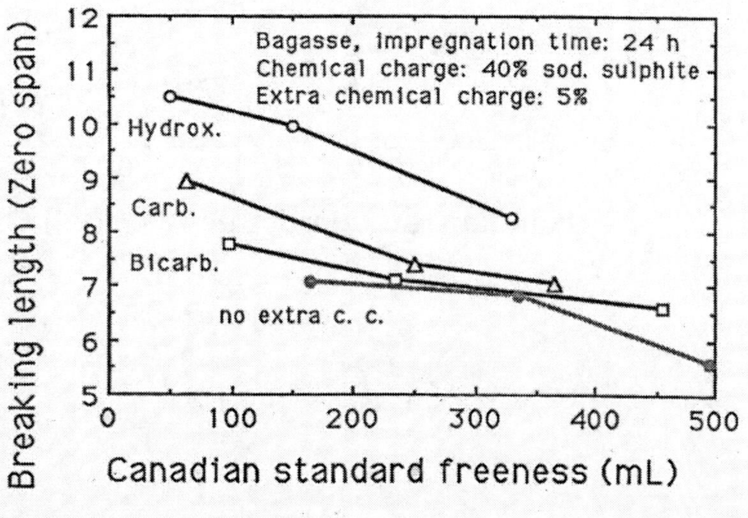

Fig.10 Zero span breaking length as a function of chemical and CSF for bagasse pulp.

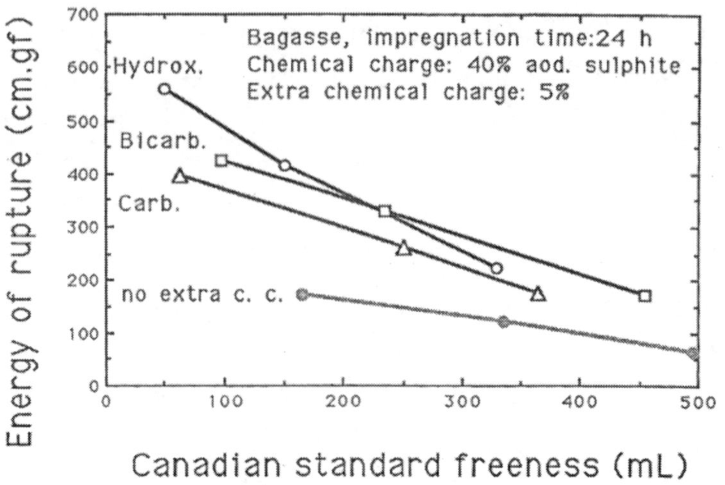

Fig.11 Energy of rupture as a function of CSF and nature of chemicals for bagasse pulp.

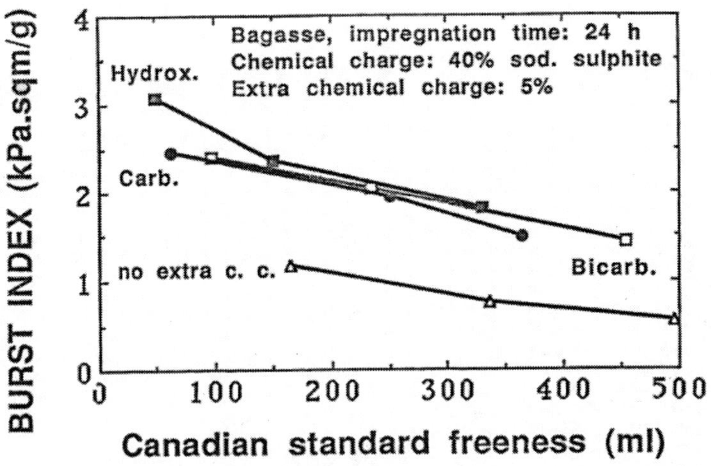

Fig. 12 Effect of chemicals and CSF on burst index for bagasse pulp.

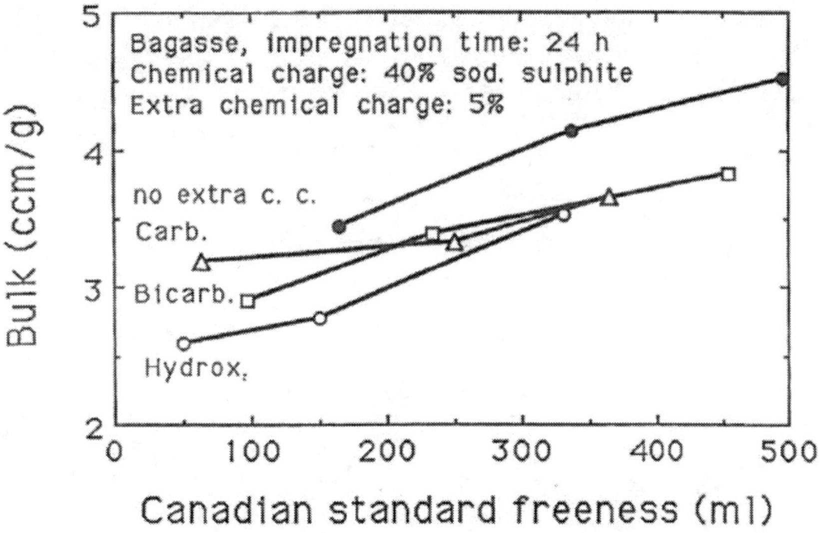

Fig. 13 Effect of chemicals and CSF on bulk for bagasse pulp

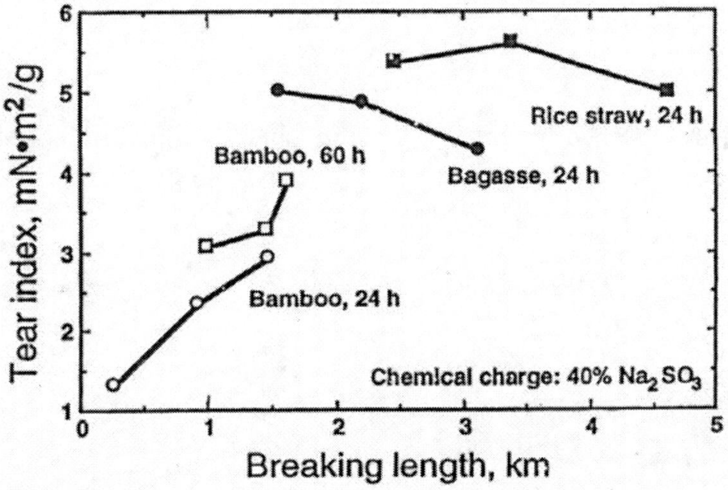

Fig.14 Breaking length –tear index interrelationship for different species and impregnation times

The variation of tear index and breaking length as a function of Canadian Standard Freeness (CSF) with percent sod. sulfite as impregnation charge on bagasse is shown in **Figs. 15 and 16** for 12 h impregnation time. Higher the chemical charges, higher are the values of strength properties. Average breaking length and tear values exceed 4.3 Km and 5.2 respectively at 100 CSF for 40% sodium sulfite impregnation charge. Higher values of breaking length or tear index could not be obtained even after protracted time of impregnation. The effect of time of impregnation on strength relationship does not give coherent behaviour.

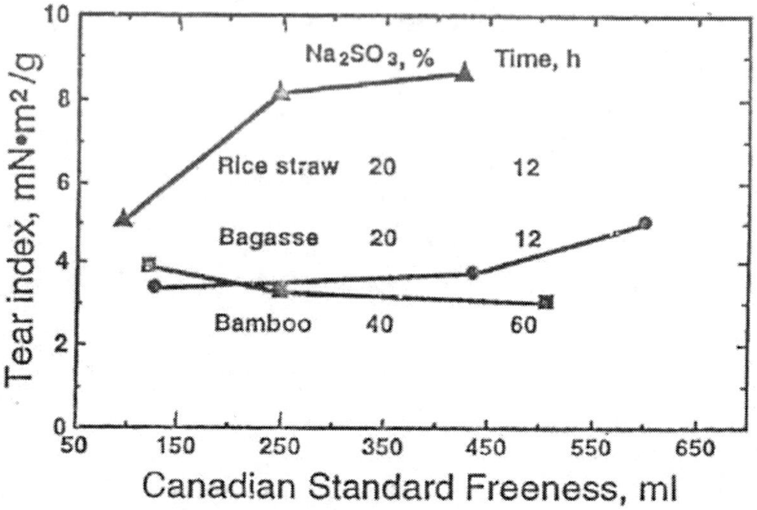

Fig.15 Variation of tear index with CSF for different species and various impregnation charges and times

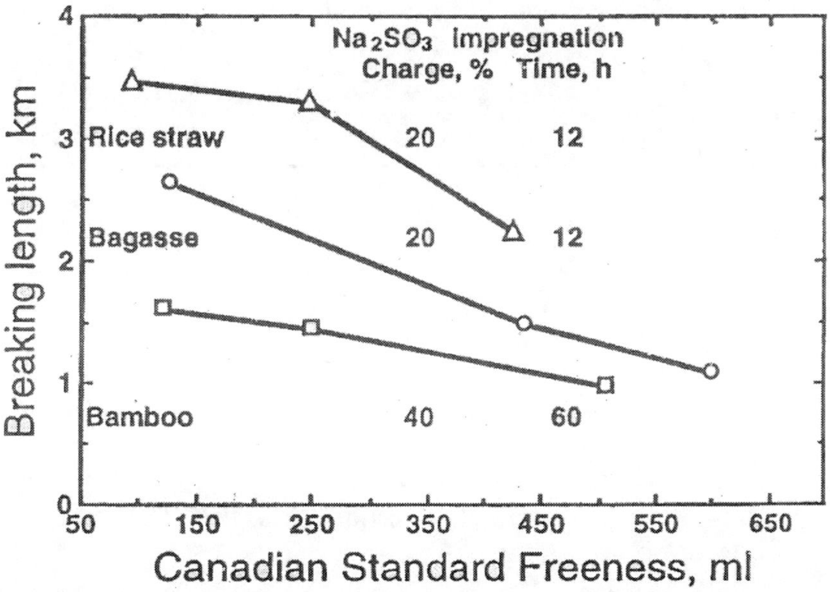

Fig.16 Variation of breaking length with CSF for different species and various impregnation charges and times

Effect of Explosion:

Effect of explosion on the properties of pulp has been explained for various pulp properties such as breaking length for 18 and 24 hr impregnation time, tear index and bulk , for rice straw pulp. These are shown in Figures 17-21. The exploded pulp exhibits higher properties than unexploded pulp in breaking length but lower values for tear index and bulk values. These are expected behavior.

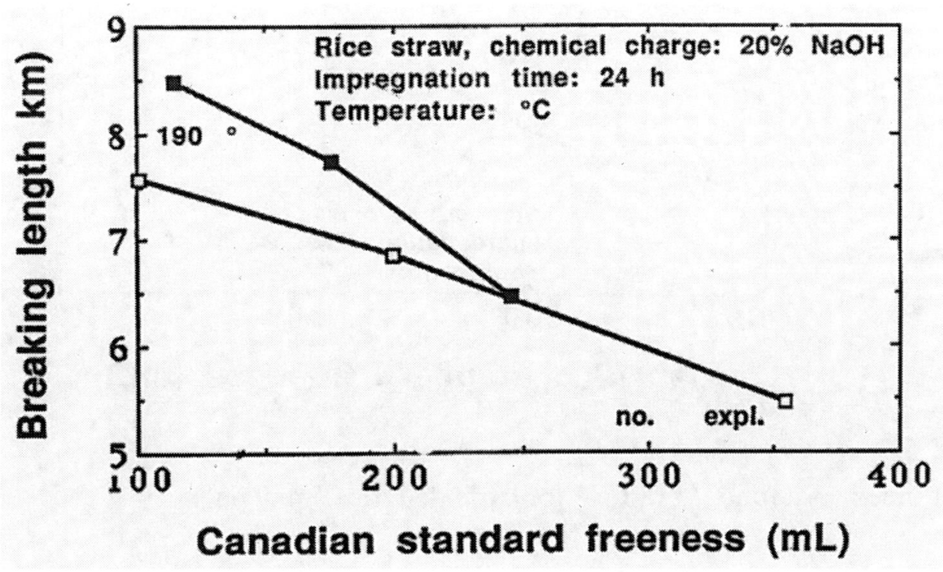

Fig.17 Breaking length as a function of CSF for exploded rice straw pulp

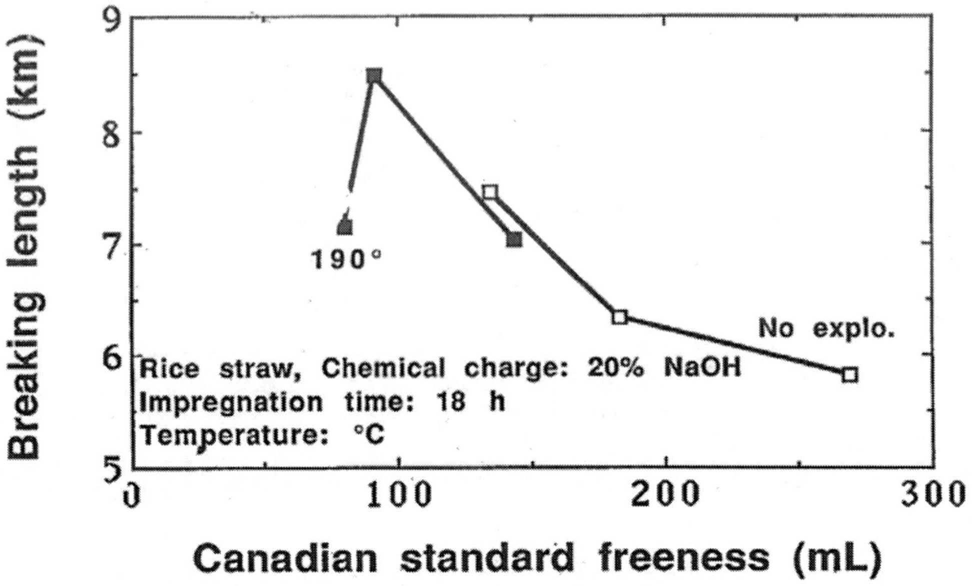

Fig.18 Breaking length as a function of CSF for exploded rice straw pulp

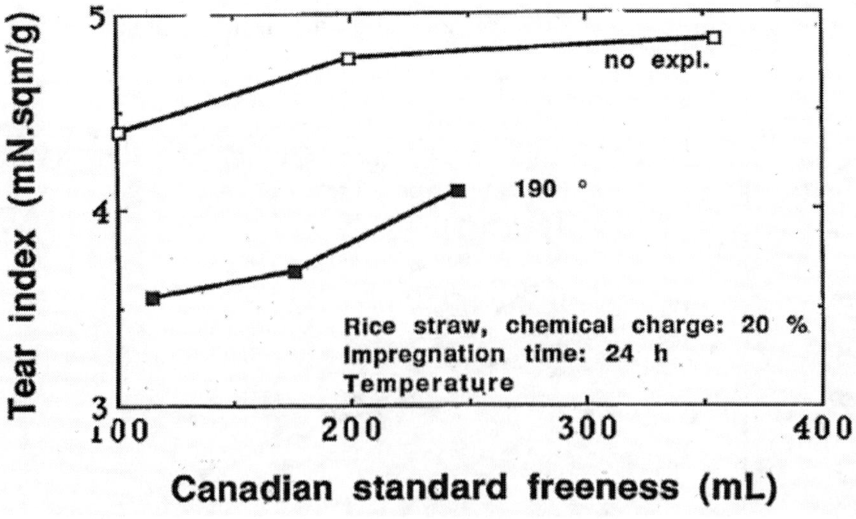

Fig. 19 Tear index as a function of CSF for exploded rice straw pulp

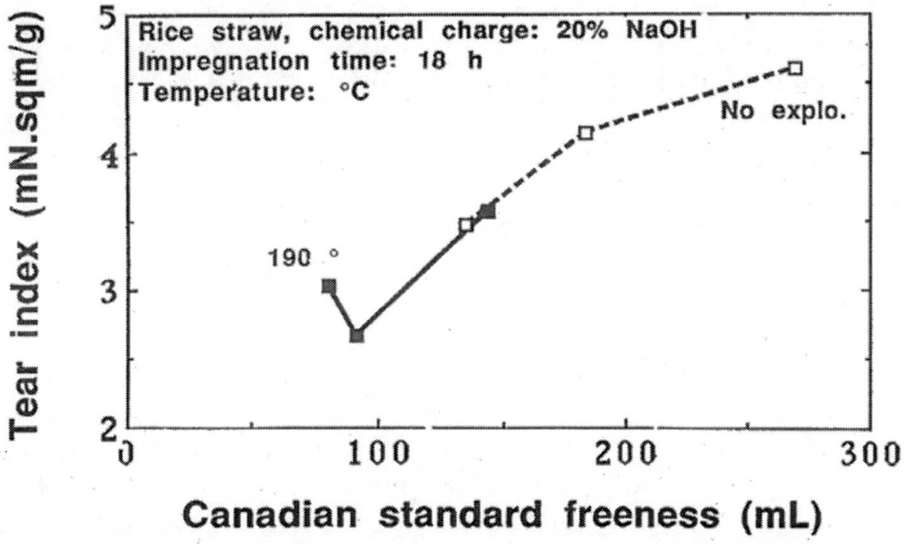

Fig.20 Tear index as a function of CSF for exploded rice straw pulp

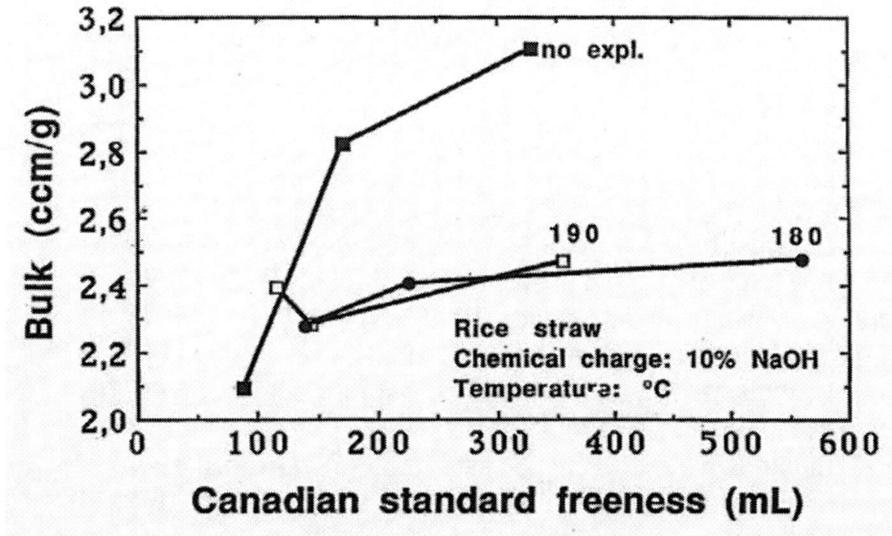

Fig.21 Bulk as a function of CSF for exploded rice straw pulp

Liquor Characteristics:

Some typical values of liquor characteristics such as change of pH, % SO_2, electrical conductivity and residual active alkali (if any shown in **Table 4.** The data indicates that change of SO_2, and electrical conductivity can give more reliable parametric changes than original liquor pH or its change from stage to stage in steam explosion process. The data on rice straw and bagasse explosion pulp properties are compared in **table 5** with those from other investigators.

5. CONCLUSIONS

Using all the raw materials ,namely bamboo, bagasse and rice straw, manufacture of newsprint and offset grade pulp is possible by medium pressure (1.25 MPa) steam explosion technique at 190^0C with primary chemical (sodium sulphite) impregnation charge. It conforms to the critical specifications stipulated for this pulp specially in terms of strength properties,opacity, brightness and scattering coefficient. Better strength properties can be developed by applications of additional chemical charges like sodium hydroxide, sodium bicarbonate and sodium carbonate but at the cost of slight lowering of optical properties. The yield is decreased with the increase of pH of secondary chemicals. The properties can further be improved by applications of only sodium hydroxide between 10 to 20% of raw material during impregnation. However, the yield will be drastically reduced. This pulp conforms to a chemical grade pulp but with higher yield than the conventional and with much lower time. The experiment further indicates that to produce chemical grade pulp, complicated equipment as used Mamers et. al. of CSIRO of Australia is not necessary. Stake Tech. Make reactor can give acceptable quality of pulp if proper conditions are choosen.

The drainage characteristics of sulphite pulps are comparable. Addition of secondary chemicals decreases the drainage rate to a greater extent compared to only sulphite-based pulp.

Liquor characteristics suggest to take in to change of % SO_2 and electrical conductivity for pulp characterization rather than change of pH in alkaline medium. Detailed investigation on influence of ash in pulp on strength properties is necessary. Any kind of pulps can be produced by Stake Technology make reactor. Effect of explosion has positive effect on breaking length and burst but decreases the value of tear index and bulk.

REFERENCES

1. Kokta, B.V., Can. Pat., 1,230, 208 (Dec. 15, 1987); Can. Pat., 1287705, (August 20, 1991)
2. Kokta, B.V., U.S. Pat., 4, 798. 651 (a.m. 17, 1989).
3. Mamers, H., Watson, A.J. and Grave, N.C. Appita 29 (5): 356(1976).
4. Mamers, H., Menz, D.J. and Yuritta. J.P. Appita, 33(3): 201 (1979).
5. Mamers, H., Yuritta J.P. and Menz, D.J. Tappi, 64(7): 93 (1981).
6. Mason, W.H., U.S. Pats. 1, 618 (January, 1928).
7. Mamers, H., Watson, A.J., and grave, N.C. Appita 29(5): 356 (1976)
8. Mamers, H., Appita, 32(2): 124(1978)
9. Mamers, H., and Rowney, J.E., U.S. pat. 4,163, 687(August 1979)
10. Mamers, H., Menz, D.N.J., and Yuritta, J.P., Appita 33(3): 201(1979)
11. Mamers, H., Yuritta, J.P. and Menz, D.J., Tappi 64(7): 93 (1981)
12. Mamers, H., Menz D.N.J., Yuritta, J.P., Appita 34(4): 281(1981)
13. Mamers, H., and Menz D.N.J., Appita 37(8): 644(1984)

14. A.K. Ray, J.J. Garceau, F. Carrasco & B.V. Kokta- Steam Explosion Pulping of Indian bagasse- Proceedings of the 57th Annual Convention of the Sugar Technologists Association of India, New Delhi, 13-15th October, 1995, General & By Product –pp 595-623, Vol., 57, 1995.
15. Ray A.K.; Garceau, J.J.; Kokta, B.V. and Carrasco, F. -"Explosion Pulping of annual Fibres"- Proceeding Bio-Mass Conversion held at Interlaken, Switzerland. May (1992)
16. Ray A. K.; Carrasco, F.; Kokta B.V. and Garcear, J.J.-"Optimizing Straw Based High Yield Pulping Process"-Second International Non-Wood fibre pulping and paper making conference held at Shanghai, PRC, April 6-9 (1992), Vol,2 pp115.101169.
17. Ray A.K.; Kokta, B.V.; Carrasco,F; and Garceau, J.J.-"Steam Explosion High Yield Pulp from Mill Wet Bagasse-A prelude to further Study" ibid Vol,1 pp.1131-1150.
18. Ray A.K.; Kokta, B.V.; Carrasco, F; and Garceau, J.J.-" High Yield Pulp from Bamboo to Steam Explosion Process-A Preliminary Investigation "ibid Vol,1 pp.446-476, 1992.
19. Ray A.K.; Carrasco, F.; Kokta, B.V. and Garceau, J.J.-"High Temperature Vapor phase Short time Cooking of Rice Straw and Bagasse"- Proceedings of TAPPI Pulping Conference. held at Atlanta, Georgia, U.S.A, November, (1993).
20. Ray A. K., B. V. Kokta, Felix Carrasco, and Jacques J. Garceau - A new look towards high temperature vapor phase short time pulping of Indian bagasse - Presented and published in IPPTA held at Calcutta between 19 - 29th November, 1999.

Table 1 Proximate and chemical analysis of Indian Bamboo, Bagasse and Rice straw

	Bamboo	**Bagasse**	**Rice straw**
Parameters		Value, %	
Solubility:			
Cold water	2.2	-	26.3
Hot water	5.1	6.3	23.2
1 % NaOH	23.5	30.0	47.0
D.C.M.	1.756	0.86	1.7
Klason lig.	25.8	23.5	11.8
Ash	4.30	1.8	18.1
Silica	2.9	-	12.0

Table-2 Experimental conditions fro bamboo, bagasse and straw

Raw material: Liquor = 1:6
Impregnation temperature = 333K
Presteaming time = min

Variables	Bamboo	Bagasse and Straw
1. Chemical charge		
% Chemical/O.D.R.M.		
Primary	Sodium sulfite	Sodium sulfite
%	40-75	20-30
Secondary	Sodium hydroxide	Sodium hydroxide
%	**0,2.5,5,10**	**0,2.5,5**
	and 25(without sulfite)	
2. Impregnation time, h	24,60	12,18,24

3. Cooking temperature, K	463, 473	463
4. Steam pressure, Mpa	1.25-1.55	1.25
5. Cooking time, min	2-8	2
6. Other chemicals	Sodium meta bisulfite	S.M.S.
	40%	40%, 32, 16

Table 3 Comparison of Typical values of properties of nonwood based explosion pulp

Properties	Bamboo		Bagasse	Rice straw
	A	B	C	D
Yield, %	68.5	76.7	87	71.6
CSF, ml	95	67	70	76
Drainage time, s	9.2	-	35	663
Bulk, ccm/g	2.7	4.1	3.1	3.1
Breaking length, km	5.2	4.4	4.4	3.87
Tear index, mN (sq.m)/g	11.4	7.4	4.97	4.95
Burst index, KPa (sqm)/g	2.8	2.3	1.67	1.74
Elongation, %	2.4	2.4	1.21	1.04
Scat. Coeff., sq.m/g	29.4	27.4	31.8	35.1
Opacity, %	98.4	96.4	92.9	98.5
Brightness, %	21.2	26.4	38.8	27.2
Porosity, ml/min	1470	456	63.9	405
Roughness, um	8.2	8.7	7.5	6.8
Zero span B.L., km	14.3	10.4	7.5	6.2
Stiffness, mN.m	0.35	0.19	0.27	0.17
Sulphonat, mmol/kg	6.85	87.2	112.8	53.3
Carboxylate, mmol/kg	181.7	158.7	127.1	176.3
Total ionic, mmol/kg	188.6	245.6	239.9	229.6
Energy of rupture, cm.gf	-	-	296.6	216.8
Ash in pulp, %	1.53	2.79	-	-

A: 25% NaOH, 60H;B: 40% sod. Sulfite+2.5% NaOH; C: 40% sod. Sulfite, 12h; D: 40% sod. Sulfite, 12h

Table-4 Comparison or characteristics of liquor streams

	Bamboo		Bagasse	Rice straw
Chemical:	NaOH	Na-sulfite +NaOH	Na-sulfite	Na-sulfite
%	25	40+2.5*	40	40
Time, h :	60	60	12	12
Parameters			Values	
pH				
Original	12.7	12.4	9.2	9.2
Aft. Imp.	12.6	10.0	7.6	8.0
Aft. Exp.	12.4	8.9	6.8	7.6

SO_2, %				
Original	0.0	3.9	3.9	3.9
Aft. Imp.	0.0	2.9	2.8	2.6
Aft. Exp.	0.0	0.24	0.67	0.47
Electrical Conductivity, us				
Original	160	47.6	54	54
Aft. Imp.	86	37.7	46.7	21.5
Aft. Exp.	15-22	11.9	22.2	16.8

* Residual active alkali in liquors were in trace amount

Table 5 Comparison of yield, fibre size distribution and drainage characteristics of exploded bagasse and rice straw pulps in present investigation with those of Mamers et al. (3-13)

Pulps	Yield	CSF	Drain. Time	Bauer-McNett Screen, mesh,		Classification Wt. %	
	%	ml	s	+30 +100	-30 +200	-100	-200
RX-1 (2.4 %)	48	200	14.6	-	-	-	-
RX-2* (5%)	65.3	170	145	16.2	33.55**	13.9	36.4
BB-1	30.8	430	5.3	30.8	35.2	13.0	21.0
BX-1	52.5	551	4.8	18.2	38.6	18.9	24.3
BX-2	52.3	467	5.1	11.5	38.6	21.2	28.3
BXS	68.5	330	10.3	26.6	44.6	18.2	10.6

Note:
RX-1 (18); BB-1 (Batch, Soda, 13% NaOH, 443K, 1h cooking time), BX-1(9% soda, 3.4 M Pa press., 473K, 5 min cooking time, due to Mamers et al.), BX-2 (8.9% NaOH,13.8 M Pa press., 473K, 5 min cooking time, due to Mamers et al.) (present, 40% sod. sulfite +5% NaOH, 1.25 M Pa, 463K, 2 min cooking time)

Captions of Figures

Fig.1

A: 40% Sodium sulfite
B: 40% Sodium meta bi sulfite
C: 40% Sodium sulfite +2.5% NaOH
D: 40% Sodium sulfite +5% NaOH

(192q) EFFECT of Recycling ON OPTICAL PROPERTIES of Bleached Bagasse PAPER Blended with Radiata PINE Wood Pulp and Retention AIDS

Shishir Sinha, Department of Chemical Engineering, Indian Institute of Technology, Roorkee, Roorkee, India and **Parmod Gupta**, Chemical Engineering, S.J.P. Polytechnique Damla, Yamuna Nagar, India

The recycling of bagasse based papers has become an essential part on the basis of pollution free and economic point of view to fill up the gap between the demand and supply of the present and future requirement of the paper. The quality of the bagasse fiber is superior to the wheat and rice straw fibers. There is no use of bagasse in the sugar industry other than burning it as a fuel, as it is a waste product of sugar industry. Due to the depletion of forest raw material bagasse fiber has become the important source of fast growing raw material to the paper industry. Bagasse being an agriculture waste contains short fiber length and low degree of polymerization it has low recycling potential and requires blending of long fibered pulp to produce required strength development in the paper produced. In the present study the recycling of bagasse paper blended with log fiber pulp was attempted with the help of varying doses of quality improving chemicals. The effect of blending of these chemicals and pulp under the various recycling stages on optical properties of paper were examined. The decreasing trend in various optical properties was tried to improvised by the optimal blending ratios of chemicals and long fibered pulp.

See more of this Session: Poster Session: Sustainable Forest Bioresources Engineering
See more of this Group/Topical: International Congress on Energy 2011

(192r) Modeling of Lactic Acid Production From Lignocellulosic Residues Using a Hybrid Cybernetic Approach

Philipp Adler[1], Hyun-Seob Song[2], Marzena Gerdom[1], Doraiswami Ramkrishna[2] and Benno Kunz[1], (1) Institute of Nutrition and Food Sciences, University of Bonn, Bonn, Germany, (2)School of Chemical Engineering, Purdue University, West Lafayette, IN

Over recent years lactic acid has received considerable attention as a potential raw material for several industrial chemicals (e.g. ethyl lactate, propylene oxide, acetaldehyde, acrylic acid) and as a feedstock monomer for biodegradable lactic acid based plastics (PLA). Lignocellulosic residues are considered to be attractive sustainable raw materials for the microbial production of lactic acid because they are renewable, inexpensive and available in large quantities. However, the composition of monosaccharide of lignocellulosic biomass varies considerably depending on its plant origin, which makes substrate screening and process optimization expensive.

In this work, we present a comprehensive model for lactic acid production, which incorporates metabolism of all quantitatively relevant sugars (glucose, galactose, mannose, arabinose, xylose) present in lignocellulosic biomass. As a model microorganism we selected the lactic acid bacteria *Pediococcus pentosaceus* DSM 20336 due to its ability to ferment pentoses. In our approach we used the hybrid cybernetic framework (HCM) (Kim et al., 2008; Song et al., 2009) which views the uptake fluxes of substrates being distributed among different elementary modes (EMs) of a metabolic network in a regulatory manner. We use this feature to predict simultaneous and sequential consumption of mixed substrates and lactic acid yield. We constructed a metabolic network for *Pediococcus pentosaceus* based on literature data reported for lactic acid bacteria and pathway information obtained from internet databases. The kinetic parameters of the HCM were estimated from batch-fermentation experiments using the respective monosaccharides as carbon sources. The model shows close agreement with various data sets of mixed-sugars fermentation experiments. Thus, the model can be a tool for time- and cost-saving evaluation of any lignocellulosic substrate and could also be employed for process optimization. By this, we attempt to contribute to the implementation of lignocellulosic residues in industrial lactic acid fermentation.

Aknowledgement

We gratefully aknowledge financial support by Max Buchner Research Foundation (MBFSt 2824) and the German Academic Exchange Service (DAAD).

Kim J.I., Varner J.D. and Ramkrishna D. 2008. A Hybrid Model of Anaerobic E. coli GJT001: Combination of Elementary Flux Modes and Cybernetic Variables. Biotechnol. Prog. 24(5): 993-1006.

Song H.S., Morgan J.A. and Ramkrishna D. 1-8-2009. Systematic Development of Hybrid Cybernetic Models: Application to Recombinant Yeast Co-Consuming Glucose and Xylose. Biotechnol. Bioeng. 103(5): 984-1002.

See more of this Session: Poster Session: Sustainable Forest Bioresources Engineering
See more of this Group/Topical: International Congress on Energy 2011

(228a) Synthesis of CuInS2 On Paper by Successive Ionic Layer Adsorption and Reaction

Chih-hung Chang, Oregon State University, Corvallis, OR

Synthesisof $CuInS_2$ on Paper by Successive Ionic Layer Adsorption andReaction

Yu-WeiSu, Wei Wang, Seung-Yeol Han, Chih-hung Chang

LeiKerr

Successive ionic layer adsorption and reaction(SILAR) technique was first reported by Y.F. Nicolauin 1985. Numerous studies of using SILAR technique have been carried out sincethen. The SILAR technique is based on sequential reactions at the surface byalternate immersion of the substrate into cation andanion precursor solutions. SILAR offers a simple and quick operation to grow avariety of thin films from solution phase at room temperature and ambientpressure. In principle, SILAR process resembles atomic layer deposition whichcan control thin film growth precisely via a layer-by-layer growth process.$CuInS_2$ is direct bandgap semiconductorwith a high absorption coefficient that is capable of absorbing a significantfraction of the incident solar radiation. We will report our progress in thesynthesis of $CuInS_2$ on paper using SILAR technique.

See more of this Session: Innovation of Forest Products
See more of this Group/Topical: Forest and Plant Bioproducts Division

The ECOwall Systems
Using Recycled Paper for Interior Partition Systems In Commercial Buildings

Scott A. Johnston, RA
Department of Architecture and Interior Design,
Miami University, Oxford, OH

Abstract

An interdisciplinary design and research team involving faculty and students from Miami University's departments of Architecture and Interior Design, Paper and Chemical Engineering, Manufacturing and Mechanical Engineering, and Marketing worked for a year on the development of new innovative approaches for using recycled paper in building related products and systems. This work was supported by an award from the Environmental Protection Agency, under their P3 Award Program – People, Prosperity and the Planet. The initial work concentrated on the design of modular paper based interior partition systems for commercial buildings. While this application does have considerable potential, the ECO-wall system is just one idea for new cost effective ways to use post consumer waste to create large format paper based products. Building systems and components that hold promise for future research include the following.

Interior furnishings or furniture components using molded high and intermediate density formed paper shapes,

- Suspended ceiling tiles where more rigid tiles may be desired to create articulated coffered shapes or where better acoustic separation is desired,

- Decorator interior wall tiles of standard patterns or customized designs using computerized numeric control (CNC) software to cut the mold faces,

- Molded interior cornice details and decorative shapes,

- Acoustic sound isolation panels for installation above suspended ceiling systems to control sound transmission between building zones or spaces

Energy and Sustainable Benefits of the ECO-Wall System

Sustainability

The proposed ECO-wall system would reduce the amount of post-consumer paper waste deposited in landfills each year by turning this material into useful products for the construction industry. Paper that would otherwise be disposed of in landfills is given a second use in products that have a considerably longer useful life than the printed materials and packaging for which it was originally created. The ECO-wall board can be made of mixed post-consumer waste paper that has a lesser value for other recycling applications. Energy is also saved because the structural ribs of the Eco-wall panels are molded paper, replacing the steel studs in conventional construction. There is a 10:1 reduction in the embodied energy needed to produce the ribs-vs-studs[1].

The ECO-wall panels are modular so the walls can be easily removed and re-installed by two people with a modest amount of training. In commercial office buildings the "churn" or turnover rate for interior spaces can vary between 7% (Cincinnati)[2] and 30% (eg. New York investment firms)[3] of the floor area per year. The ease of relocation for the ECO-wall panels eliminates most of the waste from these office renovation projects since the walls are simply moved, not demolished as is done with steel stud walls. The scrap that is produced, from damaged or field cut panels, can be re-pulped to produce the same walls over and over again.

Economic Benefits

The economic benefit from the ECO-wall project is related to the growing movement to develop new markets for recycled materials. The idea of using re-cycled paper in building products is not new. Cellulose has been used as a loose-fill insulation for many years, and has been used as a material in wall-boards as well. The ECO-wall is distinct from the prior "art", both in design and in the material compositions that were studied. For our applications the quality and homogeneity of the waste products is less important than in the printing business, cutting the material costs for the pulp by half.

In the pulp and paper industry, there is currently a greater supply of waste paper than demand for it, producing a glut of contaminated waste paper in the U.S. There are considerable incentives, both environmental and economic, to increase the demand for waste paper by developing new domestic markets. We currently export about half of the waste paper that is generated. As the supply and demand for waste paper start to equalize the price for waste paper and, hence, the incentive for collecting and recovering waste paper will increase. Our preliminary estimates indicate that the cost of producing the ECO-wall panels will be in the range of $25/lin.ft. for the basic panel[4]. This figure would allow for a 28% profit margin for the ECO-wall to still be price competitive with conventional construction at about $35/lin.ft. The ECO-wall system is just one idea we hope to pursue in looking for new cost effective ways to use post consumer waste in the creation of large format paper based products. This kind of product diversity is needed to help revive the slumping paper industry in the U.S.

Reuse and Recyclability

Buildings are responsible for two fifths of the world's material energy flows. The ECO-wall system contributes to sustainability on two levels. First we are substituting recycled paper for first generation materials. In addition the system was designed to maximize the functional longevity and the re-cyclability of the component parts when it does reach the end of its service life. The system achieves the following stated goals of the original Phase I proposal.

- The system can be easily reconfigured to meet changing interior space needs. The system has greater functional flexibility than other full-height modular interior wall systems currently on the market.

- The system is designed so the components can be easily refinished and refurbished. The stock ECO-wall panels (with flat mold face finishes) can be patched and repainted on site using the same materials used in patching gyp. board.

- The system is designed so that it can be deconstructed into the individual homogeneous materials from which it is made. This allows the components to be recycled into new materials of the same quality.

Market Applications

In the ECO-wall project we proposed to develop a new approach to the re-manufacturing of recycled materials with a focus on the use of recycled paper as the primary material in wall systems designed for the commercial building market. Instead of focusing on intermediate products – loose fill cellulose or cellulose board products - we proposed to develop end-use products directly from recycled materials. By looking at the design of a product for a specific end-use, we were able to explore material/product configurations that are unique to the application. Previous wallboards made with wood or paper waste have been designed for a broad market, and for this reason tended to focus on standard sheet sizes and thickness. While we don't discount the potential of this market, by concentrating on finished wall partition systems, we hoped to exploit other characteristics of recycled paper products. We feel there is a natural fit between the moldablility of paper pulp and the design latitude afforded in the development of wall partition systems intended for interior spaces. Pulp can potentially be formed into an infinite variety of wall shapes, configurations and surface qualities.

Paper-based products can be made impact resistant, fireresistant[5], water resistant, and aesthetically pleasing by using additives and/or coatings. In addition, we anticipate the paper-based ECO-walls will provide better sound control than lighter interior wall panels on the market. The fact that most interior partition systems are non-structural – only self-supporting – allows additional freedom in the design of the wall system.

Research Outcomes

The outcomes from Phase I of the ECO-wall project fall into four categories:

A. Definition of Target Markets
B. Integrated System Studies
C. Panel Design and Detailing
D. Preliminary Wall Board Testing
E. Evaluating Broader Environmental Impacts

A. Definition of Target Markets

In the early stage of the project a representatives from IDEO[6] came to Miami to lead a two-day design charrette, or brainstorming session, to identify potential markets for the paper based sustainable walls we proposed to develop, and to generate systems concepts aimed at those markets. Early in the project we also met with representatives from Steelcase Industries[7], Armstrong[8], and US Gypsum[9] to learn about the work they are doing in related areas of the market.

After considering more than 100 different functional use types it was determined that one of the initial markets we had identified – office structures – was, in the final analysis, still one of the most likely target markets. The following characteristics were identified as important to the market for the proposed system.

- The panel systems should be applicable for <u>new or existing construction</u>, but in either case is best suited for situations where interior walls are likely to be moved many times over the life of the structure.

- <u>Full-height modular re-locatable panel</u> systems were seen as the target market as a flexible, movable substitute for standard interior wall construction.

- Though the <u>first cost should be competitive</u> with other systems on the market, the modular re-locatable design of the <u>system provides significant life cycle cost benefits</u>. The modular Steelcase I-Line system, for example, costs 60% more than standard steel stud construction, but pays for itself the first time the wall must be moved [10].

<u>B. Integrated System Studies</u>

In designing the ECO-wall system, other interior system components (floors, ceilings, lighting, electrical, mechanical) were taken into account early in the design process. We took advantage of systems on the market and incorporated existing practices where appropriate. The ECO-wall system can stand-alone or can be integrated with other modular wall systems (eg. the Steelcase I-line system) or with standard construction.

Suspended ceiling systems are common practice in the construction industry, because they are quick to install, and are cost effective. Though slightly more expensive than their other product lines, the Armstrong Silhouette and Trim Lock systems have exposed connection slots running along the underside of ceiling track members. This feature makes it possible to attach fasteners to the ceiling track without getting into the ceiling cavity or even removing the ceiling tiles. The ECO-wall panels are designed for use with the Silhouette or Trim-Lock systems, but can be anchored to the underside of drywall ceilings as well.

In order to maximize compatibility with other systems, we installed a continuous slotted channel (1) along the top of the panel that allows for two types of anchorage at the top of the panel. A special nut can be inserted into the slot and is then threaded into the bottom of the Trim-lock system at any point along the track. This provides secure anchorage at the top of the panel as would be used in most private offices. A second option involves installing an inverted “T” to the underside of the ceiling with the middle leg of the “T” section pointing down. It can be fastened to the underside of the Armstrong Trim-Lock ceiling tracks or screwed into a conventional drywall ceiling. Positioned at a slight angle, the panel is then tilted up-right into vertical position so that the vertical "down" leg of a ceiling T-bracket slides into the slotted channel in the top of the panel. The ceiling bracket provides a pressure fit connection for panels that are likely to be moved intermittently. Feet at the base of the panel have leveling screws that raise the panel to fit snugly up to the ceiling (3).

Figure 1

Exploded Axonometric of Wall And Hardware Pieces

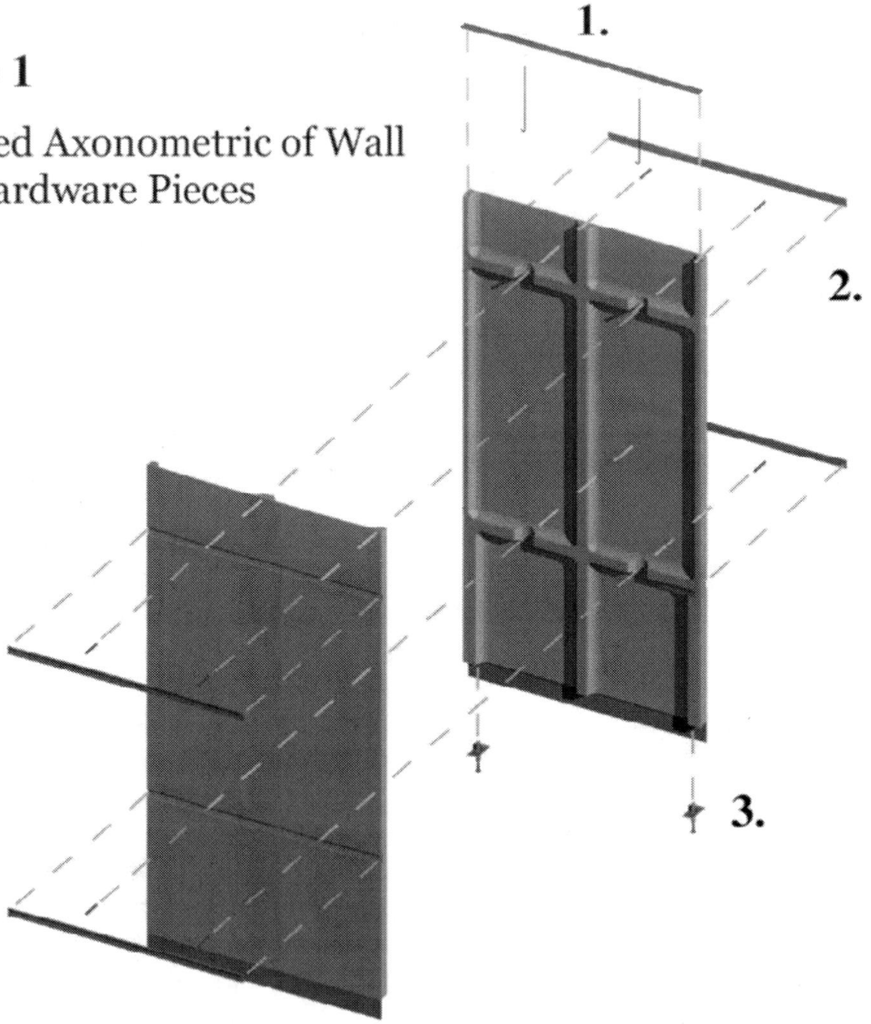

C. Panel Design and Detailing

The nominal width for the ECO-wall panels are 2' and 4'. The system also has a solid 2' wide "trim" panel that can be cut on site to any width and edge shape needed to fill in the non-modular or odd dimensions in an interior plan. Wall height can vary so that it extends to the underside of the ceiling in spaces where they will be installed.

The panel shown in Figure 1. also has channels running along both faces at desk height and door head height (2). These channels are not required in the manufacture of installation of the basic ECO-wall panel, but are an aspect of the overall system design that dramatically increases its' utility. They can provide the anchorage for shelves, flipper boxes, picture hanging, lighting fixtures, and electrical duplex outlets.

Other panel shapes and dimensions could be fabricated using the same molding techniques described below. The module for the ECO-wall system is not a limiting factor of the design. Rather it represents the first target market for a pressure based paper molding approach we feel has broad applications both in the building construction industry and in the manufacture of other products.

For a complete description of the design and construction of the ECO-wall system see the Quicktime and system details on our website *http://www.muohio.edu/ecowall*.

D. Wall Board Testing

Students in Paper Science and Engineering took the lead in conducting a series of tests on paper molding techniques. The pulp used in our wallboard samples was mixed post-consumer waste from Miami Universities recycling program. A series of fabrication studies were conducted in which varying mold shapes, sample thicknesses, and densities were made under pressures ranging from "manually" applied forces to 50 psi applied using a hydraulic press made for this purpose (see Figure 2 in Executive Summary). Starting with a honeycomb shape we progressed to flat panel shapes with thicker reinforced edges. The manually formed boards were relatively stiff for their light weight but did not have the surface finish needed to compete with gypsum board.

We were able to produce small (4" x 8") flat wallboard samples using a hydraulic press capable of exerting pressures of up to 1500 pounds. The press was limited to these dimensions, but varying thicknesses, sample configurations and pulp mixes were tested using the device. The press also allowed us to test the fabrication of wallboards with pre-made paper finishes. Test samples ranging from 1" and 3/8" in thicknesses were produced using newsprint, mixed paper, magazines, office paper, corrugate cardboard, and mixtures of each.

Though several of these tests are still in progress, our preliminary tests indicate the small (4" x 8") samples at thicknesses in the range of common gypsum wall board equaled or exceeded the characteristics of gyp board in several categories. Weight per unit volume is equal to gypsum at approximately .7 g/cc. Preliminary tests on three point break strength put our product at double the strength of standard "sheetrock" gyp board. This test was conducted using 1 inch wide strips, though and results may differ over wider areas. The gap being spanned for the 3 point test is 6 inches, with the breaking force applied at the center of the sample, 3 inches from each end.

We have not yet completed the crushing strength tests but early indications are that crushing strength will be greater than 75 psi, possibly significantly higher. Screw anchorage tests have also not been completed, though the fibrous nature of the material would suggest the ECO-wall will compare well in these tests as well, particularly when the pressure is applied perpendicular to the surface.

E. Evaluating Broader Environmental Impact

We have attempted to estimate the potential significance of the ECO-wall system by looking at existing office buildings that we perceived to be the market for our system in Cincinnati. We then re-scaled our data to estimate the impact if such a system were to replace conventional interior partitions in office buildings throughout the city of New York. We first identified 5 buildings in the Cincinnati area that would be representative or our market and reviewed the floor plans to identify high, medium and low density plan configurations. We then determined which of the walls in these floors could be constructed with the ECO-wall panels. Fire rated egress walls, service core walls, or demising walls between tenant spaces would not be a market for the unique aspects of our system. We were able to get an estimate of the percentage of walls in more open or more closed plan layouts that were a potential market for the ECO-wall system. The linear feet of wall for high, medium, and low density interior construction was then used to estimate the amount of construction waste created each year from interior office renovations in Cincinnati and New York. The environmental benefits of the ECO-wall system, compared the steel stud and gypsum walls they would replace, are listed below. (See our website at http://www.muohio.edu/ecowall for a summary of the office real estate study done as part of this grant)

1) Reduction of material from standard construction that is put into landfills.
   In Cincinnati we estimated that 60-95 cu.ft. of drywall and 397-630 pounds of steel would be land-filled each year as the result of renovation on a single floor of a downtown office - 311 Elm Street[11]. The cost to take the debris to the landfill would be between $1000 and $2000 dollars and the cost for steel stud replacement walls would be between $7000 and $12000 assuming 7% of the floor area is renovated each year. At this rate the yearly renovation costs for the whole building would be around $230,000. This cost would not be incurred with the ECO-wall system fully implemented.

   If we scale up the Cincinnati example to the entire commercial office building market in New York, where churn rates are between 15-30%, and assume that half of this turnover involves moving hard walls; a total of 3,904,436 cu. ft. of drywall and 25,719,198 lbs. of steel from office renovations must be disposed of each year. Or 26,000 dumpsters of construction waste[12].

2) Reduction of energy used to produce the steel studs in conventional construction.
   Steel is 10 times more energy intensive to produce than paper[13]. While the energy and production costs for the surfaces of the ECO-wall panels is comparable to Gypsum, the steel studs would be ten times more energy intensive to produce. Furthermore, in the New York example, the total embodied energy in the office renovation debris taken to the land fill each year is 140 million therms

for gypsum and 3.8 million therms for the steel studs – and this figure is repeated each year.

3) Reuse of paper that would otherwise go into the landfill.
The cost factors noted above could help increase the market for recycled paper. The higher profit margin for end use products will allow ECO-wall manufacturers to pay higher prices for the lower quality paper. In our Cincinnati building study[14] we found that if ECO-walls were substituted in the 312 Elm Building, in all locations where they would apply, a total of 6440 cu. ft. of post-consumer mixed waste paper would be re-used in the construction of these panels. This volume of waste would not repeat in subsequent years since the panels could be expected to have a 10-15 year useful life when maintained and refinished as needed.

The broader relevance of this project, is the case it makes for the process our team went through. We first looked at a waste material and then tried to determine what qualities it had that could be exploited in new applications. By trying to find the inherent potential in the problem itself, the process of conceiving of the solutions is reversed. Mixed post-consumer paper waste is the answer. What is the question? In this model, paper is not seen as waste, but rather as a resource; as a preprocessed fiber that still has all the characteristics it had in it's virgin state except homogeneity. It's strong, but not pretty; what can it be used for?

Discussion, Conclusions, Recommendations

The ECO-wall design builds on existing technology in a number of areas, by identifying which aspect of conventional construction are sound and which are most in need of re-conception. The interdisciplinary team we assembled explored a wide range of alternative target markets, systems configuration strategies and detailing approaches. Design breakthroughs in one area informed decisions regarding research directions to be pursued in others. This "collaboration" was neither seamless nor consistently successful. In less than one academic calendar year we were challenged with assembling twenty-three students and seven faculty members from six different academic disciplines into a functional design/research team to come up with new approaches to problems the building industry has spent decades trying to solve. While at times the size of the team structure was a liability, in the long run the diversity of the team kept the design process from maximizing the design from the perspective of one discipline over others. The final design was strengthened as much by the options we ruled out – based on feedback from one or the other of the disciplines involved – as by our success in developing the options we chose to pursue.

In the end the system we created is a "novel new approach for incorporating recycled paper products as the principle material ingredient in wall components designed for use in interior partition systems for commercial buildings," as stated in the goals of the original proposal. The system uses modular relocatible, panels formed in two pieces. The weight and shape of the panels allows them to be installed, removed and reinstalled in another location by two people without special tools. We anticipate that the panels could be relocated in this manner many times without damage. Without any special additives we are able to reach strengths in the panels that match gypsum walls.

References:

[1] Karolites, Alexis, "An Introduction to Green Building: Part 2," RMI Solutions, http://rmi.org/images/other/GDS/DO3-16-EnviroSenstv.pdf

[2] Dan Krueger, Duke Realty, 4555 Lake Forrest Drive, Suite 400, Cincinnati

[3] Keith Barr, Facilities Corporate Real Estate Division, Bloomberg LP, New York

[4] see *http://www.muohio.edu/ecowall*. - Eco-wall Cost Analysis

[5] Flame Stop is the trade name for a post treatment indoor fire retardant produced by Flame Stop Inc, located in Fort Worth Texas (http://WWW.FLAMESTOP.COM). Flame Stop I contains no toxic chemicals (eg. polybrominated diphenyl ethers, sodium bromide or formaldehydes used in other fire retardants). It pyrolitically self-extinguishes fire with minimum smoke. It has been tested to the following standards: ASTM E-84, NFPA 255, UL 723, U.S. testing #LA 62192, CA Reg. #C-14401, NFPA 701.

[6] Bob Adams, Sustainability Division of IDEO, http://www.ideo.com

[7] Marc D. Milkie, Steelcase Technical Sales Consultant for Architectural and Technology Products, 1122 Prospect Ave. Cleveland, Ohio
Todd Koenig, Steelcase Market Manager, 5225 Springborough Pike, Dayton, Ohio

[8] Monty Gillespie, Armstrong Distribution Sales Manager
3737 Redthone, Amelia, Ohio

[9] Joe Pesci, U.S. Gypsum, Project Engineer for Engineering, Gypsum Ohio Plant
Matt Montgomery, U.S. Gypsum, Paper Mill Project Engineer, Gypsum Ohio Plant

[10] I-Line system by Steelcase Inc., http://www.steelcase.com

[11] see *http://www.muohio.edu/ecowall*. - Commercial Office Renovation Analysis

[12] ibid

[13] US Gypsum publication titled "Solutions in Construction", sustainability table. Embodied energy data referenced to http://www.strategicdata.co.nz/mukhtar/sam3.htm

[14] see *http://www.muohio.edu/ecowall*. - Commercial Office Renovation Analysis

Elemental Hg Vapor Adsorption on Pyrolyzed De-inked Paper Sludge

By

Na Qin and Catherine Almquist*

Miami University

Chemical and Paper Engineering Department

Oxford, Ohio 45056

* almquic@muohio.edu; 513-529-0767

Motivation for Research

- Industrial Motivation– DPS is most often landfilled, creating both an economic and environmental burden

- Utility Motivation - Coal-fired power plants must reduce Hg emissions by 70% to comply with EPA's Clean Air Mercury Rule by 2018.

? Can we use an industrial waste (DPS) to create a value-added product that can solve a problem that has social, political, environmental, and economical implications?

MIAMI UNIVERSITY 1809 • 2009

Research Questions

- Can pyrolyzed DPS adsorb Hg?
- Will the impurities in DPS increase the adsorption capacity for elemental Hg over commercial activated carbon?
- How will pyrolysis temperature affect the properties of the resulting adsorbent for elemental Hg?

Characterization of DPS

Ultimate Analysis	
Carbon	33.3%
Hydrogen	4.2%
Nitrogen	0.9%
Chlorine	0.3%
Sulfur	0.3%
Oxygen	29%
Ash	44%
Heating Value	5400 Btu/lb_m

Element	Extractable Metals[1] (mg/kg)	Fixed Metals[2] (mg/kg)
Al	1050	12300
B	4.6	
Cd	0.27	
Ca	106000	106000
Cu	16	
Fe	610	1600
Li	1.1	
Mg	1990	890
Mn	25	
Mo	67	
K	170	890
Si	-	16200
Ag	0.12	
Na	900	530
Sr	68	
S	930	710
Ti	7.3	2700
V	2.1	
Zn	540	1700

- Calcium is the predominant metal species in DPS. It makes up approximately 10% of the raw DPS.

SEM Raw and Pyrolyzed DPS

- The cellulose fiber form is present even after pyrolysis. However, as pyrolysis temperature increases, the DPS appears more "gritty".

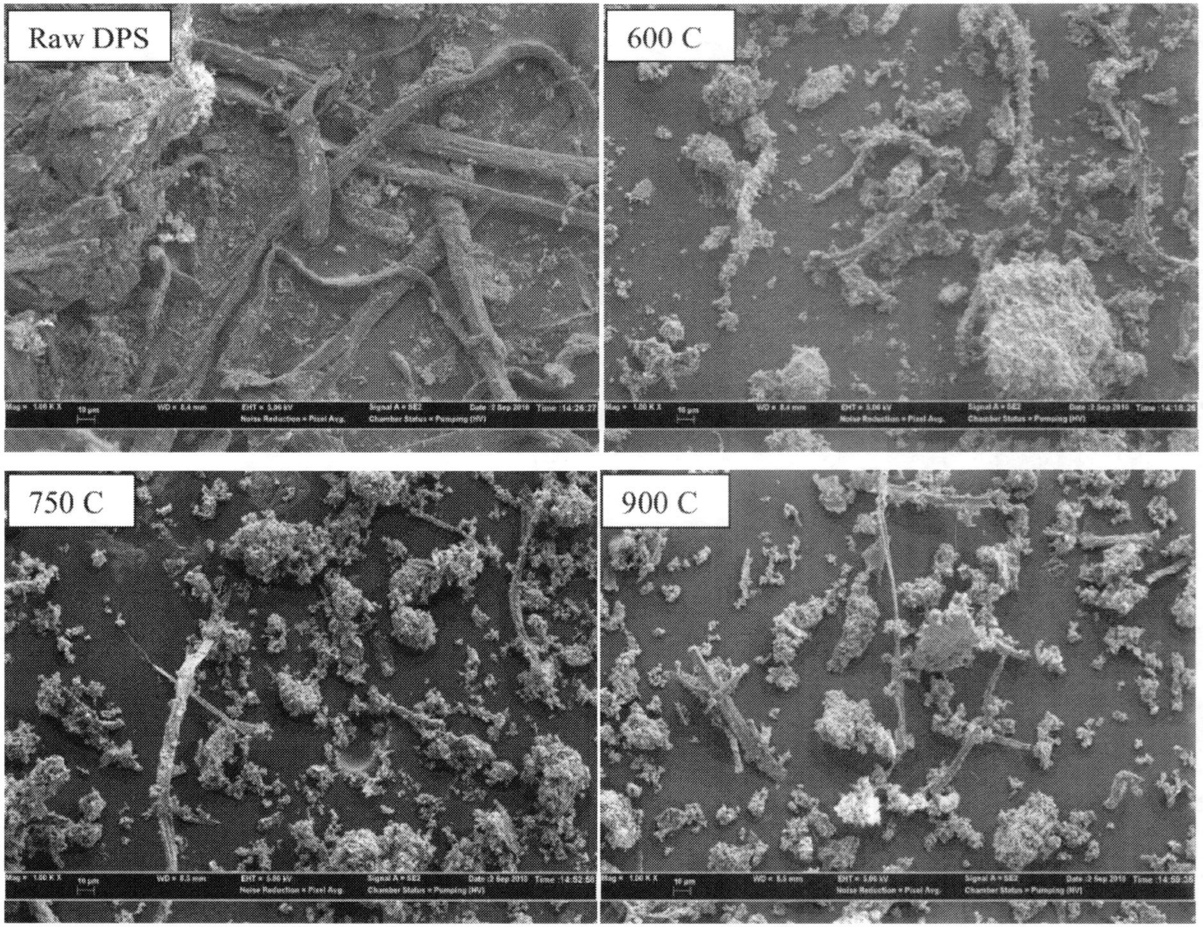

BET Surface Area and Pore Volume

- Pyrolyzed DPS has similar pore volume as BPL carbon, although its surface area is an order of magnitude lower.

Sample	Bet Surface Area (m^2/g)	Total Pore Volume (mL/g)
Raw DPS	4.8	0.024
600 C Pyrolyzed DPS	50	0.065
750 C Pyrolyzed DPS	49	0.102
900 C Pyrolyzed DPS	67	0.083
BPL	1100	0.075

- The pore size distributions are markedly different. Pyrolysed DPS has pores predominantly > 20 nm, whereas BPL has pores predominantly < 6 nm.

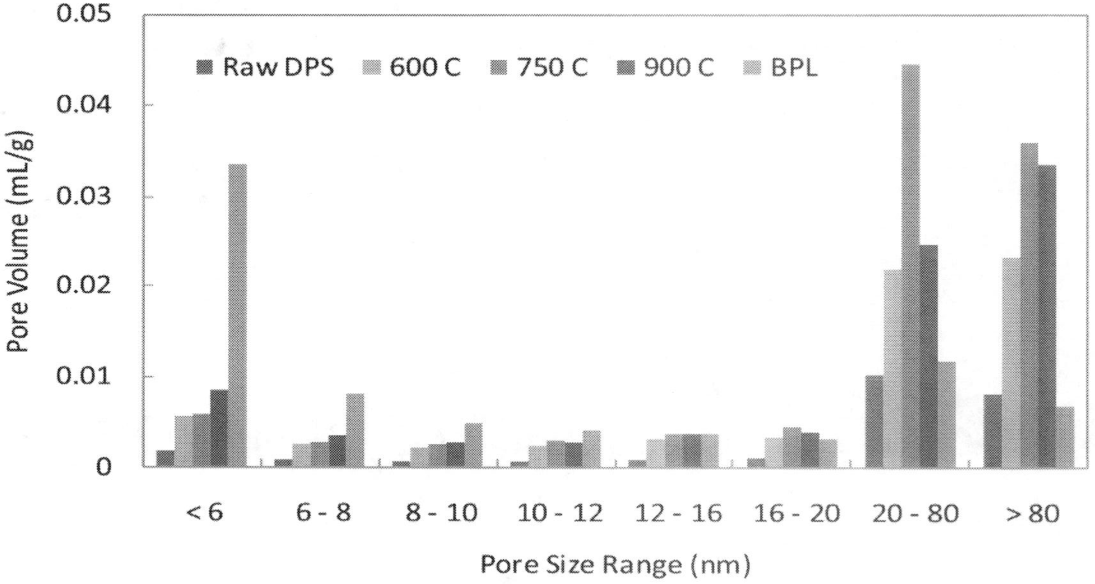

Raw and Pyrolyzed DPS Characterization

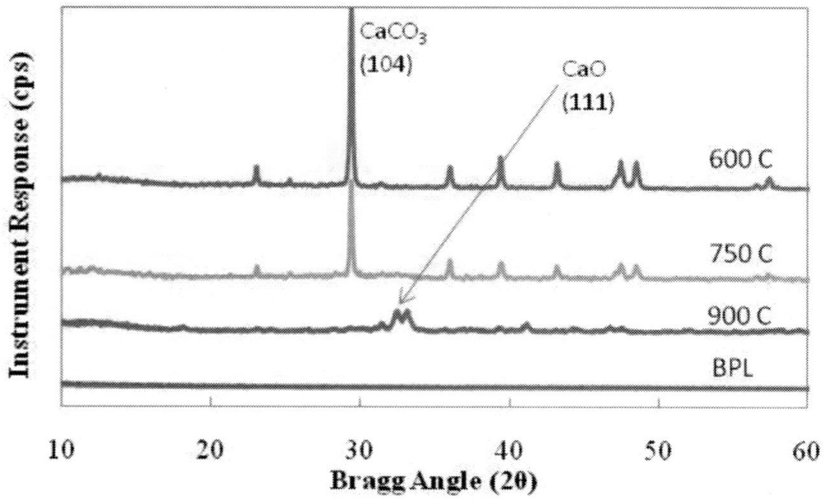

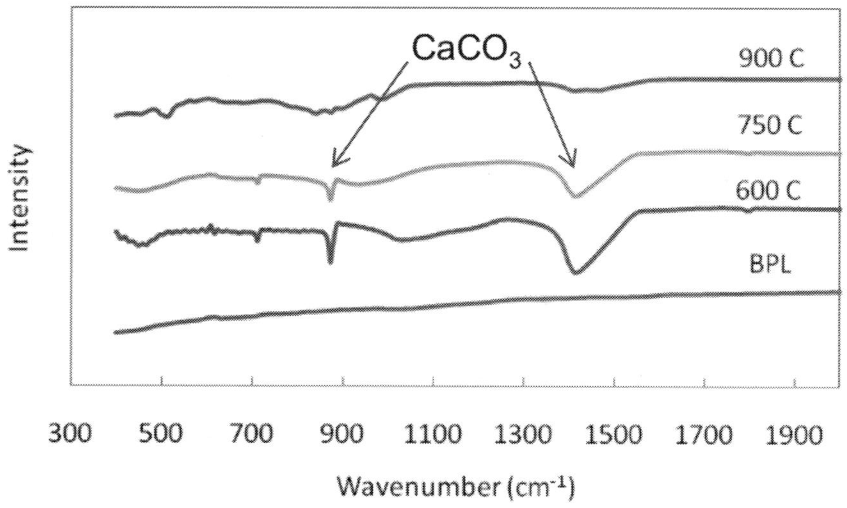

- Calcium carbonate is present in DPS pyrolyzed at 600 C and 750 C. The $CaCO_3$ has decomposed in the DPS pyrolyzed at 900 C, as noted by the disappearance of the $CaCO_3$ structures in XRD and FTIR spectra.

- BPL carbon has no observable crystal structures.

Raw and Pyrolyzed DPS Characterization

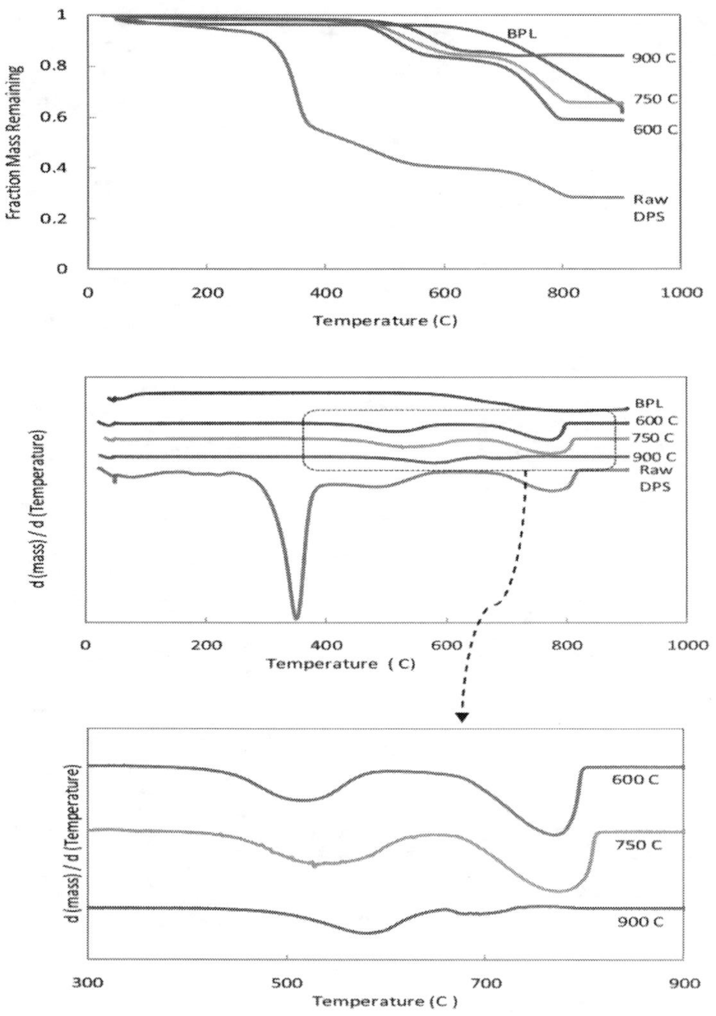

- Raw DPS loses most of its mass at ~350 C, when cellulose and hemicellulose degrade.

- At temperatures between ~500 C and ~600 C, lignin compounds degrade. As pyrolysis temperature increases, the degradation of the lignin also occurs at a higher temperature, suggesting that some lignin compounds remain in the pyrolyzed DPS, even after calcination at 900 C.

- Calcium carbonate degrades at ~770 C. DPS pyrolyzed at 900 C does not show this feature, thus supporting XRD and FTIR data.

Hg Adsorption Test System

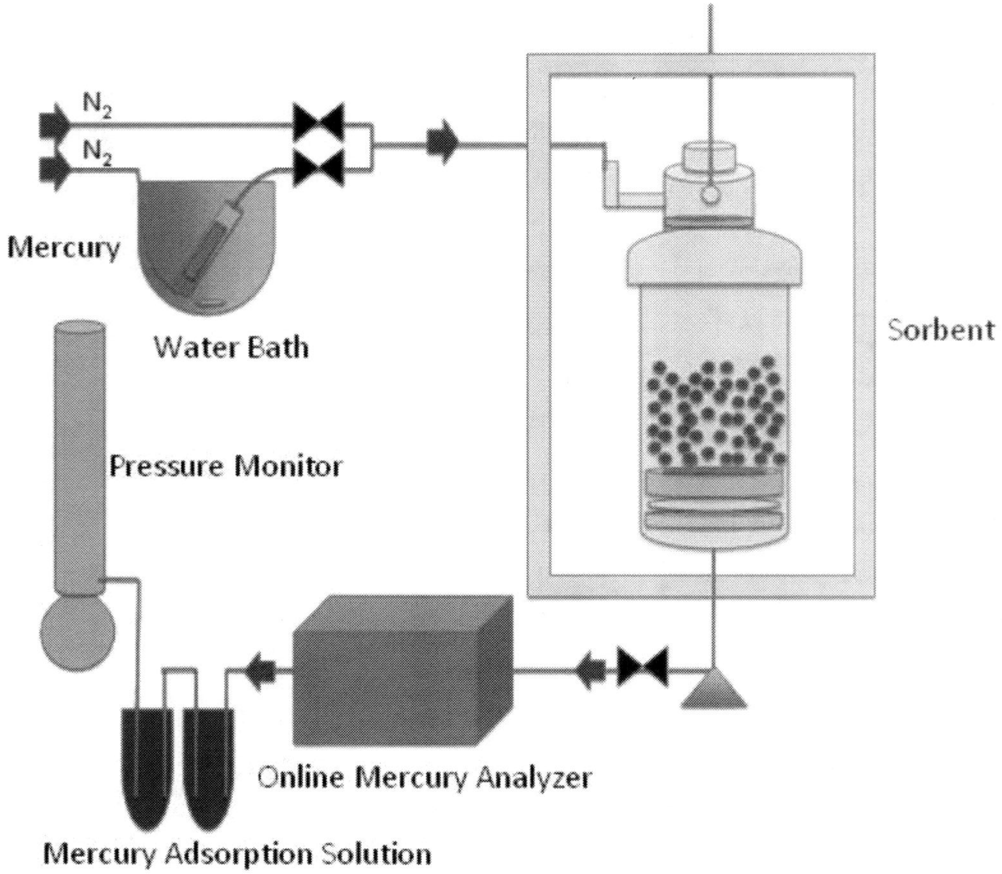

Hg Adsorption Results

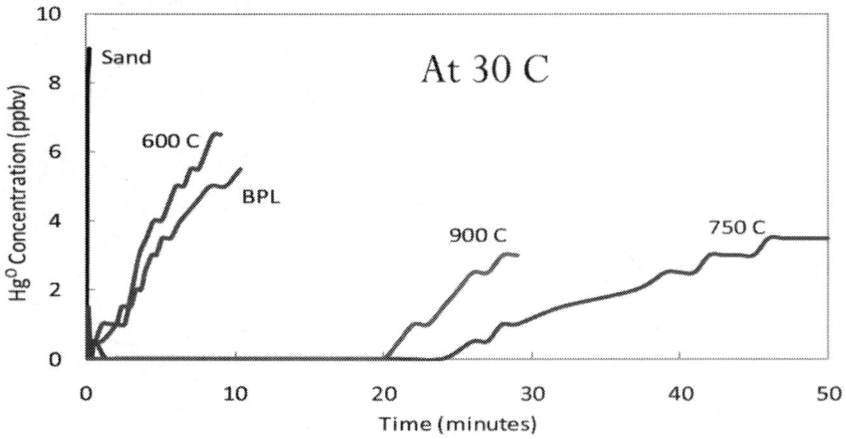

Adsorption Capacity At 30 C (ug/g)	DPS pyrolyzed at 600 C	DPS pyrolyzed at 750 C	DPS pyrolyzed at 900 C	BPL
	12	130	87	14

- Immediate breakthrough at 140 C suggests physical adsorption of elemental Hg under conditions of tests

- DPS pyrolyzed at 750 C and 900 C have much higher adsorption capacities than BPL at 30 C, despite BPL's much higher surface area

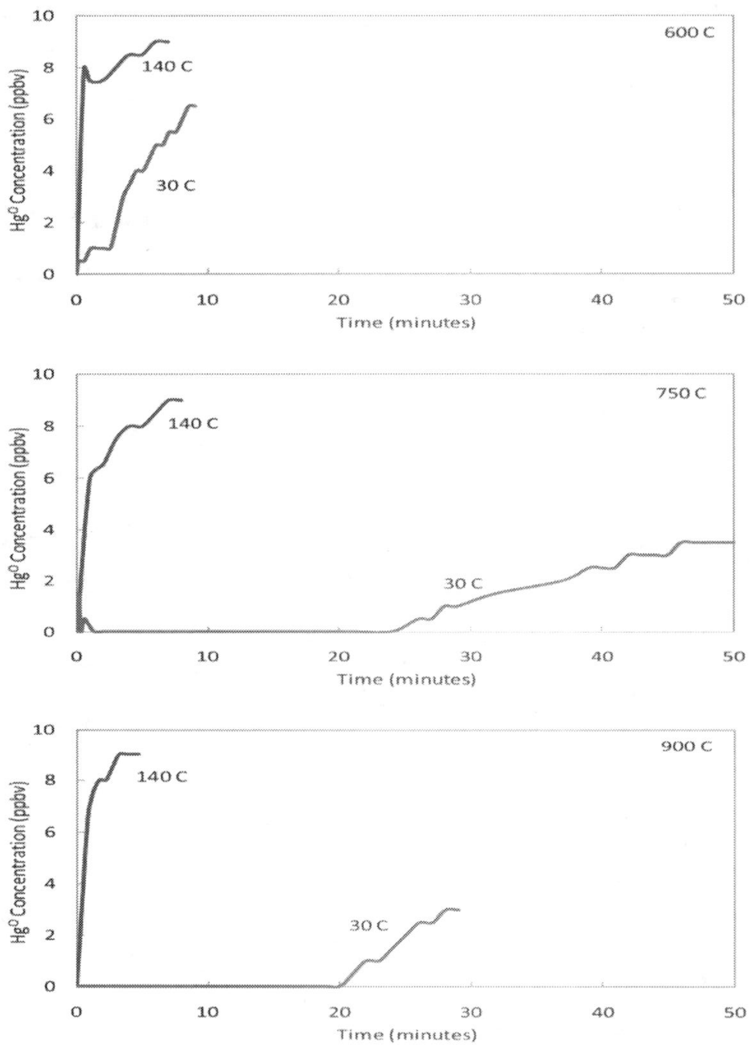

Conclusions

MIAMI UNIVERSITY 1809 • 2009

- Although both surface area and pore size impact adsorption capacity. It appears that pore sizes >20 nm are necessary for elemental Hg adsorption.

- Pyrolysis temperature impacts pore-size distribution, likely due to the degradation of lignin compounds and due to the decomposition of $CaCO_3$.

- Initial hypothesis regarding impurities in DPS will increase adsorption capacity over BPL activated carbon is not necessarily true under the conditions of our tests, except for the fact that the impurities may dictate pore structure.

To answer the research question:

- Pyrolyzed DPS can be used as an adsorbent for elemental Hg. However, to better assess its effectiveness as an adsorbent for Hg in coal-fired power plants, it should be re-tested in more representative conditions (eg. in the presence of acid gases at 140 C).

Support from Literature

MIAMI UNIVERSITY 1809 • 2009

- Takaoka, et al. (2007) – adsorption of elemental Hg is via a chemical reaction, which is influenced by impurities such as sulfur and chlorine on the surface of activated carbon.
- Olson, et al. (2000) – carbon sorbents are ineffective at temperatures in range of 100 C to 150 C in the absence of acid gases in the feed stream.
- Skodras, et al. (2004) microporous structures are more efficient at adsorbing elemental Hg, but mesopores allow for the faster uptake of Hg.
- Vidic, et al. (1997) – larger specific surface area and larger fraction of bigger pores are important factors for improving the performance of mercury sorbents.

References

- M Takaoka, T Yamamoto, N Takeda, K Oshita, T Tanaka, and T Uraga (2007). Determination of chemical states of mercury on activated carbon using XANES. *AIP Conference Proceedings*, 882, pp. 283-285.

- ES Olson, GE Dunham, RK Sharma and SJ Miller (2000). Mechanisms of mercury capture and breakthrough on activated carbon sorbents. *ACS Fuels,* 45(4), pp. 886-889.

- G Skodras, I Diamantopoulou, P Natas, T Orfanoudaki, and GP Sakellaropoulos (2004). Production, characaterization, and application of advanced carbon materials in removal of polluting substances. *Advanced Materials and Nanotechnology, Workshop of Chemical Process Engineering Research Institute, Thessoloniki, Greece, December 2004*, pp. 91-94.

- RD Vidic, W Liu, and TD Brown (1997). Development of novel activated carbon-based adsorbents for control of mercury emission from coal-fired power plants. *Conference paper: Advanced Coal-Based Power and Environmental Systems, Pittsburgh, PA July, 1997.*

BIOREFINERY BASED ON INDIAN DISTILLERY

A.K.Ray* and Prodosh Sanyal**

***Indian Institute of Technology Roorkee, Saharanpur, India**
**** National sugar Institute, Kalyanpur, Kanpur, India**

Distillery in India principally produces industrial alcohol/ rectified spirit. Alcohol is used not only portable purposes, but also is a renewable source of energy required primarily in IC engines, a supplier of energy through efficient fuel cell construction and power alcohol for substitute of gasoline. This can also be employed for production of a number of chemicals. Some of the chemicals manufactured in India are: ethylene based- such as ethylene-diamine,PVC, polystyrene, ethylene oxide/glycol, polyethylene and acetaldehyde based, namely butadiene, PV acetate, acetic acid, acetic anhydride, ethyl, butyl and other acetates, vinyl acetate monomer,pyridine,2-ethyl hexanol,n-butanol,monochloroacetic acid , propionic acid etc. In this paper the status of some of the above chemicals in regards to their installed capacities, demand and market potential are discussed.

Ethanol is usually produced by fermentation of molasses (generally 5% on cane) with a yield, 250.7 liter per tons of molasses with yeast *Saccharomyces cerevisiae* or by strains, *Zymomonas mobilis*. Uses of cane juice, jaggerry, mohua flower are also used in some installations. An alternative feedstock for ethanol production is ligno-cellulosic biomass;bagasse ,rice straw, cotton straw, groundnut shells, wood and forest residues, waste paper products, paper and pulp mill wastes using thermophilic anaerobic bacteria, *Clostridium thermocellum*. Cellulose enzyme produced by aerobic mesophilic fungus, *Trichoderma reesei, also* releases glucose from plant biomass and converts to ethanol by simultaneous saccharification and fermentation.

Acid hydrolysate of hemicelluloses extracted from bagasse and other agriculture wastes by fermentation with the yeast *Saccharomyces cerevisiae* has also been a potential feedstock.

Starch based natural produce like wheat, rice, millets, potatoes, corn, tapioca can also be a potential and economic source of ethanol production. In this paper potential for profitable production of alcohol by various sources are reviewed.

Economics of alcohol manufacture depends strongly on the economy of steam consumption as it contributes a significant fraction (40%) of the total cost. Most of the Indian distilleries produce alcohol with 1.5-2 times (of the order of 2.5 kg/kg of alcohol production) higher steam consumption than the actually required. Attempts have been made to highlight how production can be improved by various means with lower consumption of energy and the factors influencing it. The absolute alcohol is manufactured from rectified spirit, by various means such as azeotropic and extractive distillation, using dehydrant, molecular sieves or by membrane separation. The status regarding the production of power alcohol and as gasoline substitute as well as chemicals in Indian distilleries is introduced.

Spent wash and spent lees, the two liquid waste streams can also be employed for manufacture of chemicals and energy, especially biogas. Performances of installations for handling distillery spent wash using downflow stationary fixed film type anaerobic filter and other designs in Indian distilleries are reviewed.

Absolute alcohol is produced from fermentation of molasses in India using two distillation columns, namely analyzer and rectifier. From rectified spirit power alcohol can be manufactured through the following processes:

- Dehydration by chemical reaction
- Dehydration by formation of ternary azeotropes with Entrainer like Benzene or Trichloroethylene or Cyclohexane
- Dehydration by countercurrent extraction, called extractive distillation, glycerine with 25 % diluted K_2CO_3, ethylene glycol, molten eutectic mixture of K or Na acetate, $CaCl_2$ etc.
- Use of molecular sieves
- Membrane separation

INTRODUCTION:

Due to variety socio-techno-economic reasons , sustainable development of sugar industry and distillery, to supplement the depleting resources of fossil and petroleum based fuel, to avoid the extensive use of petroleum based chemicals for manufacture of polymer and other products and especially for environmental protection, concept of biorefinery based on bio-ethanol needs to be used. In order to be economically strong, a biorefinery built on integrated sugar-paper- chemical-energy complex has to be optimally designed. As bio-resource is of vegetable origin which is replenish able, renewable and abundant in India, it can be utilized both as a substitute of production of high volume low cost bio-fuel as well as value added small volume important chemicals.

There are many Industries in India where perfect biorefinery has already been established at least to manufacture many valuable chemicals from sugar industry byproducts, typically Kolhapur Sugar factory, Somaiya Organic chemicals Ltd, Dhampur Sugar mills, Ashok Alcochem limited, Ashok organic industries Ltd., to name a few. A list of the industries producing alco-chemicals without mentioning distilleries producing power alcohol is shown below: They have complex of a sugar mill, paper mill (optional), distillery, bio-gas plant, and chemical plant based on various byproducts of sugar Industry. Sugar Industries like The Salem Co-Operative Sugar Mills Ltd., Amaravathi Co-operative Sugar Mills Ltd., Bannari Amman Sugars Ltd., and many others are on the way to establish sugar-alcohol –chemical complex to accrue economic benefits of the Industry. The Simbhaoli Sugar Mills Limited, of capacity 9500 TCD, having India's largest integrated sugar refinery complex at Simbhaoli (Ghaziabad, Uttar Pradesh) is also planning to set up a biorefinery in near future.

The first true bio-refinery in India [1, 2]:

The Godavari Sugar Mills Ltd., incorporated in 1939, has been demerged to Godavari Biorefineries Ltd. as per High Court order dated 20.03.2009. The Company has recently commissioned Sugar Refinery of 300 TPD and is capable of producing 75,000 tons per annum of refined sugar of international quality (EEC grade II).Currently, GSML reports that it can produce 95 million liters/year of ethanol from non-cellulosic sources. It has plants located in the states of Karnataka and Maharashtra in India & has diverse interests in sugar, power, industrial alcohol, heavy organic chemicals, specialty chemicals, bio-fertilizers & agricultural Research. It manufacturers more than twenty products from renewable resources, thereby forming an entire value chain right from sugar cane to sugar to other value added products as already indicated. The Company has an approved Agricultural Research Institute K. J. Somaiya Institute of Applied Agricultural Research (KIAAR) at Sameerwadi, which works closely with other sugarcane research institutes & the farming community in developing better and newer varieties of cane. The company has two Council of Scientific & Industrial Research(CSIR) recognized Research centers working on areas of process technology, product improvement, new product development, recovery improvement & cost reduction. Developed by Godavari Sugar Mills Ltd in Sameerwadi, Karnataka, the biorefinery will use sugarcane bagasse to produce ethanol. The project, which is being undertaken in cooperation with the National Chemical Laboratory, Council of Scientific and Industrial Research, hopes to be able to process about 5,000 tons of raw materials within four years.

Many Indian Sugar Industry has a lot of potential for establishing a perfect biorefinery like the above which consists of exportable surplus power by using cogeneration concept, production of rectified spirit, potable alcohol, power ethanol, biogas particle board, many value added chemicals especially, the alco-chemicals. There is a lot of information in this regard [1-46]. It has been estimated that conventional mills can give 3500 MW whereas energy efficient mills yield upto 5000MW [30].

Sugar Industry discharges various by-products and co-products. Table 1 shows the various by-products / wastes produced in a sugar mill.

Table 1: Sugar Industry by-products:

For every 100 tons of Cane the products in tons are as follows [11, 27]:

Sugar of 98.5o Pol, tons	11.0- 12.5
Cane top	18.4
Leaves and straw	42.9
Final molasses 88^0 Brix	3.0- 3.6
Filter mud 77% -80% moisture	3.0- 3.4
Bagasse 50% moisture	27.5
Flue gases (180°C)	120
Electricity, kWh	1500

Surplus bagasse of 50% moisture content may be of the order of 4 tons.

The production of molasses for the last several years is shown in the Fig.1. The production rate as well as growth rate increase linearly. With the growth of the sugar industry the molasses production will also be increased. The final molasses 88^0 Brix has been reported of the order of 3.0- 3.6 percent. It is however prudent to assume that 40 kgs of molasses is produced per ton of cane from which10 litres of ethanol can be obtained. If the sugarcane is directly and fully used in ethanol manufacture, the yield of ethanol is 70 liters/ per ton of cane [34] .Further, average data of molasses % cane based on several years (1994-2004) has been found to be of the order of 4.493% [34]. In many process engineering calculations the value of molasses percent cane has been assumed of the order of 4.3% [32].

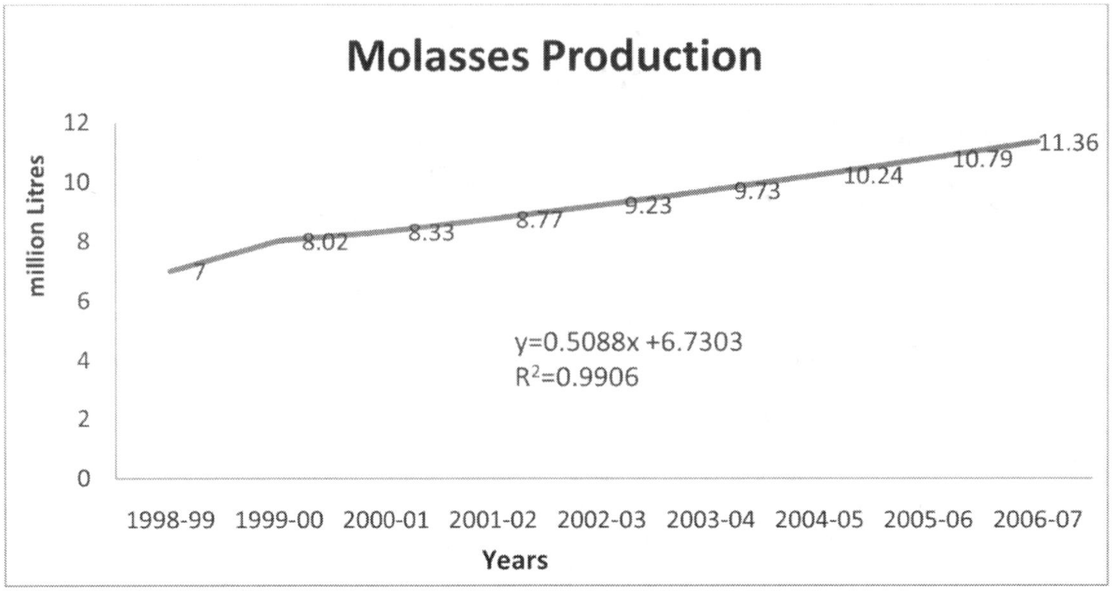

Fig.1 Production of molasses from sugar cane industries in India.

The molasses is utilized in India, mainly for two purposes, to produce rectified spirit or industrial alcohol or production county liquor. As these are produced by biochemical route, this may also be termed as bioethanol unlike synthetically manufacture from ethylene or other compounds.

Uses of bio- ethanol:

(a) Alcohol for portable purposes
(b) Distillery for production of rectified spirit, and power alcohol
(c) To manufacture various valuable chemicals:
Ethylene based- Ethylene diamine, PVC, polystyrene, Ethylene oxide/glycol, Polyethylene
Acetaldehyde based- Butadiene, PV acetate, acetic anhydride, Ethyl and other acetate, 2-ethyl hexanol, n-butanol, monochloroacetic acid etc.
(d) Utilization of distillery effluents ,Spent wash and Spent Lees for Biogas generation
(e) In Fuel Cell applications

Bio-ethanol is a thus renewable source of energy and can be used for substitute of gasoline. Hence Sugar factories in India can produce energy in terms of liquid fuel (ethanol), biogas, and electrical power and supplement to I.O.C. and grid. As indicated above, ethanol is also an important feedstock for a number of industrially important chemicals, besides, being a relatively clean fuel and possessing potable value [14].

ETHANOL PRODUCTION, CONSUMPTION AND ECONOMICS:

In India there are 300 distilleries producing alcohol using molasses as raw material, out of which 115 distilleries are attached to sugar factories. The total installed capacities of all the distilleries is 2964million litres in the year 2008 with 67.5% capacity utilization [31].The growth of alcohol production in India over the years is shown in fig. 2 which is linear

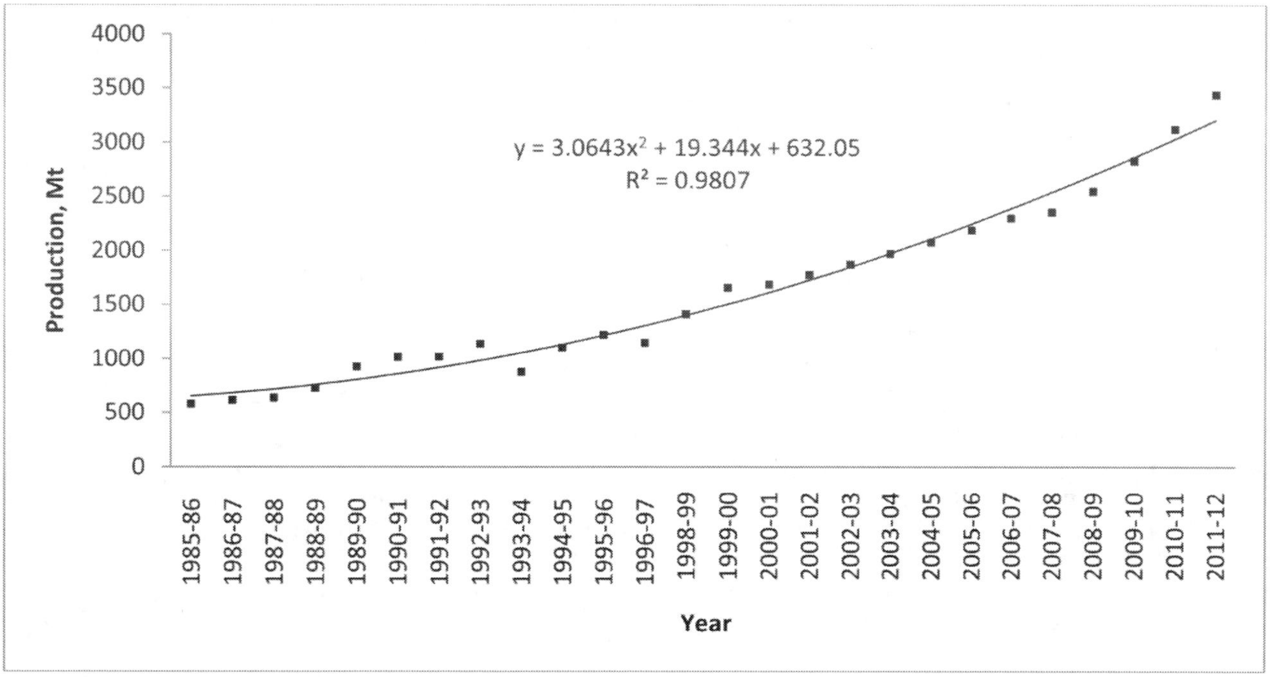

Fig.2 Production of alcohol in India

India is surplus in alcohol as on today as it produces more than 28 million tons of cane sugar and 2074.5 million litres of alcohol per annum through its nearly 300 distilleries [32, 33]. The annual total licensed capacity is 3287.491 million litres out of which 1321.585 million litres are produced from the distilleries attached with sugar factories. On the other hand total annual installed capacity for the same are 3198.314 and 1360.627 million litres respectively.

India is producing indigenously around 33 million tons of petrol whereas our annual requirement is around 120 million tons. This may go up to 365 million tons in the year 2025 [21, 33]. Presently India imports 75% fuel and will rise to 90% by 2020. The present scenario of sugar, fossil fuel stock, and power shortage, it is imperative to go for non-conventional bio liquid fuel and power production to cope up with the present and future demand. As a World's largest producer of cane molasses, India should take initiative to use biofuels, to overcome the mounting pollution problem in the metros [21]. With a normal av. production rate of 1,900 million litres a year, India is the World's fourth largest producer of ethanol after Brazil, The United States and China [34]. Production and Consumption pattern of ethanol in some countries are shown in the following tables 2 and 3[34].

Table 2. Production of Ethanol

Countries	Production of Ethanol, 10^{-3} million litres	
	2007	2017
Brazil	18	28
USA	26	58

Europe	3.0	15
India	2.2	4.0

It is evident that the ethanol production in 2007 was around 49.2 billion litres and in 2017, it will increase up to 105 billion litres. It is also clear that US is the major contributor, Brazil followed the next and unlike others, the amount is expected to be double in US and India.

It is expected that in 2017 nearly 4.0 billion litres of alcohol will be produced in India. In India 94-96% ethanol is produced from sigar cane based molasses and only 3-4% from grains. Production wise India's share is only 4%. The consumption pattern is quite different in India compared to global situation. This is shown in the following table:

Table 3: Consumption of Ethanol

Type of consumption	Global, %	India,%
Potable	10	50-55
Chemical	10	30-40
Motor grade	80	5-10

It is evident that generally consumption of ethanol in India for potable, chemical and motor grade is of the order of 50-55%, 30-40% and 5-10% respectively compared to 10%, 10%, and 80% worldwide **[33].**

BLENDING OF ETHANOL PROGRAMME IN INDIA:

Comparison of projected demand for petrol and ethanol requirements from the year 2005 to 2012 is shown in the Fig.3.The plot indicated gasoline, alcohol requirements for 5%, 10% and 25% respectively [34]. Similarly the demands for diesel and ethanol requirements for blending are shown in table 4.

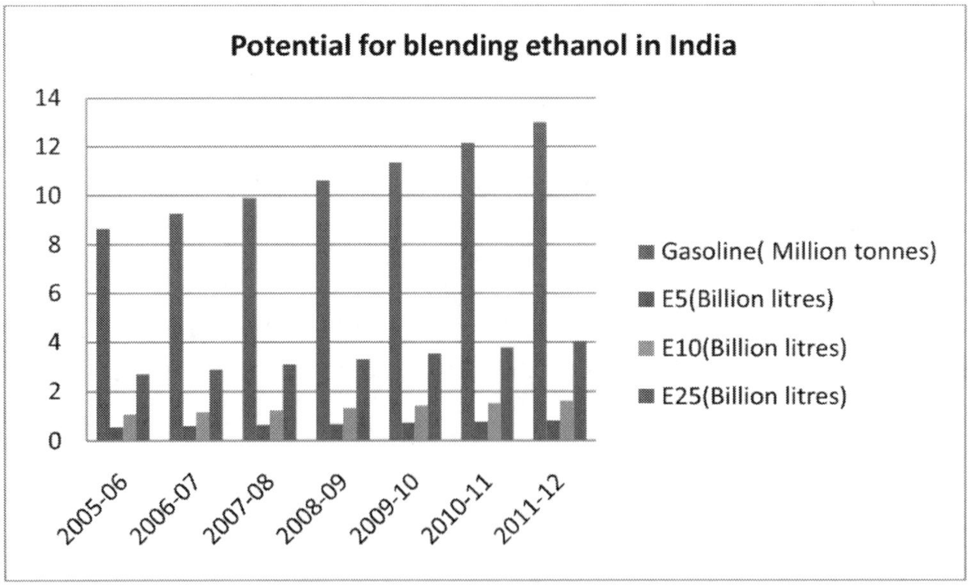

Fig.3 Potential for blending ethanol in India

Surplus quantity of ethanol available after catering the needs of other sectors like potable, chemical industries, etc. is just sufficient for 5% blending only. The government is planning to increase the blending ratio to 10% in which case the production of ethanol through molasses route alone will not be probably sufficient to meet the demand unless yield of cane per hectare for sugar industry is significantly improved with more area available for cultivation. This might be a difficult situation in India. Hence option of producing ethanol from secondary juice (even without considering cellulosic ethanol) will increase the production without affecting the economics of the Industry [32, 34].

Table 4 Demand of diesel and requirements of ethanol for blending purpose as fuel

Year	Diesel Demand, Mmt	Ethanol Blending,Mmt		
		5%	10%	20%
2006-07	52.32	2.62	5.23	10.46
2011-12	66.91	3.35	6.69	13.38
2016-17	83.58	4.18	8.36	16.72

Agarwal *etal.*[7] ,and Mathur[26], Uppal[25] discussed about the details of the quality as a blend with gasoline in IC engines whereas Sharma [19] discussed about the profitability of sugar industry in connection with utilization of ethanol as a blend of gasoline if alcohol is to be produced in traditional means of molasses fermentation process. Production of sugar, molasses, production, consumption, and export(3-4% of domestic production) potential of indigenous alcohol, capacity utilization(40-45%) and its availability for blends with petrol were highlighted.

PRODUCTION OF ALCOHOL FROM VARIOUS INDIGENOUS SOURCES:

In India, alcohol is usually produced from molasses, though some distilleries also use cane juice, deteriorated cane juice, gur, jatropha, mohwa flowers etc. Alcohol can also be produced from starchy materials, wheat, rice, millets, corn, tapioca and potatoes, rice bran, sulphite waste liquor, cellulosic materials such as bagasse, pith, and even woody biomass.

There are enough quantities of nonfood and nonfodder agricultural waste available in India, currently at 250 million tons /year. Besides rice straw, 284.99 million tons and wheat Straw 159 million tons may be a potential source of production of alcohol. The approximate yield of alcohol per metric ton of raw material can be shown in table 5 as under:

Table 5 Yield of Ethanol from some substrates

Substrate	Ethanol,kg/mt subst.	Substrate	Ethanol,,kg/mt Substrate
Molasses	200-220	Rice husk	113
Bagasse	150	Wheat straw	57
Jute Stick	42	Rice straw	42

The alcohol yields from corn and tapioca was found 352 and 370 litres at 24 hr and 411 and 437 litres per ton, at 72 hr, respectively with diazyme (fungal amyloglucosidase) and *S. cerevisiae* during fermentation as compared to about 225 -250 litres per ton of molasses normally obtained in Indian distilleries operating on conventional batch fermentation process [13,18].The acid hydrolysate of hemicelluloses extracted bagasse with the yeast *S.cerevisiae* for fermentation can produce ethanol after concentration and supplementation 8.3% alcohol with fermentation efficiency(FE) of 83.3% compared to 7.8% for molasses medium of 28°Bx with FE of the order of 83.3% [13]. The National Institute of Agriculture Research, Hyderabad has advocated the extraction of ethanol from Sweet Sorghum .There has been some efforts by some sugar factory in Chittoor in Andhra Pradesh to utilize sweet Sorghum(*Andropogan saccharatus*), a short duration 5 months crop , for production of ethanol with the help of Praz, a Pune based distillery machinery manufacturing company [31].

Attempt to increase of the production of alcohol in Indian Research Laboratory:

There are all round efforts in various research laboratories (National Sugar Institute, Govt. of India, Vasantdada Sugar Institute, Sugarcane breeding Research Institute etc.) to increase the productivity and efficiency of production of alcohol from molasses. Attempts are also being made by In-House R&D laboratories of Sugar mill and distilleries. Higher demands of ethanol increased the search for better fermentation conditions and efficient ethanol producing organism. Some of the strains [1, 2] which have been used to increase alcohol production at various conditions of temperature are :Lambanog yeast, *Saccharomyces cerevisiae*, and *Candida hypolytica, Rhodotorula rubra, C. lypolytica* and *C. tropicales.* Addition of gibberellic acid and rice bran has been found to increase the ethanol content. In India efficiency in the production of alcohol from molasses has been attempted to increase by using several thermotolerant mutants of *Sacchaomyces cerevisiae* which can enhance ethanol levels [9], by pretreatment with 0.3% sulphuric acid increasing additional 1.5% , supplementation with 0.2 skim milk and addition of 0.4 % fungal mycelium [10],by supplementing with urea and phosphate, chitin and yeast extract, addition of various carbohydrates and alumina beads, by physical means, lowering of vapor pressure of alcohol at low temperature[14]. Jain [41] reported the advantage of ethanol production by bacterium *Zymomonas mobilis* over the conventional yeast fermentation of various substrates such as cane and beet molasses, sugarcane juice, syrup, and starchy materials like corn hydrolysates and cassava starch hydrolysates. The limitations of *Zymomonas* process with respect to utilization of commercial substrates is being overcome with the development of modified strains of *Zymomonas mobilis*. In addition to batch fermentation , high ethanol productivity has been obtained in continuous culture, using flocculating or immobilized cells of *Zymomonas mobilis.*

Padmaja *etal.*, [40]investigated on the potential of various cellulosic substances such as bagasse , rice straw, cotton straw and ground nut cells for the production of alcohol, using thermophilic anaerobic bacteria , *Clostridium thermocellum* at 60°C and pH 7.5 under continuous sparging of nitrogen to make the medium oxygen free. Mild alkali treatment given to the substrate enhanced their utilization rate and ethanol yield. The yield of ethanol was found to be of the order of 0.20-0.12 gm/gm of substrate degraded. Abha Singh [39] used both unhydrolysed and hydrolysed molasses for batch fermentation using modified strain of *Zymomonas mobilis* at optimum conditions of 30°C and pH 6.0.Conversion efficiencies in excess of 90% have been obtained from a molasses yielding 10 to 11 % (v/v) ethanol within 17-20 hours.

Many others have reported about the production of alcohol from molasses [32-37].

COST OF ALCOHOL PRODUCTION FROM MOLASSES IN INDIA

In this regard there are diverse reports available. One report indicated [34] that the cost of molasses in India varies widely across different states from Rs.200 per ton to Rs.2000 per ton. A sizable part of the cost is central excise duty, Rs.75 per ton, sales tax(VAT) 30% on basic price, and excise duty , transport cost etc. The international price of molasses which was $50.0 per ton in 2004 and has now doubled to $100/ton. Assuming a molasses price of Rs.2000/ ($45.0) and a yield of 220 litres of ethanol per ton, Ravi [34] has given a detailed cost for a plant for production of ethanol and found the cost of ethanol per liter should be of the order of Rs. 17.18-17.22.

The distilleries which are not attached to sugar factories, procure molasses from sugar factories at a price varying between Rs.3500-Rs. 5000 per ton, the cost of production in these units is approximately Rs.22-25 per litre and sales price including excise duty is of the order of Rs.25-28 per litre [31].

ECONOMICS OF ABSOLUTE ETHANOL PRODUCTION FROM MOLASSES:

Economics of alcohol manufacture also depends strongly on the economy of steam consumption as it contributes a significant fraction (40%) of the total cost. Most of the Indian distilleries produce alcohol with 1.5-2 times higher steam consumption than the actually required.

The absolute alcohol is manufactured from rectified spirit, which is produced from fermentation of molasses in India using two distillation columns, namely analyzer and rectifier. From rectified spirit power alcohol can be manufactured through the following processes:

- Dehydration by chemical reaction
- Dehydration by formation of ternary azeotropes with Entrainer like Benzene or Trichloroethylene or Cyclohexane
- Dehydration by countercurrent extraction, called extractive distillation, glycerine with 25 % diluted K_2CO_3, Ethylene glycol, molten eutectic mixture of K or Na acetate, $CaCl_2$ etc.
- Use of molecular sieves
- Membrane separation

For production of 100 litres of absolute alcohol, approximately 105 litres of ethanol is required. Though many distilleries in India produce absolute alcohol using azeotropic distillation based on benzene cyclohexane technology, a few mills in the country are however, manufacturing, employing membrane technology for special purposes [31].

Estimated cost of manufacturing anhydrous ethanol using molasses as feedstock with the above processes assuming average recovery of 220 litres per ton of molasses with benzene/cyclohexane technology were examined. The membrane technology process was used by M/s Dharamsi-Morarji Chemicals Co. Ltd., at their plant at Roha for the manufacture of absolute alcohol. It has been shown that the consumptions of steam, electric power, cooling water as well as loss of alcohol in the membrane process (per evaporation process) are much less.

Molecular sieve (needing replacement of molecular sieves every after 3-4 years accounting to 70-80% of the total investment) and membrane technology were considered higher investment proposal (with 40 % more in case of membrane technology based plants). The molecular sieves and membranes are imported items. The approximate cost of molecular sieves imported from China is of the order of 69.50 rupees per kg [42].

Status of production of Power alcohol in India:

In the absence of a well knit policy in the past for purchasing and blending ethanol, not many distilleries have been producing ethanol [1]. Only three distilleries attached to sugar mills had war years' experience, and were able to gear themselves up to supply ethanol immediately. During 2003-04, about 11 factories in Uttar Pradesh will be adding facilities to produce about 75 million liters of anhydrous alcohol by end-September; 7 units in Tamil Nadu (production capacity of 62.5 million liters of anhydrous alcohol); 8 in Karnataka (anhydrous alcohol production capacity of 66.5 million liters); and 4 units in Andhra Pradesh (capacity of over 40 million liters). Similar steps have also been taken up by the cooperative sector units in Maharashtra, Punjab and UP. By the end of the year it was estimated that about 300 million liters capacity would have been created for the production of anhydrous alcohol. In the Indian Sugar Mill Association, this matter was recently examined and it was concluded that instead of taking up the scheme on a state-wise basis, it would be appropriate to take it up in metropolitan and other cities where environmental pollution is a major concern [2]. The blending should be taken up to 10% and introduced selectively to make a better impact on the environment, as no changes in the engine or carburetor are required, and other countries are already carrying this out successfully. Now, there are 131 industries in India producing absolute alcohol of annual installed capacity of 2240 million litres whereas 311 industries are engaged in producing rectified spirit of 3875 Mt. capacity.

DISTILLERY BASED CHEMICALS

In the recent past, UNIDO mentioned that India is the world leader in alco-chemistry products with 40 plants (1991) producing an entire range of products, starting with acetaldehyde, acetic acid, acetic anhydride, acetone, butanol, butyl acetate, cellulose acetate carboxymethyl cellulose, diethyl oxalate, dioctyl phthalate, ethyl acetate, ethyl silicate, ethyl acrylate, ethyl amine, ethyl ether, ethylene dichloride, ethylene oxide, ethylene glycol, ethylene glycol monoethyl ether (Cellosolve), glyoxal, monochloro acetic acid, pentaetrythritol, polystyrene,picoline, the metal acetates, vinyl chloride, vinyl acetate monomer etc. There are demands of these chemicals for various applications such as , polymers, plastics, rubber and film, textiles, paints, dyes, purification of opium to heroin, solvents ,pesticides, plasticizers, perfumery as well as numerous pharmaceutical products such as ephedrine, pyridine. A short account of these chemicals is shown in Fig.4[3]. Over 2 million tons per year [3] is being produced in India, the majority being acetic acid, acetaldehyde and acetic anhydride. However, bulks of many such chemicals are mostly imported. Actually Alcohol based chemicals is more than 20000 million Rupees Industry in India which is the biggest alco-based industry in the World [8, 17]. As molasses is the major raw material required, sugar factories are taking necessary steps in developing sugar Industry into a chemical complex. Basic criteria for selection of products from huge possibilities are its demand and supply position in Indian as well as global market. Market status of certain alco-chemicals for which indigenous technologies are available is given in the following table 6 [8].

Table 6 Demand –Supply position of some Alcohol based value added chemicals

Chemicals	Capacity,TPA	Supply,TPA	Market Demand
Acetic acid(ethanol base)	2,82,460+50,000*	131,700	3.22 lakh t
Acetic anhydride	68,320	33,900	60,000
Ethyl/butyl acetate	34,968/15,800	-	-
Vinyl acetate monomer(VAM)	50,000	23,700	44,000
Pyridine	6050	4100	6600
2-Ethyl hexanol	36,600	77,000	

*Methanol from M/s GNFC, Assam Gas Co., Rama Petrochem.

Annual production capacity(TPA) of 27 number of Alco-chemicals Industries are documented by Das[3].Further a detailed documents of various manufacturing process , industrially important process, percent conversion, important uses of 85 number of alco-based chemicals, their manufacturers, 292 Indian distilleries, 31 industries manufacturing acetic acid, 307 chemical industries(excluding sugar and distilleries) ,the basic raw materials for value added chemicals were cited[3].Some of the first generation Indian refineries using molasses as raw materials are shown in the following table 7.

Table 7 Sugar Mill / Distilleries and Alco-Chemical products

Sl.No.	Sugar Mill and Distillery In India	Alco-Chemicals
1	The Andhra sugars Ltd	Industrial alcohol, acetic acid, acetic anhydride, ethyl acetate
2	Mysore sugars Ltd	Acetaldehyde, acetic acid
3	SM Mohite Patil SSK Ltd	Acetaldehyde ,acetic acid
4	Somaiya Organic chemicals Ltd.,Sakharwadi	Acetaldehyde, acetic acid, acetic anhydride,ethyl acetate,butyl acetate
5	Somaiya Organic chemicals Ltd., Barabanki	Acetaldehyde, acetic acid, butanol, ethyl acetate, butyl acetate, crotonaldehyde, paraldehyde
6	The Kohlapur Sugar factory	Acetaldehyde, acetic acid, n-butanol,butyl acetate, Crotonaldehyde
7	The Sanjivani S.S.K.Ltd.	Acetic acid
8	Vasantdada shetkari S.S.K.Ltd., Sangli	Acetaldehyde, acetic acid, acetic anhydride
9	Trichy Distilleries & Chemicals Ltd.	Alcohol, acetaldehyde, acetic anhydride, ethyl acetate, butyl acetate, acetic acid
10	Daurala Sugar works Ltd	Phenyl acetic acid, benzyl acetate, cinnamaldehyde, cinnamyl alcohol, oxalic acid

11	Krishna Sahakari Sakhar Karkhana Ltd.	Acetone
12	Padamshri Dr. Vittalrao Vikhe Patil S.S.K.Ltd.	Acetaldehyde, ethyl acetate, glyoxal
13	Dhampur Sugar Mills and Alco-Chem Ltd.	Acetaldehyde, acetic acid, acetic anhydride, ethyl acetate
14	Sanjivani(takli)SSKLtd.	Industrial alcohol, acetic acid,acetic anhydride, ethyl acetate, butyl acetate, diethyl oxalate, ethyl aceto acetate
15	Shree Satpuda Tapi Parisar SSK Ltd	Diethyl oxalte
16	Rajarambapu Patil S SK Ltd.	Acetone
17	Walwa Taluka Sahakari Sakar Karkhana Ltd., Maharastra	Acetone
18	Gujchem Distillers(I) Ltd	Industrial alcohol, acetaldehyde, acetic acid, monochloro acetic acid, methyl cellulose, methyl & ethyl esters of mono chloro acetic acid ,2-4 D, sodium carboxy methyl cellulose

Majority of sugar factory managements which have distilleries attached to their units have established ethanol based chemical units producing value added chemicals like acetic acid , acetic anhydride, salicylic acids etc. which gives much higher profits compared to the primary product, sugar itself. There are significant developments in the chemical market in India over the years due to steep hike of prices of crude oil, imported and petro-based chemicals and improvement of supply of indigenous ethanol for the conversion of ethanol in to feed stocks for polymers and many other products such as acetaldehyde, and acetone which would give value added downstream products by dehydrogenation, dehydration, and oxidation. Carbonylation reaction when applied to ethanol can produce propionic and higher carboxylic acids with Rh or Ir complex catalyst with aqueous HI as a promoter [12] .The product has also been developed by NCL, Pune [3]. Many other new products are being developed by in-house R&D's or in National laboratories. Examples are: development by NCL of Albene technology of one step alkylation of benzene with dil. alcohol using specially structured zeolite catalyst to produce ethyl benzene which is in operation in the polystyrene plant of McDowell Ltd. in Vizag since 1989[3], the recovery of products, divinyl benzene and ethyl vinyl benzene from the byproduct stream by Polychem Ltd., Mumbai, conversion of Ethyl alcohol to C3+ hydrocarbon over silica catalyst by NCL, developments of catalysts for manufacture of many chemicals by projects & Developments(I) Ltd.

Subsequently further conversion of one of the downstream products from alcohol, acetone into some high value chemicals with long shelf life(ranging from 6 months to 3 years} such as diacetone-alcohol(DAA),methyl-isobutyl-ketone(MIBK),Isophorone(IP),bisphenol-A(BP),methyl methacrylic acid monomer (MMA), acetyl acetone(AA) in the Indian context has been possible. Some of the chemicals are indigenously manufactured by a number of public and private sector companies such as National organic chemicals industry limited(NOCIL), Mumbai, Herdilla chemicals limited, Mumbai, Hindustan organic chemicals limited, Rasayani, dist. Raigadh etc. Gujrat State fertilizers corporation-GSFC manufacture MMA from IPCL, Vadodara for the supply of a crucial and highly poisonous ingredient HCN required for manufacture MMA. However, the above three companies manufacture acetone from petroleum based starting materials viz, cumene or isobutylene. M/s Piramal organics,Cibatul and Alpha chemicals are some manufacturers of BP. Co-operative sugar mills like, Krishna SSK Ltd., Rethare, Dist. Satara, Rajarambapu Patil SSK Ltd.,Walwa, Dist.Sangli manufacture acetone from ethanol. In order to be competitive (less expensive) with the above three big companies, efforts have been made to produce low volume high value chemicals. A large number of options are available in this regard. These small volume but high value chemicals will add to the concept of biorefinery in India. The market survey involved collection of data on the various alcohol or acetone based chemicals from professional bodies like Indian chemicals Manufacturer's Association (ICMA), All India Alcohol –based Industry's Development Association (AABIDA). Growth ratio, future projections, demands –supply, price and margins (ranging 7-12%) have been the main focal points to be considered [8]. Increase in demands is found to be on the order of 10-30% per annum. In terms of demand DAA appears to be the fastest moving chemicals followed by MMA, then BP as moderately fast moving but MIBK and AA are rather slow moving chemicals. These chemicals are mostly imported from Japan and other countries. The industrial use of IP in printing inks, inks for metal containers, solvent for vinyl compounds whereas the same for AA is for paints, pharmaceuticals and explosive industries [17].

PRODUCTION ECONOMICS FOR MANUFACTURE OF ALCO-CHEMICALS IN INDIA:

The stoichiometric amount (tons) of acetone based chemicals per ton of the products well as their production viability are shown in table 8 [8, 16,17]:

Table 8 Important valuable product economics

Product per ton	Acetone reqd., ton	Viable plant cap., TPA	Investment ratio
BP	0.25	20,000	32.5
AA	0.57	-	-
MMA	0.63	10,000	17.5
DAA	1.00	3,000	5
MIBK	1.14	-	-
IP	1.24	2,000	1

Present cost of some alco-chemicals per kg [42] is given in table 9 as follows:

Chemicals	**Cost, Rs./kg**	**Chemicals**	**Costs, Rs./kg**
Diacetone-alcohol	115(India),Imp: 71.54	Bisphenol-A(BP)	Imported :.107
MIBK	135(India); Imp. 82.46	Methyl methacrylic acid monomer(MMA)	124.41(India);Imp.110.41
Pyridine	Rs.210	Acetic acid	51(India);Imp.26.16 (excluding china Rs.40.20)
Acetone	97(India);Imp.44.31	n-butanol	75.15
2-Ethylhexanol	Imp. 90.84	-	-

It has been found that in all the cases, imported chemicals are found to be cheaper than indigenously manufactured.

ECONOMICS FEASIBILITY ANALYSIS:

Basis: Assuming selling prices of various by-products are net of duties and other levies, Capacity: 2500 TCD, 150 days season as suggested [11], and 4.5 % molasses on cane, the amount of molasses required and alcohol produced will be 16875 tons & 4.23 million liters respectively. The following data are considered:
Yield of ethanol per ton of cane ranges between 110-120 litres [31]
Acetic anhydride from acetic acid (ketene):85-89% yield, whereas 70-75% yields from acetaldehyde [3]
One ton particle board from 3 tons of bagasse [45]
Cost of alcohol/litre: Rs.21.5 as per ref. [31] whereas Rs.17.30/ litre is due to ref. [34]
Cost of acetic acid: Rs.51/kg, and acetic anhydride:Rs.80.75/kg. The annual turnover is given as under:

Table 9 Preliminary economic analysis for a biorefinery in sugar mill (million Rs.)

Sugar :@ 10% sugar recovery and Rs. 29, 000/MT net realization Annual Gross turnover in Sugar Mill	1087.5
Particle Board:@ 4% surplus bagasse(50% moisture) on cane 15,000 t mill wet bagasse shall give about 5,000 m^3 of particle board. Annual turnover @ Rs. 350 per sq. meter of 8 mm Thickness	218.75
Alcohol:@ 4.5% molasses on cane ,16875 tons molasses cane yield 4.23 million litres alcohol.@250.67 litres/ton of molasses Annual turnover @ Rs. 21.5 per litre	90.95
Acetic Acid @ 4.23 million litre alcohol can yield 3384 tons Acetic acid (assuming conversion is 80%) Annual turnover @ Rs. 51,000 per ton	172.6
Acetic Anhydride:@ 3384 tons Acetic Acid yield about 2970 tons ,Annual turnover @ Rs. 80,750 per ton	239.8

Total annual turnover is approximately Rs.1546.0 million, 42.2% more gain in terms of profit, of which 22% alone for acetic anhydride (without considering particle board manufacture). More gain can be expected if acetaldehyde and butanol are manufactured.

BYPRODUCTS OF DISTILLERY:

The other byproducts of distillery are: carbon-di-oxide, fuel oil, and recovered yeast [3, 27]. Theoretically 48.8 kg of CO_2 is produced for 100 kg of fermented sugar. Thus, for each litre of alcohol, 0.76 kg of carbonic gas is produced which can be used in carbonated beverage industry, in clarification of sugar juice by carbonation process, dry ice, as preservatives of vegetables, fruits and other food products, in the automobile industry in the soldering process, synthesis of polymers and co-polymers, including polycarbonates, production of urea, methanol, carbonates, ammonium, calcium, sodium, potassium and magnesium bicarbonates.
Vam Organic Chemicals Ltd.is one of the producers of both liquid and solid CO_2 (dry ice) in their distillery.

Fusel oil, consists of isoamyl alcohol, isobutyl alcohol, amyl alcohol, and n-propyl alcohol small amounts of aldehydes, esters, and other alcohols. The mixture of amyl alcohol is employed as artificial essence, perfumes and laboratory reducing agents with sodium and in milk testing kits. The fusel oil is produced in amount of 4 litres per 1000 litres of rectified spirits.
The recovered yeast is mainly used for animal feed [3].
The effluent spent wash; spent lees can be utilized for biogas generation as discussed later.

UTILIZATION OF DISTILLERY EFFLUENT FOR FUEL/ BIOGAS PRODUCTION

Establishment of Biomethanation Plant in India:

Anaerobic digestion of distillery effluent (spent wash) to produce biogas is a well established technology with around 100 commercial plants operating in the country [24]. On an average about 15 to 20 litres of spent wash of a very high pollution load of 45,000 mg/l (ranging 40,000-60,000 mg/l) BOD is produced per one litre of spirit [23, 28].
Thus approximately an average distillery in India producing 30,000litres of alcohol per day will generate effluent in the range of 5 to 6 lakhs litres per day [29]. Agarwal *etal.*, [29] treated this effluent with ammonifying bacteria in Somaiya Organic chemicals distillery , Sakarwadi, M.S., S.K.G.Sugar Limited, Lauriya Distillery, W.Champaran Distt. of Bihar.

In order to increase the efficiency ,Sanjeev Kumar *et al.*, [24] took several trial run with additive, Biologic-SR2 in a biogas plant of m/s Manganga S.S.K.Ltd, Sangli for a upflow CSTR design with a cell recycle and in M/s Kopargaon S.S.K.Ltd, Ahmednagar in a downflow –fixed film reactor configuration with effluent recycle system produce gas production of 18000m^3/day(with 25% more production) with 64.4 % methane (with 7.3% more)and 9,100 m^3/day(21.3% increase) with 58.4% methane were obtained.

Vishwanathan *etal.*,[43,44] of Western Biosystem limited ,Pune in collaboration with Sulzer Brothers Ltd., Switzerland put up about 28 plants in India (either operating or under commissioning) for higher efficiency and productivity. Depending upon the pollution load and other factors, the process yields a COD reduction of 65%to 70% and BOD reduction of 85%to 90%. Biogas production is about 0.4 to 0.5 m^3/kg COD reduced with methane content 60 to 70%. The plant has to maintain temperature, 35°C to 38°C, pH =7.0(>6.8) with continuous recycling at prescribed rate and agitation.
Devendra Kumar [6] indicated new anaerobic process which enabled to reduce the BOD up to 90% and COD load by 70% with production of biogas containing about 70% methane gas by optimizing relevant parameters. In conjugation with sludge recycling technique. Enhanced biogas production with admixture of cattle waste and distillery effluent were tried by Malik *etal.*, [15,20].
A process of distillery waste water treatment by UASB reactor methanogenic biphasic reactor was used by Gurpal Singh *etal.*, [22] in The Simbhaoli Sugar Mills Ltd., Uttar Pradesh. The plant has shown 70-75% reduction in COD and 90-95% reduction in BOD in continuous operation for several years. The biogas evolved in the methanogenic reactor consists of 65% methane and 34% carbon dioxide with a calorific value 21840 kJ/m^3 of biogas and is used as fuel in the distillery. The process has resulted in a daily saving of 60-65% as fuel for 30 kL /day alcohol plant in terms of rice husk and has become a profitable recycling resource in pollution management.

Performance of Installations for Handling Distillery Spent Wash Using Downflow Stationary Fixed Film Type Anaerobic Filter in three Indian distilleries are shown [4,5] in the following table 10.
(Assuming steam requirement = 2.5 kg/kg of alcohol production)

Table 10 Performance of Biomethanation plant from Indian distilleries

Items	Parameters	ASI	KSSK	BACARDI
A. Distillery	1. Alcohol, Capacity kl/d	15	30	100-120
	2. Spent Wash, m^3/d	200-225	450-500	1500-1700
	3. BOD, gpl	40-45	40-45	36-42
B. Reactor	1. DiaXmedia htXTotal ht.	18X6X12	24.8X6.1X9.0	36.6X9.1X12.8
	2. Media Vol. m^3	1526	300	9600
	3. Area, m^2/m^3	90	100	90
	4. Loading, kg COD/m^3/d	10.5	12.5	12.5
C. Bio-Gas	1. Methane, %	55-60	55-60	50-55
	2. Rate, m^3/kg/COD	0.5	0.52	0.56
	3. Gen. Rate, m^3/d	6800	13600	48,000
	4. Energy Potential, kJ/d	138.6X106	268.8	1075.2
	5. Steam Demand, t/d	40.4	80.8	350
		37.5	75	300
D. Reactor Performance	1. BOD Removal, %	80-82	80-85	75-80
	2. COD removal, %	67-70	65-70	65-70
	3. HRT, d	8-10	8-10	8-10
Note: ASI= Andhra Sugar Ltd., KSSK = Kopargaon Sahakari Sakhar Karkhana Ltd., Bacardi Corporation; Kesar Sugar Ltd.				

ALCOHOL BASED CO-GENERATION OF POWER

Almost all Indian large sugar mills are running cogeneration plant for production of power and exhaust steam for process heating using bagasse as a fuel. These are extensively documented in literature [4,5, 38]. There is a possibility of establishing alcohol/biogas based co-generation plant for self- sustaining through gas turbines. Estimates of the potential for such schemes have shown that power generation can be doubled compared to conventional steam turbines. This alcohol or biogas can be coupled with fuel cell for co-generation.

Alcohol/Biogas-Fuel Cell Based Co-Generation in Sugar Mill

The fuel cell has the components like an anode ,an electrolyte and a cathode and has the advantages: low power consumption, low nutrient demand and lesser sludge generation.

Types Of Fuel Cell: Alkaline fuel cell (AFC);Polymer Electrolyte fuel cell (PEFC);Phosphoric Acid Fuel Cell (PAFC);Molten carbonate fuel Cell (MCFC);Solid Oxide Fuel Cell (SOFC).

Various Reaction Schemes and operating conditions in Fuel Cell

- Reforming Reaction (Ni catalyst, at 650°C -700 °C, with 1.5 times the stoichiometric amount of steam): $CH_4 + H_2O = CO + 3H_2$ $\Delta H = +223$ kJ/mole
- Two – stage water-shift conversion (CO, a poison to Pt electro catalyst ;tolerable limit = 1.4 vol. % at PAFC operating temperature of 190 C):$CO + H_2O = CO_2 + H_2$, $\Delta H = -41$ kJ/mole
- Electro-chemical Oxidation in the PAFC System (190 °C):
  $H_2 \rightarrow 2H^+ + 2e^-$ (Anodic Reaction)
  $2e^- + (1/2)\ O_2 + 2H = H_2O$ (Cathodic Reaction)
  $H_2 + (1/2)\ O_2 = H_2O$ (Overall Reaction)

In high temperature shift, iron Oxide catalyst at 450°C LTS, CuO/ ZnO catalyst on A1203 at 250°C. Electro-chemical Oxidation in the PAFC System (temperature: 190°C)

Technical Parameters For HRAD-PAFC System designed in India [38].

Basis: Conversion Rate: 0.4 m^3 bio-gas /kg dry matter;
Dry matter conversion: 40%; CH_4 = 75 % in bio-gas

- System Heat Rate: 10710 kJ/kWh
- Bio-gas to electricity conversion: 0.5 Nm^3/kWh
- Plant Utilization factor = 0.6
- Power output ~ 300kW
- Heat Recovery Rate ~ 250 kW at 160 °C

CONCLUSIONS:

There is an enough potential in India to establish biorefinery in a sugar mill complex consisting of a distillery, power alcohol plant, plant for cogeneration of power, a biogas plant with sugar mill effluent, as well as with distillery waste, a chemical plant and a bagasse based paper plant . Molasses can provide a unique source for a large number of downstream products , particularly alcohol based chemicals like oxalic acid, acetic acids, acetaldehyde, acetic anhydride etc. pharmaceutical products, plastics , synthetic rubber and innumerable other products. There is an opportunity to increase the production of alcohol from molasses making it a surplus one, opening avenues either for export or for producing value added products. Use of more amount of absolute alcohol (10% or more)as substitute of gasoline and rectified spirit for biodiesel will be possible by increasing the efficiency of alcohol production from molasses and sugarcane juice. Efficient energy storage system like fuel cell can also be designed. For sustainability point of view the sugar industry with distillery and chemical plant producing only acetic acid / acetic anhydride can have an attractive turnover of Rs.1546.0 millions equivalent to 34 million US dollars. Therefore, the molasses should not be let out of the mill control , rather mill be encouraged by government regulatory bodies to put three industries, sugar-alcohol- chemicals, together in chain to improve Indian economy , and to enhance export potential of both chemicals and energy. In order to achieve these targets, government policies regarding levies, central tax, state tax and VAT must be relooked and reexamined.

REFERENCES:

1. http://search.conduit.com/?SearchSource=10&ctid=CT2776682

2. http://www.google.co.in/From Wikipedia, A sugar industry perspective & ethanol production
3. Das A.K., Alcohol based chemical industries in India, Akay Publications, January ,1999
4. Ray A.K. Energy production and conservation techniques for building and industries, Proceedings of All India Seminar on"Emerging trends in Energy Conservation and Management"The Institution of Engineers (India) Roorkee Local Centre, Roorkee, India, pp.1-34.

5. Ray A.K.,Proceedings of the refresher course of Sugar Technologists and Engineers, 2003, National sugar Institute , Govt. of India,2003.

6. Devendra Kumar, Anaerobic treatment of distillery spent wash, optimization of technology, Proceedings of the 51th Annual Convention of STAI, Kanpur, 14th -16th October,1988,pp.G60-G65
7. Agarwal P.K., Renewable source of energy, Use of Alcohol as motor fuel in internal combustion engines,ibid pp.G67-G76

8. Kulkani A.D.,S.P.Phandis and S.J.Jadhav,Current Status certain Acetone based chemicals in India, G-17-G 22.,Proceedings of the 54th Annual Convention of STAI, Bangalore, 17th -19th January,1992.

9. LeelaWati ,D.S.Dahiya ,S.S.,S.S.Damija and B.S.Yadav, Ethanol production by strains of *Sachharomyces cerevisiae* at high temperature, G9-G14, Proceedings of the 55th Annual Convention of STAI, New Delhi, 21st -23rd August,1993.

10. Yadav Ajay, S.Sharma, and D.S.Dahiya, Effect of pretreatment and supplementation of cane molasses on ethanol production by yeast strains, G1-G7, Proceedings of the 55th Annual Convention of STAI, New Delhi, 21st -23rd August, 1993.

11. Sharma S.K.,Increasing potential of Sugar Mill by-products,pp.G667-672, Proceedings of the 57th Annual Convention of STAI, New Delhi, 13th -15th October,1995.

12. Patil R.P.,Production of propionic acid via ethyl alcohol carbonylation reaction,pp.G6657-G 666, Proceedings of the 57th Annual Convention of STAI, New Delhi, 13th -15th October,1995.

13. Lehri Madhu, Vartika Saxena and P.K. Agarwal, Preliminary studies on ethanolic fermentation of acid hydrolysates of hemicelluloses-extracted bagasse,pp.G587-G593, , Proceedings of the 57th Annual Convention of STAI, New Delhi, 13th -15th October,1995.

14. Sudarsan Y.,and P.K.Agarwal, Some factors affecting quality of molasses for alcohol production,G 573-G578, Proceedings of the 57th Annual Convention of STAI, New Delhi, 13th -15th October,1995

15. Malik R.K.,and Dalel Singh, Biogas generation from Anaerobic digestion of cattle waste and molasses,ibid,pp.G 551-G558

16. Patil G.D., and B.N.Shinde,Chemical nature of spent wash,Chemical nature of spent wash/spent slurry- press mud compost,pp.G565-G572,ibid

17. Kulkani A.D.,S.P.Phandis and S.J.Jadhav,Status of important Alco-Chemicals: Now and into The 21st Century,G33-G42, Proceedings of the 58th Annual Convention of STAI, New Delhi, 14th -16th January,1996.

18. Lehri Madhu and P.K.Agarwal,Studies on Alcohol production from corn &Tapoica,G85-G-97, Proceedings of the 58th Annual Convention of STAI, New Delhi, 14th -16th January,1996.

19. Sharma, S.K.,Global trends towards bio-fuels, G23-G26, Proceedings of the 58th Annual Convention of STAI, New Delhi, 14th -16th January,1996.

20. Malik R.K.,and Dalel Singh, Biogas production from cattle waste and distillery effluent mixed slurry, Proceedings of the 59th Annual Convention of STAI, Goa, 26th -28th September,1997,pp.Co-product 37-41.

21. Sharma S.K. and D.K.Goel,Ethanol:An ecofriendly biofuel bio-fuel from sugar industry, ibid,Co-Product 3-23.

22. Gurpal Singh, B.Chatterjee,K.A.Prabhu, and Luckshendra Kumar,Biomethanation process in pollution abatement and energy recovery in distilleries, Proceedings of 61st Annual Convention of The Sugar Technologists of India, pp.Co-Product-64-72,1999

23. Patil, P.U.,and V.S. Damankar,Performance of Biomethanation process for reduction of pollution load in spent wash at different distilleries, Proceedings of 61st Annual Convention of The Sugar Technologists of India, pp.Co-Product-35-39, 7th -9th September,1999.

24. Sanjeev Kumar , Ravindra Singh and Sanjay Patil,Performance evaluation of Biologic-SR2 additive for primary anaerobic digester treating molasses based distillery effluent, Proceedings of 62nd Annual Convention of The Sugar Technologists of India,pp.Co-Product G27-G38.,2000

25. Jai Uppal, Gasohol An ecofriendly transport fuel for the twenty first century, ibid,Plenary,17-27.

26. Mathur H.B.,Ethanol-a bio-solar fuel substitute for meeting the twin challenges of vehicular fuel oil scarcity and pollution,ibid,Plenary10-16

27. Paturau, J. Maurice,Byproduct of the Cane Sugar Industry, Elsevier Science Ltd, 1969,1982

28. Lehri Pradeep and Baldev Singh,Continuous treatment of high load distillery effluent by ammonofying bacteria process, Proceedings of All India Seminar on Pollution Control in Sugar Industry,, National sugar Institute,Sept,25,1992.pp.89-90

29. Agarwal, P.K., G.L.Shukla, and Santosh Kumar, Treatment of Distillery Effluent by ammonifying bacteria on large scale,ibid,pp.56-68.
30. Natu, S.C., Bagasse based cogeneration, India marching ahead,International Sugar Journal,Vol.107,International Sugar Journal,1279, pp.416-424, www.internationalsugarjournal.com
31. Ramaiah, N.A., Production of ethanol from cane –A profitable process and economics,pp.3-27,Sissta J.,2008
32. Bakthavatchaalam,N., Multi Energy from Sugar Cane by effective utilization to its full potential,ibid, pp.29-30(Source:Ministry of petroleum)
33. Ramasamy P.,Ethanol production in sugar complex,ibid, pp.33-39
34. Ravi S., Ethanol: A solution to global fuel demand,ibid,pp.41-52
35. Bakthavatchaalam,N, Ways to improve economics of sugar Industry by effective utilization of sugar cane to its full potential, Sissta 2007,pp.172-175
36. Bhagat J.J., and Dilip Jain, Indian Sugar Industry-an overview, International sugar Journal,Vol.109,No.1304, pp.505-509
37. Uppal Jai, International sugar Journal,Vol.109,No.1304, pp.189-193
38. Kurup ,G. Rajendra Chawla S.K.,Mathur A., Biogas –fuel cell based cogeneration in the pulp and paper Industry, IPPTA convention issue ,1992,pp.25-30.
39. Singh Abha and V.K.Jain, Batch fermentation of cane molasses for ethanol production by Zymomonas mobilis,Indian Chemical Engr., Section A, Vol.37, No.3 ,July – Sep.1995,pp.94-98.
40. Padmaja A, Venkateshwar S, Seenayya G and Bhagavanth Rao M, Evaluation of cellulosic substrates in the conversion of biomass to ethanol,Indian Chemical Engr., Section A . Vol.37 No.3, Jul-Sep(1995),pp.140-144.
41. Jain, V.K., Present status and prospects of Bacterium *Zymomonas mobilis* for ethanol production,Proc. of the Eleventh National convention of Chemical Engineers, sept.28-29,1995,pp.III-1- III-5.
42. Chemical weekly,July 5,2011pp.226-237,239-273
43. Banka S.N. ,L.Viswanathan, and V.S.Raju,Some thoughts on secondary effluent treatment after biomethanation,, Proceedings of All India Seminar on Pollution Control in Sugar Industry,, National sugar Institute,Sept,25,1992.pp.13-15
44. Viswanathan, L.,V.S.Raju and A.Patil, SULZER'S " an –OPUR" biogas plant,ibid, pp.1-4
45. Thangavelu S., By-products of bagasse-A review, SISSTA,2007,pp.168-171
46. Jain S.L., India and its impact on the World sugar market, International Sugar Journal,Vol.111, No.1316,pp.455- 459.

PRINCIPAL ALCOHOL - BASED CHEMICALS

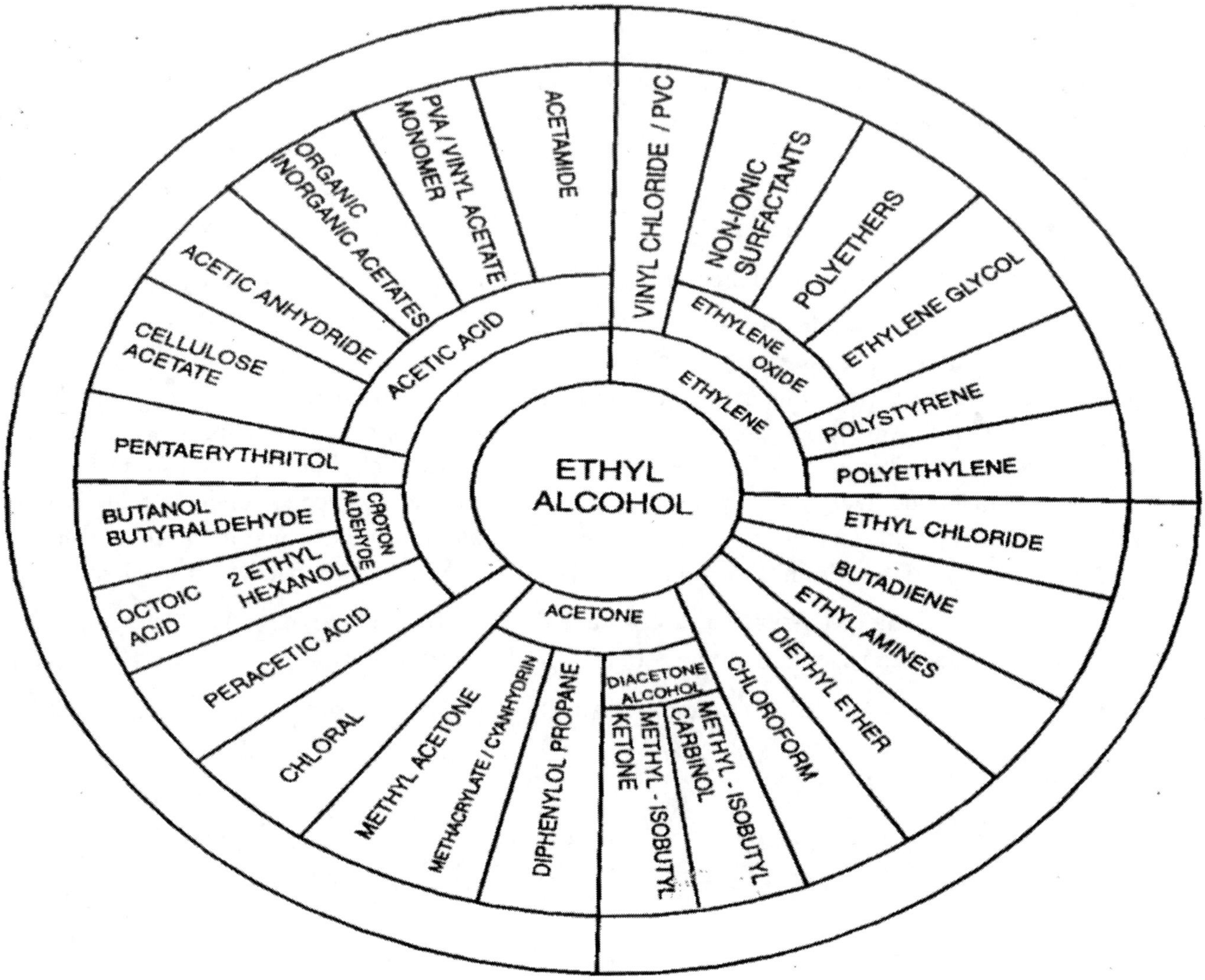

Fig.4 A brief account of value added chemicals from ethanol

(228e) Dye Sensitized Solar Cells On Paper Substrates

Bo Wang, Chemical and Paper Engineering, Miami University, Oxford, OH and **Lei L. Kerr**, Paper and Chemical Engineering, Miami University, Oxford, OH

This study reports for the first time in the literature, a dye sensitized solar cells with 1.21% efficiency (V_{oc}= 0.56 V, J_{sc}= 6.70 mA/cm^2, F.F.= 0.33) on paper substrates. The current dye sensitized solar cell technology is based on fluorine doped SnO_2 (FTO) coated glass substrates. The problem with the glass substrate is its rigidity and heavy weight. Making DSSCs on paper opens the door for both photovoltaic and paper industries. The potential of using mature paper making and coating technologies will greatly reduce the current PV cost. Paper substrate based DSSCs not only offer the advantages of flexibility, portability and lightweight but also provide the opportunities for easy implantation to textile. In this study, a low temperature process is developed to coat uniform nickel on paper substrate as the metal contact to replace the traditional expensive FTO. The Ni paper showed excellent conductivity of 8-10 Ω/. It is found that the control of the metal oxide electrode morphology is critical to the solar cell performance. The TiO_2 film has the tendency to crack on Ni coated paper which resulted in the shunt of the device and no solar cell efficiency was obtained. ZnO film on the other hand had good morphology tolerance on Ni coated paper and yielded solar cell efficiency of 1.21% (V_{oc}=0.56 V, J_{sc}= 6.70 mA/cm^2 and F.F.= 0.33) under AM 1.5 (Activation area is 0.16 cm^2). The control sample of ZnO solar cell on FTO glasses has the efficiency of 2.66% (V_{oc}=0.64 V, J_{sc}=9.97 mA/cm^2, F.F.=0.42).

See more of this Session: Innovation of Forest Products
See more of this Group/Topical: Forest and Plant Bioproducts Division

Dewatering Characteristics of Indian Nonwood and Wood BASED PAPER Mill ETP Sludge with Chemical Conditioning

Amiya kumar Ray, Rahul Meena, Department of Paper Technology, Indian Institute of Technology Roorkee, Saharanpur, India , and **Harish Kumar Dhingra**, Software development Division, Stanislaus County Superior Court, Modesto, CA and V.K. Mahendru, Retired Scientist, Central Pulp and paper Research Institute, Saharanpur, India

ABSTRACT

Sludge handling and its disposal from waste water treatment plant is one of the major constraints in any chemical industry in India, pulp and paper industry in particular. This problem relating to sludge handling can be overcome by applying efficient sludge dewatering techniques. Chemical conditioning of sludge is a commonly applied practice to improve the dewatering characteristics of sludge to result in improved dried sludge. While doing chemical conditioning of sludge, the understanding of sludge and flocculant characteristics is very much needed for better economical dewatering. However, these chemicals are numerous and generally expensive; thus it increases the cost of operation significantly. Therefore, it is absolute necessary to evaluate the viability of using the chemicals at plant or laboratory scale on the basis of the balancing the gain against cost.

In India effluent treatment plant (ETP) of paper mill are generally based on either nonwood or wood fibers or combination of the former two which generate large amount of primary and secondary sludges in aqueous form. Nonwood based ETP sludges generally become more detrimental during handling. These problems can be overcome by efficient dewatering of sludge for which study of characteristics of sludge and the effect of chemical conditioning is very much needed. The later method is used to improve dewatering of sludge by flocculating the sludge with organic/inorganic flocculants.

An attempt has been made in this present investigation to study in-depth by laboratory scale experiments, the dewatering characteristics of both primary and secondary sludges of two different Indian mills using different kinds of raw materials, wood and non-wood. The effects of various parameters like, pH, time of agitation, and storage on dewaterability of sludge from wood and nonwood based paper mills during conditioning with inorganic and organic flocculants in terms of specific cake resistance and other filterability parameters..The present investigation also provides a step by step procedure to evaluate the effectiveness of flocculants from the economic standpoint.

KEY WORDS: Primary and secondary sludge, dewatering, chemical conditioning, flocculants, rapid and slow mixing time, pH, storage, optimum dose, under dose, economic analysis.

INTRODUCTION

Environment pollution due to industrial activity is one of the major challenges for any industry today, pulp and paper industry in particular.It is well known that sludge from effluent treatment plant (ETP) of a paper mill is one of the major streams affecting environment to a considerable extent and is also difficult to handle. The amount of ETP sludge from small paper mill using agro-waste is of the order of 180 kg per tonne of paper (15% of total). Its disposal is therefore, one of the major constraints relating to environmental degradation. The first stage in handling of sludge is obviously efficient dewatering of sludge. Basically dewatering helps to reduce transportation volume, to make it suitable for landfill, or to prepare the sludge for incineration or other value added byproducts. ETP Sludge can be mainly classified as primary sludge and secondary sludge.

Dewatering is usually accomplished by mechanical means using vacuum filter, belt filter press, screw presses, centrifuge, pressure filter or double wire belt washer. Table-1 shows the achievable cake dryness of primary sludge by various mechanical dewatering equipments mostly being used **(1).**

Table-1: Achievable cake dryness by mechanical dewatering equipments

	Achievable Cake Dryness	
	Mean (%T.S.)	Range (%T.S.)
Vacuum Filter	20	15-22
Vacuum Filter and V-press	35	30-40
Belt Filter press	35	30-40
Screw presses	45	40-50

All the above mechanical dewatering devices consume huge electrical power and other costs .It is therefore an imperative necessity to reduce the same.

Of late various paper mills have installed secondary treatment plant which has led to significant increase in generation of secondary sludge (biological sludge). Efficient dewatering of mixed secondary (biological) and primary sludges have received attention because biological sludge is generally considered more difficult to dewater than primary sludge. There is a growing tendency of the Industry to dewater in the existing devices more efficiently by increasing cake dryness as much as possible for its efficient utilization using various means.

The problem pertaining to sludge handling can be overcome by efficient dewatering of sludge for which chemical or biochemical conditioning is very much essential. The biochemical conditioning is however in research stage. It is reported that short exposure of pulp and paper mill sludge to cellulose enzymes leads to rapid shortening of the fibers, which provides higher cake solids when the sludge is dewatered **(2)**.On the other hand the chemical conditioning is more often practiced. The Chemical conditioning demands the use of efficient and economic flocculants. The selectivity of the chemicals in terms of its power to flocculate the sludge is necessary to estimate only through experimental means.

Unfortunately, the performance of a particular flocculants and the optimum dose required are dependent on sludge characteristics which are diverse in nature. It is important to mention that these vary from sludge to sludge in a same mill and also from mill to mill.

There are few other operating parameters in addition to conditioning by flocculants and sludge characteristics, like pressure drop, speed of the agitator and time of mixing, pH, concentration and nature of ions present in sludge, temperature, and duration of storage of sludge also greatly influence the performance of chemicals to flocculate the sludge. This in turn affects dewatering efficiency.

It is therefore, important from both cost and performance considerations, to optimize not only the addition of flocculant for sludge conditioning but also the above parameters in order to achieve efficient sludge dewatering for better handling.

To handle such a complex situation with innumerous variables for optimal results laboratory scale experiments are generally performed before taking plant scale trials. Even during the routine operation in plant, to control the parameters laboratory scale tests are necessary to get the desired values. However, the information in this regard is scarce and scatteredly available. No set procedure is followed.

In this present investigation, a systematic step by step procedure is developed for evaluating different flocculants for efficient dewatering of pulp and paper ETP sludge. It includes the development of methodology, mathematical formulae, computational procedure, experimental results and their interpretations. The various experiments have been performed with the sludge samples obtained from wood based paper mill and the result has been also compared with nonwood based paper mill. Various design parameters has also been determined. More detailed work, for example the mathematical analysis, effect of operating variables and a non detailed economical feasibility is available elsewhere **(1).**

In the present investigation the dewatering characteristics of both primary and secondary sludges of two different Indians mills using different kind of raw materials were studied in the laboratory. The influence of various flocculants, both organic and inorganic in nature on dewatering was also studied at room temperature. Further, studies have been performed to study the effect of various operating variables like time of agitation, pH, storage etc. the step by step procedure and the relevant mathematical equations employed with a preliminary cost viability analysis is presented elsewhere **(1).**

PRINCIPLES

The most important parameter for identifying the dewatering characteristics of ETP sludge is usually the specific resistance to filteration (SRF), compressibility of cake, mass concentration of cake and loading factor. These are defined as under.

r (Specific Resistance to Filteration) (3-5)

The basic filteration equation derived form the Poiseuille and Darcy' law is written as-

$$dV/dt = \Delta P\ A/\mu\ (R_C + R_m) \qquad (1)$$

Eq^n (1) can also be written by inserting values of R_C and R_m

$$dV/dt = A^2\ \Delta P/\ \mu\ (V+AV_f)\ \mu \qquad (2)$$

The physical significance of parameter r can be appreciated if in Eq^n (2) medium resistance R_m is neglected (which is permissible in engineering estimates). Solving for r,

$$r = PA^{2}/\ [\mu cV\ (dV/dt)\ 1 \qquad (3)$$

From Eq^n (3) it follows that r is numerically equal to the pressure difference (applied vacuum $_\Delta$p) required to produce a unit rate of filtrate flow (i.e.dV / dt = 1.0) through a unit mass of cake (i.e. cV = 1.0) and a unit filter area (A = 1.0), if filtrate viscosity is unity (μ = 1) or r = p if dV /dt = 1.0, cV = 1.0, μ = 1.0, and A = 1.0. Thus, the specific resistance r measures the ability of the sludge to be filtered; higher the value, the more difficult is the filtration. In the usual range of operating conditions, the value of r can be related to pressure drop $_\Delta$p by empirical equation

$$r = r_0\ (\Delta P)^S$$

Integrating Eq^n (2) assuming constant pressure over time, if at t =0, V = 0 and at t = t, V = V, integration of Eq^n (2) yields.

$$t / v = (\mu\ r\ c / 2\ PA^2)\ V + \mu\ R_m / AP \qquad (4)$$

From Eq^n (4) it follows that a plot of t / V vs, V yields a straight line, values of specific cake resistance r and medium resistance R_m are evaluated from the slope and intercept of this line, respectively.

$$r = (2\ PA^2 / \mu\ cV)\ s_1 \quad (5)$$

$$R_m = i\ A\ P\ \mu \quad (6)$$

Eq^n (5) is the required equation for specific resistance to filtration.

Coefficient of compressibility

Coefficient of compressibility was determined by slope of line when a graph is plotted between r and ΔP on log-log scale based on the following equation.

$$r = r_0 * p^S \quad (7)$$

The value of s, the coefficient of compressibility lies between 0 and 1. The larger is s, the more compressible is the sludge. When s = 0, the specific resistance is independent of pressure and the sludge is incompressible and Eq^n (3) yields.

$$r = r_0 = \text{Constant}$$

Relationship for c (Mass of Cake Deposited Per Unit Volume of Filtrate) (3)

The cake deposited per volume of filtrate, c, can be represented as

$$c = c_c\, c_i / 100\ (c_c - c_i) \quad (8)$$

Primary and secondary sludge samples were collected from effluent treatment plant of two Indian paper mills A and B. Mill A is an integrated Pulp and Paper Mill which employs mixture of Hard Wood integrated Pulp and Paper Mill which employs mixture of Hard Wood (15%), Soft wood (5%), Bamboo (75-80%) and produces both bleached and unbleached variety. Mill B is a small paper mill which uses nonwood raw material (Bagasse and Sarkanda) for producing unbleached grade paper. The pulping process being followed by Mill A is entirely based on Chemical Soda Sulfate Process while the mill B is chemical soda process and chemi-mechanical process. Both the paper mills use activated sludge process as secondary treatment of eh effluent.

The secondary sludge samples were thickened after collection from mill and allowed to settle. The settled samples with comparatively high solid content were used while the primary sludges were used as such.

The flocculants selected for the study were inorganic chemicals viz. ferric chloride, ferric sulfate, alum and lime and also organic polyelectrolytes (one cationic and two anionic).

Fresh 10% w/w stock solutions of inorganic flocculants and 0.1% w/w stock solution of polyelectrolytes were prepared and used whenever experiments were conducted.

A mechanical stirrer was used to mix sludge samples. After adding the required flocculant dose to the samples, the same were mixed for 1 min, at 150 and 250 rpm to generate homogeneous samples and for 2min, at 40 and 70 rpm (lower value for primary sludge of higher T.S. concentration and higher value for primary sludge of higher T.S. concentration) to promote the flock formation. Parameters for rpm and time combination have been selected based on preliminary laboratory scale trials and literature information.

Specific resistance to filtration (SRF) is calculated from equation 5 after getting filtrate volumetric flow rat by Buchner Funnel test **(6-7).** All the experiments have been performed at constant vacuum (375 mm Hg) except during study of the effect of pressure drop. Also, the same Buchner funnel was used throughout the study. In other words the area for filteration has remained constant.

Coefficient of compressibility was determined by the slope of the graph plotted between r and P on log-log scale based on Eq^n 7. The step-by step procedure is as followed.

EXPERIMENTAL INVESTIGATION

The step by step investigation procedure to evaluate the dewatering characteristics of sludge with or without the addition of flocculants is as follows.

1. Fiber content of sludge- required for the classification of the sludge in terms of its fiber content in different mills.
2. Determination of SRE of unconditioned sludge
3. Impact of pressure drop on filteration of unconditioned sludge
4. Evaluation of effect of solid content of sludge
5. Conditioning of sludge and its SRF values
6. Effect of pressure drop on conditioned sludge
7. Determination of optimum dose of various flocculants

RESULTS AND DISCUSSION

Based on the experiments conducted in the laboratory, data on Specific Resistance to Filtration (SRF) as a function of various parameters have been estimated using mathematical models. The data have been interpreted in various figures. These are described in the following paragraphs:

1. Fiber Content of Sludge

This is the first step in evaluating the dewaterability of primary sludge and gives a comparative idea of determining the amount of flocculant required. In general, the more is the amount of fines, the more difficult is the dewatering and hence more flocculants is required to achieve the same degree of dewaterability. The different fiber fraction of the primary sludge has been determined by Bauer-McNett fractionation. The result has also been compared with fiber fraction of nonwood paper mill primary sludge. The results are presented in **Table-2.**

Table-2: FIBER CLASSIFICATION AND ASH CONTENT OF PRIMARY SLUDGE OF MILL A AND MILL B

Mesh Specification	% Fraction (Weight/Weight)	
	Mill A	**Mill B**
+30	11.09	4.28
+50	16.81	3.95
+100	13.20	6.95
+200	2.80	3.80
-200	56.10	81.02
Ash %	**21.25**	**20.56**

The fines content i.e. material passings through 100 mesh screen are more than 80% (on O.D. total material basis) for mill B compared to only 60% for Mill A. Hence, it was more difficult to dewater the primary sludge form mill B than that of a mill A and demanded higher dosage of flocculants in the later stage. This is expected as mill B is a non-wood based mill which handles short fibers even from the beginning.

2. Determination of SRF of Unconditioned Sludge

Fiber fractionation of sludge gives only an qualitative idea for determination of dewaterability and amount of flocculants required for conditioning. The more exact determination of dewaterability and flocculant requirement is found by evaluating the values of specific resistance of cake to filtration for primary and secondary sludges. These values for mill B and Mill A sludges without flocculants are shown in histogram

(Fig. 1) for better comparison. It is evident that sludges of mill B have high SRF value in comparison to the higher amount of fines in primary sludge of Mill B. The reason for higher SRF of secondary sludge of Mill B may be attributed to carry over of more fines into the secondary treatment system resulting in more difficult dewatered sludge.

Similarly, on comparing the primary and secondary sludges from the same mill, it is clear that secondary sludges do have larger SRF values compared to primary sludges, and thus are found more difficult to dewater. The reason for larger SRF of secondary sludges may be due to presence of vary fine biological solids, and also more hydrous in nature.

3. Impact of Pressure Drop on Filtration of Unconditioned Sludge and the Resulting Compressibility Coefficient

The influence of vacuum, the driving force for filtration on SRF values of unconditioned primary and secondary sludges of Mill A have been estimated. Some experiments are also tried on conditioned primary sludge of the same Mill a (which will be shown later). The comparisons are made in **figure 2.** It is clear that there is increase in SRF values with increase in vacuum. Almost linear relationship between average SRF values and pressure drop has been found. The coefficient of compressibility which has also been estimated sludges and primary. It reflects that the cakes are highly compressible in nature. However, the primary sludge has been found to be more compressible than secondary activated sludge, though the difference is very small.

4. SRF of Unconditional Sludge as a Function of Solid Content

Specific resistance is supposed to be independent of solids content of sludge. But, experimentally, it is evident that SRF increases with increase in solid content with a slight tendency of non-linearity in behavior as reflected **in figure 3** for unconditioned primary and secondary sludges respectively. The increase in SRF with solid contents. This characteristic trend coincides to those given in **literature (8)** for secondary activated sludge. This also agrees with the finding of Coackley that the SRF decreased with decreasing solids content. It therefore supports the present investigational findings.

5. Determination of SRF of Conditioned Sludge

The effect of various flocculants (both inorganic and polymers) on the SRF values at varying dosage of flocculants is experimentally determined. Knowing the SRF values of unconditioned sludge and type of flocculants, the different dosage range are determined. For example in case of organic polymer, the starting dose must be very low compared to inorganic flocculants.

The typical plots of SRF as a function of dosage of different flocculants for both primary and secondary sludge for mill A and mill B have been shown in **figures 4 to 11**. Similar kinds of plots in different trials can also be shown **(1).** Comparing the data for the effect of inorganic flocculants (namely Alum, Ferric chloride, Lime and Ferric Sulfate) on primary sludge and secondary sludge, it has been observed that all the flocculants improved the sludge dewatering characteristics as indicated sludge. However, Ferric Chloride displayed the best and the lime the poorest.

Further experimental results on the cationic polymers and inorganic flocculants shows that the dose required for cationic polymer to reduce the SRF value by same magnitude is much low in comparison to inorganic polymers.

On comparing the effects of cationic and anionic polymers in conditioning primary sludges for dewatering i.e., on SRF values, it is reflected that only cationic polymer has improved the dewatering rate by lowering SRF significantly and anionic polymers has not been found effective at all.

6. Effect of Pressure Drop on conditioned Sludge

Attempts have also been made to see the effect of a conditioned sludge. A typical plot between vacuum and SRF of Ferric Chloride conditioned primary sludge and pressure drop is shown in **figure 12.** It is reflected that the conditioned primary sludge is much more compressible than unconditioned primary sludge. It is interesting to note that the compressibility coefficient exceeds 1.0 for $FeCl_3$ conditioned primary sludge. Though normal value of compressibility coefficient the between 0 – 1.0, it has been mentioned in literature, the value may be greater than 1.0 in quite a few cases.

On the contrary a value of 0.8 has been reported for coefficient of compressibility for pulp and paper activated sludge conditioned with 2.5% Ferric Chloride **(9).**

7. Determination of Optimum dose of Various Flocculants

The optimum dose of various flocculants is defined as the dose which gives maximum reduction of SRF values. Adding higher amount than optimum dose of flocculants, gives almost no further decrease in SRF value of any improvement of sludge characteristics. Besides the dosage must be economically viable for a special flocculants. Comparison of optimum dose of different flocculants for both primary and secondary sludges from both the mills has been shown in **figure 13**. The optimum dose, however, has been found to be sludge dependent. The slight variation in optimum values have also been observed for the same sludge but collected at different lime intervals.

It further reveals that the optimum dose requirement for sludges of Mill B is higher compared to sludges of Mill A in the case of all the flocculants used. However, Ferric Chloride exhibit different behaviour. The optimum dose requirement is minimum for cationic polymer followed by Ferric Chloride, alum / Ferric sulfate and lime for both primary and secondary sludges from both the mills.

As expected, the optimum dose requirement of secondary sludge is found to be higher than those for primary sludge of the same mill under treatment of different flocculants.

8. Other Parametric Effects:

The parametric effects on flocculation efficiency are described hereunder.

8.1 Effect of Mixing Time during Conditioning with Flocculants

It is a known fact that for such a process mixing speed has a significant role. However, the optimum mixing speed depends upon total solids concentration (if difference in concentration is appreciable).

There are two periods of mixing, namely rapid and slow to control such situation. The main aim during the rapid mixing period must be to disperse the flocculants properly but limited to a certain time. However, just mixing of chemicals through rapid mixing will not improve the filterability and hence, slow mixing is a must to promote floc formation, which also should be gentle and for a short period. The purpose is that the floc should not break.

The total mixing time has been maintained to 3 min (composed of 1 min. of rapid mixing and 2 min. of slow mixing) in all the cases during the experiments.

The effect of rapid mixing time has been determined by measuring SRF at various mixing times. The change in total mixing time was due to change in rapid mixing portion of the mixing cycle. Ferric chloride, alum and cationic polymer have been used as flocculants at around their optimum doses which has been found earlier (1). The representative results for mill A and B are plotted in **figures 14 and 15.**

These results indicate that with the increase in rapid mixing time beyond 1 -1.5 min or with decrease in rapid mixing time below 1 min, increase in SRF value has been observed indicating the deterioration of dewatering rate. The total optimum mixing time has therefore been found to be 3 min. (composed of 1 minute of rapid mixing and 2 min. slow mixing) in the cases. For an example, with 10% ferric chloride to condition secondary sludge of Mill A, at total cycle time of 2 min (0+2), zero minute rapid mixing, and 2 minutes slow mixing), 3 min (1+2), 4 min (1+3) and 5 min (1+4), the values of SRF have been found to be 4.31x109, 4.34x107, 5.54x107 and 1.11x108 s^2/g respectively.

Similarly, the effect of slow mixing time has been determined by keeping rapid mixing time of 1 min constant and changing slow mixing time. The results are shown in **figures 16 and 17** for mill A and mill B respectively.

The results indicate that if the slow mixing time is very low (less than 2 min) there has been decrease in dewatering rate indicated by high SRF value. However, longer time (more than 2 min.) does not affect the results significantly provided gentle mixing is maintained.

For example, at a total cycle time of 1 min (1+0, 1 min. rapid mixing and zero min. slow mixing), 2 min (1+1), 3 min (1+2), 4 min (1+3) and 5 min (1+4), the SRF values of 2% cationic polymer conditioned secondary sludge of Mill B are of the order of 3.01x1010, 2.09x108, 4.63x107, 4.63x107 and 7.02x107 s^2/g respectively.

8.2 Effect of Varying Raid Mixing Time during Conditioning with Different Dose (Less than the Optimum and Optimum) of Flocculants (Interaction of Dose and Mixing Time)

The experiments have also been conducted at low dose (less than optimum dose found) of ferric chloride with varying fast mixing time. The results have been compared with the results obtained at optimum dose of ferric chlorides with varying fast mixing time. The results are plotted in **figure 18.** it is clear that at low dose (under dose) there is more pronounced effect of mixing time on SRF values compared to values obtained at optimum dose i.e. there is more increase in SRF values at less than or more than 3 min mixing time with under dosage.

For example, for secondary sludge conditioned with 6% and 10% ferric chloride, the values of SRF at 5 min total cycle times have been found to be around 12.5 times and 2.6 times more of SRF value at 3 min cycle time respectively.

This reflects that rapid mixing time has paramount importance particularly during conditioning with underdose of flocculants.

8.3 Effect of pH Variation during Conditioning with Flocculants

The effect of pH on SRF has been determined for different flocculants at their optimum doses and the results are shown in **figures 19 and 20** for secondary and primary sludges of mill B respectively. It is observed that with increase in pH, SRF increases. This increase is very sharp after pH of 7.5, but very less pronounced at a pH between 5.5 and 7.5. Results of lime were exceptional in case of primary sludge.

For example in case of secondary sludge conditioned with cationic polymer the values of SRF at a pH of 6.5, 8.5 and 10.5 have been found to be respectively 2.9, 166.5, and 3726.8 times more than the SRF values at 4.5 pH. In case of sludge conditioned with lime, the values at same pH levels are respectively 4.7, 38.9, and 98.7 times more than the SRF value at 4.5 pH. While in case of primary sludge conditioned with lime, SRF values at above mentioned pH values have been found to be 0.4, 1.6, and 2.88 times more of SRF values at 4.5 pH. Similarly, in case of cationic polymer treated primary sludge, the values are 2.1, 11.2, and 308.8 times more respectively compared to SRF value at 4.5 pH.

8.4 Effect of pH during Conditioning with Different dose (Less than the Optimum and Optimum) of flocculants (Interaction of Dose and pH)

Effect of pH on conditioning of sludge has been also determined at low dosage (lower than optimum dose) for Ferric Chloride and cationic polymer. The results have been compared with that obtained at optimum dose and are depicted in **Figure 21.** It is evident that in case of under dose of flocculants, the SRF values increases at greater proportion compared to increase in SRF values of optimally conditioned sludge with the increase of pH. For example with primary sludge treated with 4% and 10% ferric chloride, the values of SRF at 6.5, 8.5, and 10.5 pH are respectively 11.5 and 4.3 times, 72 and 7.9 times, 414 and 39 times more than that the SRF value at 4.5 pH.

This shows that pH is of much more importance during conditioning with under dose of flocculants.

8.5 Effect of Storage on Dewatering

The influence of secondary sludge storage on its dewatering performing test on the stored sludge. The results are plotted in **figures 22, 23 and 24** with Ferric Chloride, Cationic polymer and lime respectively.

The increase in SRF values of stored sludge (SRF without flocculants) compared to fresh sludge samples is reflected. For example SRF values of fresh sludge and stored sludge have been found to be 6.71×10^{10} and 1.00×10^{11} s^2/g respectively. It indicates that stored sludge is difficult to dewater compared to fresh sludge. An effect of storage on the sensitivity of conditioning with flocculants was nevertheless evident for sludge with all the flocculants used (FeC13, Lime, Cationic Polymer). With storage, under dosed sludges tends to be increasingly difficult to dewater having higher SRF values. SRF for the sludge conditioned optimally with flocculants is not affected by storage. Thus, it is possible to achieve the same filterability of the optimally conditioned sludge. For example, in case of Ferric Chloride treated fresh and stored sludge, the SRF values at 4%, 6% and 10% (optimum dose previously found) dose have been found to be 13.7 and 10.9 times, 139.8 and 100.5 times, 1455.5 and 1607 times lower respectively compared to initial SRF value without adding flocculants. However, there was slight increase in optimum dose requirement in this case, and this increase is different for different flocculants. Minimum increase is found in case of Ferric chloride and maximum in case of lime. For example, SRF values of 10% Ferric Chloride conditioned sludge are $4.61x10^7$ and $6.21x10^7$ for fresh and stored sludge respectively while in case of 16% lime treated, the SRF values for fresh and stored sludge have been found 7x107 and 1.58x108 respectively. The optimum dose of lime in case of stored sludge has been found to be 22% and 24% compared to 16% with fresh sludge.

CONCLUSION

Primary sludge has lower SRF values compared to secondary sludge whether it is unconditioned or conditioned by various flocculants. This is applicable for any kind of mill is it is wood based or non-wood based. However for the same kind of sludge, the non-wood based mills gives higher SRF values resulting in poor dewaterability due to high fines content in the sludge.

With the increase of solid content, the SRF values increases almost linearly.

The compressibility values indicate that both the sludges are compressible in nature.

With the increase of vacuum, the SRF values increases almost linearly.

Flocculants diminish the SRF values thereby increases the sludge dewaterability.

However, conditioning depends on type of flocculant used. However, the results depend on type of sludges. The optimum dose therefore found to be different for different flocculants for the same constant other parameters like Vacuum, solids content, pH, mixing time and storage.

Among the inorganic flocculants ferric chloride displayed the best results whereas the lime the least. Alum and lime exhibited characteristics which are strongly influenced by the sludge.

Among the cationic and anionic polymers, the cationic polymers give better results whereas anionic does not affects so much.

Data has been extended to a 100 TPD mill and indicate a promise in future for conditioning sludge.

Based on the results obtained experimentally, the following conclusions can also be drawn:

With increase in pH, increase in SRF values is noticed with all the flocculants used except in case of lime treated on primary sludge. However, increase was not very sharp between 5.5 to 7.5 pH values; the amount of dose required will be increased for a particular flocculants for the better dewatering capability. Otherwise, with the optimum dose the results will be poor.

From the results of effect of fast mixing time, it is found that it is very necessary that flocculants must be properly mixed with sludge (higher SRF at no fast mixing or fast mixing for ½ min.). However, too prolong fast mixing will deteriorate the conditioning (higher SRF at longer fast mixing time) because the fibers will not get the time to flocculate and some small flocs which are just developed will also break. Similarly, from the results of effect of slow mixing time on conditioning, it is found that there is particular time for which slow mixing (flocculation) is to be carried out. If the slow mixing time is much low than the optimum required, the SRF of sludge will be higher due to poor flocculation. Longer slow mixing time (upto 4-4.5 min), however, has not shown severe effect on SRF in the present study. But excessive variation may severely deteriorate the dewatering capability.

Optimum value of operating variable (pH, rapid/slow mixing time etc.) is of paramount importance during conditioning with underdose of flocculants.

Operating variables have influence in almost equal manner in case of both wood and non-wood ETP sludge.

The stored sludge dewaters poorly. However, it can be dewatered to the same level at its optimum dose or at dose slightly higher than the optimum dose used for fresh sludge. The underdosage of flocculants will not improve the results and much over dosage will not also provide any extra benefits.

DESIGN PARMETERS EVALUATED & FORMULA USED

For estimation of the design parameters, certain additional calculation is necessary. These are for the loading rate (L), flocculation paddle, air suction rate and power requirement for a 100 TPD plant. The design equations are well known. The equations and the value estimated are given below.

Sludge	**Unconditioned**	**Conditioned**
Vacuum filter loading rate, $kg/m^3/h$	0.75	1.29
Filter area, m^2	250	20
L(actual)= $(2\ Pck_f / urt_C)^{1/2}$		
A(TS to be proceed/L)	15ft x12 ft	10ft x 7ft
Storage(2 days) Tank requirement(Sludge Flow Rate/Storage period), m^3	160	160

Conditioning Tank Requirement(Sludge Volume to be handled per day x retention time)(40% extra),m^3	-	0.55
Power requirement of pump required for transportationof sludge from storage tank, $P = (QP_dGh) / (1000xft)$,kW	0.4	0.4
Area required for flocculant Paddle, $P = 2\ P/C_d\ Pd^{V3}$,m^2	-	0.04
Air Suction Rate and Power requirement of vacuum pump volume of air per unit time=weight of cake per unit time $x(k_a\ \mu r/k_f \mu_a 2 r_a c)$,m^3/h	33.18	22.57
$P = G\ W_{ad} / 3600 * 1000$ $W_{ad} = [k/ (k-1)]\ p_1 V_1\ [(p_2/p_1)^{(k-1)/k-1]}$ kW	0.73	0.31

REFERENCES

1. Dhingra H.K.," Dewatering Characteristics of Waste Water Treatment Sludges and Effects of Flocculants", M.E. Dissertation submitted at the institute of Paper Technology, University of Roorkee, India 1994.
2. Banerjee Sujit," Energy from Pulp Mill Sludge",Inpaper International,October-December,2010, PP.10-12
3. Vesilind, P.A., Treatment and disposal of wastewater sludges, 2nd Ed. Ann Arbor Science Pub., Ann Arbor Michigan, 1979.
4. Peters, M. S. and Timmerhaus, K.D., Plant design and economics for chemical engineers, 2nd Ed., McGraw Hill Kogakusha Ltd., Tokyo, 1968.
5. McCabe, W.L. and Smith, J.C., Unit operation of chemical engineering, 2nd. Ed., McGraw Hill Kogakusha Ltd. Tokyo, 1967.
6. Gale, R.S., Wat. Poll. Cont., "Filtration theory with special reference to sewage sludges", 66: 622-632 (1967).
7. Kavanagh, B.V., Wat. Poll. Cont. "The dewaterability of activated sludge: Measurement of specific resistance to filtration and capillary suction time", 79:388-398 (1980).
8. National Council for Air and Stream Improvement, Full scale operational experience with filter presses for sludge dewatering in the North American P & P industry, Technical Bulletin No. 299, October, 1971.
9. Ekenfelder, W.W. and Conner, O., Biological waste water treatment, Pergamon Press, 1961. LMO

NOMENCLATURE

V = Volume of filtrate delivered in time t (m^3)

dV/dt = Rate of filtration

ΔP = Driving force = pressure difference across filter (P is numerically equal to applied vacuum in gauge, because $P = P_{vacuum} - P_{atmospheric}$ Hence, it is indicated as P = applied vacuum)

R_C = Cake resistance to the flow of filtrate

R_M = Medium resistance to the flow of filtrate

A = Area of filtrate (Ns/m^2)

μ = Viscosity of filtrate (Ns/m^2)

r = C/P_C and is known as specific cake resistance (m/kg) or (k/g)

r_0 = A constant

s = A constants, known as coefficient of compressibility of the cake.

s_1 = Slope of the straight line.

i = Intercept of the straight line.

c_C = Cake solids concentration, %

c_i = Feed solids concentration, %

L = Filter loading rate $g/m^2/s$.

Weight of dry cake per unit time per unit area)

c = Mass of dry cake deposited per unit volume of filtrate (g/m^3)

k_f = Fractional filter drum submergence

P = Power (in KW) required by the motor of a single stage compressor

G = Kg of gas per hour

P_1 = Initial pressure

P_2 = Final pressure

W_{ad} = Theoretical amount of work done (in j / kg) by a single stage compressor in the Adiabatic (isentropic) compression of 1 kg of a gas

k = Adiabatic exponent equal to the ratio C_P / C_V (the specific heat capacities at constant pressure and constant volume)

V_1 = Specific volume of gas in the initial conditions i.e. at a pressure of P_1 and a Temperature of T_1 (m^3 / kg)

μ_a = Viscosity of air at temperature and pressure of surroundings

r_a = C' /P_C and is known as the specific air suction cake resistance

Q = Volume of output (delivery) through pump

P_d = Density of slurry being pumped,

g = Acceleration due to gravity 9.81 m /s^2

H = Total head developed by the pump (m)

∩ = Overall efficiency

P = Power requirement for mixing, (Watts)

A = Area of paddle, m^2

v = Relative velocity of paddle in fluid, (m /s) usually 0.7 – 0.8 of paddle tip speed, v_P

C_D = Coefficient of Drag of flocculator paddles moving perpendicular to the fluid. For rectangular paddles this is about 1.8.

P_d = Mass fluid density (kg / m^3)

Fig:-1 Specific Resistance to filtration (SRF) values of unconditioned primary and Secondary Sludge.

SRF (Sec^2/g)

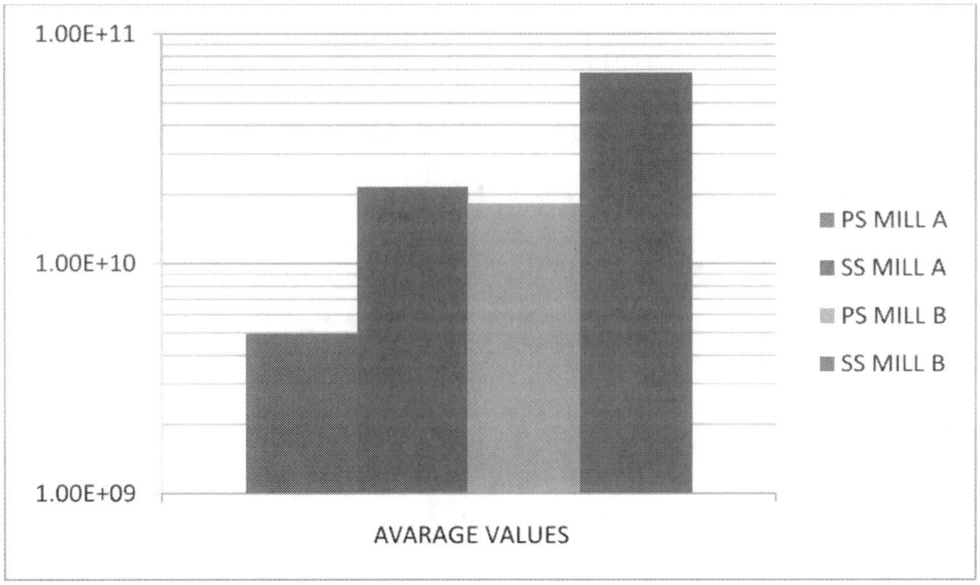

Fig:-2 Effect of Pressure Drop on SRF of Sludge (Mill A)

SRF (Sec^2/g)

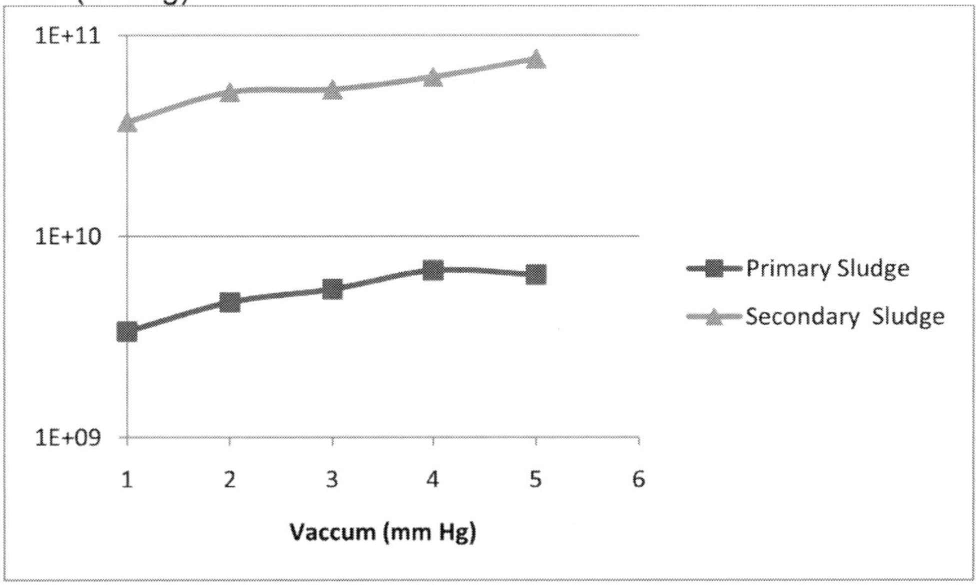

Fig:-3 Effect of Consistency (TS) of sludge on SRF

SRF (Sec^2/g)

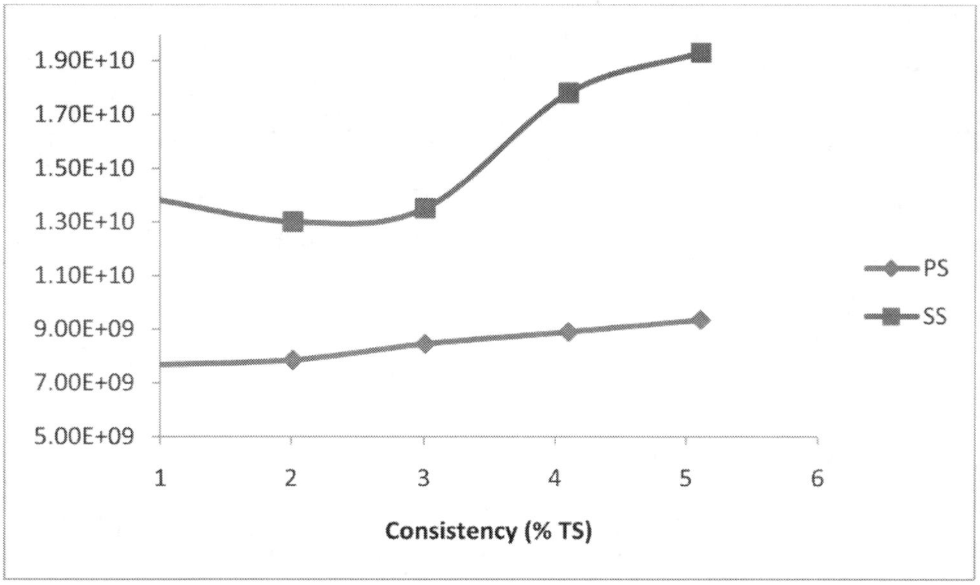

Fig:-4 Primary Sludge of Mill A

SRF (Sec^2/g)

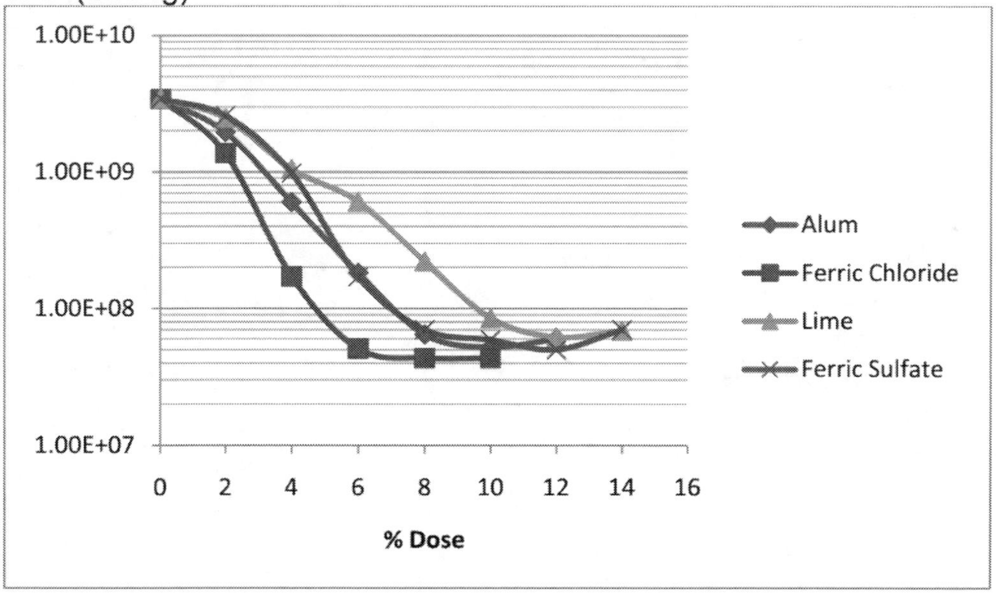

Fig:-5 Secondary Sludge of Mill A

SRF (Sec^2/g)

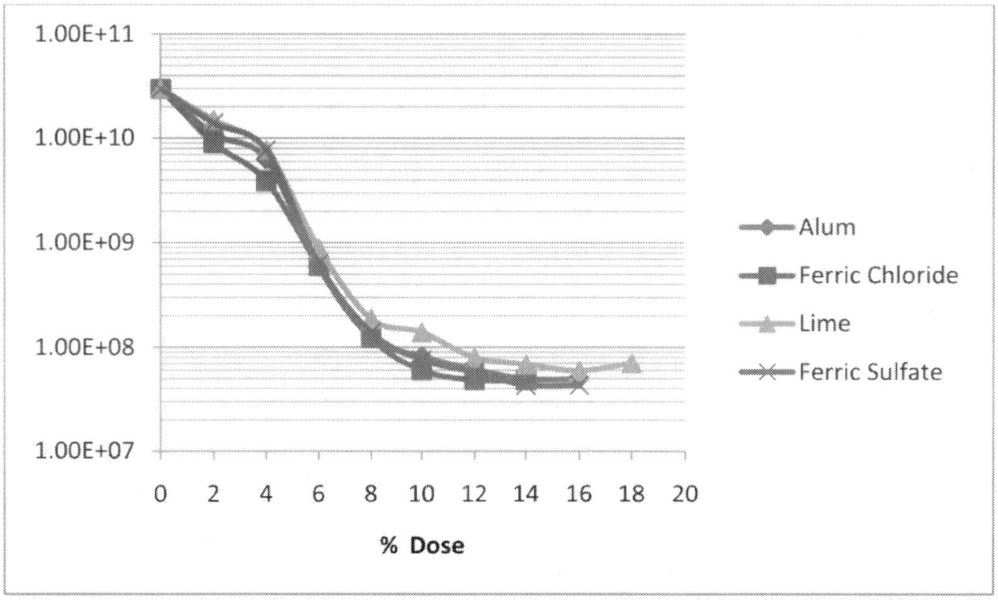

Fig:-6 Primary Sludge of Mill B

SRF (Sec^2/g)

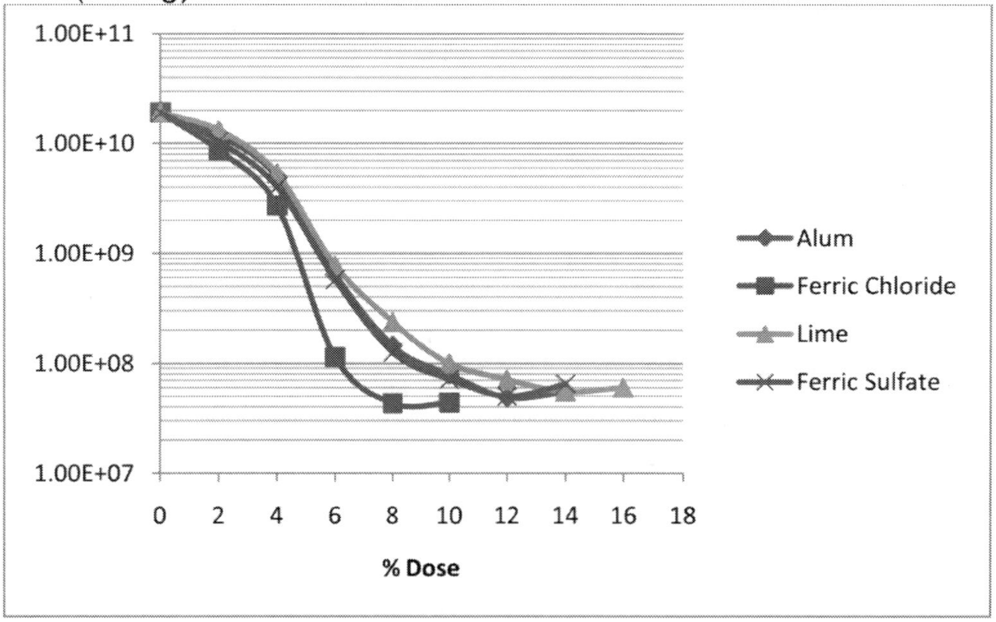

Fig:- 7 Secondary Sludge of Mill B

SRF (Sec^2/g)

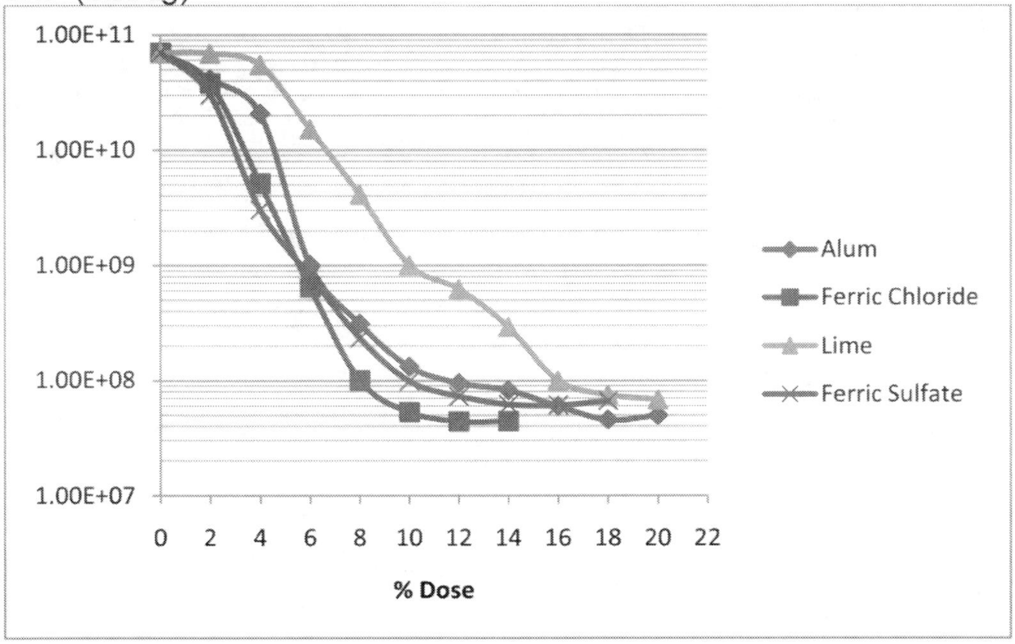

Fig:-8 Primary Sludge of Mill A

SRF (Sec^2/g)

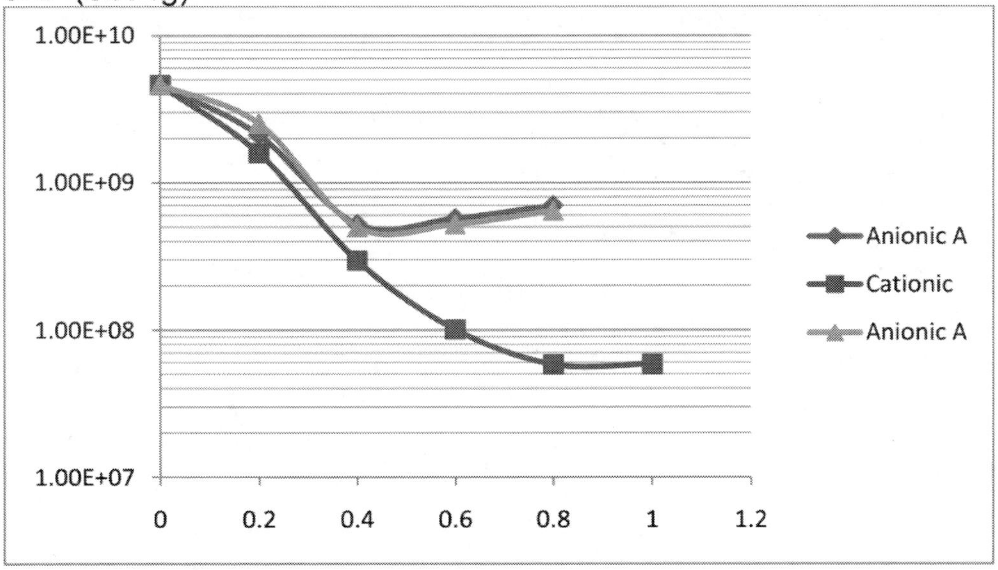

Fig: - 9 Secondary Sludge of Mill A

SRF (Sec^2/g)

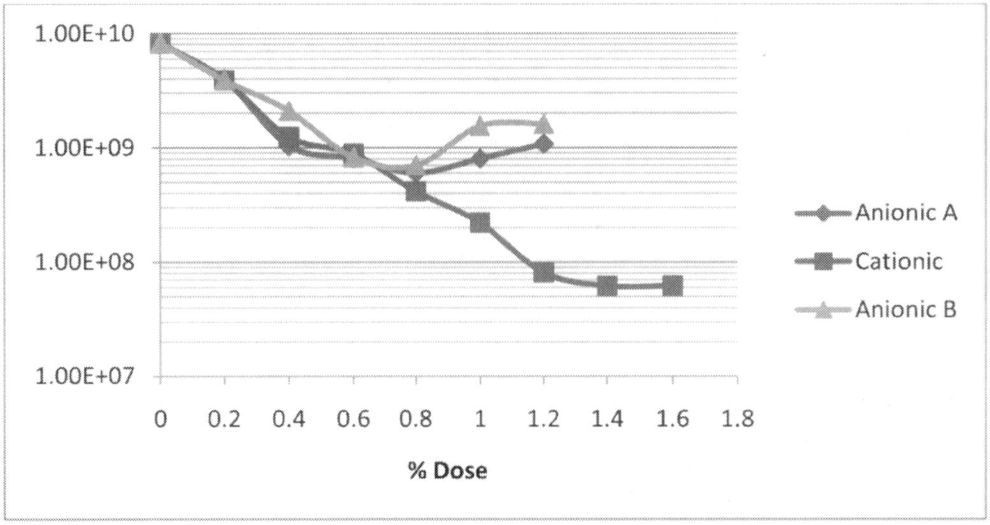

Fig :-10 Primary Sludge of Mill B

SRF (Sec^2/g)

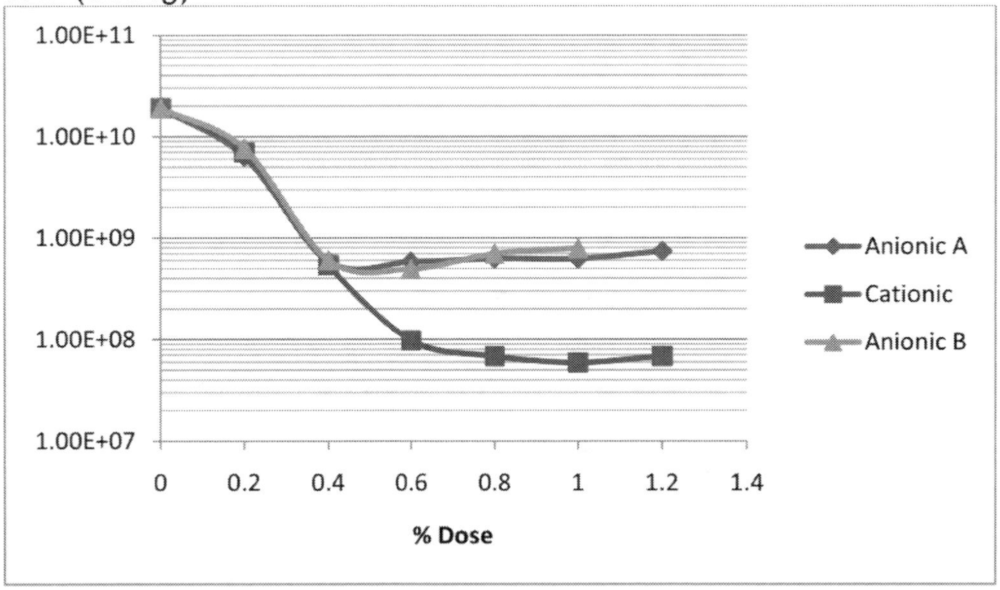

Fig :-11 Secondary Sludge of Mill B

SRF (Sec^2/g)

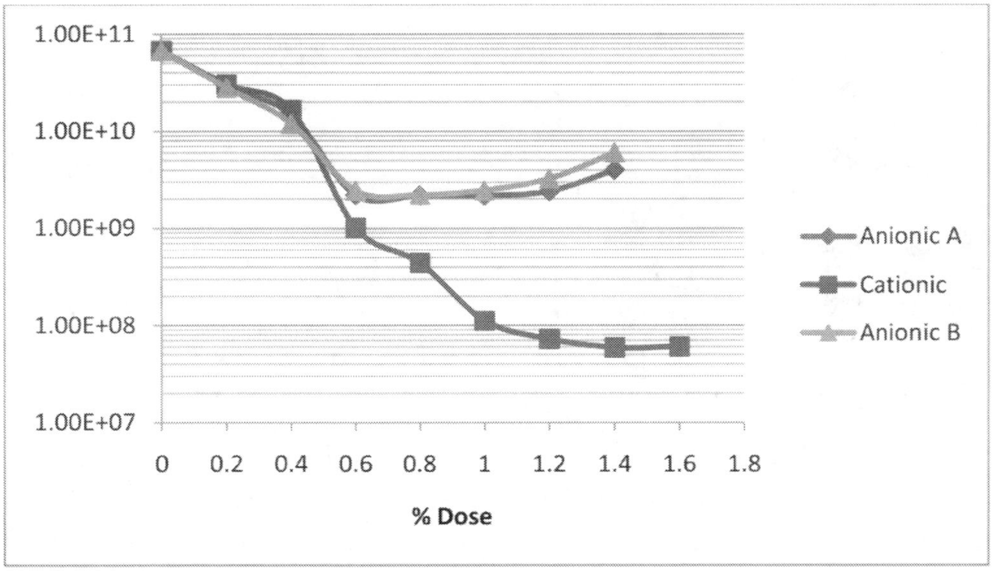

Fig :-12 Effect of Pressure Drop on SRF ($FeCl_3$, Conditioned Primary Sludge)

SRF (Sec^2/g)

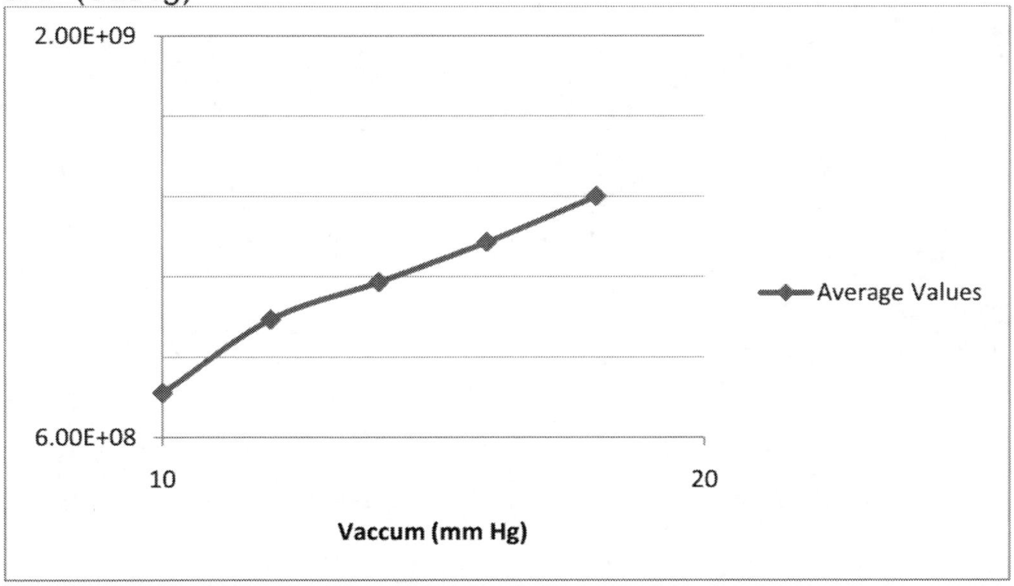

Fig: - 13 Optimum Dosage of Different Flocculants for Different Sludges.

% Optimum Dosage

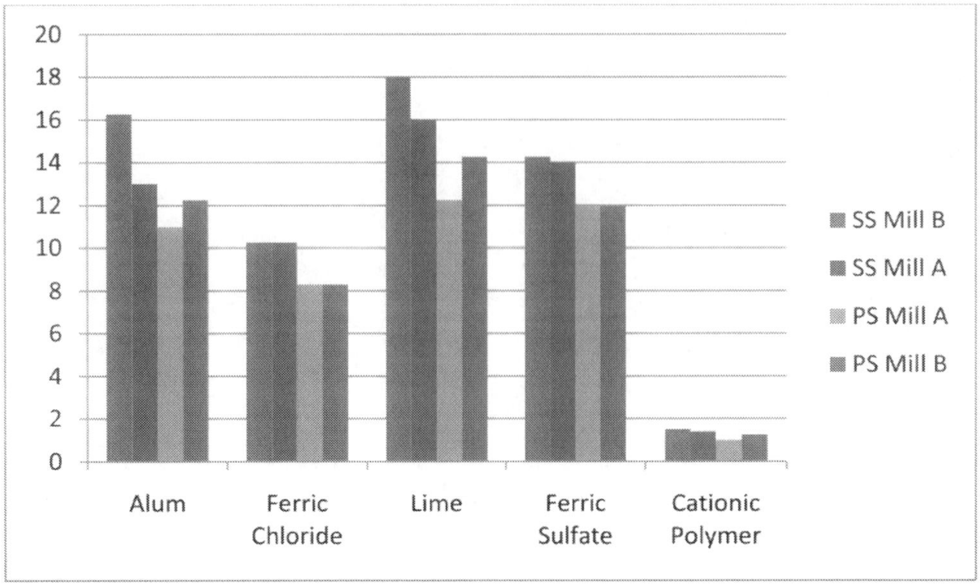

PS = Primary Sludge

SS = Secondary Sludge

Fig:-14 Effect of varying Rapid Mixing Time on SRF during conditioning with Flocculants (Slow Mixing Time two min. kept constant) (Secondary Sludge Mill A)

SRF (Sec^2/g)

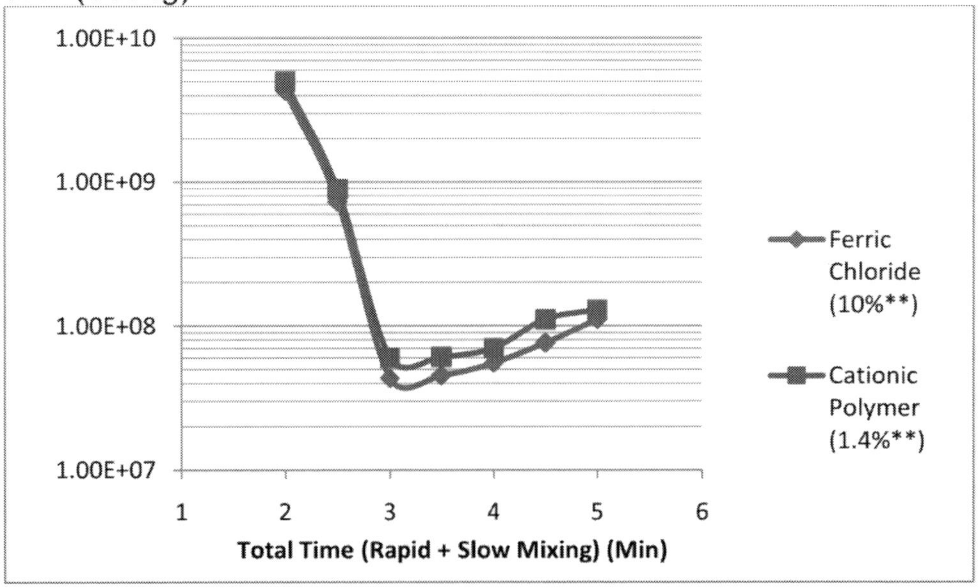

Fig:-15 Effect of varying Rapid Mixing Time on SRF during conditioning with Flocculants (Slow Mixing Time two min kept constant) (Secondary Sludge Mill B)

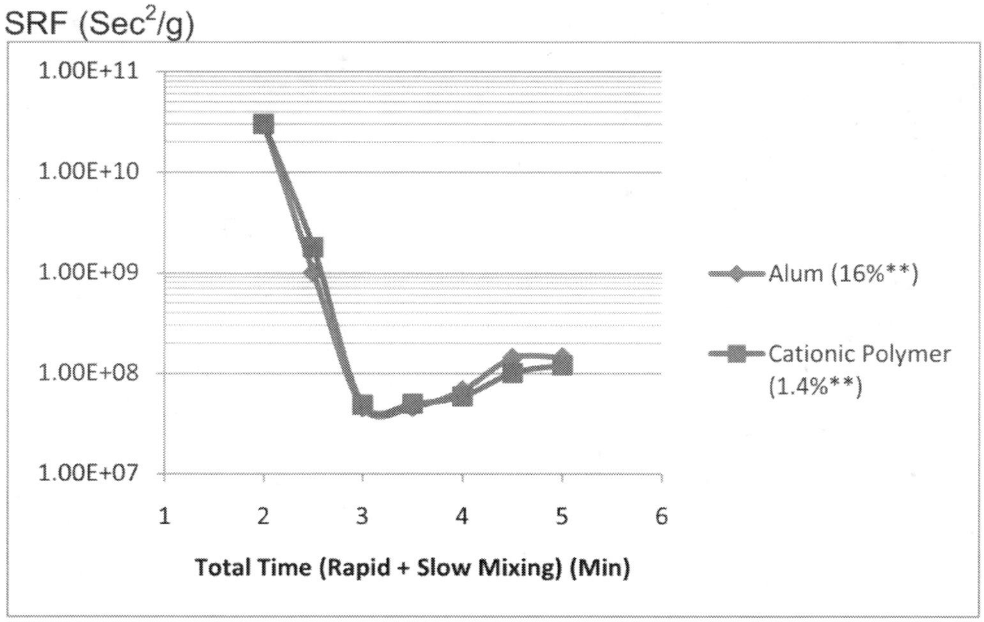

Fig:-16 Effect of varying Slow Mixing Time on SRF during conditioning with Flocculants (Rapid Mixing Time one min kept constant) (Secondary Sludge Mill A)

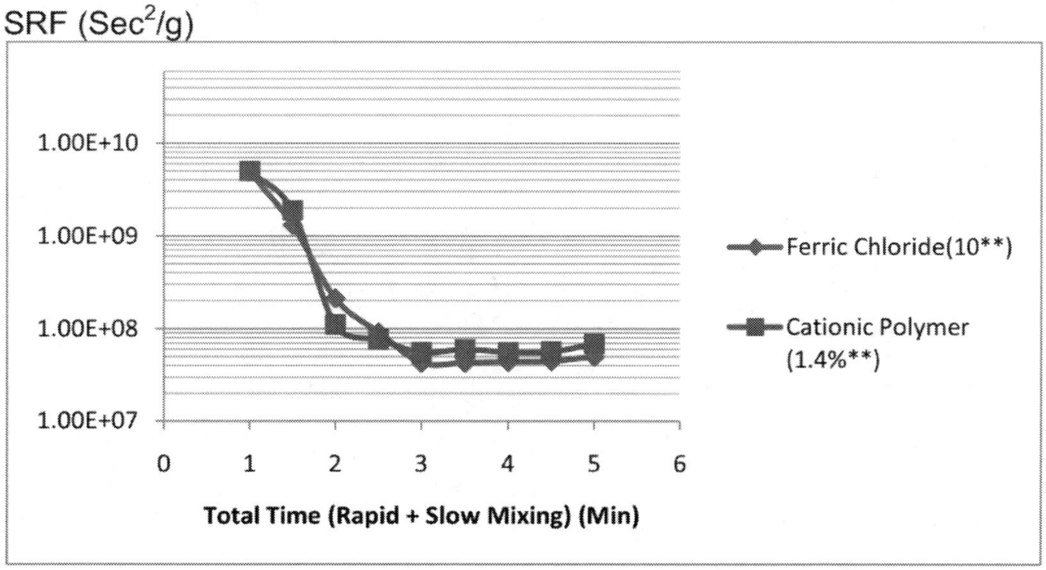

Fig:-17 Effect of varying Slow Mixing Time on SRF during conditioning with Flocculants (Rapid Mixing Time one min kept constant) (Secondary Sludge Mill B)

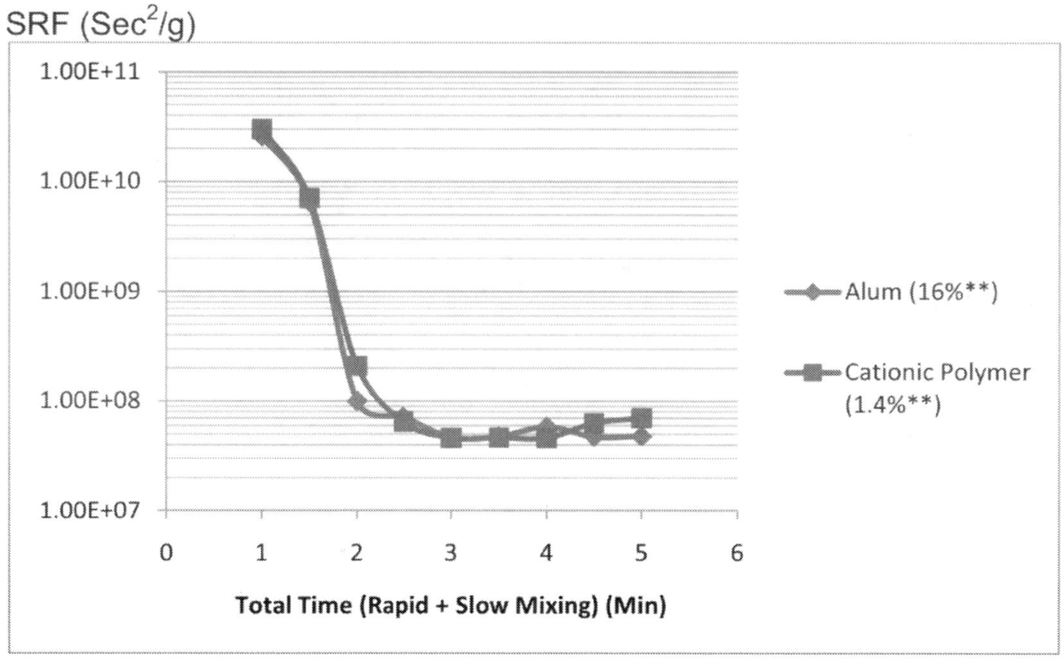

Fig:-18 Effect of varying Rapid Mixing Time on SRF during conditioning with Different dose Flocculants (Optimum* and Less than Optimum**) (Secondary Sludge Mill A)

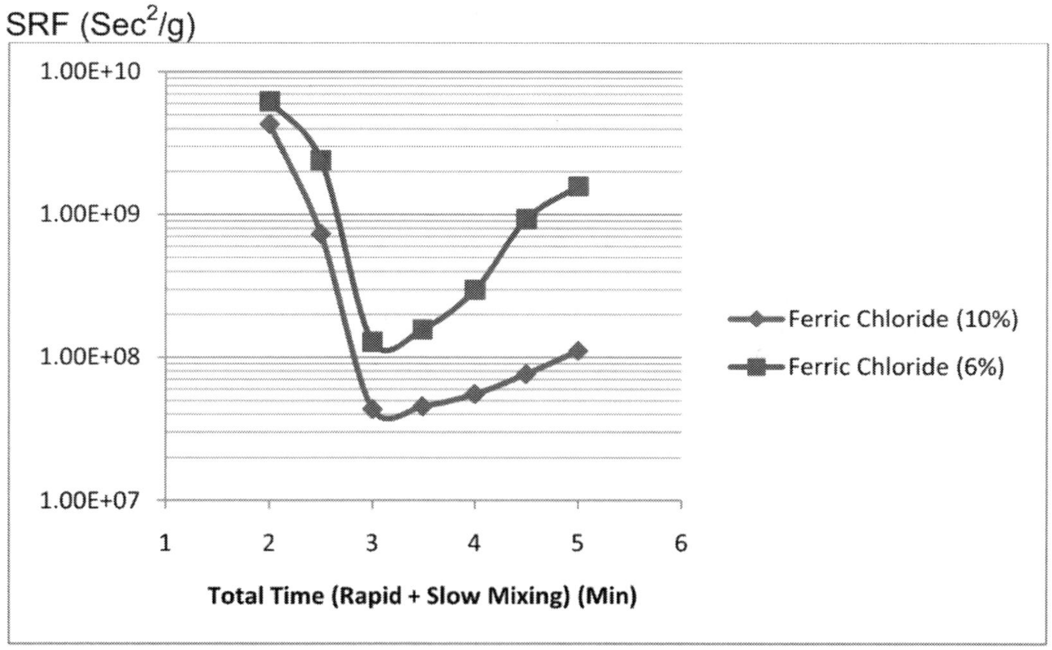

Fig:-19 Effect of varying pH on SRF during conditioning with Flocculants. (Secondary Sludge Mill A)

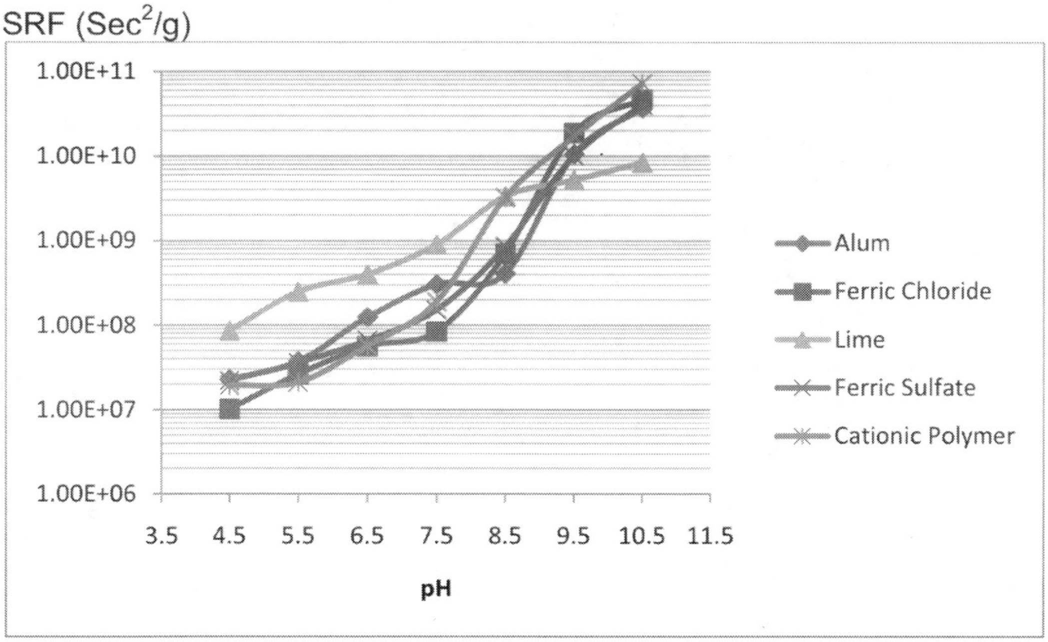

Fig:-20 Effect of varying pH on SRF during conditioning with Flocculants. (Primary Sludge Mill A)

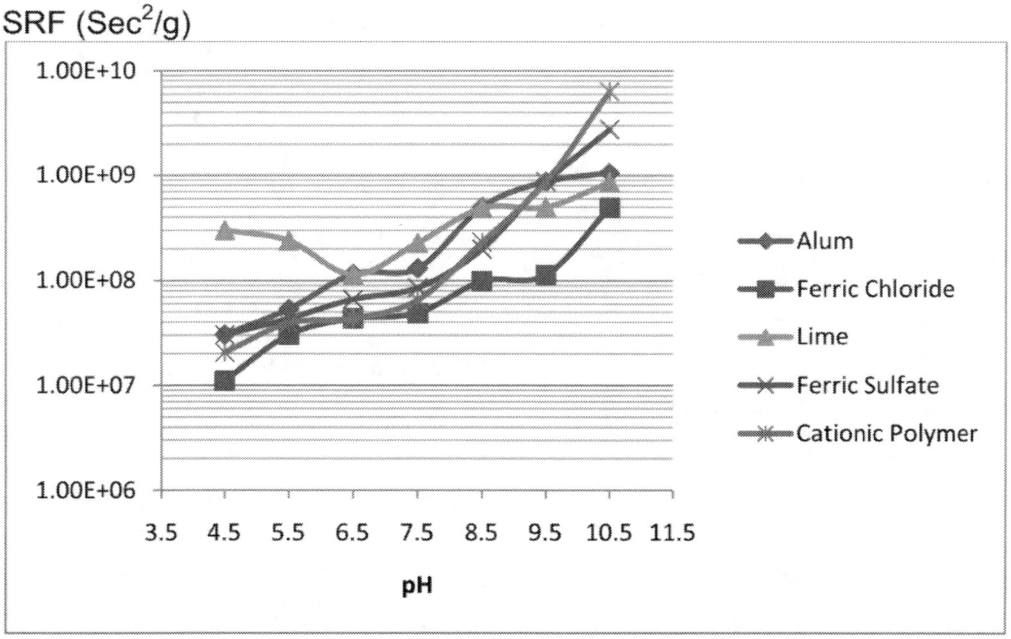

Fig: - 21 Effect of varying pH on SRF during conditioning with Different dose Flocculants (Optimum* and Less than Optimum**) (Secondary Sludge Mill A)

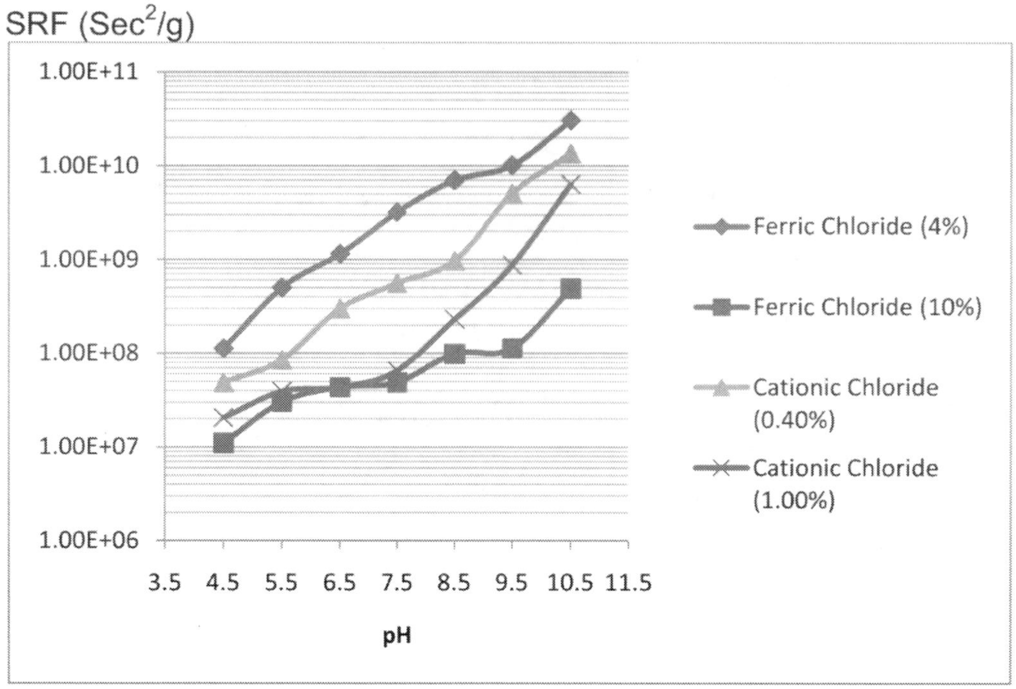

Fig: - 22 Effect of Storage on Dewaterability and on conditioning with different dose of Ferric Chloride (Secondary Sludge Mill B)

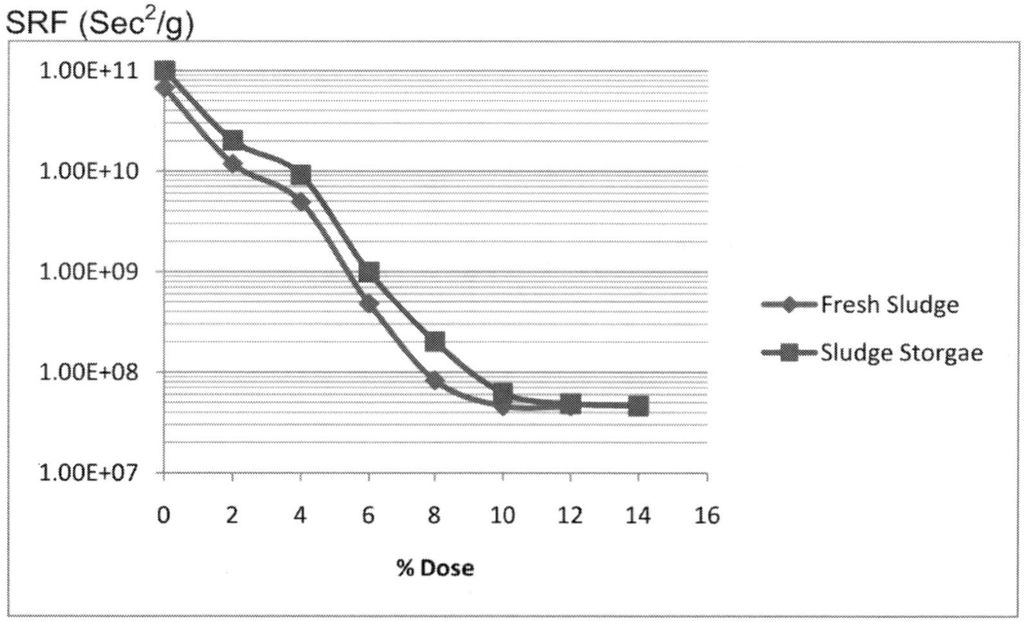

Fig:-23 Effect of Storage on Dewaterability and on conditioning with different dose of

Cationic Polymer (Secondary Sludge Mill B)

SRF (Sec^2/g)

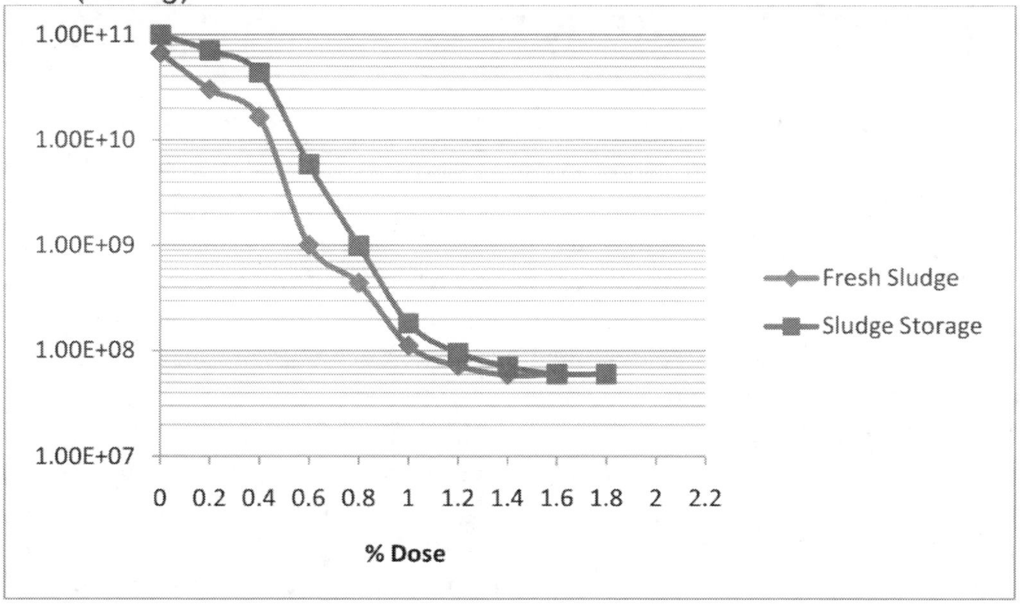

Fig: - 24 Effect of Storage on Dewaterability and on conditioning with different dose of Lime (Secondary Sludge Mill B)

SRF (Sec^2/g)

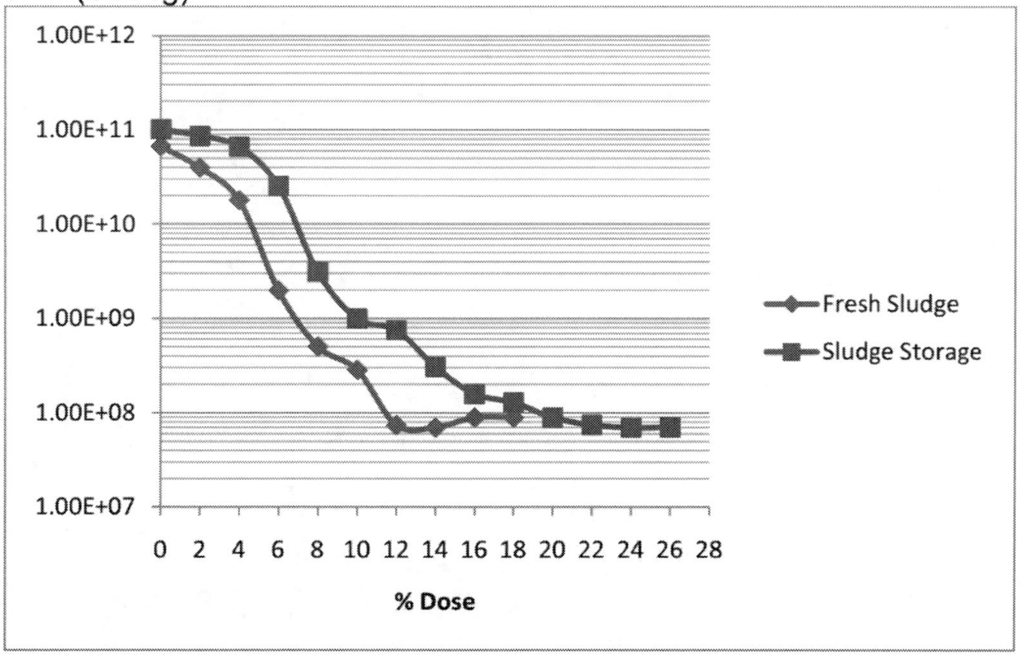

(264a) Enhanced Food Waste Hydrolysis for Integrated Biofuel Production

Jingwei Ma[1], Liang Yu[2], Yubin Zheng[1] and Shulin Chen[3], (1)Department of Biological Systems Engineering, Washington State University, Pullman, WA, (2)Washington State University, Pullman, WA, (3)Biological Systems Engineering, Washington State University, Pullman, WA

In recent years, there has been an increasing interest in using anaerobic digestion as a means of reducing the volume of municipal solid waste (MSW) and generating methane biogas for energy recovery. In order to increase process efficiency of food waste treatment and obtain valuable product, hydrolysis of food waste were enhanced by optimizing pH and total solids (TS), then two strategies of volatile fatty acid (VFA) utilization for biofuel production were proposed in this study. One of the strategies was methanization of VFA abundant liquid in high rate anaerobic digester, such as up-flow anaerobic sludge bed reactor (UASB) as the methanogenic reactor. Another strategy was production of single cell oil (SCO) or microbial lipid by oleaginous yeast *Cryptococcus curvatus* to be used as biodiesel feedstocks. High concentration of VFA production was the optimizing objective during the hydrolysis of food waste. After 20 d incubation time at pH 6.0 and 12% TS, VFA concentration, VFA yield and volatile solids (VS) reduction were reached to 53.9g/L, 0.38gVFA/gVS and 59.2%, respectively. Biofuel production of two strategies was compared and the mass balance of two systems was calculated. At steady-state, methane production rate of 2.28L/L/d and lipid production of 1.1g/L/d were achieved in two systems, respectively. Ammonia and phosphor recovery were added into these two systems to produce valuable products and alleviate inhibition.

See more of this Session: Sustainable Fuel From Renewable Resources I
See more of this Group/Topical: Environmental Division

(264b) Biodiesel Production by Utilizing by Products From Fish Oil Concentration

Martin Ernst, R&D BioFuels, BDI BioEnergy International AG, Grambach/Graz, Austria

Due to the health-promoting properties of polyunsaturated fatty acids (PUFA) shown in numerous medical studies, omega 3 concentrates based on fish oil are enjoying increasing popularity in the food industry. According their occurrence in natural fish oils, concentrates with high concentrations of PUFAs of Eicosapentaenoic acid (EPA) and Docosahexaenoic acid (DHA) are used. The applications range from nutritional supplements and food supplements up to pharmaceutical preparations. For large scale industrial production, high-quality fish oils with concentrations about 30% in total in EPA/DHA are commonly used for the production of EPA/DHA ethyl ester and triglyceride concentrates. For concentration to desired EPA/DHA amounts the fish oil is transesterified to ethyl esters. Adjacent to the transesterification the EPA and DHA esters are concentrated by molecular distillation. During the concentration step more than 60% of the starting ethyl ester amount forms the so called ethyl ester light fraction which is low in content EPA and DHA. Due to the low content of poly unsaturated fatty acids, this light fraction is usually thermal utilized. However, since the quantities of high-quality fish oils are limited, there is consequently an increasing trend also using fish oils with considerably lower concentrations of these omega-3-fatty-acids. Naturally by converting fish oils which are low in EPA/ DHA content also the amount of ethyl ester light fraction will increase. Therefore a further upgrade of this light fraction to higher value products is important to improve the overall economics.

Target of the ongoing research is the evaluation and necessary treatment of ethyl ester light fraction for utilization as alternative fuel. Therefore fish oil with lower content of polyunsaturated fatty acids was transesterified to ethyl ester. By molecular distillation this ester was used to produce marketable ethyl ester concentrates. The resulting ethyl ester light fraction with low contents of PUFAs was evaluated for its ability to produce biodiesel according ASTM standard. Compared to conventional methyl ester several parameter of the light fraction has to be improved to fulfil the ASTM standard. Especially the quality parameters of free glycerine, carbon residue and atmospheric residue are standard parameters out of specification. By utilizing this side stream, a renewable fuel could be generated, offering also the opportunity to recover remaining PUFAs in the light fraction.

See more of this Session: Sustainable Fuel From Renewable Resources I
See more of this Group/Topical: Environmental Division

(264c) A Sustainable Energy Campus

Earl R. Beaver, Practical Sustainability, LLC, Chesterfield, MO

The Sustainable Energy Campus combines multiple technologies from different industries to create a truly sustainable "waste-to-profit" campus where the waste from one technology being used becomes the input feedstock to another technology that utilizes that waste to create a full spectrum of highly profitable fuel products. All the while, greenhouse gases are reduced, materials are recovered, energy is optimized and virtually no emissions are released to the environment. Further, the reality is highly profitable immediately and for the long term. Thus, genuine waste-to-profit.

Technologies from the rubber, chemical and petroleum industries are at play in the Sustainable Energy Campus. A Carbon Black recovery production line takes scrap automobile and truck tires and uses a vacuum pyrolysis process to convert those tires into a product called Carbon Black Alternative and a stream of pyrolysis oil. Carbon Black is present in 75-85% of all new tires produced worldwide. A new cracking and reaction technology combined with a sophisticated separation process converts the pyrolysis oil into raw materials for gasoline, kerosene, diesel, naphtha, jet A fuel base and liquified low molecular weight hydrocarbons (propane, butane and pentane). While each step can be profitable independently, combining them in a single site with cogeneration allows for energy and materials optimization as well as maximizing revenue. Further, minor components are captured and isolated which have extremely high value. The key materials are cyclic terpenes and high been recovered at high yields and high purities.

Since a single scrap tire weighs approximately 21 pounds and one tire is produced per year per person in industrialized countries, the Sustainable Energy Campus provides a means of producing massive amounts of valuable products which otherwise would be produced from incremental fossil fuels and raw materials at significant cost to society and the environment. A full scale tire pyrolysis unit utilizing this teechnology has been in operation for more than one year and has produced more than ten million pounds of carbon black along with the co-products cited above. The data related to mass of products, energy savings and net environmental impact will be presented and discussed.

See more of this Session: Sustainable Fuel From Renewable Resources I
See more of this Group/Topical: Environmental Division

(264d) Air Quality Impacts of Achieving U.S. Renewable Fuel Mandates

Kristina M. Wagstrom[1], Christopher Tessum[1], Jason Hill[2] and Julian Marshall[1], (1)Department of Civil Engineering, University of Minnesota - Twin Cities, Minneapolis, MN, (2)Department of Bioproducts and Biosystems Engineering, University of Minnesota - Twin Cities, St. Paul, MN

The current US Renewable Fuel Standard (RFS2) requires increasing amounts of biofuel production through 2022. In this analysis we use lifecycle assessment, air dispersion modeling, and health risk assessment to estimate air pollution and related health impacts of the production and use of an additional 7.5 billion gallons of corn grain ethanol and 5 billion gallons of corn stover cellulosic ethanol.

Specifically, we considered three potential increases in ethanol volumes associated with moving from the Renewable Fuel Standard – Phase 1 (RFS1) to Phase 2 (RFS2): (i) a 5 billion gallon increase in cellulosic (corn stover-derived) ethanol, (ii) a 7.5 billion gallon increase in corn-derived ethanol, and (iii) a combination of the two (5 billion gallons of cellulosic ethanol plus 7.5 billion gallons of corn ethanol).

We use the Greenhouse Gases, Regulated Emissions and Energy Use in Transportation Model (GREET) to estimate the life cycle emissions associated with the production and use of both the ethanol fuels and an energy equivalent amount of gasoline. Temporal and spatial information was also utilized in order to allocate the lifecycle emissions as necessary for photochemical modeling. The Sparse Matrix Operator Kernel Emissions (SMOKE) model was used to calculate the base case emissions. We use the CAMx reactive photochemical air quality model to estimate the resulting changes in ozone and $PM_{2.5}$ concentrations, relative to a non-RFS scenario (i.e., relative to consuming an energy-equivalent amount of gasoline), for the entire United States using 2005 as a base year. Finally, we use BenMAP to evaluate the potential health impacts from the anticipated air quality changes.

Preliminary results for February and August indicate that population-weighted ambient $PM_{2.5}$ concentration increases are up to 189% higher for RFS than for the non-RFS scenario; analogous values for nitrate $PM_{2.5}$ are 15%-150% higher for RFS. Concentration impacts vary in space and time; predicted concentrations exhibit greater spatial variability for RFS than for the non-RFS scenario. Preliminary findings from the $PM_{2.5}$ health risk assessment suggest greater impacts from RFS2 scenarios than from an energy equivalent amount of gasoline.

See more of this Session: Sustainable Fuel From Renewable Resources I
See more of this Group/Topical: Environmental Division

(264e) Multi-Objective Footprints-Based Synthesis of Integrated Bioethanol Production Systems

Lidija •u•ek and Zdravko Kravanja, University of Maribor, Faculty of Chemistry and Chemical Engineering, Maribor, Slovenia

Abstract

Energy consumption in the transport sector depends almost exclusively on fossil fuels – oil. The forecasts show a growing tendency of this sector in the following years. The price of oil on global markets continues to remain high. Peak oil production is predicted to start shortly or even that it has been already started (Narodoslawsky, 2007; Robelius, 2007; The Oil Drum). At the same time, a significant rise in carbon dioxide emissions from the burning of fossil fuels is raising the earth's temperature and is causing climate change (Hamilton, 2008). Therefore it is important to take actions towards energy efficiency, emissions reductions, fuel supply security, and low-carbon economy.

Biomass is renewable only if production is sustainable. Also, different research studies have showed that lower carbon footprints are achieved by utilizing biomass, when compared to energy generated from fossil sources. On the other hand higher amounts of nitrogen are emitted into air and water (Cherubini and Strømman, 2011; Bauer, 2008; •u•ek et al., 2011).

This contribution focuses on the simultaneous integration of different technologies for converting starchy and lignocellulosic raw materials to bioethanol. Different paths are selected: biochemical, thermo-chemical and thermo-biochemical and integrated in one superstructure so that operating costs, energy and agricultural land are minimized and profit is maximized. The ethanol production is assumed to be centralized. Raw materials are assumed to grow on the agricultural area of the maximum of 50 000 hectares. The fossil energy used in the supply chain of bioethanol production is also calculated and analyzed. The degree of (un)sustainability of the bioethanol production system is evaluated by the carbon and nitrogen footprint and compared to gasoline supply chain. Multi-objective optimization is performed on bioethanol supply chain where net footprints are minimized. The optimization of the system is formulated as a mixed-integer non-linear programming (MINLP) problem. The problem is solved by the use of the Mixed-Integer Process Synthesizer MIPSYN using mass and energy balances and other constraints with profit maximization as the optimization criterion.

References

Bauer C., 2008, Life Cycle Assessment of Fossil and Biomass Power Generation Chains. An analysis carried out for ALSTOM Power services, PSI-report No. 08-05, Paul Scherrer Institute, Villigen, PSI,

Switzerland

Cherubini F., and Strømman A. H., 2011, Life cycle assessment of bioenergy systems: State of the art and future challenges, Bioresource technology, 102, 437-451

•u•ek L., Klemeš J. J., and Kravanja Z., 2011, Overview of footprints and relations between carbon and nitrogen footprints, Chemical Engineering Transactions, 25, 923-928, DOI: 10.3303/CET1125154

Hamilton L., 2008, Firewood and Woody Biomass and their Role in Greenhouse Gas Reduction, Agriculture Notes < www.homeheat.com.au> Accessed: April 2011

Narodoslawsky M., 2007, Sustainable Processes – The Challenge of the 21st Century for Chemical Engineering, Proceedings of European Congress of Chemical Engineering (ECCE-6)

Robelius F., 2007, Giant Oil Fields -The Highway to Oil, Giant Oil Fields and Their Importance for Future Oil Production, Acta Universitatis Upsaliensis, Uppsala, Sweden

The Oil Drum <www.theoildrum.com> Accessed: April 2011

See more of this Session: Sustainable Fuel From Renewable Resources I
See more of this Group/Topical: Environmental Division

(264f) Dilute Acid Pretreatment of Farm-Waste Cultured Algae and Its Effect On Enzymatic Hydrolysis of Cellulose

Rui Chen, Zhengbo Yue, Lauren A. Deitz, Jin Wang, Yan Liu and Wei Liao, Biosystems and Agricultural Engineering, Michigan State University, East Lansing, MI

The objectives of this study are: (1) to determine the optimal conditions of hydrolysis of farm-waste cultured algal biomass using dilute acid method, and (2) to investigate the effects and correlation of algal hydrolysate and enzyme loadings on enzymatic hydrolysis of pretreated lingocellulose. We statistically studied the effect of reaction conditions of acid concentration, temperature, and reaction time on the accumulation of total sugar released in dilute sulfuric acid treatment of AD effluent cultured algal biomass. Based on the regression model, the optimal conditions for sugar accumulation were 4% H_2SO_4 (w/w), 116°C reaction temperature and 30 min reaction time. Under these conditions, sugar concentration reached 28.18% of total biomass (DW), which is 99% of total sugar in algal biomass. Furthermore, the algal hydrolysate was applied as medium in the enzymatic hydrolysis of alkaline pretreated AD fiber, it demonstrated a significant increase ($P < 0.001$) in glucose yield compared to the control of hydrolyses using citrate buffer (conventionally employed in enzymatic hydrolysis) and deionized water. This study presented a potential pathway of integrated utilization of both liquid (algal cultivation) and solid (enzymatic hydrolysis) wastes from aerobically digested dairy manure for biofuels production. Implementation of such process will provide a mutually beneficial solution for agricultural operations, bioenergy production, and environmental protection.

See more of this Session: Sustainable Fuel From Renewable Resources I
See more of this Group/Topical: Environmental Division

(264g) High Quality Biodiesel Fuel Production From Crude Jatropha Oil without Upstream and Downstream Processing

Naomi Shibasaki-Kitakawa, Kei-ichi Kanagawa, Kazunori Nakashima and **Toshikuni Yonemoto**, Department of Chemical Engineering, Tohoku University, Sendai, Japan

1. Introduction

Jatropha curcas oil has beenreceived much attention as a renewable source of biodiesel fuel with no competing food uses. Crude *Jatropha* oil hashigh contents of free fatty acid (FFA) andwater. FFA and water inthe raw oil significantly decrease the yield and quality ofthe product biodiesel in the industrial production process using the homogeneousalkali catalyst. Thus, the present production process is usually accompaniedby a refining processto remove FFA in the raw oil to less than 0.5 wt% and a dewatering process of the raw oil. In addition, the downstream purificationprocess to remove the by-products, such as soap and glycerin, from the biodieselis required to satisfy the fuel's quality requirements. These upstreamand downstream processes raise the production cost of the biodiesel fuel.

In this study, the bench-scale productionsystem connected the expanded-bed reactor packed with the cation-exchangeresin and that packed with an anion-exchange resin in series was constructed. The cation-exchangeresin has a catalytic ability to convert FFA to biodiesel by esterification reaction without inhibition by water[1]. The anion-exchange resin has acatalytic ability to convert glycerides to biodieselby transesterification reaction without soapformation[2] as well as an absorbent ability to remove the impuritiessuch as glycerin, water and pigment contained in the product[3]. Thus, high quality biodiesel fuelwould be produced from the crude *Jatropha* oil without any pretreatment of the raw oil and the purification of the products.

2. Experimental methods

Awater-jacketed column packed with the cation-exchangeresin, Diaion PK208LH, and two columns packed with theanion-exchange resin, Diaion PA306S, were connectedin series in the bench-scale production system (Fig.1). The temperature of each column was keptconstant at 50 °C using a hot-water recirculatingsystem. The mixed-solution of thecrude *Jatropha*oil (FFA content of 2 wt%, water content of 3300 mg/kg) and methanol at the stoichiometric molar ratio of methanol to the total fattyacid residue in the oils was supplied to the bottom of the first column at theconstant flow rate. The effluent solutions from the top of thecolumns were collected, and the concentrations of the reactants and products inthe solution were determined.

3. Results and discussion

Figure 2 shows the HPLCchromatograms of the feed (a) and the effluents from the first column (b), thesecond column (c), and the third column (d). Several peaks of the FFA, diglyceride (DG), and triglyceride (TG) were observed in thefeed. In the effluent from thefirst column, the peaks of the FFA completely disappeared and the peaks of thefatty acid methyl ester (FAME) appeared. The peaks of diglyceride and triglyceridebecame smaller in the effluent from the second column and completelydisappeared in the effluent from the third column. Thus, the FFA and glycerides in the crude *Jatropha* oil were completelyconverted to FAME.

Table 1 shows the analyticalresults of the product biodiesel compared with standards of USA (ASTM D6751)and Europe (EN14214). The acidvalue and the contents of free glycerin and total glycerin in the product fullysatisfied with both standards without downstream processing. In addition, the contents of remainingreactants (monoglyceride, diglyceride,triglyceride), water and FAME also satisfied with thestandards of Europe. The glycerinand water as well as FFA and pigment were confirmed to be removed by theadsorption on the resin. Therefore,the proposed system permitted the production of high quality biodiesel fuelwithout upstream and downstream processing.

4. Reference

1) N.Shibasaki-Kitakawa*et al., Energy Fuels,* **24**, 3634 (2010)

2) N.Shibasaki-Kitakawa *et al., Bioresour. Technol.,* **98**, 416 (2007)

3) T.Tsuji *et al., Energy Fuels,* **23**, 6163 (2009)

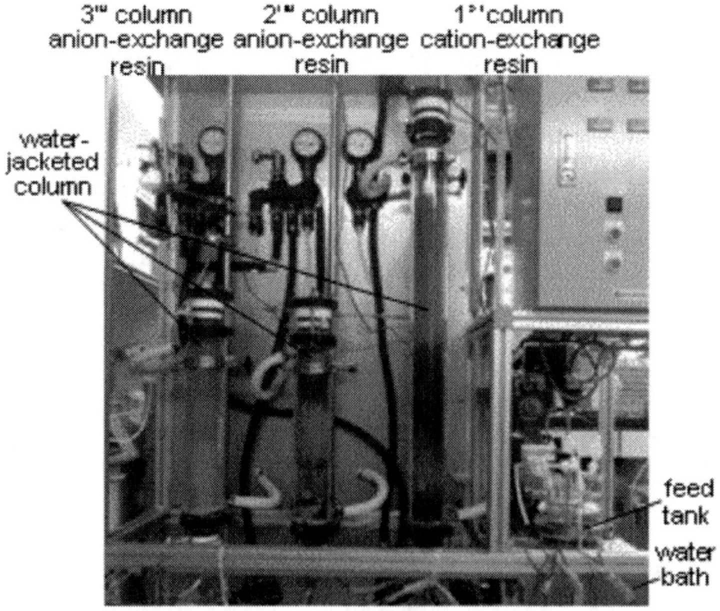

Fig. 1 Photograph of bench-scaleproduction system with cation- and anion-exchangeresins

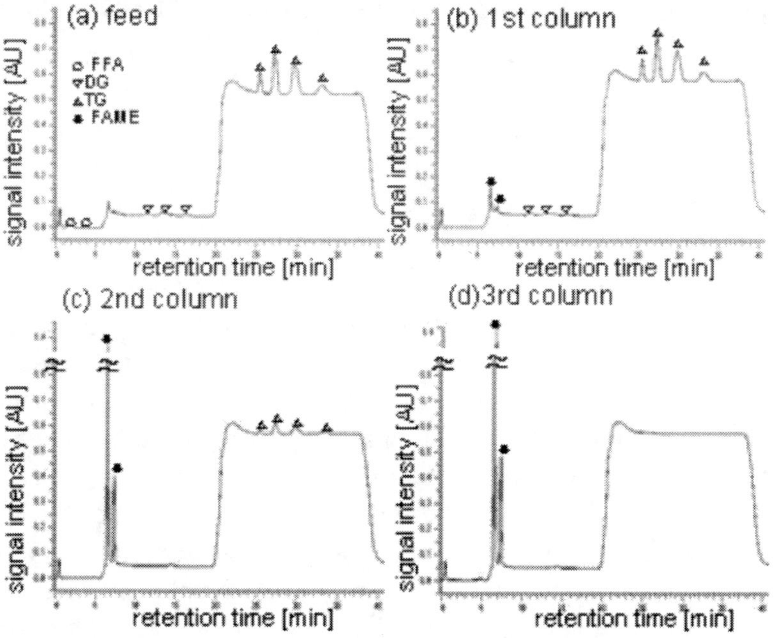

Fig. 2 HPLC chromatograms of feed andeffluents from each column

Table 1 Comparison ofproduct properties with standards ofbiodiesel fuel

property	units	product	standards	
			ASTM D6751	EN14214
acid value (FFA)	[mg-KOH/g]	0.03	$\leq$0.8	$\leq$0.5
free glycerin	[wt%]	0.01	$\leq$0.02	$\leq$0.02
total glycerin	[wt%]	0.03	$\leq$0.24	$\leq$0.25
monoglyceride	[wt%]	0.06	-	$\leq$0.80
diglyceride	[wt%]	0.01	-	$\leq$0.20
triglyceride	[wt%]	0.10	-	$\leq$0.20
water	[mg/kg]	148	-	$\leq$500
FAME	[wt%]	99.1	-	$\geq$96.5

See more of this Session: Sustainable Fuel From Renewable Resources I
See more of this Group/Topical: Environmental Division

(267a) Pyrolysis Vs. Gasification: Effect of Oxygen In Reaction Atmosphere On Biochar Properties

Catherine E. Brewer, Center for Sustainable Environmental Technologies, Iowa State University, Ames, IA and Robert C. Brown, Mechanical Engineeering, Iowa State University, Ames, IA

Since biochar properties can vary widely depending on feedstock and processing conditions, making meaningful comparisons between biochar produced by different processes (pyrolysis, gasification, kiln carbonization) and at different locations can be difficult. Biochar characterization methods can provide some useful metrics such as van Krevelen diagrams, fixed carbon to volatiles ratios, and aromatic ring cluster size. One of the key parameters shown to influence biochar properties is the highest heating temperature (HHT) reached during the reaction; clear trends can be observed in the characteristics of slow pyrolysis biochars over the 200-800°C HHT range. These trends do not hold up as well, however, when biochars made at the same HHT by fast pyrolysis or gasification are considered. For low and high temperature gasification, the presence of oxygen in the reaction atmosphere creates biochars with higher O/C molar ratios and higher apparent HHTs than analogous slow pyrolysis reactions. This study used a large sample size thermogravimetric analyzer (TGA) to produce corn stover biochars reaching different HHTs under slow pyrolysis conditions with varying levels of oxygen present in the flow gas. The biochars were characterized by proximate analysis, CHNS elemental analysis and ^{13}C solid-state nuclear magnetic resonance spectroscopy (NMR) to understand the combined effects of HHT and oxygen on biochar properties. The goal of the study is to determine if the presence of oxygen in the reaction atmosphere at a given HHT would be beneficial for the creation of oxygenated functional groups on biochar surfaces similar to biochars that have "aged" in the soil environment.

See more of this Session: Thermochemical Conversion
See more of this Group/Topical: International Congress on Energy 2011

(267b) Pretreatment of Loblolly Pine for Fast Pyrolysis Via Torrefaction

Jiajia Meng, Forest Biomaterials, North Carolina State University, Raleigh, NC

Topical 6: Lignocelluosics: Biorefineries and Sustainable Energy

T6007 Thermalchemical Conversion I

AIChE 2011, October 16~21, 2011

Minneapolis, MN

Pretreatment of Loblolly Pine for Fast Pyrolysis via Torrefaction

Jiajia Meng, David Tilotta, and Sunkyu Park*

Department of Forest Biomaterials, North Carolina State University

*Corresponding author: sunkyu_park@ncsu.edu

Fluidized-bed fast pyrolysis was performed on torrefied loblolly pine with differing degrees of torrefaction severity. Compared to the pyrolysis of non-torrefied loblolly pine, which has a bio-oil yield of 65%, the bio-oil produced from thermally processed loblolly pine has a 10-30% lower yield depending on torrefaction severity. In order to increase the liquid recovery efficiency from the pyrolysis of intensely treated wood, optimal biomass particle size and pyrolysis temperature were investigated in this study. GC/MS analysis was applied to quantify several important chemical markers in the bio-oils, such as levoglucosan, glycolaldehyde and acetol. It was found that the decomposition of torrefied wood with different thermal treatments leads to varied content of these chemical components in the bio-oils. This result suggests that torrefaction may exert power on the pyrolysis pathway of the major components in the woody biomass. For instance, the opportunity for ring scission and the formation of levoglucosan from cellulose may be altered. Also, the decreasing amount of guaiacols and guaiacyl derivatives in the bio-oil, with respect to the torrefaction temperature, implies the reduction of methoxy groups either in the torrefaction process or in the condensation process occurring in fast pyrolysis. To understand the effect of torrefaction pretreatment on the variation of bio-oil composition, py-GC/MS and MBMS analysis of both torrefied wood sample and the residues from acid hydrolysis of torrefied wood was utilized. In addition, the stability and the acidity of the bio-oils produced from torrefied wood were examined to identify the bio-oil quality improvement gained from torrefaction. These results indicate that fast pyrolysis of the torrefied wood may be an effective way to produce high quality bio-oil.

See more of this Session: Thermochemical Conversion
See more of this Group/Topical: International Congress on Energy 2011

(267c) Characterization of Biochar, the by-Product of Energy Production From Biomass Via Thermochemical Processes

Ellen M. Knapp, Chemical Engineering, Villanova University, Villanova, PA and Justinus A. Satrio, Department of Chemical Engineering, Villanova University, Villanova, PA

Characterization of Biochar, the By-Product of Energy Production from Biomass via Thermochemical Processes

Ellen M. Knapp and Justinus A. Satrio

Chemical Engineering Department, Villanova University, PA 19085

Biochar is the solid product obtained through thermochemically processing biomass, namely through fast pyrolysis, slow pyrolysis, and gasification. The use of biochar has been shown to hold great potentials to aid in the sequestration of carbon dioxide and to act as a soil restoring agent. The thermochemical processes for energy production from biomass that stores carbon as biochar can be considered carbon negative due to high recalcitrant property of biochar. As a soil restoring agent, biochar has been shown to improve the quality of soils by increasing microbial activity, lowering soil acidity and density, and improving nutrient and water retention. It has also been shown that during thermochemical processing of biomass, a significant portion of mineral nutrients contained in the original biomass are recovered in biochar.

It is believed that all biochars are not created equal and will have different physical and chemical properties depending on their biomass origins and the thermochemical processing used for producing the biochars. The purpose of this study is to evaluate the effects of biomass precursors and process conditions of thermochemical processes on the chemical and physical properties of the produced biochars. For this study, biochars are synthesized by fast pyrolysis and slow pyrolysis processing of selected biomass materials using a small tubular reactor furnace at selected process conditions. A variety of biomass materials, commonly available in the state of Pennsylvania, particularly in the region where Villanova University is situated, are used for the study, which include wheat straw, switch grass, woodchips and spent mushroom substrate (SMS). The last material, SMS, is the waste product from mushroom growing industries and is produced in very large quantities in Chester County, which is known as the "mushroom growing capital of the world." Currently, Chester County accounts for approximately 37% of the mushroom production in the U.S.A. Biochar produced from various biomass materials by various process conditions will be analyzed for their physical and chemical characteristics/ properties. Analysis will include proximate analysis, elemental analysis, heating value, surface area/ porosity, and composition of mineral matters. Focus will be placed on the comparison of the chars formed from each biomass precursor at different process conditions. In particular, emphasis will be placed on comparing the characteristics of SMS char versus wheat straw char. SMS is a derivative of wheat straw that has been combined with various manures and grains. The comparison of chars produced from these two biomass precursors in particular will provide insight into the effects of additional materials added to the straws in the SMS on the properties of chars.

See more of this Session: Thermochemical Conversion
See more of this Group/Topical: International Congress on Energy 2011

(267d) Pyrolysis of Cellulose with the Use of Ionic Liquid for Efficient Production of Levoglucosenone

Shinji Kudo[1], Zhenwei Zhou[2], Koyo Norinaga[3] and Jun-ichiro Hayashi[3], (1)Research and Education Center of Carbon Resources, Kyushu University, Fukuoka, Japan, (2)Department of Applied Science for Electronics and Materials, Kyushu University, Fukuoka, Japan, (3)Institute for Materials Chemistry and Engineering, Kyushu University, Fukuoka, Japan

Pyrolysis is a most practical method to directly convert cellulose into condensable organic product, which is in general termed tar, with high yield. The tar consists of a variety of organic compounds, such as anhydrosugars and furan derivatives, and this is due to progress of various reactions in parallel and in series. For selective production of valuable chemicals, thus, the pyrolysis is currently less suitable compared to indirect methods via saccharification. A compound, levoglucosenone, can be produced by means of pyrolysis more easily than by general chemical methods, but it is difficult for any methods reported so far to convert cellulose to this particular compounds with either high selectivity or yield. This study proposes the pyrolysis of cellulose with the use of a type of ionic liquid (IL). ILs have widely been used as solvents for catalytic conversion of cellulose due to their characteristic properties such as solvent powers for cellulose, low vapor pressures and thermal stabilities. The proposed pyrolysis employs a type of IL, 1-butyl-2,3-dimethyl imidazolium trifluoromethanesulfonate ([BMMIM][OTf]), with an expectation of its high thermal stability occurring from alkyl substituent to 2-carbon. TG/DTA analyses in fact demonstrated its perfect stability at temperature up to 350 °C, where the most of the ILs underwent self-decomposition. The anion of [BMMIM][OTf], i.e., [OTf], is the conjugated base of a superacid, trifluoromethanesulfonic acid ($pK_a < -14$), and it was therefore expected catalysis of this IL in the pyrolysis of cellulose. Mixed with [BMMIM][OTf] at a mass ratio of 1 and temperature of 60 °C, cellulose was swollen. The swelling was maintained even after the mixture was cooled down to temperature below the melting point of [BMMIM][OTf], 41 °C. A TG analysis showed that the pyrolysis of cellulose in the mixture started at 160 °C, caused rapid and slow mass releases at 200–275 °C and at higher temperature up to 350 °C. Neither decomposition nor evaporation of [BMMIM][OTf], if any, was detected in the course of heating. The products from the pyrolysis at 300 °C were 32% char, 30% tar, 33% water, and 3–5% gas on a cellulose mass basis. Most importantly, levoglucosenone accounted for more than 70% and 27 mol-% on tar mass and cellulose monomer bases, respectively. Formation of levoglucosenone during the pyrolysis of a phosphoric-acid-treated cellulose was reported previously, but its yield was less than half of that from the cellulose-IL mixture. Such a high yield of levoglucosenone seems to be attributed to a catalytic role of catalysis of the [OTf] anion. Repeated use of [BMMIM][OTf] was also examined. The IL was separated from the char by washing the mixture with ethanol, and then recovered by means of evaporative removal of ethanol. The IL recovery was complete within the margin of experimental error. TG analysis of the used IL and also its repeated use in the subsequent pyrolysis of cellulose confirmed little or no change in the property of the IL in the processes.

Modeling of the formation and reaction properties of biomass

Karsten Müller, Olga Lobanova, Liudmila Mokrushina, Wolfgang Arlt
Friedrich-Alexander-University Erlangen-Nuremberg,
Chair of separation science and technology

Biomass as an alternative source of energy to replace fossil fuels has drawn much attention in recent years. Many of the biomass utilization processes developed in the past later turned out to be ineffective in terms of their energy balance. For example, some publications indicate that the amount of energy needed for the production of biodiesel and bioethanol exceeds the energy derived from the respective fuels.

A systematic development of new processes, avoiding the development of energetically inefficient processes, requires knowledge of the thermodynamic data of the biomass used. Especially for complex ingredients like polysaccharides and lignin, these data are not available from the literature, yet.

This research project aims at the determination of the formation and reaction properties of such biomass compounds as well as other thermodynamic properties like heat capacity.

In a succeeding step thermodynamics of reactions utilizing biomass as a feedstock have been analyzed. A group of model reactions, such as the conversion of saccharides or lignin to biofules like methane or hydrogen have been evaluated. For an appropriate thermodynamic modeling valid substance data are necessary. In the course of the project it could be shown that the free energy of formation is linear function of the degree of polymerization. For details see [1].

By balancing enthalpy and Gibbs' energy of the respective reactions, the equilibrium yields and conversions have been estimated. The first reaction has already been implemented as a high temperature process. Since in terms of energy, it is desired to carry out reactions at mild conditions, the feasibility of doing this was researched from the thermodynamic viewpoint.

It could be shown that e.g. the production of gaseous products such as hydrogen or methane from saccharides at low temperatures is thermodynamically possible. The equilibrium yield can be further increased by lowering the system pressure. This allows the gaseous products to leave the system and therefore shifts the equilibrium position towards product formation. The addition of inert gas can also be used to increase the formation of hydrogen.

Many biomass compounds are not, or only limitedly, soluble in water. Hence, solvents have to be used. Calculations showed that the usage of an appropriate solvent does not lower equilibrium conversion and can even help increasing the conversion.

This project has been supported by the Deutsche Forschungsgemeinschaft (Projektnummer: AR 236/34-1).

[1] O. Lobanova, K. Müller, L. Mokrushina, W. Arlt, „Estimation of Thermodynamic Properties of Polysaccharides", Chemical Engineering & Technology, 2011, 32, No. 6, 867–876

(267f) Effect of Pressure and Heating Rate On Biomass Devolatilization and Gasification

Pradeep K. Agrawal, School of Chemical & Biomolecular Engineering, Georgia Institute of Technology, Atlanta, GA and Kristiina Iisa, National Renewable Energy Laboratory, Boulder, CO

The goal of present study is to characterize the effect of pressure and heating rate on the devolatilization behavior of different biomass species. Three biomass species (southern Pine, corn stover, and switchgrass) were studied. A combination of two complementary gasification devices are utilized in this study: (i) pressurized entrained flow reactor (PEFR) provides high heating rates ($\geq 10^3$ °C/sec) and short residence times (1-20 sec), and (ii) pressurized thermo-gravimetric analyzer (PTGA) provides lower heating rates (0.1-1 °C/sec), but larger residence times (5-100 min). The pressurized thermogravimetric analyzer connected to a mass spectrometer and a Fourier-transform infrared analyzer provides information on the evolution of gaseous species during pyrolysis.

The char yield appeared to increase with an increase in pyrolysis pressure. The yields of the lighter gas components (CO, CO_2, H_2, CH_4, H_2O) increased as the pressure was increased. The increase in the amount of char and lighter gases was accompanied by a decrease in the formation of tar species. This was supported by an observed decrease in furan and furfural and higher hydrocarbon fragments with increasing pressure. There were clear differences in the evolution of gaseous species between the woody and herbaceous and agricultural residue feedstocks. The evolution of char morphology is strongly influenced by pyrolysis pressure and heating rates. The results suggest a complex role of pyrolysis conditions and biomass ash content on the char reactivity.

See more of this Session: Thermochemical Conversion
See more of this Group/Topical: International Congress on Energy 2011

(267g) Technoeconomic Analysis of Biomass Gasification System Utilizing Geothermal Energy for Processing

Sudhanya Banerjee[1], Yanan Zhang[1], **Song-Charng Kong**[1], Robert C. Brown[1] and Brian Anderson[2], (1) Mechanical Engineering, Iowa State University, Ames, IA, (2)Chemical Engineering, West Virginia University, Morgantown, WV

The use of biomass as a source of energy has been studied extensively inrecent years, and special attention has been paid to biomass gasification. In this study, ASPEN Plus is used to conduct technoeconomic analysis of a large biomass gasification system. A biorefinery scenario consisting of 2,000 dry metric tonnes per day of corn stover was developed using the model. In the original setup, steam is purchased and used in various processes. In this study, we explore the use of geothermal energy by substituting the purchased steam with steam derived from geothermal resources. Geothermal energy is in in theform of steam with temperature of 150°C. The flow rate of the geothermalsteam is in the range of 40 to 60 kg/sec and the cost is estimated at $3 – $7/MMBTU. This technoeconomic analysis assesses the use of geothermal steam instead of the purchased steam for various processes. Results are reported including the feasibility of the process, product characteristics, and the technical and economic aspects of using geothermal steam for biomass gasification system.

See more of this Session: Thermochemical Conversion
See more of this Group/Topical: International Congress on Energy 2011

(274b) Biodegradation of Hardwood Lignocellulosics by the Western Poplar Clearwing Borer, Paranthrene Robiniae (Hy. Edwards)

Jing Ke, Dhrubojyoti D. Laskar and Shulin Chen, Washington State University, Pullman, WA

Plant lignin is a source of useful chemicals, and also the major barrier for saccharification of lignocellulosic biomass for ethanol conversion. The enzymatic lignin degradation/modification could bypass the need for chemical pretreatment and thereby facilitate bioprocess consolidation. Herein, we reveal our new discovery on the elucidation of the prototype of hardwood lignin modification/ degradation in clearwing borer system, *Paranthrene robiniae*. The wood-boring clearwing borer, *P. robiniae*, effectively tunnels hardwood structures during the larval stage, and its digestion products from wood components has not yet been investigated. Subsequent solid state ^{13}C cross polarization magic angle spinning (CP/MAS) nuclear magnetic resonance (NMR) spectroscopy, attenuated total reflectance fourier transform infrared (ATR-FTIR), pyrolysis-gas chromatography/mass spectrometry (Py-GC/MS), and thermogravimetric (TG) analysis on tunnel walls and frass produced, provided evidence for structural alterations and lignin degradation during such hardwood digestion process. Our results strongly suggest that the structural alteration of lignin primarily involved a preferential degradation of syringyl units accompanied with oxidation on the side chains of lignin guaiacyl moieties. This study also further indicated that, unlike the wood-feeding termite, the clearwing borer does not target cellulose as an energy source, and thus its lignin degradation ability should provide potential information on how to disassemble and utilize hardwood lignin. Overall, this new biological model with an efficient lignin disruption system will provide the new insights into novel enzyme system required for effective plant cell wall disintegration for enhanced cellulose accessibility and production of value-added lignin derived products.

See more of this Session: Biological and Chemical Conversion Processes In Forest Biorefineries I: Enzymatic and Other Biological Processes
See more of this Group/Topical: International Congress on Energy 2011

Improved Ethanol Production through Anaerobic Thermophilic Co-Culture From Various Carbohydrates

Lei Xu*, Ulrike Tschirner

Bioproducts and Biosystems Engineering University of Minnesota, St. Paul, MN 55108, USA

1. Introduction

Affordable bio-fuel has become popular due to an imbalance between supply and demand of existing energy sources as well as the problems associated with recent fuel refinery processes (1). Lignocellulosic biomass offers a great potential as a bio-fuel resource due to their abundance and high conversion potential.

Consolidate Bioprocess Process (CBP) fermentation with proper microbes is a promising solution offering the potential for higher efficiency and lower production cost compared to other current conversion processes (2). The major attempts of this study were to evaluate the advantages of integrating both *C.themocellum* and *C.thermolacticum* into the CBP fermentation on the ethanol and acetate formation under extreme pH (5-10) on various saccharides (3). In order to optimize the co-culture, by overcoming the inhibitor effect of high pH and high ethanol concentrations, allowing for an effective merging of pretreatment and fermentation processes to bring the technology closer to economically feasible biofuels from non-food based biomass.

2. Materials and Methods

Both the *C. thermocellum* (ATCC 27405) and *C. thermolacticum* (ATCC 43739) were cultured with media described by He et al. (2009) under controlled conditions (80%N_2/20%CO_2). The conversions of selected materials including cellulose, hemicellulose and their mono- or di-saccharide intermediates are investigated. The major saccharides studied include glucose, cellobiose, and xylose. A low initial substrate/medium ratio (1% w/v) was used and a 5 mL homogeneous microbe culture with a cell weight of 5 g/L was transferred into 100 mL anaerobic bottles containing 50 mL of medium prepared under anaerobic conditions. Cultures incubated at 57 °C without shaking at pH 5 to pH 10 were analyzed for the end-products (ethanol and acetate). The impact of the ethanol production and accumulation within the medium on total solvent production was studied by adding various levels of ethanol at the beginning of the fermentation. The addition of 0 g/L of ethanol represents the beginning stage of the fermentation when no ethanol was produced. By adding ascending amounts of ethanol from 0.5 g/L to 4 g/L at the start, this study simulated the ethanol accumulation progress as a sequence of the continuous fermentation. The influences of pH, ethanol concentration and substrate types were further compared between *C. thermocellum/C. thermolacticum* mono-cultures and co-culture fermentation by analyzing the ethanol and acetate production using HPLC.

3. Results and Discussion

The innovative co-culture is extremely efficiently in fermenting separate components of lignocellulosics, especially MCC, for ethanol production due to enzymatic characteristics of the strains (4, 5). Almost all the ethanol yields obtained from co-culture fermentation are above any of its mono-culture fermentation yields, often almost tripled as compared to the mono-cultures under identical conditions on all the substrate. In addition, this co-culture exhibited a shorter lag period (48 h) than any of their mono-culture before the initiation of the selected sugar fermentations. However, the increase of ethanol yield caused by applying a co-culture was observed not to correlate to ethanol/acetate ratio improvement.

Based on ethanol and acetate yield after 10 d fermentation, statistical analysis indicated that the solvent yields were significantly related to the growth conditions, including the interactions among application of co-culture, initial ethanol level, pH value and substrate type. Fermentation yields for both strains were sensitive to pH changes. The sensitivity towards pH change was enhanced with the increase of initial ethanol concentration. When the starting ethanol concentration reaches 4 g/L, the pH influence has the maximum input on ethanol production for all substrate conversions and for all strains. Alkaline conditions, especially pH equal to 9, is preferred for *C. thermocellum*/*C. thermolacticum* co-culture fermentation if the goal is high solvents production. Almost all of the ethanol yields obtained from co-culture fermentation at this pH were higher than the sum of their mono-culture fermentations. Also, the co-culture produced the largest amount of acetate during fermentation under high initial ethanol levels (4 g/L), and the optimum pH of 9. Under this condition, the ethanol yield from xylose was increased from 0.1 g/L for *C. thermolacticum* and 0 g/L for *C. thermocellum* to 4.5 g/L using a co-culture, which almost approached 90% of the theoretical xylose fermentation efficiency. For MCC fermentation, improvement was detected to be from 1.4 g/L for *C. thermocellum* and 0.1 g/L for *C. thermolacticum* to a maximum of 3.8 g/L for co-culture. This represents 1.5 mol of ethanol per mol of sugar utilized for co-culture fermentation (around 75% of the theoretical efficiency) (6), almost three times the ethanol yield of the combined mono-cultures. Hemicellulose fermentation appeared to be slower and less effective than conversion of other fermentable sugars due to the lack of effective enzymes (4, 5). In this case, by applying the co-culture of *C. thermolacticum* and *C. thermocellum* to lignocellulosic fermentation, optimization of the conditions should focus on the degradation conditions for cellulosic material and the hemicellulose fragments, mainly xylose. Furthermore, in order to archive higher solvent productions, initial mild alkaline condition, prompt removal of acetate from the system, or pH-control is essential.

4. Conclusion

In the present work, a detailed analysis of the co-culture fermentation with various carbohydrates under extreme pH with existing of ethanol was carried out. The results demonstrated the technical feasibility of using our innovative simultaneous *C. thermocellum/C. thermolacticum* co-culture to converting components of lignocellulosics directly into ethanol and acetate because of shorter lag time and higher solvent production. An interaction among pH, substrates and initial ethanol level to the solvent production could be shown. To obtain maximum ethanol yield, providing a comparatively higher starting pH and higher initial ethanol level are preferred in this co-culture anaerobic fermentation.

Acknowledgements

The authors acknowledge the support of the University of Minnesota and the Buckman Fellowship fund for supporting this research.

References

1. Kumar P, Barrett DM, Delwiche MJ, Stroeve P. Pretreatment of lignocellulosic biomass for efficient hydrolysis and biofuel production. Industrial & Engineering Chemistry Research. 2009;48(8):3713-3729.

2. Perlack RD, Wright LL, Turhollow AF, Graham RL, Stokes BJ, Erbach DC. Biomass as feedstock for a bioenergy and bioproducts industry: The technical feasibility of a billion-ton annual supply. [Internet]. 2005:May 20th, 2011. Available from: http://feedstockreview.ornl.gov/pdf/billion_ton_vision.pdf.

3. Willaert RG, Baron GV. Gel entrapment and micro-encapsulation: Methods, applications and engineering principles. Rev Chem Eng. 1996;12(1-2):1-205.

4. Raman B, Pan C, Hurst BG, Rodriguez M, McKeown KC, Lankford KP, et al. Impact of pretreated switchgrass and biomass carbohydrates on *clostridium thermocellum* ATCC 27405 cellulosome composition: A quantitative proteomic analysis. PLoS ONE 4. 2009;4(4):e5271.

5. Pellerina P, Gosselina M, Lepoutrea J, Samaina E, Debeire P. Enzymic production of oligosaccharides from corncob xylan. Enzyme Microb Technol. 1991;13(8):617-21.

6. Kim YJ, Weigand AW. Experimental analysis of a product inhibited fermentation in an aqueous two-phased system. Appl Biochem Biotechnol. 1992;34-35(1):419-30.

(274d) Thermostable Cellulases and Xylanases

Aditya Bhalla[1], Venkat Kunreddy[1], Sudhir Kumar[1], Kenneth Bischoff[2], Lew P. Christopher[3], Stephen R. Hughes[4] and Rajesh K. Sani[5], (1)Chemical and Biological Engineering, South Dakota School of Mines & Technology, Rapid City, SD, (2)Bioproducts and Biocatalysis Research Unit, USDA-ARS, NCAUR, Peoria, IL, (3)Center for Bioprocessing Research & Development, South Dakota School of Mines & Technology, Rapid City, SD, (4)USDA-ARS, NCAUR, Peoria, IL, (5)Chemical & Biological Engineering, South Dakota School of Mines & Technology, Rapid City, SD

Thermostable cellulases and xylanases offer several advantages including greater rates of hydrolysis for lignocellulosic substrates, lowered risk of contamination, and increased flexibility with respect to process design. We have isolated several thermophilic cellulose- and xylan-degrading cultures belonging to the genera *Bacillus*, *Clostridium*, and *Geobacillus* using soil samples collected from the Homestake Gold Mine, SD and a local compost facility. One of the isolates, *Bacillus* sp. (DUSELR13 strain) produced xylanase which had optimum pH and temperature of 7 and 75^{0}C, respectively. Other isolate *Geobacillus* sp. (WSUCF1 strain) secreted endoglucanase having maximum activity at pH 5 and temperature 70^{0}C. The thermal stability of enzymes secreted by DUSELR13 and WSUCF1 was tested by determining the enzyme activity remaining after incubation of the enzymes at different temperatures (50 -100^{0}C) for a period of 40 days. WSUCF1 endoglucanase showed remarkable residual activity (>50%) after prolonged incubation for 25 days at 60^{0}C compared to others reported in literature. DUSELR13 xylanase retained >50% residual activity after incubation for 35 days at 60^{0}C. The effect of various lignocellulosic feedstocks on the endoglucanase and xylanase secretion by these isolates was also examined. K_m values for DUSEL R13 xylanase and WSUCF1 endoglucanase were calculated using Lineweaver–Burk plots. In case of DUSELR13 xylanase, the K_m value with xylan from birchwood was 1.4 mg/ml while for WSUCF1 endoglucanase with carboxymethylcellulose this value was 1.08 mg/ml. These thermostable enzymes were overexpressed extracellularly using *E.coli* as a heterologous host. Comparison studies were carried out between recombinant enzymes and native enzymes. These thermostable cellulases and xylanases are currently being used to develop a novel cost-effective and more efficient form of the simultaneous saccharification and fermentation process for biorefineries.

See more of this Session: Biological and Chemical Conversion Processes In Forest Biorefineries I: Enzymatic and Other Biological Processes
See more of this Group/Topical: International Congress on Energy 2011

(274e) Ionic Liquid Pretreatment of Wood Biomass Leading to Enhanced Enzymatic Delignification

Muhammad Moniruzzaman, Department of Material & Energy Science, Okayama University, Japan, Tsutomu Ono, Department of Material & Energy Science, Okayama University, Hiroshi Kanzaki, Department of Chemistry and Bioactive Compounds, Okayama University and Kenji Okada, Department of Life Science, Kurashiki University of Science and the Arts, Japan

The biological pretreatment of wood biomass for degrading lignin, a complex aromatic polymer, has received much attention as an environmentally safe or “green” process. Generally, enzymes isolated from naturally occurring fungi, or with enzymes produced by genetically engineered fungi have been used for wood biodegradation. However, this approach for lignin degradation in aqueous solution has been found to be very slow mainly due to the difficulties in enzyme accessibility to the hemicelluloses and the poor solubility of the lignin. To address these issues, this presentation will describe a new approach for enzymatic delignification of wood biomass using room temperature ionic liquids (RTILs) as pretreated agents or/and (co)solvents. It is well known that ionic liquids (ILs), a potentially attractive “green” recyclable alternative to environmentally harmful organic solvents, have been increasingly exploited as solvents and/or (co)solvents and/or reagents in a wide range of applications including pretreatment of lignocellulosic biomass. The pretreatment of wood biomass comprises using the hydrophilic IL [emim] [OAc] (1-ethyl-3-methylimidazolium acetate) that possesses very high solubility of wood biomas and lignin, which makes biomass very suitable for enzymatic degradation. It has been shown that the cellulose rich materials obtained from combination effects of IL and biological pretreatment contained significantly lower amounts of lignin as compared to the amounts found when each method applied alone. The produced cellulose rich materials were characterized by acid hydrolysis, Fourier-transform infrared spectroscopy (FTIR), Scanning electron microscopy (SEM), Thermalgravimetric analysis (TGA) and X-ray diffractometry (XRD). SEM and XRD revealed considerable microstructural and crystallinity index changes in the pretreated cellulose rich materials. We believe that this newly developed environmentally friendly process will play a great role in converting cellulosic biomass, the most abundant renewable biomaterials in the world, to biomaterials, biopolymers, biofuels, bioplastics and so on.

See more of this Session: Biological and Chemical Conversion Processes In Forest Biorefineries I: Enzymatic and Other Biological Processes
See more of this Group/Topical: International Congress on Energy 2011

(274f) The Evaluation of Inhibition of Enzymatic Hydrolysis Following Dilute Acid Pretreatment of Herbaceous, Softwood and Hardwood

Mahdieh Aghazadeh, Chemical and Biochemical Engineering, University of Maine, Orono, ME and G. Peter van Walsum, Chemical and Biological Engineering, University of Maine, Orono, ME

Presenter: Mahdieh Aghazadeh

The Evaluation of Inhibition of Enzymatic Hydrolysis following Dilute Acid Pretreatment of herbaceous, softwood and hardwood

Peter Van Walsum* and Mahdieh Aghazadeh

Department of Chemical and Biological Engineering

University of Maine, Orono, Maine 04496

Phone: 207-581-2226

Fax: 207-581-2323

Email: pvanwalsum@umche.maine.edu

ABSTRACT

The concept of severity combining residence time and temperature is often used in the pulp and paper industry. The influence of the severity of dilute acid hydrolysis of lignocellulosic biomass on sugar yield and on the subsequent enzymatic hydrolysis of the hydrolysate by commercial enzyme preparation (from Novozymes) was investigated. A dilute acid pretreatment was performed at 0.7 % (w/v) solid concentration with 100 mL Accelerated Solvent Extraction (ASE Dionex Corp.). The pretreatment conditions (residence time, temperature) were treated as a single parameter, s*everity factor* (log R_0), with constant acid concentration. The decrease in the yield of oligosaccharides coincided with the maximum concentrations of furfural and 5-hydroxymethylfurfural (5-HMF). With the further increase in s*everity factor*, the concentrations of furfural and 5-HMF decreased while the formation of formic acid and levulinic acid increased. The effect of acetic acid, formic acid, levulinic acid, furfural, and 5-HMF on enzymatic hydrolysis was investigated. In former studies the effect of the sugar produced from the hydrolysate itself had not been studied - in this study the different enzyme inhibition effect on three major types of biomass, herbaceous (corn stover), soft wood (pine) and hardwood (maple/oak) were observed.

See more of this Session: Biological and Chemical Conversion Processes In Forest Biorefineries I: Enzymatic and Other Biological Processes
See more of this Group/Topical: International Congress on Energy 2011

(274g) Role of Enzyme Blend Composition, Biomass Particle Size and Solids Loading On Pretreated Biomass Saccharification Efficiency

Shishir PS Chundawat[1], Nirmal Uppugundla[1], Mary Lipton[2], Margaret Magyar[1], Dahai Gao[1], Venkatesh Balan[1] and Bruce Dale[1], (1)Chemical Engineering and Material Science, Michigan State University, Lansing, MI, (2)Biological Sciences, Pacific Northwest National Laboratory, Richland, WA

One of the limitations of screening enzyme combinations using microplate based assays (Gao et al. 2010, Bioresour Technol 101, 8:2770) is that these are typically carried out using milled biomass (<500 mms) at low solids loading (<1%, w/v). Biomass particle size and solids loading can both influence the enzyme set necessary for efficient hydrolysis. In this study, three different harvest varieties of AFEX pretreated corn stover were hydrolyzed at 0.6-18% solids loading using multiple combinations of cellulases, hemicellulases and other accessory enzymes for extended hydrolysis time periods. Three particle size ranges were tested (<100 mm, <500 mm, >1 mm). The crude glycosyl hydrolase mixtures were also analyzed using a high-throughput proteomics approach to identify enzyme activities necessary for efficient hydrolysis. We show that proteomics based spectral counting approach is a reasonably accurate and rapid analytical technique that can be used to determine protein composition of complex glycosyl hydrolase mixtures that also correlates with the specific activity of individual enzymes present within the mixture. We also find that both particle size and solids loading can significantly impact the optimal combination of hydrolytic enzymes necessary for efficient hydrolysis. In summary, we found that optimal cellulase-hemicellulase mixtures can achieve near-theoretical glucan and xylan yields at greater than fourfold lower total protein loading than using individual commercial enzyme blends alone.

See more of this Session: Biological and Chemical Conversion Processes In Forest Biorefineries I: Enzymatic and Other Biological Processes
See more of this Group/Topical: International Congress on Energy 2011

(338a) Comparing Resource Consumption and Sustainability of Biobased Materials Using Eco-LCA

Erin F. Landers and Bhavik R. Bakshi, William G. Lowrie Department of Chemical and Biomolecular Engineering, The Ohio State University, Columbus, OH

Ecosystem goods and services are essential to all human activity. In spite of this importance, the Millennium Ecosystem Assessment has found that many ecosystem services are highly degraded. In many engineering analyses, these necessary goods and services are undervalued, or even ignored completely. It is therefore crucial to develop and modify existing methods to allow these services to be valued and accounted for in decision making. Conventional life cycle assessment (LCA) tends to only look at energy consumption and emissions. However, Ecologically Based LCA (Eco-LCA) is a thermodynamically based LCA that allows for many ecosystem goods and services used throughout our economy to be quantified, in addition to the traditional LCA results of energy consumption and emissions.

To create the Eco-LCA model, a nationally based economic model must first be obtained, to which the resources will then be added. Eco-LCA uses the national input-output economic model that the Bureau of Economic Analysis provides every five years. This current model, based on the 2002 national economy, shows the interactions between different industries and is used to track the use of ecosystem goods and services through the economy, which will be compared with previous research done to create an Eco-LCA inventory for the 1997 economic model. Resource use must then be added to the model; this is done by finding values for resource use from government agencies and various research groups. The thermodynamic properties of exergy and emergy allow for the many different units of ecosystem goods and services to be converted to the common basis of thermodynamic work. This common basis means the results of Eco-LCA analyses can be aggregated and directly compared in metrics such as Cumulative Exergy Consumed and Percent Renewable. After adding the ecosystem goods and services to the national economic model, resource intensities are calculated for every resource and economic industry by taking into account the resources entering the economy, the interactions between industries, and the total monetary activity of every industry. This then provides a measure of individual and total resource use per dollar of economic activity in each industry. These new resource intensities can be directly compared to the resource intensities from the 1997 Eco-LCA model and provide insight as to changes in both resource use and interactions between industries from 1997 to 2002.

A case study will be presented using Eco-LCA to evaluate biobased materials such as biofuels and biopolymers. The sustainability of biofuels will be discussed by investigating the resources required and emissions produced for various feedstocks, including changes since 1997. This will allow for a more dynamic view of the progress towards the most viable, sustainable biofuels. Resource consumption to make biopolymers will be compared to traditional polymers, and conclusions will be drawn as to the most sustainable polymer materials.

See more of this Session: Sustainability and Biobased Materials
See more of this Group/Topical: Forest and Plant Bioproducts Division

(338b) Sustainable Green Composites

Amar K. Mohanty, Department of Plant Agriculture & School of Engineering, Guelph, ON, Canada and Manju Misra, School of Engineering and the Department of Plant Agriculture, University of Guelph, Guelph, ON, Canada

In progress

See more of this Session: Sustainability and Biobased Materials
See more of this Group/Topical: Forest and Plant Bioproducts Division

(338c) Integrated Biorefinery Operations: Biomass Pretreatment As It Relates to Biochemical and Thermochemical Products

Brian L. Cooper, Jeff London and Robert Mellon, Hazen Research, Inc., Golden, CO

Biomass based chemical refining represents many new opportunities in engineering and process design. The population of companies seeking to develop their technology is large and diverse, from pure liquid fuel producers to integrated multi-product refineries. A common flow sheet among many of these organizations shows a hybrid biochemical/thermochemical process, whereupon the carbohydrate values in a biomass are removed and the resultant lignin portion is utilized for co-products or process heat. As a company with capabilities in both areas, Hazen Research is conducting a series of experimental programs that generate data for a hypothetical process for the biorefining of Lodgepole Pine (Pinus contorta) and Ponderosa Pine (Pinus ponderosa) into chemical values and thermal chemical products. Initial studies demonstrated a clear correlation between biomass pretreatment conditions and the resultant pyrolysis/gasification products generated from the hydrolysis residue. The presentation will focus on new data generated in a second round of experiments, which include optimizing carbohydrate yields in hydrolysis as well as running a continuous thermal processing in place of batch experiments.

See more of this Session: Sustainability and Biobased Materials
See more of this Group/Topical: Forest and Plant Bioproducts Division

Green Poly (lactic acid) Composites Reinforced with Sugarcane Bagasse Residues

Letian Wang, Zhaohui Tong

Agricultural and Biological Engineering Department, IFAS, University of Florida,

Rogers Hall Museum Road, PO Box110570, Gainesville, FL, 32611-0570

Email: ztong@ufl.edu

Poly (lactic acid) (PLA) composites reinforced by sugarcane bagasse residues were prepared using twin-extrusion. By using near zero-value bio-based residues, these composites are not only 100% renewable and biodegradable, but also have the competitive cost in comparison with petroleum-based composites. The sugarcane bagasse residues (10-30 wt.%), containing different content of fibers and all the remaining lignin, were used to prepare lignocelluloses reinforced PLA composite with desirable physical and chemical properties. In this study, the sugarcane bagasse residues include the residues originally from the sugar extraction process, after pretreatment, and after fermentation of biorefinery processes. The varieties of different bagasse residues include the composition, particle size and its surface reactivity during the chemical and biological (enzymes, microbes) treatments that the biomass is subjected to. The effect of these varieties on the morphologies, thermal properties, mechanical properties and the molecular weight of PLA composites was studied. Furthermore, a coupling agent (Desmodur® VKS 20) was used to increase the interfacial bonding of PLA with lignocelluloses. Desmodur® VKS 20 is a mixture of diphenylmethane-4, 4'-diisocyanate (MDI) with isomers and higher functional homologues (PMDI). The microstructure and the properties of the composites in the presence of couple agents were also investigated in this study. It was found that this coupling agent could extend the polymer chain and therefore improved the mechanical properties of the composites.

ALTERNATIVE USES OF SUGARCANE MOLASSES AND JAGERRY FOR PRODUCTION OF VALUE ADDED FATTY ACIDS, ESPECIALLY OXALIC ACID

P.Sanyal*, S. Gupta, S. Chattopadhyay and A.K. Ray
Department of Paper Technology
Indian institute of technology roorkee
Saharanpur campus, Saharanpur-247001, India
*National Sugar Institute, Ministry of Food and Civil Supplies, Govt. of India, India

ABSTRACT

Sugarcane molasses and jaggery provide a unique source for large number of downstream products which include livestock and human food sources, potable alcohol and absolute alcohol, alcohol based chemicals like oxalic acid, citric acid, tartaric acid, lactic acids, acetic acid, diethyl oxalate, ethyl acetate, acetic anhydride etc., which are raw materials for manufacture of pharmaceuticals, plastics, synthetic rubber, base for yeast, lysine, protein and innumerable other products. In this present study, status and economics of Indian industries who utilize molasses to produce organic acids through both chemical and biochemical routes like lactic, citric, tartaric, aconitric and oxalic acid etc. are reviewed with particular reference to oxalic acid production and updated research findings. The oxalic acid or its derivatives have widespread industrial applications such as textiles, leather, oil refinery, metallurgy ,nuclear power plant, dyes, catalysts, explosives, straw bleaching, printing, marble polishing, metal and cloth cleaning, pharmaceuticals and antibiotics. It is also a very important chemical in petroleum, rare-earth, ink, rust, corrosion inhibitor, and dental adhesive processing.

Oxalic acid can be produced from a number of sources from sugar itself, from sugarcane or sugar industry wastes, sugarcane trash, wastes from paper industry by oxidation of their main components of carbohydrates e.g. sucrose, glucose, and other sugar bearing raw materials, from dry powder of parthenium hysterophorus L., and other cellulosic wastes, wood dust, wood cellulose etc. Sugar cane or sugar beet molasses, sugarcane jaggery, deionized cane juice can be a potential source from sugar industry wastes. Both chemical as well as biochemical process may be followed to produce this important chemical. In chemical route generally catalytic oxidation of the above ingredients by nitric acid or by formed nitrogen oxides in batch or continuous stirred tank reactors produces oxalic acid, acid sachharate, and tartaric acid. Production of oxalic acid by fermentations of carbohydrates using *Aspergillus niger* and *Pseudomonas fluorescens* and radiation processing of carbonate solutions and molasses are reported. Production of tartaric acid and citric acid under certain conditions can also be produced. However, the majority of work involves oxidation of sugar bearing materials by nitric acid to produce oxalic acid. In this present study, an attempt has also been made to optimize the processes of producing oxalic acid from sugarcane molasses and sugar cane jaggery by oxidation with nitric acid in presence of dehydrant and catalyst with or without the presence of catalyst promoters. Parameters affecting the reaction were: air flow rate, temperature, the amount of V_2O_5, catalyst promoters and the concentrations of reactants and dehydrant. The maximum yields obtained by different processes were compared. Treatments of NO_x vapour by absorbing in a suitable medium (in soda ash or sulphuric acid), to avoid air pollution (and also to produce value added by-products, glauber salt and sodium nitrate), mother liquor and catalyst recycling were attempted to enhance the economy of the plant.

INTRODUCTION:

India is the largest producer of sugar in the world [1-4]. In terms of sugarcane production, India and Brazil are almost equally placed now a days. Two to three years' before it was the second largest producer of sugar in the world next to Brazil. At present, total world production is of the order of 110-116 million tons from both beet and cane sugar.India is expected to produce round 27.5 million metric tonnes (MMt)and 30.5MMt in 2011-12 and 2016-17 respectively.

It is wellknown that one of the important by- product of the Sugar Industry is the cane molasses, which is a strong building block for many chemical industries. In spite of producing cane molasses in large quantity in the country, several molasses based products are being imported in substantial quantity[4]. Department of Science and technology in India submitted a report which detailed the demand supply scenario and status of the Indian molasses based industry concerning the five products namely Lactic Acid, L-Lysine, Gluconic Acid, Glutamic Acid/Monosodium Glutamate and Itaconic Acid. It is necessary that India should set up facility for the production of the above molasses based products for the reasons [4]such as to utilize the molasses availability, to avoid the import of the molasses based downstream products, to export the molasses based downstream products in the global market. To achieve breakthrough in the fermentation technology for the molasses based products, as molasses is the byproduct of the sugarcane industry it is important to highlight regarding the status of this industry in this country.

STATUS OF SUGAR CANE INDUSTRY IN INDIA:

Sugar industry, worth Rs 60,0000 millions is also one of the largest agro-based industries in India-a country that happens to be also the largest consumer of white plantation sugar.[1-2] It has manufactured 26 million MT in 2007-08 which has declined over the following years due to varieties of reasons, 14.7 million tons 2008-2009 and again increased to 16.0 million tons 2009-2010. As already indicated in the coming years, the production of sugar is likely to go up steadily in the country [1-3].With approximately 25 million tons of sugar production in 2010-11 and 24 million tons of consumption, both sugar production and consumption are bound to increase lineally. The growth rate of sugar production in India is shown in Fig.1 and the same for molasses is given in Fig.2.

Indian sugar consumption is growing at a CAGR of 3.75% .Sugar production in India has occupied a place of pride in the Indian economy being one of the five largest industries comprising of more than 571 units spread all over the country.

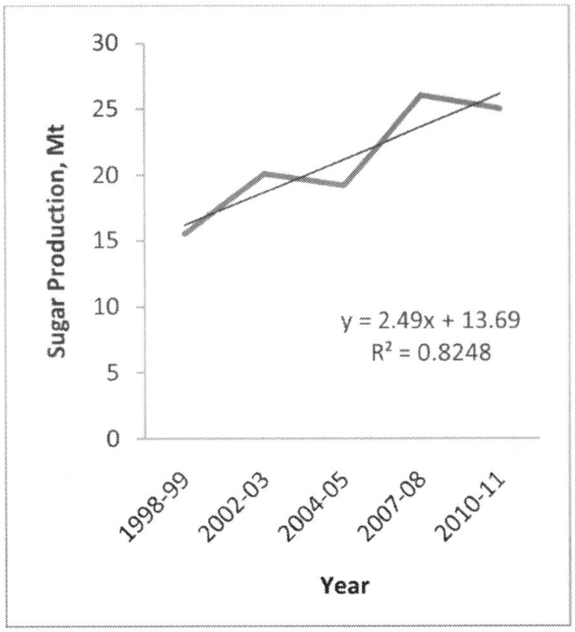

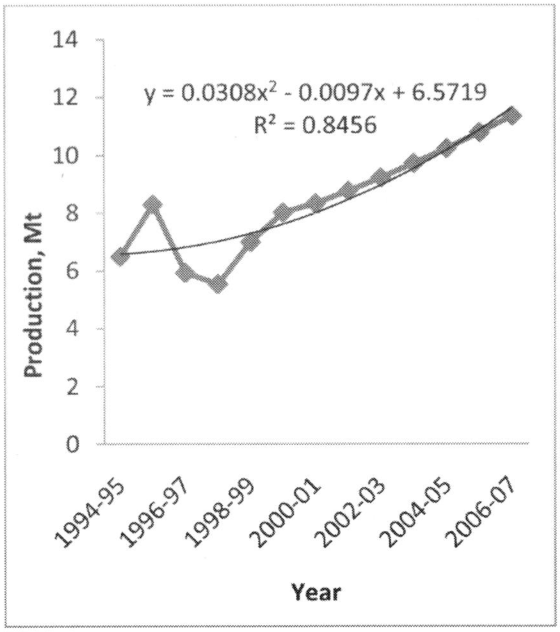

Fig.1 Production of white plantation sugar. Fig. 2 Production of Molasses

The average yield of cane in India is around 70 -77 t/ha in tropical states to about 52t/ha in subtropical stases[1] though in UP(which produces 40% sugar) the yield has decreased from 55 MT/ha to 45 MT/ha in the recent past. High yielding new varieties, namely Co 238, Co 239 and Co 118 identified can be used to increase the productivity of sugar cane and sugar. The area under sugarcane is presently less than 2% of total cultivable area in the country and about 3% of the irrigated area. There is considerable scope for increasing the area under sugarcane considering the fact that it is more profitable compared to other crops.

INDIAN MOLASSES CHARACTERISTICS AND ITS UTILIZATION:

Molasses is the uncrystallizable syrup which remains as final residue in process of manufacture of sugar. Its composition varies due to a number of factors, including cane maturity, variety, agricultural conditions, industrial processes etc. A typical composition of final molasses(all in %) in Indian Sugar Mill is as under:

H_2O	-	17-25	
Sucrose	-	30-40	} 50% Saccharine matter
Glucose	-	4-9	
Fructose	-	5-12	
Other reducing sugars	-	1-4	
Total reducing sugar	-	10-25	
Carbohydrates, Gums, Starch	-	2-5	
Water soluble carbonates	-	7-5	
Nitrogen	-	2.5-4.5	
Others	-	0.5-1.5	

Waxes, sterols phosphorus	-	0.1-1.0
Vitamins	-	varying amounts

As molasses contains substantial quantity of sucrose as well as reducing sugars, it must produce many chemicals acids as pure sucrose does as shown in the following diagram[1]:

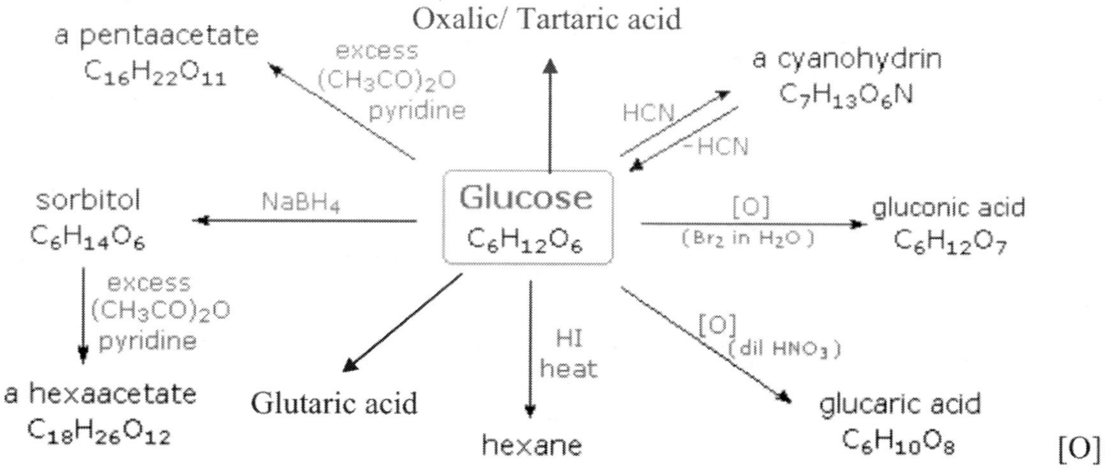

Most of the chemicals are however imported in India.

Use of Molasses

(a) Direct utilization: Fertilizer, animal feed
(b) Distilling industry: Ethyl alcohol, rectified spirit, power alcohol, alcohol derivatives, potable liquor, such as beverages
(c) Other fermentation industries: Vinegar and acetic acids, acetone and butanol, citric acid, lactic acid, glycerine, yeast,
(d) Other chemical industries: Oxalic acid, and tartaric acid
(e) Miscellaneous industries: Aconitic acid, monosodium glucamate, dextran,L-Lysine, Xanthenes gum, 2-keto gluconic, 5-keto gluconic , a-ketoglutaric, fumaric acids , glycolic acid,kojic acid and glucaric acid etc.

Besides the above, molasses can be used for many other value added products.Some of the important value added chemicals (mostly imported) are briefly described below:

Molasses is the only raw material available in large quantities in India for an economic production of alcohol until petrochemical source develops. Almost all the commercial molasses in the world is cane molasses because the beet-sugar producing countries usually consume it domestically. The most significant use of final molasses in many industrialized, countries is in animal feed. Other uses of molasses are as the raw material for alcohol production and as the base for yeast, lysine and citric acid production and may others. The chemicals derived from molasses could be produced cost effectively and be profitably exported [3,5-40].

Molasses contains small amount of Lysine, or L-lysine, an essential amino acid, 2,6-diaminohexanoic acid, building blocks of protein, required for proper growth but the body can't manufacture it; lysine has to be obtained from food. Lysine it plays an essential role in the production of carnitine, a nutrient responsible for converting fatty acids into energy and helping to lower cholesterol. Lysine appears to help the body absorb calcium, and plays an important role in the formation of collagen, a substance important for bones and connective tissues including skin, tendon, and cartilage. Glucaric acid is a member of a larger group of compounds known as sugar acids – and more specifically, aldaric acids. Glucaric acid has gained attention because it was identified as one of the top 12 renewable building block chemicals.

Fumaric acid

The principal industrial use of fumaric acid is as a mordant in textile dyeing. It appears that it could find application as a replacement for tartaric acid in leavening agents, as well as in the manufacture of resins and plastics for special purposes. Itaconic and fumaric acids, if sufficiently cheap, could flow into plastics. The same is true of citraconic acids which may be derived from citric acid. In addition, this group is outstanding as acidulating agents in foods. Citric, lactic, and gluconic acids are known to be safe ingredients of the diet and are extensively employed in foods, soft drinks, confectionary and pharmaceuticals.

Itaconic acid

Itaconic acid, 2-Ketogluconic acid is produced by the action of various species of bacteria of the genus *Pseudomonas* on glucose in aerated cultures in the presence of calcium carbonate. This is formed by *Aspergillus itaconicus* and is now available commercially. It has the possibilities in the preparation of resins and plastics because it is a substituted acrylic acid. It is now possible to get 37-percent weight yields in 12 days at about pH 2.2. With submerged fermentations, yields as high as 45% have been obtained in 5 days at a pH of 1.8.Weight yields of 90 percent can be obtained in a 30-hour submerged fermentation. The keto acids can be converted to d-araboascorbic acid which is a good antioxidant for food use. In this capacity, it can serve in the preservation of flavors in the canning industry and in preventing the development of oxidation flavors.

Kojic acid

It is very chemically reactive compound and many derivatives have been prepared from it. It can be prepared readily in yields of more than 50 percent by molds growing on a variety of carbon sources like glucose and xylose, with the use of members of the *Aspergillus flavus, A.oryzae* group.

5 keto gluconic acid is produced by various species of *Acetobacter* in weight yields exceeding 90 percent. When oxidized by air in the presence of vanadium pentoxide, it is converted to d-tartaric acid with good yield. A ketoglutaric acid can be produced in a 25 percent yield when fermentation of glucose by *Pseudomonas fluorescens* is allowed to proceed beyond the 2-ketogluconic acid stage to the point where no more reducing sugar is observed. Good yields of bionic acids such as maltobionic and lactobionic acids could be obtained by the action of *Pseumonas graveolens* on maltose and lactose. The pentonic acids such as d-arabonic, and d-

ribonic acids are formed with good yields during oxidation of d-arabinose , d-xylose and d-ribose by the genus *Pseudomonas*.

Lactic Acid:

Five tons of molasses would be needed to produce one ton of lactic acid.

Lactic acid is used primarily as an acidulant and preservative agent in the pharmaceutical, chemical, and food industries like beverages, beer, jelly, etc. Described as ``the chemical of the future" because of its use in the manufacture of polylactic acid, it is reckoned that polylactic acid could substantially replace plastic bags, since it is bio-degradable [9-10]. The current interest lies in making biodegradable plastics used for general and medical applications as a delivery vehicle for vaccinations.

Lactic acid production parameter optimization[9] was performed using cane molasses by the design of experiment(DOE).Eight factors with three levels studied were yeast extract, $CaCO_3$, $MnSO_4$, pH, temperature, molasses, urea, and Tween 80. Lactic acid production was significantly affected by interactions of two factors such as temperature-urea, pH-$CaCO_3$, temperature- Tween 80, but individually they have minimum impact on lactic acid production. Individually pH, yeast extract, molasses, and urea are the most significant factors in lactic acid production. Lower amount of $CaCO_3$ and $MnSO_4$ and Tween 80 enhanced lactic acid production. The expected yield of lactic acid under optimal condition was 98.66% and the actual yield was 96.8%. Overall 6% improvement in yield has been observed when compared with same (mutant) species grown on cane molasses.Lourdes R. Fragante [5] has conducted fifteen experiments using *Lactobacillus delbreukii* in aqueous molasses and molasses/coco water solutions containing 9, 10, 11 and 12% sugar as invert. The additives used were 2% nitrogen, 1% phosphate and 8.3% invertase per 35g. molasses. The incubation temperature was 45°C. Highest yields were obtained at 10% sugar level for all substrates. Coconut water was found to support lactic acid production. Substrate-Molasses in coco water solution with invertase further increased the yields but only up to 10% sugar level. The presence of higher glucose/fructose levels seemed to inhibit lactic acid production. All substrates gave decreasing yields at 11 and 12% sugar.

Citric Acid

Seven tons of molasses would be required to produce one ton of citric acid.

This chemical is mainly used as a preservative in food processing, pharmaceutical and cosmetic industries. It is also used to manufacture sodium citrate, used in the production of detergents. Industrial production of citric acid by fermentation is one of the most efficient examples of fungal fermentation technology[7-8].The food industry is thought to comprise as much as 70% of this market; the remainder being largely pharmaceutical produce, though the metal cleaning, agriculture and detergent industries would also be of notable importance.

Tartaric Acid:

Tartaric acid is used above all in many applications in the food and textile industries. For example, tartaric acid is used as a food additive, especially as an antioxidant and acidification agent. In addition, tartaric acid is used as a building material additive for modification of the setting characteristics of cement and serves as a starting substance for numerous chemical reactions, especially for chiral synthesis.

A method of preparing tartaric acid in which 5-ketogluconate is oxidized in a carbonate buffer at an alkaline pH preferably in the range of 8 to 10 using vanadate has been described [11-13]. One way of producing tartaric acid is to recover it from tartar, a substance which is produced during the production of wine.

However, the crude tartar which results from the production of wine is as a rule contaminated by organic materials so that expensive preparation is necessary to obtain the tartaric acid from the tartar.

It has been found surprisingly that under these reaction conditions, i.e. also without the addition of vanadate or another catalyst, a conversion of 5-ketogluconate to tartaric acid results whereby a conversion of 5-ketogluconic acid to tartaric acid of up to 55% is achieved. If vanadate is additionally supplied to the reaction mixture, conversions up to 75% of the 5-ketogluconate used to tartaric acid are obtained[11-12].

A further process for producing tartaric acid resides in the conversion of glucose to tartaric acid through microorganisms like *Acetobacter* or *Gluconobacter* in the presence of vanadate as trace element. It can be demonstrated that in these processes the tartaric acid production is carried out in two steps. In the first step 5-ketogluconate is excreted into the medium of microorganisms cultured in a glucose-containing medium, while in a second step the conversion of 5-ketogluconate to tartaric acid is carried out purely chemically with the vanadium salt as trace element, i.e. independently of the presence of microorganisms[13].

PRODUCTION STATISTICS AND ECONOMICS:

Although India is a major producer of molasses (fig.2). Sugar mills produce around 8 million tons (mt) of molasses, which is projected to increase to 10.5 mt over the next ten years. About 14 per cent of the molasses produced today is exported, without any value addition. There is a growing demand world over for molasses-based chemicals, which are presently imported [6]. The huge export opportunity is being lost, because, principally, the Indian industry is not fully aware of the business potential in this area. There is a possibility of manufacturing these chemicals with indigenous resources. Here again, a huge export opportunity exists manufacturing this chemical at competitive prices, further harnessing of overseas business opportunities. On the other hand, the country imports molasses-derived chemicals in very large amounts

Estimated demand for lactic acid (partly imported) is 2500 tons in the year 2015 A.D. Lactochem Ltd of the Malladi group has a 200 tpa plant in operation. Karnataka-based Munirabad Chemical Co, which has a 180 tpa plant, is another example of a successful plant. Orchem Ltd, an associate unit of Atul Drug House, attempted to put up a 300-tpa plant, but did not succeed. The company first attempted to manufacture the chemical synthetically on a laboratory scale, but ran into difficulties when getting into commercial scale operations. Another company called Sangita Bio-Chem of Orissa also attempted a venture, but is said to be held up for want of funds. Two new units are coming up; one is Prathista Industries in Andhra Pradesh, with 10000 tpa capacity and Interspice Biotech near Chennai. The latter is coming up with technology from GaLactic S.A. of Belgium. The study recommends setting up of a plant of 3,000 tons capacity, which would cost around Rs 27 crore.

In 2002, citric acid production by fermentation was estimated to be in the region of $7.0x10^5$ ton/year [8]. Such production levels are necessary to meet the high demands of the now massive market. From the design of the original fermentation process right up to the present day, citrate production has largely been overseen by Pfizer. Deviation from the market profile of the 60s and 70s has seen the once dominant pharmaceutical market demand overshadowed by that of the ever-growing food market, particularly the carbonated drinks industry. This shift has seen the price of citrate fall significantly, thus encouraging further improvements in the rate and efficiency of its production. Unlike oxalic acid, the technology for producing citric acid from molasses is established and available. The Indian industries worth mentioning are Bharat Starch Industries of the Thapar group, Citric India Ltd of Mumbai and Andhra Citrates of Hyderabad, Citurgia Biochemicals of Mumbai.

For the first time, a detailed study (commissioned by the Centre and conducted by a consultancy company called Nandini Chemical Consultancy[3] has identified some molasses-based products and the business potential for each of them. The study, which advocates setting up of plants for the identified molasses-based chemicals, makes two postulates for its recommendations. First, the price fluctuation of molasses in the country is immaterial because the value addition in the manufacture of these chemicals is so high that volatility in prices is inconsequential. Second, the target market is the world market and any entrepreneur who wants to first establish a ``good domestic base" might as well look for opportunities elsewhere.The following table indicates the approximate business statistics for three principal molasses derived acids.

Table Major Molasses based acids, their demand-Supply, growth rate, export potential

Acids	Global Demand, tons(year)/ Growth Rate,%	Demand/Installed and Production Capacity, tons	Export, tons
Oxalic acid	250,000; 3.0	30,000 /22,000	9,000
Citric acid	703,200(2001), $7.0x10^5$ (2002),1,040,900(2011)	12,000(2001)/ 38,475(2011)	4,610(0.4% of world market)
Lactic acid	70,000(2001)/100,000(2002) 181,560(2005),500,000 (2011)	560(2001)/ 1,980 (2011)	5,000(3% of world share)

Techno-Economic feasibility studies:

Department of Science and technology (DST), Government of India, has estimated the likely gap in supply by 2015 AD in India and recommended the Installed Capacity of the following chemicals. The detailed project report has been made based on number of 300 working days per annum and 24 numbers of working hours.

Lactic acid, is presently consumed 200 Mt, partially met by imports, likely to increase to 600 M t. Only M/s Lactochem India Ltd. is manufacturing it. L-Lysine, itaconic acid, gluconic acid and glutamic/ MSG acid all are imported, India will need 18,000, 1800, 1500 and 10,000 Mt respectively. Adequate electrochemical process technology has been developed for the production of Gluconates from Glucose by Central Electrochemical Research Institute, Karaikudi to produce the later two chemicals.

The Nandini Chemicals study has also said that huge export possibilities exist for other molasses-based products such as lysine, glutamic acid, itaconic acid, gluconic acid, ephedrine hydrochloride and yeast.

Oxalic Acid:

To produce one ton of oxalic acid, 2.5 tons of molasses would be required.
Oxalic acid and its derivatives have widespread industrial applications like textile, leather or tanning, oil refinery, catalysts, dyes, metallurgy, pharmaceutical, marble polishing, petroleum, nuclear engineering, explosives, detergent, plastic, film industries and antibiotics.
Other uses include dental adhesive processing, corrosion inhibition, straw bleaching, printing, precipitating agent in rare-earth mineral processing, bleaching agent in the textile activities, wood pulp bleaching, wood bleaching agent(used to lighten and brighten dingy gray wood), remover of cement residues, rust remover for metal treatment, commercial rust removers to remove rust stains from tubs and sinks, food and rust stains from kitchen countertops, grinding agent, such as marble polishing ,waste water treatment, removing calcium in water, cleaning and sterilizing home brewing equipment, as a reducing agent for photography and ink removal ,as purifying agent in pharmaceutical industry, clean rock collectors, clean mineral specimens, , plumbing fixtures and fabric.

Indian Statistics for oxalic acid manufacture:

Nandini Chemicals recommends a minimum economic size of 15,000 tons, which will call for an investment of about Rs 17 crore. In India, oxalic acid is produced from sugar, by a dozen-odd companies, most of them Mumbai-based. Two companies have attempted to produce oxalic acid from molasses instead of sugar - but both attempts failed. One was a Chennai-based company called Krebs India, which put up a 300 tpa plant at Pondicherry. The other was Punjab Chemicals and Pharmaceuticals Ltd, the largest producer of oxalic acid through the sugar route. The cost advantage can be gauged by the fact that while it takes 600 kg of sugar to produce one ton of oxalic acid, it takes 2.5 tons of molasses to produce the same.

Present cost of molasses based acids:

The present market cost of the following acids from molasses in Indian currency is as under:

Citric acid:Rs. 51.74,Oxalic acid, Rs.52,Fumaric acid(Food grade):60.72,Lactic acid:Rs.68.33,L-Tartaric acid:122.75,D-Tartaric acid :251.09,. These are in the ratio of 1:1.005:1.174:1.321:2.372 :4.853. It is reported that the ratio of prices of some important acids including the mentioned one are as under: Glycolic: Formic acid: Acetic Acid: Oxalic acid : Tartaric acid: Butyric Acid=1:1.44:1.5:1.61:3.56:3.78. From the current market study it is usual noticed that the imported chemicals are found to be cheaper than those indigenously manufactured.

R& D efforts for Oxalic acid production:
Literature Survey:

There are various methods for manufacturing oxalic acid [14-40]. Process depends upon the raw material selected. The conventional method of production generally consists of glucose or sucrose oxidation in presence of nitric acid and vanadium pentoxide is used as a catalyst. Attempts have been made to prepare oxalic acid from ethylene, ethylene glycol, propylene, lignin, jaggery, cane and beet molasses, sugarcane, sugars, flour, plant wastes, formate, carbonate and bicarbonate salts, cellulose and wood wastes by using different methods. These methods are: oxidation of pure sucrose or impure commercial sugars, cellulose and other carbohydrates by nitric acid, or oxides of nitrogen, fermentation of carbohydrates, fusion of sawdust with caustic soda, oxidation of olefins(ethylene , propylene) and glycols, radiation processing of carbonate solutions and molasses, and decomposition of formates.

Of the above the first two received much attention.Other methods such as thermal decomposition of sodium formate (Radulescu et al., 1993)have little bearing on economic production of oxalic acid.

Fermentations of carbohydrates to oxalic acid have been reported for *Aspergillus niger* (Rujiter et al., 1999) and *Pseudomonas fluorescens* (Hamel *et al.*, 1999).

Number of attempts have made for the production of oxalic acid by the nitric acid oxidation of carbohydrates such as glucose, sucrose, cellulose and commercial waste materials (Metin Gürü,*et al.*,2001,Guo *et al.*, 1997, Sattar *et al.*, 1988, Mane *et al.,* 1986; Mane *et al.*, 1988, 1986, Nowak,1986 , Sullivan *et al.*,1983,Dube *et al.*, 1982, Shukla *et al.*, 1982 ,Sanyal *et al.* 1981,82,Gupta *et al.*, 1980, Despande and Vyas ,1979,. Agrawal *et al.*, 1979, Raha *et al.*, 1976, Bose *et al.*,1971, Preen,1942)

Bretschneider *et. al,* [15] reported oxidation of sucrose by nitric acid and discussed the kinetics of the process for tartaric acid as the main product., not oxalic acid.
Berntsen and Degering(1949) oxidized glucose to produce potassium acid sacchharate with oxides of nitrogen and Chia and Lin(1962) oxidized sucrose to produce gluco-saccharic acid. Tanz and Kao (1944) reported work on the selection of catalyst and promoters for this oxidation reaction using oxides of Vanadium, Molybdenum and Tungsten. The oxidation of ethylene glycol to oxalic acid has been discussed by Vishwalingam *et al.*, (1980) and kinetics was studied. The production of oxalic acid from other sugar bearing commercial by-product materials such as jaggery (gur) , beet molasses , cane molasses and wood cellulose have also been tried by a very few investigators [14].
Raha *et al.*, [30] have described especially the method for producing oxalic acid from Gur. In most of these cases after pretreatment the raw material was oxidized by nitric acid in presence of vanadium pentoxide. Agarwal and Rao [14] oxidized cellulosic materials by nitric acid and found ammonium-meta-vanadate which is a cheaper material to be equally effective as the oxide of vanadium.
The attempts for producing oxalic acid using waste materials like final molasses have not been very economical. Bose *et al.,* [17] attempted the preparation of oxalic acid from sugarcane

molasses using nitric acid and vanadium pentoxide and obtained about 55 percent of the theoretical yield.

Deshpande and Vyas [19] have produced oxalic acid from sugar and the nitrogen oxides evolved were absorbed to produce sodium nitrite as a by- product . Gupta *et al.*, [22] reported the production of oxalic acid from deionized sugar cane juice.

Optimum conditions of reaction as developed earlier were employed in the present investigation and the raw materials (sugarcane molasses) of three different compositions were utilized.

Metin Gürü,*et al.*, [23]conducted exhaustive investigation for the production of oxalic acid from sugar beet molasses using a series of three reactors with formed nitrogen oxides in the following two reactors Parameters, air flow rate, temperature(45 °C to 58°C), the amount of V_2O_5 catalyst and the concentrations of molasses and H_2SO_4 were optimized. The process is however a noble one in respect of recovery of NOx, to prevent air pollution ,to reduce the unit cost of reactant for oxalic acid production and the equipment arrangement. The yield was reported of the order of a maximum 77.5% and 74.1% of theoretical yield obtained by using the absorption solutions with NOx.

Though the process is claimed to be Novel, definitely improvement over the earlier one, in India firstly the raw material used by Guru *et al.*, beet molasses is scarcely available(almost negligible) whereas cane molasses are abundantly available. Besides, beet molasses from refinery goes for cattle feed, hence not available. It however contains more sucrose than that of molasses but lower than jaggery. The sugar beet molasses used in the experiment contained 48.1% sucrose, 3.2% invert sugar and 19.4% water. These are compared in the following table.

Investigators	**Yield of theoretical,%**	**Raw Materials employed**
Metin Guru *et al.*,[23]	78.9/ 77.5 /74.1	Sugar beet Molasses
Sanyal *et.al.*, [33]	74	Jaggery (Gur)
Raha *et al.*, [30]	74	Jaggery(Gur)
Shukla *et al.*,[35]	48.5	Cane Molasses
Agarwal *et al.*[14]	< 50	Cellulosic materials
Sattar *et al.*,[34]	48.5	Molasses
Mane *et al.*,[27]	51.6/60	Sugar cane trash/Saw dust
Sullivan [36]	80.2	Saw dust
Bose *et al.*,[17]	55	Sugar cane molasses

Therefore it can be ascertained if final cane molasses is at all allowed to use , manufacture of oxalic acid, and tartaric acid with the three CSTR as proposed by Metin Gürü,*et al.*, with optimum process conditions of temperature, catalyst concentration etc can give the best yield.

Therefore the work done previously by Sanyal *et al.*, [33] are reproduced here under for effect of two different raw materials , namely jaggery and final cane molasses for further critical analysis;

Raw Material and Chemicals Used:

1. Sugarcane Molasses: Khandsari and final sugar cane factory molasses were used and their analyses are as follows:

Specifications on Sugarcane molasses

Sl. No.	Item	Molasses Sample		
		A	B	C
(i)	Brixs (Bx^0)	70.30	80.30	84.00
(ii)	Pol.	33.00	34.90	-
(iii)	Purity	46.94	43.46	-
(iv)	Sucrose %	29.90	28.00	39.98
(v)	Sucrose Purity	42.53	34.80	41.80
(vi	Reducing Sugar %	21.42	18.36	20.80

2. Gur(Sugar cane Jaggery):

Large quantities of gur are produced in India mostly from small scale units and the properties are likely to change depending upon the sugar cane used. Sudies were made using a single batch product with the following composition:

Specifications of Sugarcane Jaggery (Gur)

Parameters	Percentage
Sucrose	76.3
Reducing sugars	9.5
Purity	84.65
Ash	3-6
Colour	Yellow Brown
pH	6.3

1. Acids: Commercial grade Nitric Acid (60 % HNO_3) and A.R. grade Dehydrant(H_2SO_4)
2. Catalyst: LR grade Vanadium Pentoxide (V_2O_5) in powder and pellet forms.
3. Promoters: LR grade oxides of Tungsten(WO_3) Molybdenuim(MoO_3) and Iron(Fe_2O_3)
4. Soda Ash: commercial grade Soda Ash (40% Na_2CO_3).

Experimental:

The experimental set-up for the batch reactor and other accessories used has been described elsewhere [33, 35]. The reactor consisted of long glass tube vessel of 1 liter capacity fitted with internal glass cooling coil through which constant temperature water was circulated. Three side tubes were attached for liquid feed, for air bubbling and for continuous removal of the reaction gases.

Procedure:
Predetermined amounts of gur/ molasses, distilled water and catalyst 0.08 % were charged in the reactor. The air compressor was started to introduce at a rate of 9-10 l/min and the acid mixture containing measured amount of nitric acid and sulfuric acid was added slowly. Reaction was allowed to proceed for a particular batch time and the temperature was maintained at 54.0±1.0°C.

After completion of the batch whole mass was transferred to a flask and was left overnight at a room temperature of 25.0±4.0°C. Needle shaped crystals of oxalic acid dehydrate were formed which were filtered, washed and dried. This product gave the yield for first crop. The filtrate was again cooled in the freezer at 2.0±1.0°C for about 12 hrs. which gave further crystals and the product gave the yield for the second crop. The two crystal yields were combined to get the total yield.

The repeated experiments were carried out for both jaggery as well as molasses at desired air flow rates, reaction time, temperature and catalyst concentration. Product yield was defined as the ratio of oxalic acid dehydrate produced to the amount of oxalic acid dehydrate that was expected by reaction stoichiometry from a particular amount of sucrose present in the jaggery sample. For further experiments on molasses, the filtered air flow rate was kept constant and the reactant to acid ratio was varied. Oxalic acid yield was computed after separating the crystals from mother liquor.

Results and Discussions:

Experimental results on sugar cane molasses were shown in Fig.3 indicating how reactant ratio(molasses : dehydrant: HNO_3).The molasses however, has been kept in the ration as 1.

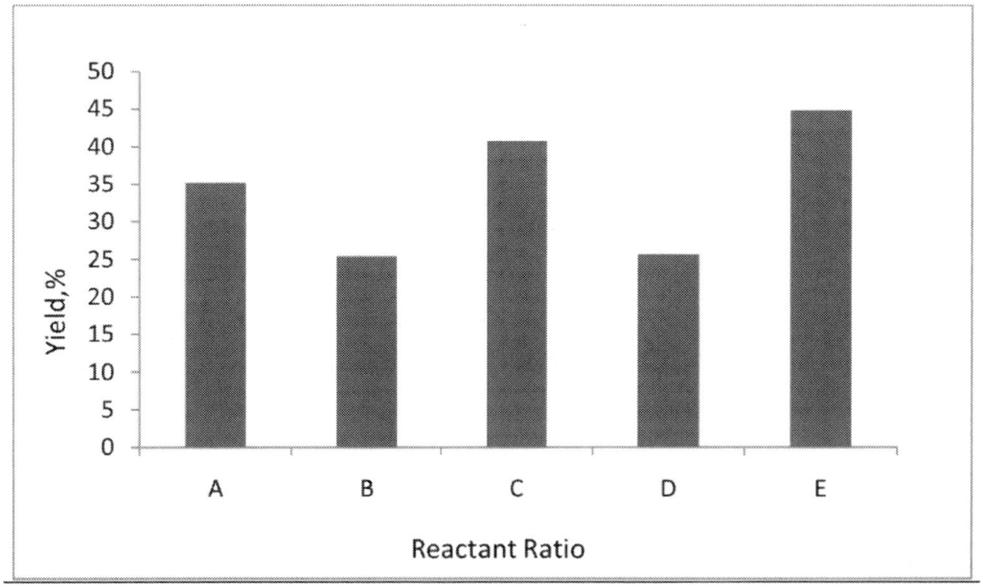

A=1:0.92:4.20, B=1:1.38:4.06, C=1:1.84:4.06, D=1: 1.895:3.60, E=1:2.134:5.41

Fig.3 Yield of oxalic acid as a function of reactant ratio

It is clear from the plot that slightly higher amount of dehydrant and nitric acid gave maximum yield of oxalic acid about 48% of theoretical yield. There was not much difference observed between C and E. considering the cost of dehydrant and nitric acid C might be chosen as the optimum yield.

Vapor Treatment: Reaction vapor products consisting of moisture and nitrogen oxides were allowed to pass through soda ash solution. Thus the gas absorption with chemical reaction took place in the absorber. After the batch completion, the solution was concentrated and cooled for crystal formation. The crystalline product was analyzed for carbonate and nitrate fractions. It is evident from the results that product ratio of $NaNO_3$ and Na_2CO_3 ranges between 1.69 -1.90 with the percentage conversion of soda ash between 60-65%.

Mother Liquor Treatment and recycling: After complete possible extraction of oxalic acid, the mother liquor was treated with soda ash, under constant stirring, till neutral solution was obtained. This neutralized solution was concentrated and cooled to get glauber salt (Na_2SO_4) mixed with unreacted soda ash (Na_2CO_3). The yields and analysis of results indicated that $CO_3^{//}$: $SO_4^{//}$ and yield of Na_2SO_4 (in percentage of theoretical) were between 1:2.70- 1:3.35 and 67.0 - 78.2 respectively. Analytical procedure given in the literature was employed [37- 38].Mother liquor recycling results gave yield (in % of theoretical yield) 18.01% -22.60%.

Product Purity: Oxalic acid crystals were tested chemically and physically using Bausch a Lomb Spectrophotometer and Erma Colorimeter. A comparison was made with I.S.I. grade sample and the product was found to be 99.5% pure.

Parametric relationship for jiggery utilization:

Effect of Reaction Time:

The effect of reaction time on yield was evaluated by varying the reaction time from 1 to 12 hours at constant 0.08% catalyst concentration as shown in Fig.4. The experimental results indicated that the yield of about 74.0% of oxalic acid was attained at a reaction time of 10.0 hrs, beyond which the yield started decreasing. This

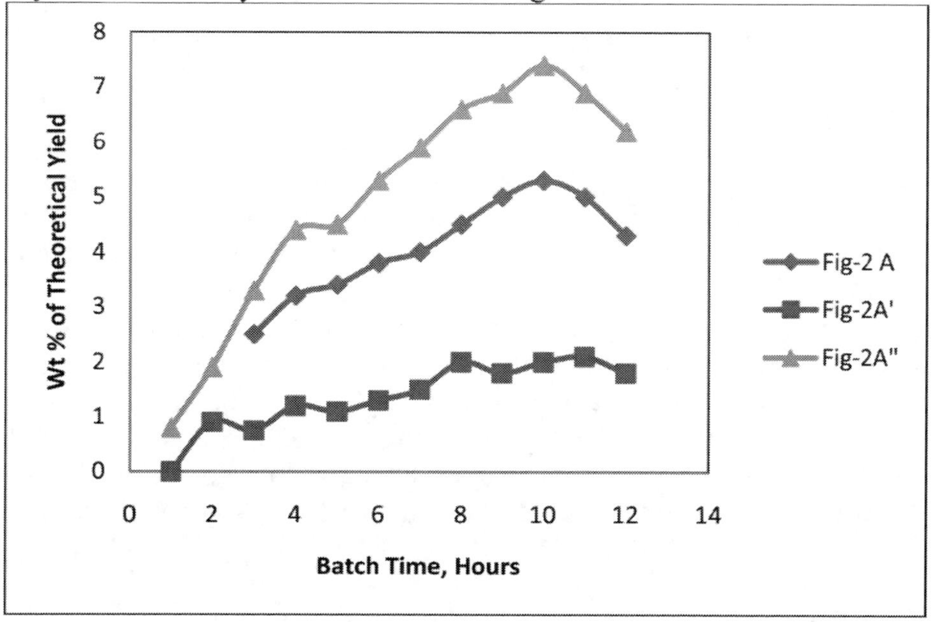

Fig.4 Percent of theoretical yield as a function of reaction time

Effect of Catalyst Concentration:

Vanadium pentoxide powder was used as catalyst. The effect of catalyst concentration on yield was studied at the reaction time of 10 hrs. Amount of catalyst in relation to the amount of gur was varied from 0.01 percent to 0.20 percent. The experimental results as shown in Figs.5 and 6 revealed that the maximum yield could be obtained at a catalyst concentration of 0.08 percent.

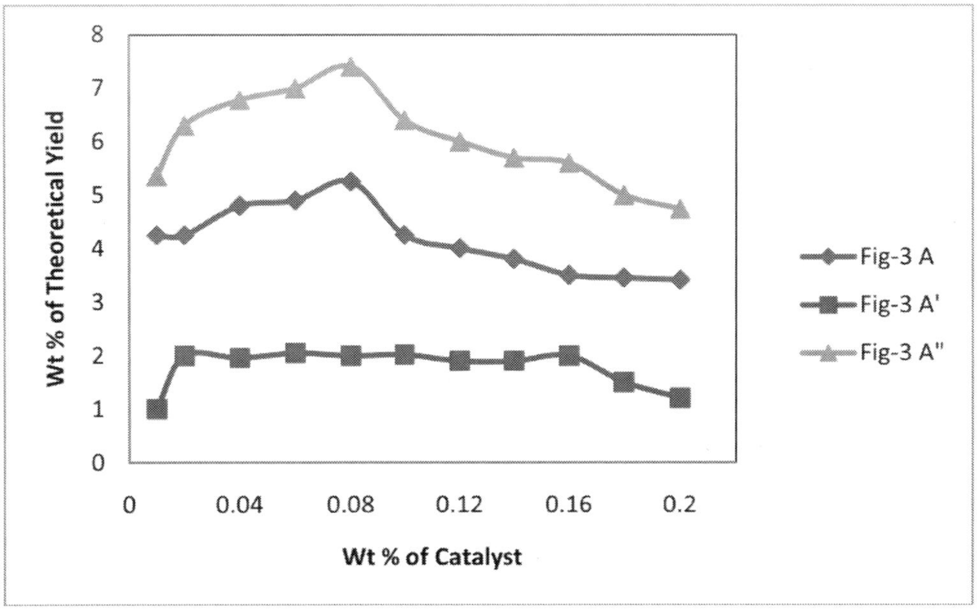

Fig. 5 Percent of theoretical yield of oxalic acid as a function of percentage of catalyst

Effect of Catalyst Promoters:

At the optimum amount of catalyst (0.08 wt. % V_2O_5), the effect of three promoters was studied using oxides of Tungsten (WO_3), Molybdenum (MoO_3) and Iron (Fe2O3). The amount was varied from 0.04 to 0.12 wt percent. The experimental results are shown in Fig. 6 reveals the following:

In case of Tungsten Oxide the yield gave a maximum at wt. 0.08% while in other two cases the yield decreased as the amount of promoters was increased. Furthermore, on comparing the results, it was observed that vanadium pentoxide alone used as catalyst gave better yield (74.4% of theoretical) than the maximum obtained in the case of Tungsten Oxide (66.0 wt% of the theoretical). It was also observed that the use of promoters added colour to the oxalic acid crystals such as reddish brown in the case of ferric oxide, and lost the aesthetic value of the product.

For sugarcane molasses the following additional experiments were performed.

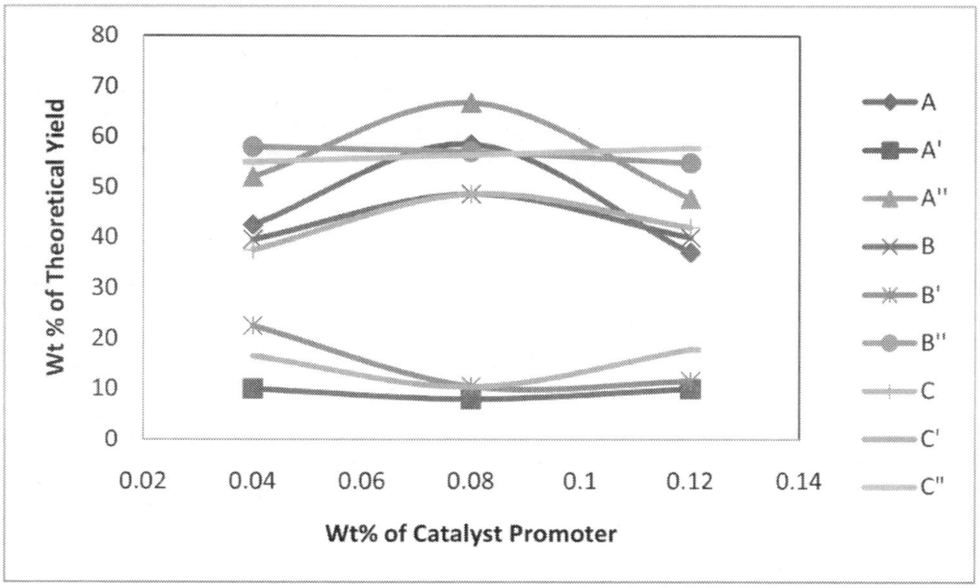

Fig.6 Percent of theoretical yield as a function of percentage of catalyst promoters.

Catalyst Recirculation: Vanadium pentoxide was impregrated in glass beads and alumina and used in place of the powder. It was observed that although at high acid concentrations there was no effect but at lower acid concentrations after the batch was over the oxide was soluble and hence it could not be recovered or recycled.

Gas Absorption: Nitrogen oxides evolved along with air and water vapor could be arrested only partially. As it is clear from the results, it is felt that a better absorber should be designed for complete removal of nitrogen oxides may cause air pollution and also offset the cost of oxalic acid production. The best way is to absorb it and recycle so as to reduce the nitric acid requirements.

Chemicals from Mother Liquor: Effluent liquid in the form of mother liquor contains unreacted acids, catalyst and other impurities originally present in the raw material (in the present case sugarcane molasses). Production of oxalic acid on industrial scale may pose the liquid effluent disposal problem and hence this treatment was tried. Effluent stream should be treated in such a manner that the pollution hazards are controlled as well as the useful chemicals are recovered to reduce the cost of production effluent treatment. Commercial grade soda ash was employed and the results are given in Table 3. It was noticed that no nitric or nitrates were present and about 73% conversion to glauber salt was obtained. In this way the left over sulfuric acid was recovered as well as the liquid effluent was treated.

Mother Liquor Recycling: Mother liquor analysis (after oxalic acid extraction) showed that a large portion contained unreacted dehydrant. Five batches were run using mother liquor. Make up nitric acid and fresh sample of molasses were added to start the batch. The results as discussed clearly show that an overall reduction in oxalic acid yield to an extant of 50% was obtained. Hence it is not instead it should be treated as discussed earlier.

CONCLUSIONS:

There are possibilities of establishing various molasses based acids in India such as lactic acid, oxalic acid, tartaric acid, ctric acid, aconitic acid, monosodium glucamate, dextran,L-Lysine, Xanthenes gum, glycerine, yeast, 2-keto gluconic, 5-keto gluconic , a-ketoglutaric, fumaric acids , glycolic acid,kojic acid and glucaric acid etc.These are mostly imported. Production are very scarce.

Projects regarding production of lactic acid, citric acid, tartaric acid and oxalic acid can be considered as competitive though there could be scope for further optimization of quality & cost standards. Also, more R&D initiatives are required to produce products of various grades, catering to the requirement of specific user. R& D work on the above acids are going on in various industries as well as research laboratories in India. Little R&D push, a unit would be able to produce oxalic acid from molasses. Producing the acid from cane molasses would be very cost-effective and India would be able to take a substantial share of the world market.

The following conclusions can be drawn from the results of investigation for producing oxalic acid dehydrates crystal using sugarcane jiggery and molasses as the raw materials.

1. The maximum yield, 74 wt percent of the theoretical could be obtained in 10 hrs. and increasing or decreasing the reaction time would decrease the yield.
2. Vanadium pentoxide at a concentration of 0.08wt% gave the maximum yield.
3. Addition of promoters like WO_3, MoO_3 and Fe_2O_3 decreased the yield and added colour to the final product, hence it is not recommended.
4. Maximum yield, 48.5 weight percent of theoretical could be obtained while in the case of gur (sugarcane jaggery) under similar condition it was 74.0 wt. percent.
5. Vanadium Pentoxide used as a catalyst has some solubility in dilute acids and hence cannot be re-cycled at this stage.
6. Nitrogen oxides can be absorbed in alkaline solution to give useful products (fertilizers) and control the air pollution.
7. Mother liquor re-cycling reduces the yield by 50% and is hence not recommended at this stage.
8. Mother liquor can be treated with alkalis to get useful products like glauber salt, ammonium sulphate etc.

REFERENCES:

1. http://search.conduit.com/?SearchSource=10&ctid=CT2776682
2. http://www.google.co.in/
3. *http://www.nandinichemical.com/publications/molassesbasedprojects.html*
4. Lourdes R. Fragante, Lactic acid from molasses, RDE abstract of completed research,Sugar and co-products utilization/molasses,ibid
5. TIFAC Report on "Technology for Production of Lactic Acid" 2001
6. Chemicals Weekly, July 5,2011 pp.239-273
7. Ali. S. et al (2002), Production of citric acid by *Aspergillus niger* using cane molasses in a stirred fermentor, Electonic Journal of Biotechnology, 5, Issue December 15th 2002

8. Ikram-ul-Haq et al (2004), Citric Acid production by selected mutants of *A. niger* from cane molasses, Bioresource Technology 93:125-130
9. Bhatt S.M. and S.K.Srivastava,Lactic Acid Production from Cane Molasses by *Lactobacillus delbrueckii* NCIM 2025 in Submerged Condition: Optimization of Medium Component by Taguchi DOE Methodology, Food Biotechnology, 22:115–139, 2008
10. Aksu Z. and T.Kutsal,Lactic acid production from molasses utilizing *Lactobacillus delbrueckii* and invertage together, Biotechnology Letters , Vol.8, No.3, pp.157-160,1986
11. James M. Church, Ruth Blumberg, Synthesis of Tartaric Acid, *Ind. Eng. Chem.*, 1951, *43* (8), pp 1780–1786
12. W. E. Barch, The Oxidation of 5-Keto-gluconic Acid with Nitric Acid in the Presence of Vanadium, J. Am. Chem. Soc., 1933, 55 (9), pp 3653–3658
13. R. Klasen, S. Bringer-Meyer and H. Sahm, Biotechnology and Bioengineering 40, 183, 1992.
14. Agrawal, Prasad, H. and Rao, M.B., 1979. "Influence of different processing conditions on the production of oxalic acid from cellulosic materials by nitric acid oxidation. Ind. J. Technol. 17, pp. 11–15.
15. Bretschneider, R.B.,Kopriva, J.. Markova and Z.Ondrackova, " Oxidation of Sucrose by nitric acid",Listy Cuker 1966-82,215-219,1967,36-39,and 227-232.
16. Bernsten,R.A.,and E.F. Degering," Oxidation of glucose with oxides of Nitrogen "Sugar Industry Abstracts,1949,11,87,258
17. Bose , S., S.Mukherjee,and A.N. Srivastava," Production of oxalic acid from sugarcane molasses", Proc.4th Joint Conv. S.T.A.I., D.S.T.A. and S.I.S.S.T.A.,G-17-21,1971
18. Chia, K.S. and C.C.Lin," Oxidation of Carbohydrates-synthesis of Glucosachharic Acid from sucrose",Chemistry, Taiwan,1962,98-104
19. Deshpande, S.D. and Vyas, S.N., 1979. Oxidation of sugar to oxalic acid and absorption of oxides of nitrogen to sodium nitrite. Ind. Eng. Chem. Prod. Res. Dev. 18 1, pp. 69–71.
20. Dube, S.K., Vasudevan, P. and Khandelwal, B., 1982. Oxalic Acid Manufacture. J. Chem. Tech. Biotechnol. 32, pp. 909–919.
21. Guo, W., Jiang, H. and Huaxue, Yu.N., 1997. Manufacture of oxalic acid from beet waste slag by hydrocarbonylation at atmospheric pressure. J. CA 45 4, pp. 240–241.
22. Gupta, K.C., Suri, S.K. and Bose, S., 1980. Oxalic acid from deionised sugarcane juice. Proc. STAI 44, pp. 119–125.
23. Gürü, M., et al., 1991. eker Pancarı Melasından Nitrik Asitle Okzalik Asite Yükseltgenmesinde Optimum artların Belirlenmesi. VII. Kimya ve Kimya Mühendisli i Sempozyumu, 2, 464, Gazima usa
24. Gürü, Metin, Ali Y. Bilgesüb and Vecihi Pamuka, Production of oxalic acid from sugar beet molasses by formed nitrogen oxides ,Bioresource Technology,Volume 77, Issue 1, March 2001, Pages 81-86
25. Hamel, R., Levasseur, R. and Appanna, V.D., 1999. Oxalic acid production and aluminum tolerance in *Pseudomonas fluorescens*. J. Inorg. Biochem. 76 2, pp. 99–104.
26. Mane, J.D., Jadhav, S.J. and Ramaiah, N.A., 1986. Production of oxalic acid from dry powder of parthenium hysterophorus L. J. Agricultural Food Chem. 34 6, pp. 989–990.
27. Mane, J.D., Modak, H.M., Ramaiah, N.A. and Jadhav, S.J., 1988. Utilisation of sugarcane trash and other cellulosic wastes for production of oxalic acid. Biological Wastes 25, pp. 171–176.

28. Nowak, P., 1986. Technology for obtaining oxalic acid from sugar by a pressure method. Przemysl Chemiczny 65 9, pp. 488–490.
29. Preen, F.H.E., "Oxalic acid from wood waste", Chemical age,1942-32,72,
30. Raha A.C., R.B.Nigam,and P.Sanyal : Production of oxalic acid from gur", Proc.S.T.A.I.,44(part 2),G-1-6,1976
31. Radulescu, G., Moise, M.-I., Cornos, L., Vapler, G., 1993. Preparation of oxalic acid by the thermal decomposition of sodium formate followed by cation exchange. Patent CA Section:45, Rom. RO 106878 B1 30: 4
32. Rujiter, G.J.G. et al., 1999. Oxalic acid production by *Aspergillus niger*: an oxalate-non-producing mutant produces citric acid at pH 5 and in the presence of manganese. Microbiology 145 9, pp. 2569–2576.
33. Sanyal, P., Ray, A.K. and Shukla, N.P., 1981. Oxalic acid from gur-effect of catalyst. Proc. STAI 45, pp. 21–26.
34. Sattar, A., Muhammad, D., Ashraf, M., Khan, S.A. and Bhatty, M.K., 1988. Oxidation of molasses to produce oxalic acid. Pakistan J. Sci. Ind. Res. 31 3, pp. 163–166.
35. Shukla, N.P., Ray, A.K., Sanyal, P., Oxalic acid from sugarcane molasses: process development and waste utilisation. Proc. 46th Convention STAI, pp. 93–101,1982
36. Sullivan, J.M., Williard, J.W., White, D.L. and Kim, Y.K., 1983. Production of oxalic acid via the nitric acid oxidation of hardwood (Red Oak) sawdust. Ind. Eng. Chem. Prod. Res. Dev. 22, pp. 699–709.
37. Scott., W.W.,Standard Methods of Chemical analysis, Fifth Ed., Vol.1, The Technical Press, London,644
38. Snell, F.D. and F.M.Biffen,Commercial Methods of analysis, First Ed., McGraw Hill Book Co., New York,1944,140
39. Tanz and Kao, Chemical Abstracts,1944,355466
40. Vishwalingam,K.,T.V.Suramaniam and B.Jagannadhaswamy," Process oxidation of ethylene glycol to oxalic acid in liquid in the presence of nitric acid" Ind. Chem.Engr.1980-22(4),11-13

(338f) Modeling the Hydrolytic Degradation and Distribution of Polymer Species In Biopolymers

Justin T. Kaffenberger, Huajiang Huang, Ulrike Tschirner, Ben Schroeder, Waleed Al-dajani and Shri Ramaswamy, Bioproducts and Biosystems Engineering, University of Minnesota, St Paul, MN

Modeling thehydrolytic degradation and distribution of polymer species in biopolymers

Justin Kaffenberger, Huajiang Huang, Ulrike Tschirner, BenSchroeder, Waleed Al-dajani, and Shri Ramaswamy

Department of Bioproducts and Biosystems Engineering,University of Minnesota, St. Paul, MN 55108

Abstract:

The US Department of Energy in its Plant/Crop-BasedRenewable Resources Report 2020 has put forth an initiative to have 10 percent ofchemical feedstock coming from plant-based renewable sources by 2020 and afurther increase to 50 percent by 2050.[1] Meeting these targets will require productionof current chemical feedstocks from renewable sources, such as the dehydrationof plant-based ethanol to produce ethylene.It will also require the development of alternative feedstocks that cansupplant the petroleum-based ones that currently predominate. One example of a chemical feedstock ofincreasing importance is lactic acid.

Lactic acid is a metabolite formed during glucosefermentation by many organisms. Due toits versatility, lactic acid has the potential to be a major renewable chemicalfeedstock. It can be esterified to formlactate esters, reduced to propylene glycol, or dehydrated to acrylic acid.[2] It can also be homopolymerized to formpolylactic acid (PLA) via condensation reaction. The resulting polymer exhibits mechanicalproperties similar to many petroleum-based polymers, including polyethyleneteraphthalate, polyethylene, and polystyrene.

The degradability of PLA is critical to its currentcommercial success. In disposable foodpackaging and service ware, the degradability and bio-absorptivity of thedegradation product of PLA makes it marketable as a green and environmentally-friendlyproduct. In biomedical applications, thedegradability and biocompatibility of PLA are essential. As PLA's ability to degrade is a major factorin its application functionality, understanding the fundamentals of thisprocess is critical to all stages of PLA manufacture and end-use application.

To this end, a mathematical model has been developed thatpredicts the time dependent change in the average molecular weight ofpolylactic acid (PLA) of varying crystallinity as it hydrolytically degrades ata given temperature. This model includes three possible hydrolysis reactionroutes: random chain scission, end scission (monomer removal from the acid endgroup, and monomer removal from the alcohol end group). Apparent rate constants for these reactionpathways were determined from experimental data. The resulting values suggest that,independent of the crystallinity of the starting PLA, acid end groups arehydrolyzed more rapidly than either alcohol end groups or random chainscission. This may help develop suitableapproaches to control the polymer properties and degradation and help developwider application for PLA products.

See more of this Session: Sustainability and Biobased Materials
See more of this Group/Topical: Forest and Plant Bioproducts Division

(338g) Experimental Study On Bast Fiber Reinforced Soy-Based Polyurethane

Shanshan Huo and Chad A. Ulven, Mechanical Engineering, North Dakota State University, Fargo, ND

Growing importance has been placed on the research of natural fibers and resins derived from natural resources as the need for reducing dependency on petroleum-derived products has increased. Natural fibers and biobased resins are potentially inexpensive, renewable resources capable of increasing the specific strength, specific stiffness, and damage tolerance of a variety of composites.

Polyurethane is one of the most versatile and intensively used industrial materials. Bio-based polyols developed from vegetable oils such as soybean oil, canola oil, palm oil and castor oil, are now being used in industry. Soy-based polyurethanes normally exhibit equivalent or improved physical and chemical properties due to the hydrophobic nature of triglycerides. Cellulosic fibers as an alternative for fiberglass have been studied by many researchers. Chemical modifications can be applied to reduce the amount of hydroxyl groups in cellulosic fiber to increase the hydrophobicity and improve the interfacial properties between fibers and matrices. Distillers dried grains with solubles (DDGS) is a by-product of corn-ethanol. Their potential use as reinforcing filler in biocomposites has been studied in thermoplastic composites, but not to a great extent in thermoset composites.

In this study, bast fiber reinforced soy-based polyurethane composites with DDGS filler are produced. The mechanical properties of soy-based PU matrix, soy-based PU matrix with DDGS, and cellulosic fiber reinforced soy-based PU composites with DDGS are investigated. The thermal properties of the matrix system and their composites are evaluated.

See more of this Session: Sustainability and Biobased Materials
See more of this Group/Topical: Forest and Plant Bioproducts Division

(338h) Plant Seed Oils and It's Derivatives Based Value-Added Products From Cenral India Region of the India

Bhalchandra Vibhute, Department of Oil Technology, Laxminarayan Institute of Technology,Nagpur (Maharashtra),India, Purna, India and Dr.Anand S. Kulkarni, Dept.of Oil Technology, Laxminarayan Institute of Technology,Nagpur(Maharashtra),India, Purna, India

See more of this Session: Sustainability and Biobased Materials
See more of this Group/Topical: Forest and Plant Bioproducts Division

(394a) Integrating Nanocellulose Production with Biofuel

Junyong Zhu, USDA Forest Products Laboratory, Madison, WI

This study demonstrated the feasibility of integrating the production of nano-fibrillated cellulose (NFC), a potentially highly valuable biomaterial, with sugar/biofuel (ethanol) from wood fibers. Commercial cellulase enzymes were used to fractionate the less recalcitrant amorphous cellulose from a bleached kraft eucalyptus pulp resulting in a highly crystalline and recalcitrant cellulose (RC). The RC is difficult to hydrolyze to sugars but very suitable for producing biobased nanomaterials through mechanical homogenization. A range of fractionation yields of RC from 10–70% can be achieved by varying fractionation duration and enzyme dosage. The crystallinity of the RC was found to be as much as 24% higher than that of the original bleached pulp. The cellulase fractionation process facilitated mechanical homogenization by significantly reducing the degree of polymerization (DP) to about 400 and the length of the fibers to about 200 mm. The hydrolyzed sugars were found to be easily converted to ethanol through yeast fermentation with excellent efficiency of 91%. Films made from nano-fibrillated cellulose were found to be optically transparent with opacity as low as 12%. The NFC films were mechanically strong and stiff with tensile strengths and moduli approximately 10 and 6 times higher than film made from fibers that had not been nano-fibrillated.

See more of this Session: Nanocellulose-Based Materials and Composites
See more of this Group/Topical: Forest and Plant Bioproducts Division

SURFACE AND AGING CHARACTERIATICS OF PAPER COATED WITH NANO SIZE ZINC OXIDE PIGMENT

Sanjay Tyagi. Renu Tyagi* , B P Thapliyal* R M Mathur* and A K Ray***

**Central Pulp and Paper Research Institute, Saharanpur – 247001*
*** Indian Institute of Technology, (Saharanpur Campus), INDIA*

ABSTRACT

Quality of paper is improved on pigment coating in terms of surface, optical properties and printability characteristics. One of the major parameter for aqueous pigment coating is the type of pigment used and also its particle size and distribution. Zinc oxide is one of the pigment s employed for special coating application. It is offlet nano size particles are playing prominent roles in several areas ranging from medicine to electronics. In paper making there is no exception. Few specialty grades of paper are already using nano size calcium carbonates both as filler and also as coating pigments. Application of water based barrier coating in paper and board in food packaging industry have proven successful. It is also important due to substantial smaller size particles (nano range) compared to macro size solid particles (in micron range) can provide better functional properties of paper in some specialized applications. However, ZnO is present in various forms nano size ZnO particles can enhance the properties significantly to improve optical and surface properties and specially the effect of aging. In this present therefore an attempt has been made to study the potential technological applications of nano zinc oxide particles on paper surface made from hardwood furnish and bamboo using 100% nano ZnO and with admixture of coating grade China clay using laboratory rod coater. All the required properties were evaluated as per ISO standard and compared with the same base paper coated with micro size ZnO pigment particles. From the detailed study it is observed that paper coated with nano ZnO particles imparts better properties as compared to bulk ZnO in terms of brightness, whiteness, gloss, smoothness and printability. Aging properties were also thoroughly studied under artificial condition of varying temperature (50% – 120%)and humidity (10% - 90%). And also UV light, paper coated with nano ZnO showed better resistance to brightness reversion and UV protecting potential.

INTRODUCTION

Paper is pigment coated to improve the quality of paper in terms of surface, optical and printability characteristics. Paper coating ingredients which generally contains pigments, binders and additives is applied to the paper surface to improve the above properties of the paper. One of the major parameter for aqueous pigment coating is the type of pigment used and also its particle size distribution [(1,2,3)]. A commonly used pigment includes china clay, calcium carbonate, silica, zeolite, titanium dioxide, alumina trihydrate and polystyrene [(2)]. Zinc oxide is one of the pigments employed for special coating application [(4)]. Presently all above mentioned conventional pigments are available in the market have their particle size in the range of 1 – 10 micron [(5,6)]. The size of these particles is too big for the preparation of coating pigments to be used efficiently in coating machines. Evenness and uniformity in surface smoothness and thickness of coating layer which is one of the two prime requirements for producing good multi color print, generally lacks with conventional micron sized coating pigments [7,8,9]. Pore structure and contact angle wettability effect print quality by preventing ink spreading. In order to prevent wicking and feathering, it is important that the thickness of coating layer be homogeneous to a scale of few microns in depth which also helps in the absorption of successive droplets of ink at high delivery rates and any water present[9,10].

Application of nano size Zinc oxide in papermaking

While papermaking is normally a statistical process, nano technology makes use of the classic process where the position of every atom is specified and controlled. Yet despite this apparent discrepancy, between two technologies, the paper industry has been practicing nano particles over the conventional pigments long before it was marketed as latest technology trend [5,6].

Nano sized calcium carbonate has been used by paper industry since long as filler due to the advantage of high opacity, high brightness and whiteness [11]. At present, it is also used in special paper products such as baby diapers, sanitary napkins, etc.

Zinc oxide is also one of the pigments finding its use for special coating applications[4]. It is offlet nano size particles are playing prominent role in several areas ranging from medicine to electronics [12]. ZnO is essential for rubber and tier industry and also used as natural rubber, synthetic rubber and latex vulcanization active agents, reinforcing agent and coloring agents. In paint industry, ZnO in addition to color strength and hiding power outside but also a light coating of preservative and agents [13]. In medical and health and food industries, ZnO has a Sida, bleeding, hygienic convergence of functions. In electronics industry, ZnO varistor is the main raw material is magnetic, optical material such as the main additive. Prepared using ZnO varistor, not only has low sintering temperature, but also improve the performance of porous sensitive resistor such as the flux, the non linear coefficient [14,15].

In the present study, an attempt has been made to study the potential technological applications of nano zinc oxide particles on the paper surface made from the pulp of hardwood furnish and bamboo using 100% nano ZnO and with the admixture of coating grade china clay by taking trials on laboratory rod coater. The Zinc oxide soluble starch nano composite was prepared by wet chemical method using Zinc nitrate and sodium hydroxide as precursors and soluble starch as stabilizing and compared with the same base paper coated with micro size ZnO pigment particles.

EXPERIMENTAL

Coating formulation

The coating trials were carried out using Laboratory rod coater. The coated sheets were dried immediately in laboratory oven at the temperature of 105 C for maximum 2 minutes. The following two coating formulations were used for coating trials.

Ingredient	Composition I	Composition II
Pigment		
• Zinc Oxide (Nano or Ordinary)	500 parts	1000 parts
• China clay	500 parts	
Binder		

• Oxidized Starch	90 parts	90 parts
• Styrene Butadiene Latex (SBR Latex)	60 parts	60 parts
Dispersant		
• Anionic polymer	5 parts	5 parts
	Total solids maintained- 65±2%	

Preparation of coating colour

The preparation method for the coating slip is as under-

(a) Starch cooking

The oxidized starch solution was prepared separately by cooking under continuous agitation in hot water at temperature 90-95 ^{0}C, for about 15 minutes, i.e. upto the point when the whole solution became fully gelatinized.

(b) Pigment preparation

Pigment slurry was made in laboratory high-shear coating mixer. The dispersant was firstly added to desired quantity of water. When homogeneous solution was formed, weighed quantities of pigments were added, under constant stirring. A thick paste thus resulted, which became uniform at high shear. Time of agitation was, however kept at about 15 minutes.

(c) Addition of binders to pigments

The freshly prepared starch solution and synthetic binder SBR latex (in required amount) were added to the pigment paste in laboratory high-shear coating mixer, under constant agitation. The mixture was kept under constant agitation at high shear for about 30 minutes to ensure uniform mixing and dispersion of all the ingredients. Finally the slurry was taken out of the mixer.

(d) Screening

The coating slurry was next screened on a 300-mesh screen to eliminate foreign matters, coagulated coating and particles of dried coating slurry etc. Thus a uniform coating slurry; free from dirt, was ready for coating.

Calendering

Coated sheets after dried in laboratory oven were calendered using the laboratory calender with softnip mode, at 40 ^{0}C temperature & 30 bar pressure. The sheets were passed twice through the softnip.

Evaluation of optical characteristics of the pigments

The two pigments- Zinc Oxide (ordinary) and Zinc Oxide (nano particles), which were used in coating trials, were evaluated also for following optical characteristics; as an extension of the detailed study.

- Brightness (%)
- Whiteness (%)
- Yellowness (%)

Evaluation of base paper and coated paper

Base paper as well the coated & calendared papers were conditioned at 27±1 ^{0}C temperature and 65±2 % relative humidity prior to testing of following properties as per standard methods mentioned below.

Property	Standard methods
Grammage (g/m^2)	ISO 534
Brightness (% ISO)	ISO 2470

Gloss (%)	Tappi 480 om-99
Whiteness (%)	ISO 2470
PPS Roughness (micron)	ISO 8791/4
Picking velocity (cm/s)	ISO 3783
Print density (g/cm^3)	Using IGT printability tester Model AIC2-5
Print uniformity index	Using IGT printability tester Model AIC2-5
Stain length (mm), oil absorbency	Using IGT printability tester Model AIC2-5

RESULTS & DISCUSSION

Optical characteristics of base paper

The base paper used in the coating trials, was procured from commercial paper mill. The base paper was made from the pulp containing furnish of hardwood pulp and bamboo. Base paper was evaluated for different optical & printing properties for comparison purpose with coated & calendered paper. The result is tabulated in Table-I.

Optical characteristics of pigments

The optical characteristics of the pigments used in coating are given in Table-II. By comparing the values of the properties obtained for nano Zinc Oxide and ordinary Zinc Oxide, it was found that the starch-encapsulated nano Zinc Oxide particles had slightly lower brightness, whiteness and was relatively more yellowish in colour as compared to ordinary Zinc Oxide particles. This is probably due to the preparation process.

Characteristics of Coated & Calendered Paper samples

The coated paper samples, after softnip calendering, were evaluated for

(a) Optical characteristics
(b) Surface characteristics
(c) Printing characteristics
(d) Aging characteristics

(a) Optical characteristics

The effect of nano size zinc oxide particles on brightness of coated and calendered paper is given in Figure 1. Results show that nano zinc oxide gives better brightness as compared to conventional zinc oxide in both cases. i.e. when used alone and when used in combination with coating grade china clay. The same trend is true with whiteness value of coated paper (Fig. 3).

The value gloss of paper coated with nano ZnO also improved as compared to paper coated with conventional ZnO (Fig. 2). However, the improvement is low as compared to brightness and whiteness values of base paper, there is a fall in their values after coating with ZnO and China clay. The reduction in values is more in case when china clay is used in coating formulation. This is probably due to the lower brightness of china clay and binder additives which results in overall reduction of optical properties of coated paper.

(b) Surface characteristics

PPS Roughness

The surface smoothness of coated paper which is an important property in deciding the print performance of paper, improved with the application of nano ZnO. The PPS roughness in case of nano ZnO is relatively lower than conventional ZnO. The same trend was obtained when nano ZnO is used with china clay. A smooth surface was obtained when china clay is used in coating furnish. This is probably due to particle shape. China clay being platelet shape flattened more during the calendaring operation in compared to ZnO which is spherical in shape (Fig. 4).

(c) Printing characteristics

Picking velocity

Picking velocity of paper coated with nano ZnO was found to be higher as compared to conventional ZnO (Fig. 5) in both cases when used alone and in combination with china clay. This indicates that paper coated with nano ZnO can run easily on high speed printing presses or at the same printing speed, less binder is required in coating formulations with nano ZnO. The excess amount of binder always has negative effect on the printing characteristics.

Print Mottle (Print Non-Uniformity)

Print non uniformity leads to print mottle which is uneven appearance of colour density in a print. This is due to the surface factors to absorb ink evenly due to variation in base paper formation or coat weight. The non uniformity printing index is lower for paper coated with nano ZnO as compared to paper coated with conventional ZnO (Fig. 8).

The variation in sheet matrix formation formed by coating layer of conventional ZnO (due to non uniform size of ZnO particles) probably the reason for this. Paper coated with ZnO alone showed lower number as compared to paper coated with a mixture of china clay and ZnO since particle size of china clay is largely varied. The sheet matrix variation can lead to print mottle non uniform compression during calendering, resulting in variation in coating layer densification. Sheets that contain binder migration, coat weight variation, inferior coating profile and heterogeneous pigment blends can all cause mottle formation on the printed sheet.

Print density

Print density is affected by ink setting/ink holdout. For good density the vehicle of printing ink should stay on the top of paper together with ink pigment. The more the dried ink is retained on the outer surface of the paper, the higher will be the ink density, accordingly, the lower will be the ink demand. Paper with large pores in coating will have more penetration of printing ink leading the lower density. This is observed in case of papers coated with conventional ZnO coating (Fig. 7). The same is true with china clay in the coating formulation. This indicates that more ink will be required to print coated papers with conventional ZnO coatings.

Oil absorbency

Oil absorbency which is an indication of rate of penetration of oil was found to be more in the case of paper coated with nano size ZnO alone and in combination with china clay. (Fig 6)

(d) Effect of Aging

The effect of aging on the optical properties of coated and calendered paper is shown in the graph (Fig. 9-11) and tabulated in table (4-6). The results show that nano zinc oxide gives excellent UV protecting strength in both cases when used alone and when used in combination with coating grade china clay. The reduction in brightness and whiteness is minimum in the case of nano zinc oxide. The lower value of indicated the higher yellowness, which incase after aging maximum when china clay is used.

CONCLUSIONS

The use of nano zinc oxide as a coating pigment improves the paper properties in terms of optical, surface and printability as compared to conventional zinc oxide. This indicates that use of nano ZnO can be a good substitute for the replacement of conventional ZnO in the case of speciality grade papers. This opens a new scope for the use of nano Zinc oxide as a coating pigment for speciality grade of papers. The application of nano ZnO also exhibits excellent UV protection characteristics when tested at different condition of humidity and temperature. Paper represents good aging resistance properties which increases the longevity of paper. This opens a new scope of manufacturing paper with much longer life which lasts for longer years paper manufactured under connectional methods

The use of nano Zinc oxide as a coating pigment results in the paper with better properties as compared to conventional ZnO in terms of brightness, whiteness, gloss, smoothness, UV protection and printability characteristics i.e. picking velocity, print uniformity, print density, and oil absorbency .

Refrences:

1. 'Y.V.Sood, S.Tyagi. R.Tyagi, P.C.Pande, R.Tandon & T.K.Roy', 'Quality of indigenous coated paper and ways to improve it', *'J.IPPTA, 2008, 20 (4), pp'*
2. 'Coating pigments: from size and shape to nanotech', *'www.thefreelibrary.com/coating+pigments:+from+size+and+shape+to+nanotech.....'*
3. 'Surface characteristics of coating layers formed by coating pigments with different particle sizes and size distribution' *'J. Coat. Tech. and Res'*
4. 'N. Vatansever & S. Polat', 'Effect of zinc oxide type on aging properties of styrene butadiene rubber compounds', *'www.sciencedirect.com, 2009'*
5. Roxare, 'Nanotechnology in paper making', *'http://pulpandpapercircle.com/2008/07/nanotechnology –in-paper-making'*
6. 'J.Shen, Z.Song, X.Ojan, F.Yang & F.Kong', 'Nano fillers for papermaking wet end applications', *'www.bioresources.com, 2010, 5(3)'*
7. G. Chinga, 'Detailed characterization of paper surface structure for gloss assessment', *'JPPS, 2004, 30 (8), pp 222-227'*
8. G. Ström, S. E. Research, A. Englund & M. Karathanasis, 'Effect of coating structure on print gloss after sheet fed offset printing', *'J. Nord. Pulp & Paper, 2003, 18 (1), pp108-115'*
9. 'G. Chinga-Carrasco, M. Axelsson, Ø. Eriksen & S. Svensson', 'Structural characteristics of pore networks affecting print through', *'JPPS, 2008, 34 (1), pp 13-22'*
10. 'D.W. Donigian, J. N. Ishley & K.J. Wise, 'Coating pore structure and offset printed gloss', *'J.TAPPI, 1997, 80 (5), pp 163-172'*
11. T. Nypelö, M. Österberg & J. Laine, 'Ultra thin coating of paper by tailor made nanoparticles', *'TAPPI International Conf. on Nanotechnology for the Forest Product Industry, 2010'*
12. 'V. Bianchard & P. Blanchet', 'Color stability for wood products during use: Effects of inorganic nanoparticles', *'www.bioresources.com, 2011, 6(2)'*
13. 'B. Wang, N. Han, D. Meng, R. Yue, J. Yan & Y. Chen', 'Filter paper- templated preparation of ZnO thin films and examination of their gas sensing properties', *'www.sciencedirect.com, 2011'*
14. 'V. Prasad, A. J. Shaikh, A. A. Kathe, D. K. Bisoyi, A.K. Verma, N. Vigneshwaran', 'Functional behavior of paper coated with zinc oxide-soluble starch nanocomposites', *'J. Mat. Proc. Tech., 2010, 210 (14), pp 1962-1967'*
15. 'Surface characteristics of coating layers formed by coating pigments with different particle sizes and size distribution' *'J. Coat. Tech. and Res.'*

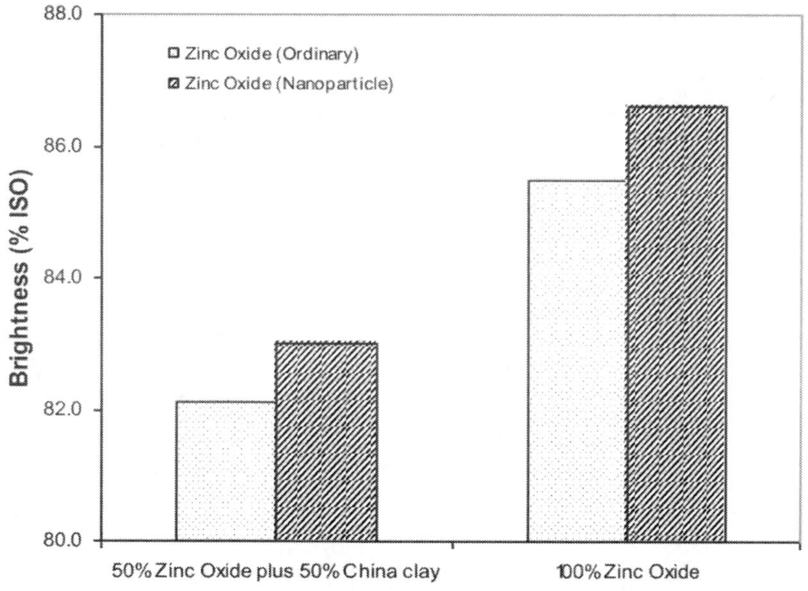

Fig. 1: Effect of Zinc Oxide nanoparticles on brightness of coated & calendered paper

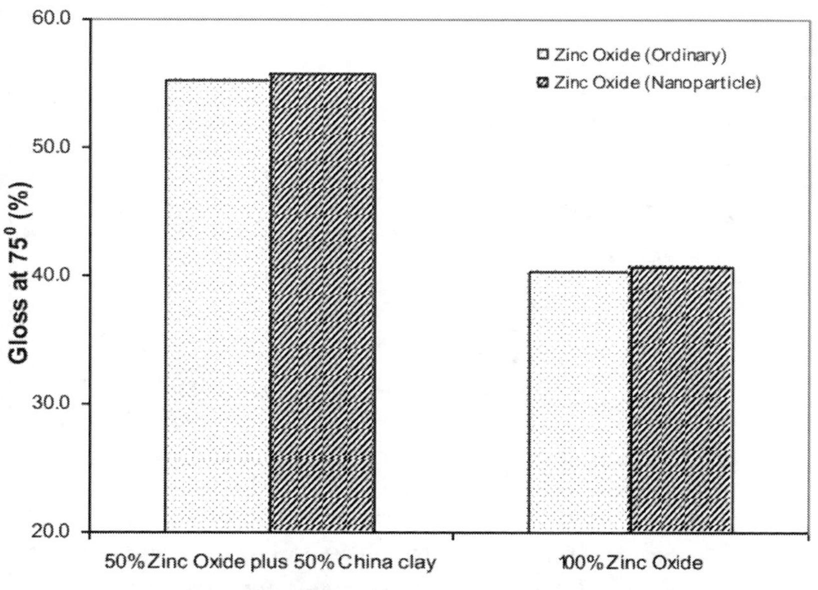

Fig. 2: Effect of Zinc Oxide nanoparticles on gloss of coated & calendered paper

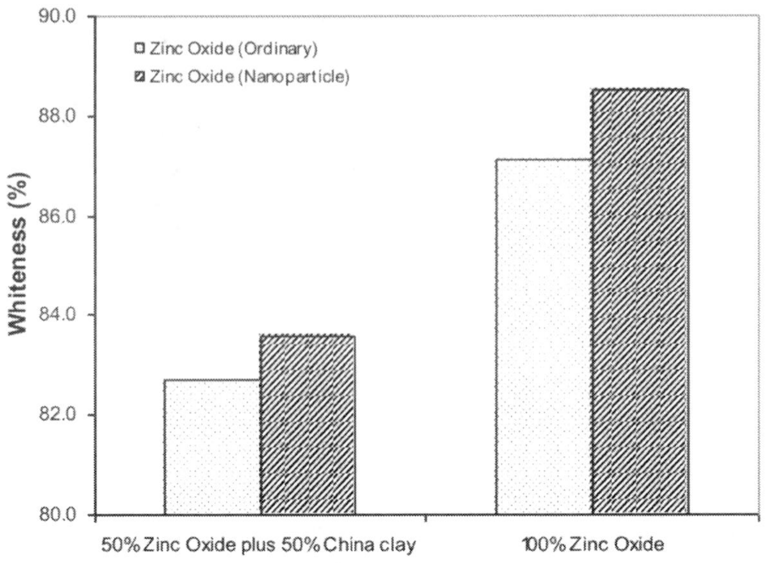

Fig. 3: Effect of Zinc Oxide nanoparticles on whiteness of coated & calendered paper

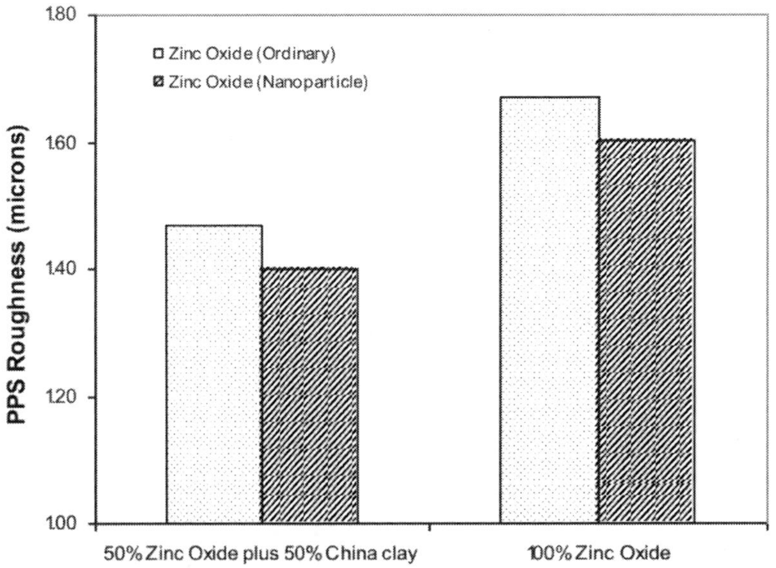

Fig. 4: Effect of Zinc Oxide nanoparticles on the roughness of coated & calendered paper

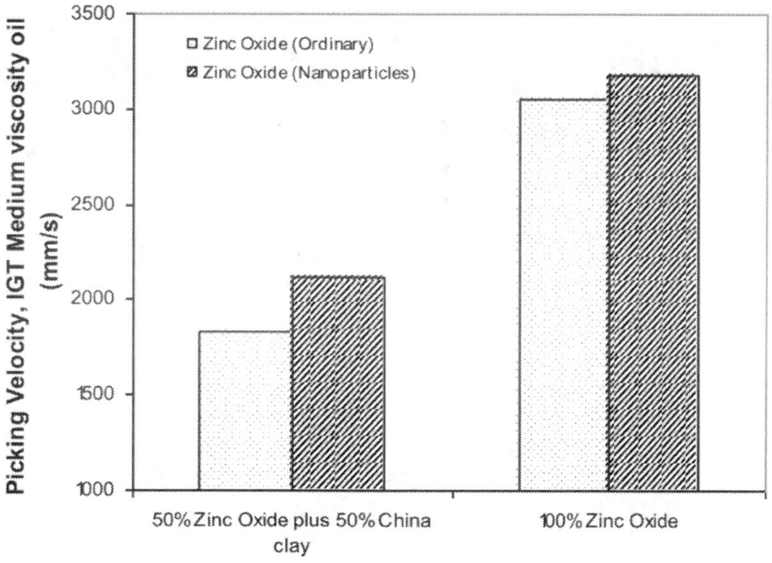

Fig. 5: Effect of Zinc Oxide nanoparticles on picking velocity of coated & calendered paper

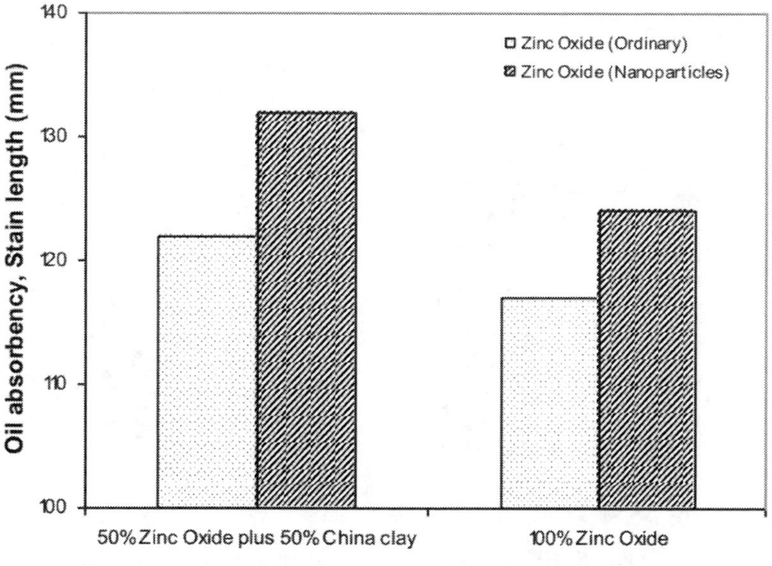

Fig. 6: Effect of Zinc Oxide nanoparticles on oil absorbency of coated & calendered paper

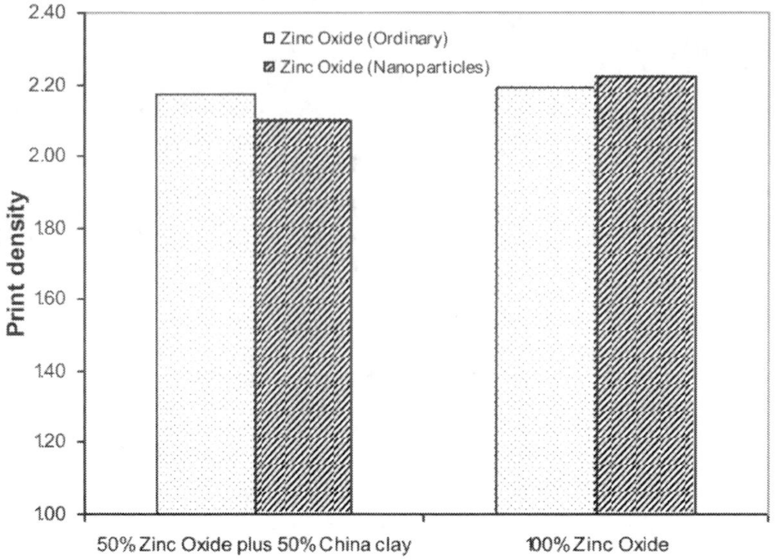

Fig. 7: Effect of Zinc Oxide nanoparticles on print density of coated & calendered paper

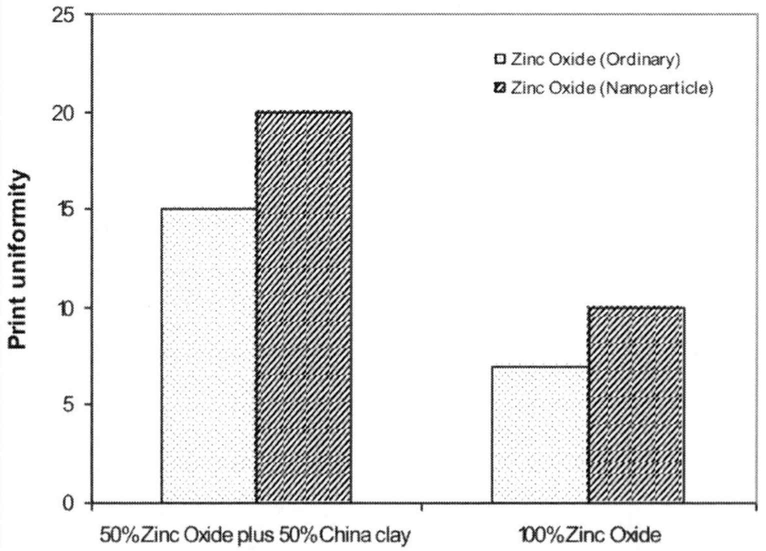

Fig. 8: Effect Zinc Oxide nanoparticles on print uniformity of coated & calendered paper

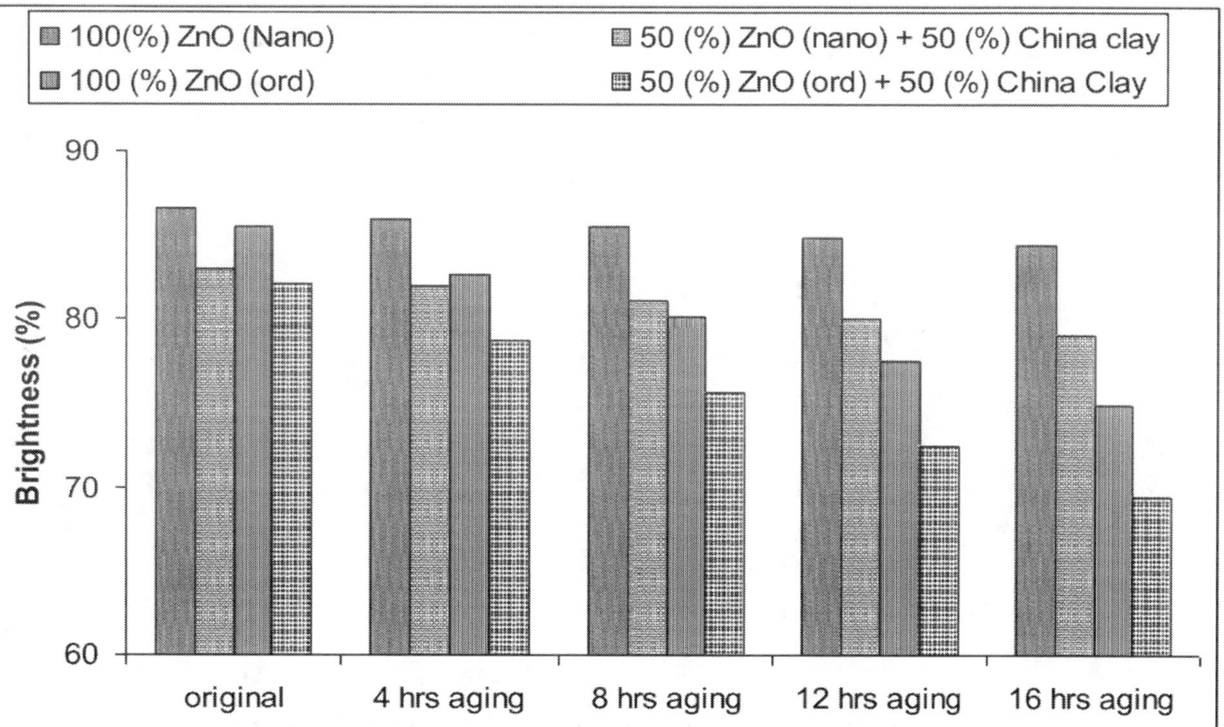

Fig. 9: Effect of zinc oxide nano particle on brightness reversion of coated and calendared paper

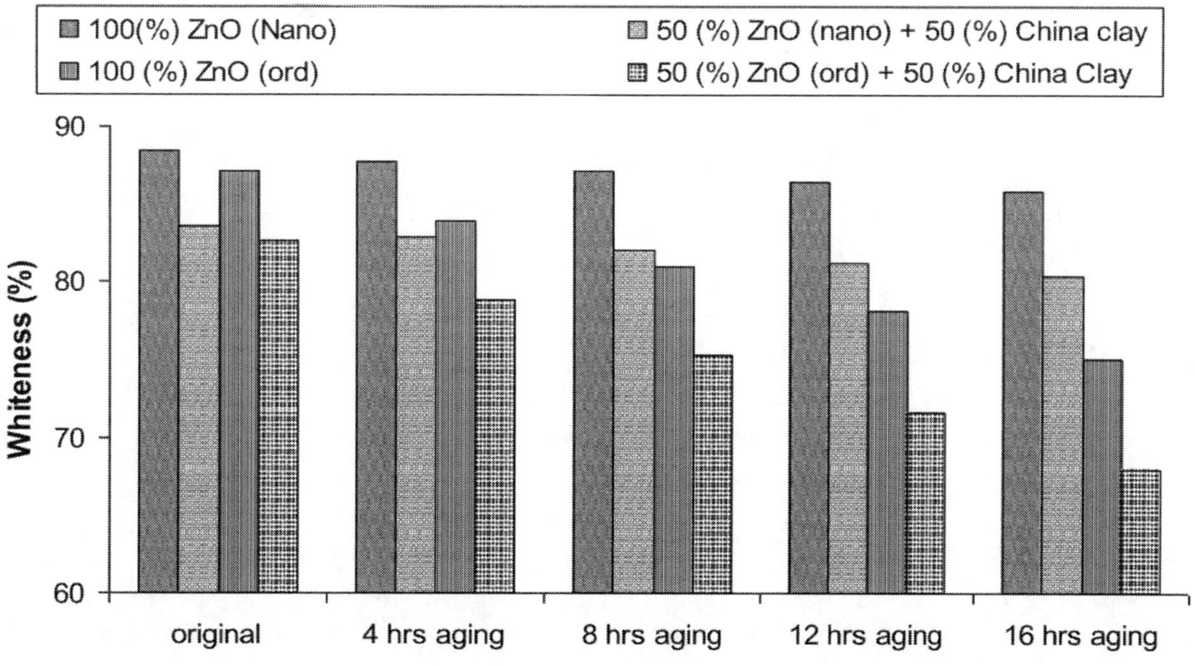

Fig. 10: Effect of zinc oxide nano particles on whiteness reversion of coated and calendared paper

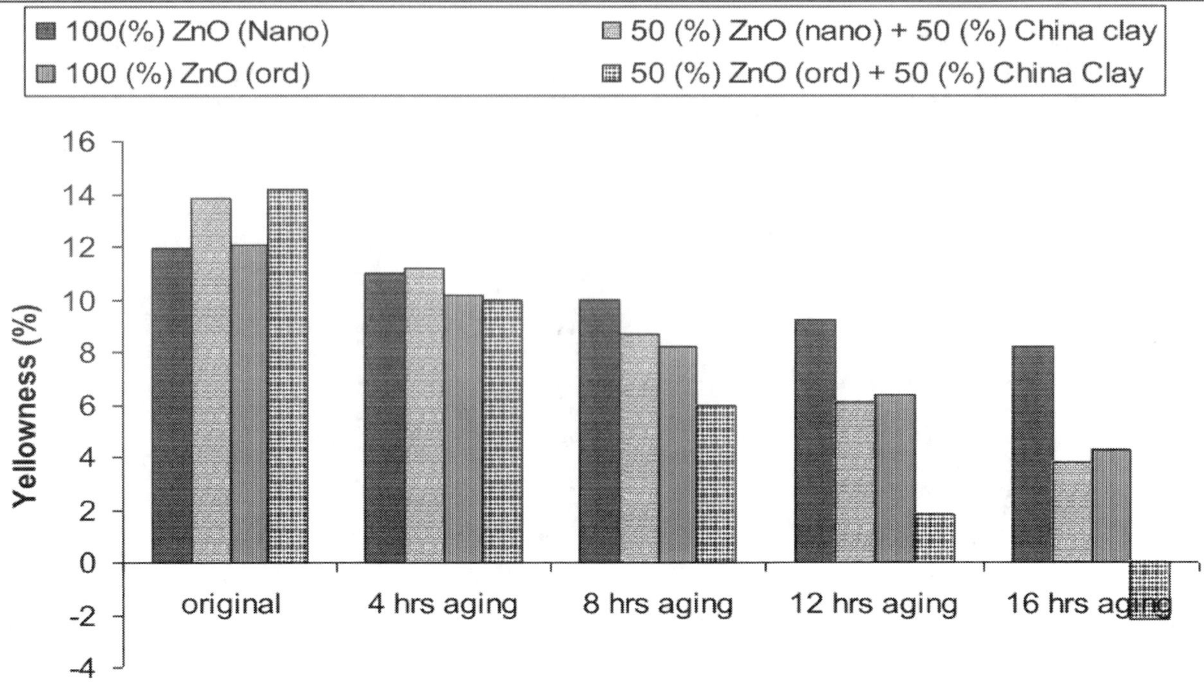

Fig.11: Effect of zinc oxide nano particle on yellowness reversion of coated and calendared paper

Table-I: Optical & printing characteristics of base paper used for coating

Property	Values obtained for the sample
Brightness (%)	88.5
Opacity (%)	80.0
Yellowness (%)	-5.13
Whiteness (%)	124.2
Gloss at 75^0 (%)	6.9
Porosity, Bendtsen (ml/min)	593

Table-II: Optical & printing characteristics of Zinc Oxide pigments used in coating

Property	Zinc Oxide (Ordinary)	Zinc Oxide (Nanoparticle)
Brightness (%)	97.8	96.2
Whiteness (%)	93.9	91.2
Yellowness (%)	1.64	1.70

Table-III: Different optical & printing characteristics of coated & calendared paper

Property	Pigment composition for coating			
	50% Zinc Oxide (Ord.) plus 50% China clay	50% Zinc Oxide (Nano) plus 50% China clay	100% Zinc Oxide (Ord.)	100% Zinc Oxide (Nano)
Coat weight (g/m^2)	12.7	12.6	12.8	12.4
Brightness (%)	82.1	83.0	85.5	86.6
Whiteness (%)	82.7	83.6	87.1	88.5
Gloss at 75°(%)	55.2	55.8	40.4	40.6
PPS Roughness (microns)	1.47	1.40	1.67	1.60
Oil absorbency, IGT Stain length (mm)	122	132	117	124

Picking velocity, IGT Medium viscosity oil (mm/s)	1830	2120	3060	3180
Picking Viscosity Velocity Product (VVP), IGT(Poise.m/s)	805	933	1346	1399
Print density	2.17	2.10	2.19	2.22
Print uniformity	15	20	7	10

Table-IV: Brightness reversion of coated and calendared paper after aging

Pigment Composition	Brightness (%)				
	Original	4 hrs aging	8 hrs aging	12 hrs aging	16 hrs aging
100(%) ZnO (nano)	86.6	86	85.5	84.9	84.4
50(%) ZnO (nano) + 50 (%) China Clay	83	82	81.1	80	79
100 (%) ZnO (ord)	85.5	82.7	80.2	77.5	74.9
50 (%) ZnO (ord) + 50 (%) China Clay	82.1	78.7	75.7	72.5	69.4

Table-V: Whiteness reversion of coated and calendared paper after aging

Pigment Composition	Whiteness (%)				
	Original	4 hrs aging	8 hrs aging	12 hrs aging	16 hrs aging
100 (%) ZnO (nano)	88.5	87.7	87.2	86.5	85.9
50 (%) ZnO (nano) + 50 (%) China Clay	83.6	82.9	82.1	81.2	80.4
100 (%) ZnO (ord)	87.1	84	81	78.1	75.1
50 (%) ZnO (ord) + 50 (%) China Clay	82.7	78.8	75.3	71.6	68

Table-VI: Yellowness reversion of coated and calendared paper after aging

Pigment Composition	Whiteness (%)				
	Original	4 hrs aging	8 hrs aging	12 hrs aging	16 hrs aging
100 (%) ZnO (nano)	11.9	11	10	9.2	8.2
50 (%) ZnO (nano) + 50 (%) China Clay	13.86	11.2	8.7	6.1	3.8
100 (%) ZnO (ord)	12.1	10.2	8.2	6.4	4.3
50 (%) ZnO (ord) + 50 (%) China Clay	14.2	10	6	1.8	-2.2

(394c) Investigation of Mass Transport Properties of Microfibrillated Cellulose (MFC) Films

Matteo Minelli[1], **Marco Giacinti Baschetti**[2], Ferruccio Doghieri[2], Mikael Ankerfors[3], Tom Lindstrm[3], David Plackett[4] and Istvn Sir[4], (1)Department of Chemical Engineering, Mining and Environment Technology, University of Bologna, Bologna, Italy, (2)Department of Chemical Engineering, Mining and Environmental technology, University of Bologna, Bologna, Italy, (3)Innventia AB, Stockholm, Sweden, (4)Solar Energy Programme, Ris National Laboratory for Sustainable Energy, Technical University of Denmark, Roskilde, Denmark

Investigation of masstransport properties of microfibrillated cellulose (MFC) films

M.Minelli, M. Giacinti Baschetti, F. Doghieri, M. Ankerfors, T. Lindstrm, I.Sir, D. Plackett

Cellulose-based nanocomposites are now emerging as promising newmaterials suitable for a variety of applications. One source for such newnanocomposites is microfibrillated cellulose (MFC), produced by delamination ofcellulosic fibers in high-pressure homogenizers. MFC is an organic andbiodegradable reinforcement in polymer nanocomposites because of its highaspect ratio, good mechanical properties and ability to form networks: promisingproperties that may make this material suitable for use in a number ofindustrial fields and products.

The structure and transport properties of a four different filmsbased on two different generations of microfibrillated cellulose (MFC), aloneor in combination with glycerol as plasticizer, were investigated throughFE-SEM analysis and sorption or permeation experiments. FE-SEM revealed the existenceof complex structures in the different samples. A porous, closely packed fibernetwork, more homogeneous in the samples containing glycerol, wascharacteristic of the surface of MFC films; while film cross-sections presenteda dense layered structure with no evidence of porosity. Water vapor sorptionexperiments confirmed the hydrophilic character of these cellulosic materialsand showed a dual effect of glycerol, which reduced the water uptake at lowwater activity while enhancing it at high relative humidity, as showed inFigure 1. The observed trends of water sorption can be described considering thatboth physical adsorption on fiber surfaces and absorption in the celluloseamorphous phase takes place and therefore a dual-model sorption model wasemployed.

The water diffusivity (in the panel in Figure 1) in dry samples wasremarkably low for a porous material (10^{-11}-$10^{-12}$$cm^2/s$), confirming the existence of complex structures below thefilm surface; however, when water is present in the system, *D* rapidlyincreases with an exponential trend. Diffusivity is also definitely affected byplasticization, being higher for glycerol-containing samples.

Similar behavior was observed in permeation experiments. Dry MFCfilms showed excellent oxygen barrier properties, comparable with those ofpolymers usually considered suitable for ultra-high barrier applications.Indeed, when dry, the cellulose network presents a particularly compact andstiff structure, causing the observed barrier effect. Furthermore, although MFCfilms are characterized by a certain porosity, FE-SEM images of sample surfacessuggests that the pores are not substantially interconnected

When the water content in the membrane is raised, however, adramatic decrease in these properties was observed (Figure 2).

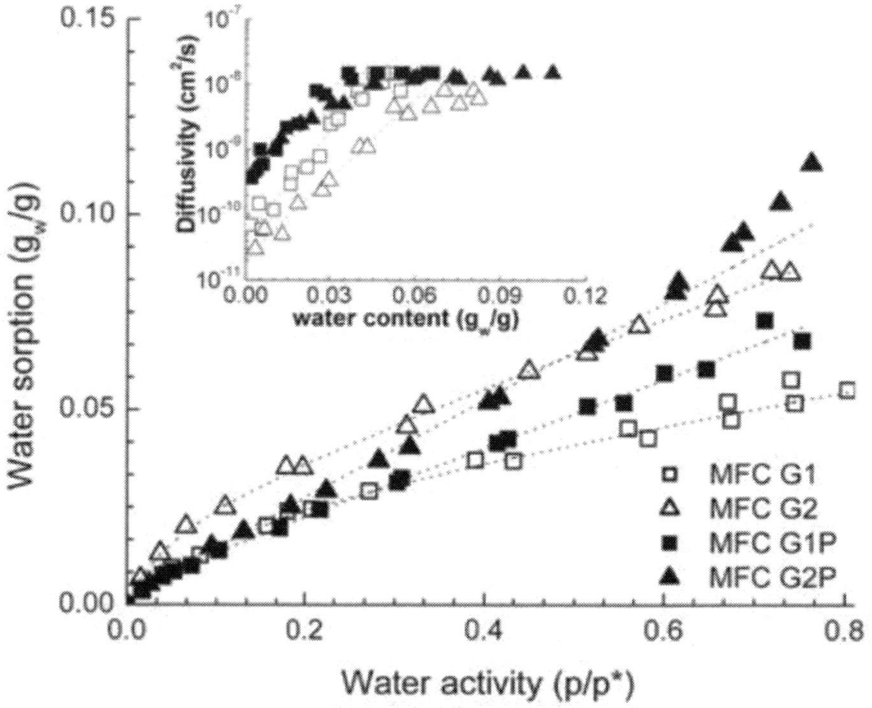

Figure 1. Water vapor sorption isothermsin MFC samples at 35°C, dotted lines are given by the best fitting of the data withthe dual-mode model equation. In the panel: water diffusivity at 35°C in the four different MFC samples as function of the average water concentration.

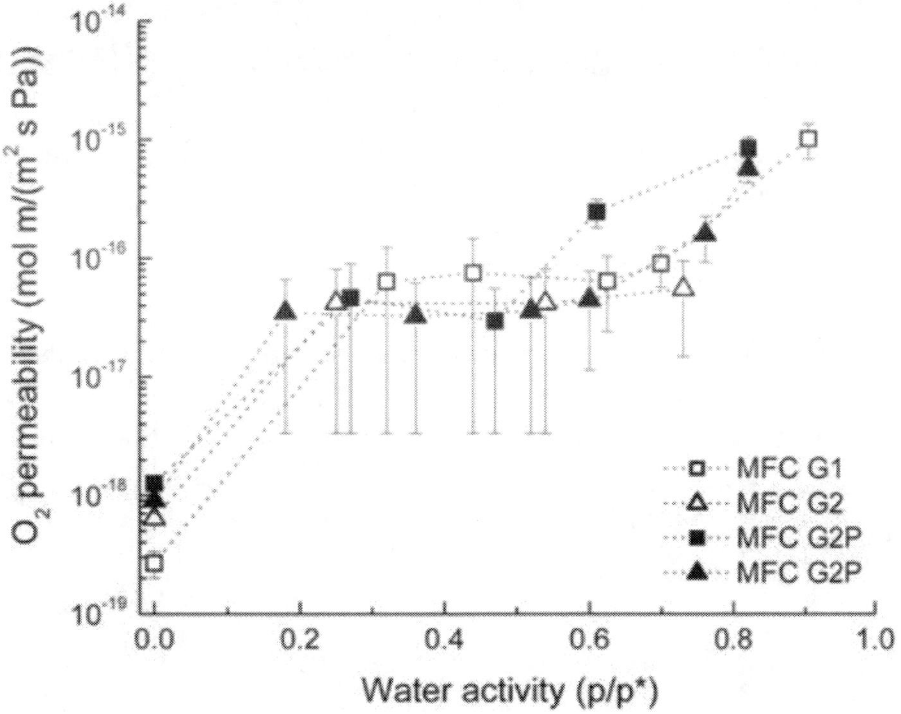

Figure 2. Oxygen permeability in thefour MFC samples at 35°C as a function of water activity.

See more of this Session: Nanocellulose-Based Materials and Composites
See more of this Group/Topical: Forest and Plant Bioproducts Division

(394d) Co-Products of Bioenergy System: Characterization of Saccharification Residuals

Han-Seung Yang, Shona Duncan, William T. Y. Tze and Jonathan Schilling, Department of Bioproducts and Biosystems Engineering, University of Minnesota, Saint Paul, MN

AIChE2011 Annual Meeting - Forest and PlantBioproducts Division

Co-Products of Bioenergy System: Characterizationof Saccharification Residuals

Han-Seung Yang[1], WilliamT. Y. Tze[1], Jonathan Schilling[1] and Shona Duncan[1]

[1]Departmentof Bioproducts and Biosystems Engineering

University of Minnesota

Saint Paul, MN 55108-6130 USA

Tel: 612-625-5200 Fax: 612-625-6286

e-mail: hsyang@umn.edu

The objective of this study is to investigate the effects ofdilute acid pretreatment and enzymatic saccharification on the reactivity,porosity, and crystallinity of solid residuals of biomass. Aspen wood flour was pre-treated with adilute acid treatment followed by enzymatic saccharification. The solid phase of the saccharificationprocess was analyzed using standard NREL protocols for amounts of lignin, pentosans,and hexosans. The functional groupsof solid residuals were analyzed using mid-infrared spectroscopy and hydroxylnumbers. The porosity andcrystallinity of the solids were also evaluated usingrespectively, nitrogen adsorption and X-ray diffraction. Cellulose nanocrystals were extractedfrom the solid residuals and verified using atomic force microscopy. The results will be discussed in thelight of reactivity, porosity, and crystallinity of saccharification residues topave the way for down-stream conversion into co-products of bioenergy(bio-alcohol) systems.

See more of this Session: Nanocellulose-Based Materials and Composites
See more of this Group/Topical: Forest and Plant Bioproducts Division

(394e) Bio-Nano Reinforcement of Polylactic Acid with Surface Modified Cellulose Nanocrystals

Jos Luis Orellana, Esteban E. Urea-Benavides and Christopher L. Kitchens, Department of Chemical and Biomolecular Engineering, Clemson University, Clemson, SC

Cellulose nanocrystals (CNC) have the potential to serve as a nanoscale reinforcement for polylactic acid (PLA), which has become a popular polymer but some of the deficiencies still need to be overcome. The main goal of this work is to improve the mechanical properties of the PLA films – especially toughness – by the addition of surface-modified CNC in the polymer matrix without sacrificing the thermal stability and processability.

CNC were isolated via acid hydrolysis from cotton and surface modified with an esterification process. Transmission Electron Microscopy (TEM) images demonstrated the isolation of the nanocrystals although aggregation can be evidently observed. CNC was further functionalized with different surfactants including decylamine (DA) and cetyltrimethylammonium bromide (CTAB). The dispersibility of the CNC was studied in different solvents such as THF, chloroform and ethyl acetate. Surface modified CNC were added in different concentrations to PLA. Tensile properties of the films were studied using an Instron universal testing machine.

It was found that the stability of the surface-modified CNC improved in organic solvents. DA greatly improved the stability of CNC in both THF and ethyl acetate; while CTAB did in chloroform. No settling was observed in these suspensions for at least three weeks. The toughness and elongation at break of PLA-CNC nanocomposites were improved by the addition of CNC by at least two-fold increase compared to plain PLA. Tensile strength and tensile modulus were not significantly affected by the addition of the crystals. The mechanical properties of the composites were compared and related to the CNC dispersion quality in the different solvents.

See more of this Session: Nanocellulose-Based Materials and Composites
See more of this Group/Topical: Forest and Plant Bioproducts Division

Jet Fuel from Air, Water, and Uranium

Charles Forsberg and John Michael Galle-Bishop
Massachusetts Institute of Technology; Email: cforsber@mit.edu
Proceedings of the American Institute of Chemical Engineers; October 16-21, 2011; Paper 235073

Abstract

Liquid fuels can be produced from air and water. We developed flowsheets for production of 6400 barrels per day of JP5, a jet fuel that is useable as diesel fuel. The feed stocks are carbon dioxide extracted from air and water obtained from the ocean. Heat and electricity to convert the feed stocks into liquid fuel are provided by either a pressurized-water reactor (PWR) or a high-temperature reactor (HTR). PWRs are the most common existing commercial nuclear power technology with peak temperatures of ~285°C. The PWR thermal heat-to-liquid-fuel efficiency is ~22% versus the typical heat-to-electricity efficiency of 33%. HTRs are advanced reactors with peak temperatures of ~700°C. The heat-to-liquid-fuel efficiency is ~31% versus a heat-to-electricity efficiency of 45%. Because the supply of air and ocean water is effectively unlimited, the flowsheet provides an upper limit for the cost of liquid fuel relative to the cost of electricity—between 2 and 3 times the cost of electricity on a per unit heat basis. On a per unit energy basis, today liquid fuel prices in the U.S. are about equal to the price of electricity The study was done in the context of a navy nuclear tanker to manufacture JP-5 at sea at a rate sufficient to supply a carrier strike group under wartime conditions. The flow sheets are essentially identical for navy-at-sea or commercial-on-land applications.

I. Introduction

The design goal [1, 2] was a U.S. navy nuclear-powered tanker that could manufacture on-board at-sea all the liquid fuel as needed for a carrier strike group. This included fuel for both the carrier aircraft and all escort vessels. The aircraft carrier has nuclear reactors for propulsion but the escort vessels use marine diesel or jet fuel for power. Such a tanker would free the U.S. Navy from most of its logistical burdens. JP5 is a refined jet fuel for navy applications that is compatible for use in marine gas turbine and diesel engines and thus could meet all U.S. Navy at-sea liquid-fuel requirements. It differs from conventional jet fuel primarily in its lower flammability—an added requirement because of the risk of fire aboard ship.

The flow sheets for liquid fuels production on a navy tanker are essentially identical to those required to manufacture unlimited liquid fuels from air and water to meet commercial liquid fuels requirements. Uranium resources are sufficient for millennia [3]. The manufacture of fuels from air and water results in no net carbon dioxide releases to the atmosphere and thus no risk of climate change. Air and ocean water are available in unlimited quantities in most countries; thus, this is an option for long-term independence from foreign liquid fuel sources.

Today oil is the preferred low-cost feedstock for liquid fuels production. Alternative feed stocks (natural gas, coal, shale oil, oil sands, biomass) are beginning to be used for liquid fuels production. Liquid fuels from air and water are the only unlimited long-term liquid-fuels production option; thus, it provides an upper long-term estimate of the costs of liquid fuels. The plant size used in this case study is ~1% of a global-scale oil refinery.

The concept of converting carbon dioxide from air and water to liquid fuels is not new. Initial studies [4] were conducted in the 1970s and recent reviews [5] report on much of the progress since then. Advancing technologies however have improved the economics and provided a much sounder basis for estimating costs. These areas are described below

II. Flowsheets

The flowsheet is shown in Fig. 1. The major process steps include:

- *Carbon dioxide feedstock.* Carbon dioxide is extracted from air by absorption with regeneration using humid air. This energy-efficient technology is being tested in pilot plants [6]. This is the most technically immature process step. There are many other options to recover carbon dioxide from air [7].

  The concentration of carbon dioxide in seawater is about 40 times larger; however, in the ocean the carbon dioxide exists in several forms including bicarbonates. The equilibrium kinetics between these various forms and dissolved carbon dioxide are slow [8]. As a consequence, the size of the equipment to remove carbon dioxide from the ocean is too large for a ship-based system. However, extraction of gaseous carbon dioxide from the ocean may be an option for a plant sited on the coast.

- *Water feedstock.* Water is purified and converted to steam—a commercial technology.

- *Syngas production.* Water in the form of steam and CO_2 are fed to high-temperature coelectrolysis cells [9] to produce syngas (H_2, CO) and oxygen. Added hydrogen [10] is made by high-temperature steam electrolysis (HTE). The equipment operates at temperatures near 800°C, is at the pilot plant stage of development, and is based on solid-oxide electrolytic fuel cells operated in reverse. HTE uses about 75% of the energy of traditional alkaline-cell electrolysis.

  HTE is potentially a key technology for integration of nuclear-wind or nuclear-solar electricity hydrogen systems [11]. In these applications, the HTE system uses steam (heat) from nuclear plants and electricity from the grid to produce hydrogen that primarily supplies industrial markets but can also be used for peak electricity production. Such advanced systems may provide an economic method to use excess renewable electricity when electricity production exceeds demand. Consequently, there are expected to be significant incentives for development and ultimate commercialization of this technology.

- *Liquid fuels production.* The syngas is fed to a Fischer-Tropsch (FT) system to convert syngas to liquid fuels. Micro-channel chemical reactors are assumed [12, 13]. The FT process using traditional chemical reactors is a commercial process whereas micro-channel chemical reactors are in the pilot plant stage of testing.

  There are large efforts underway today to improve the FT process for conversion of natural gas to syngas and then to liquid fuels. Shell Oil recently completed its Pearl Project in Qatar, the first 19-billion-dollar global-scale natural-gas-to-diesel-fuel-plant that consumes the equivalent of ~2.5% of U.S. natural gas production. Given the large price differences between natural gas and liquid fuels per unit energy, there are large financial incentives to build more such plants, improve their efficiency, and reduce their capital costs. The rapid advances in natural-gas syngas to liquid-fuels plants is directly applicable to liquid fuels from air facilities.

- *Refining*. Standard refining technology is used to produce the final JP5 liquid fuel with unusable hydrocarbons recycled through an auto-thermal reformer.

  Because of the increased demand for jet and diesel fuel, there is significant work underway [14] to develop better flow sheets to convert FT liquids from coal liquefaction plants into primarily diesel and jet fuel. The normal FT process produces a spectrum of light, medium, and heavy hydrocarbons. Narrowing the product stream to produce more high-value gasoline, diesel, and jet fuel with smaller quantities of low-value light and heavy fractions has a large impact on profitability for existing and future FT plants.

- *Auto-thermal reforming*. The FT process produces a relatively wide distribution of hydrocarbons. Refining operations can convert many of these products to JP5 but there are significant light and heavy products that are not convertible. All of these are recycled back to an auto-thermal converter and burnt with oxygen to produce a syngas for recycle.

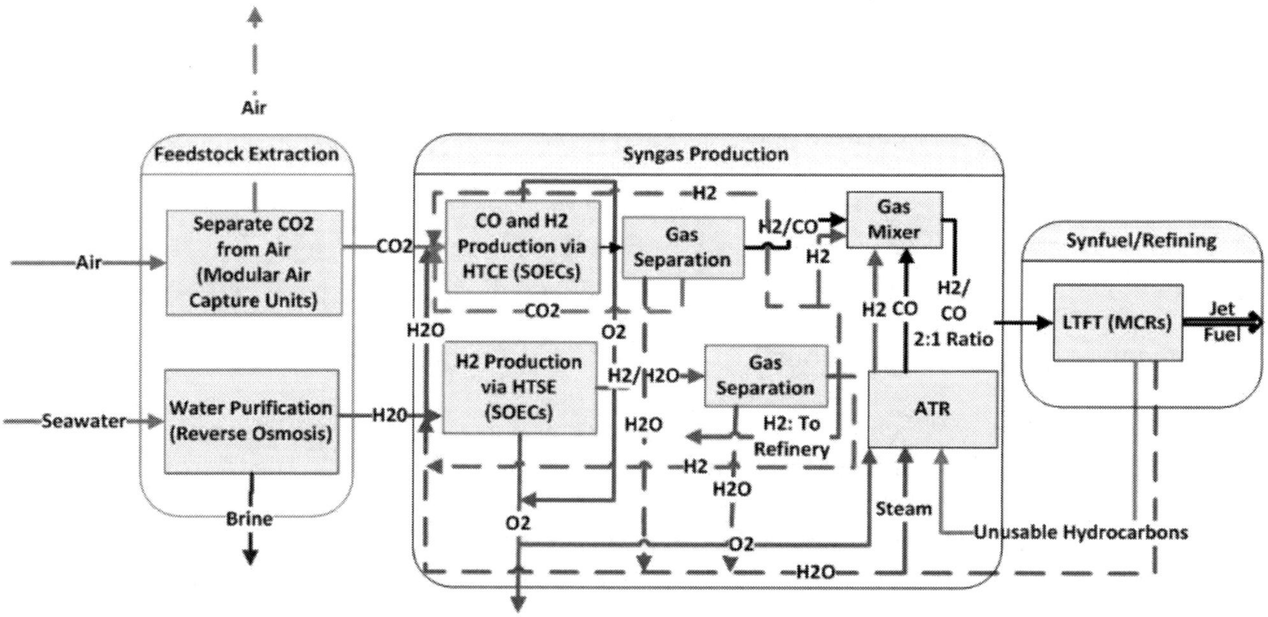

Figure 1: Top-Level Flow Sheet for Air and Water to Jet Fuel/Diesel

The technologies were chosen based on efficiency and space requirements. Space is a major constraint aboard a ship but would not be a constraint for a land-based facility. Each of the non-commercial technologies is at a point where a reasonable case for future commercial deployment can be made. For each non-commercial technology there is a backup commercial technology. For hydrogen production, it is conventional electrolysis, a technology that has been commercial for over a century. The least developed technology is the removal of carbon dioxide from air. Because most of the energy is consumed by high-temperature electrolysis and co-electrolysis systems, this is the critical technology in terms of overall plant efficiency and costs.

The process produces a single product (JP5). However, the flow sheets herein are identical to those required to manufacture unlimited gasoline, diesel, and jet fuel from air and water—except for minor changes in the last refining steps. The differences in energy consumption would be minor. The navy tanker output is ~1% of a global-scale oil refinery. Consequently, the work provides an estimate of what is required to produce liquid fuels with no net release of carbon dioxide to the atmosphere. A commercial plant would be slightly easier to build because (1) no need to restrict plant size to fit aboard ship and (2) somewhat smaller refinery subsection because there is no need to convert gasoline fractions to JP-5

III. Nuclear Reactor Choices

The process requires heat and electricity. All energy was assumed to be supplied by nuclear reactors. Two cases were evaluated. The base case assumed energy provided by pressurized water reactors (PWR)—the main commercial nuclear reactor technology today. The PWR peak temperature is 285°C. The alternative case assumed energy was provided by a salt-cooled high-temperature reactor with a peak temperature of 700°C. This is an advanced high-temperature reactor that may be available in 20 years. Resistance heating is used where required to obtain high temperatures.

IV. Energy Balances

The heat balances for a system to produce 6400 barrels per day of JP5 are shown in Table 1. The percentage of the total nuclear reactor power used in particular unit operations is shown in parenthesis. As would be expected, most of the energy input is for coelectrolysis and electrolysis that together require 590 MWe. That energy is to produce carbon monoxide from carbon dioxide and hydrogen from water for fuels production. The capture of carbon dioxide from the atmosphere is not a major energy consumer. The energy requirements for other operations are small. The smaller power requirements for ship propulsion are not shown.

If PWRs are the source of heat and electricity, the thermal heat-to-liquid-fuel efficiency is ~22% versus the typical heat-to-electricity efficiency of 33%. If HTRs are the source of heat and electricity, the heat-to-liquid-fuel efficiency is ~31% versus a heat-to-electricity efficiency of 45%.

V. Implications of Liquid Fuels from Air and Water

Converting carbon dioxide from air and hydrogen from water into liquid fuels places an upper limit on the cost of liquid fuels. By doing so, it defines the minimum technical and economic requirements for alternative technologies such as electric cars that might replace liquid fuels for specific applications. The technology could also be deployed under several different conditions.

- *Military Applications—Navy Tanker*. The original goal of this work [1] was a nuclear tanker for the U.S. Navy that could manufacture on-board at-sea all the liquid fuel as needed for a carrier strike group. This included fuel for both the carrier aircraft and all escort vessels. The carrier has nuclear reactors for propulsion. Such a tanker would free the U.S. Navy from most of its logistical burdens. The economics of navy fuel supply are significantly different than for commercial users because of the need to assure fuel is provided under war time conditions—a requirement that imposes the need for military protection of tankers.

Table 1. Heat and Energy Balances for Producing 6400 Barrels of JP5 Jet Fuel Per Day Using a Pressurized-Water Reactor or High-Temperature Reactor

System	PWR Parameter; (% Heat[1])	HTR Parameter: (% Heat[1])
Reactor Power		
Total Thermal Power (MWth)	2082 (100%)	1456 (100%)
Thermal Power to Power Cycle (MWth)	1973 (95%)	1285 (88%)
Electrical power from Power Cycle (MWe)	658 (95%)	650 (88%)
Thermal Power for Process Heat (MWth)	109 (5%)	171 (12%)
CO_2 Capture and Desalination (MWe)	21.7 (3%)	21.7 (3%)
High Temperature Coelectrolysis		
Temperature of heat input (°C)	27-285	27-655
CO_2 Feed Heating (MWth)	7.80	21.1
H_2O Feed Heating (MWth)	69.3 (3%)	102 (7%)
HTCE Heat Total (MWth)	77.1	123
HTCE Electricity (MWe)	534 (77%)	534 (73%)
High-Temperature Electrolysis		
H_2O Feed Heating (MWth)	12.0	17.6
HTE Total (MWth)	12.0	17.6
HTE Electricity (MWe)	55.9 (8%)	55.9 (8%)
HTE Electric Resistance Heating (MWe)	0.771	0.0
Auto thermal Reforming		
H_2O heating (MWth)	10.3	13.4
Hydrocarbon Heating (MWth)	2.96	7.86
ATR Heat Total (MWth)	13.2	21.2
ATR Electric Resistance Heat (MWe)	5.380	0.481
Product Upgrade		
Temperature of heat input (°C)	27-285	27-350
Product Upgrade (MWth)	6.24	8.59
Electric Resistance Heating (MWe)	2.49	0.0
Compression (MWe)	4.85	4.85
Compression Non-Refining (MWe)	34.3	34.3
Overall Cycle Performance		
Power cycle efficiency	33.1	44.6
Heat JP5 Combustion/Reactor Heat Input	21.8	31.1

[1]Percentage of total heat from the reactor

JP5 is a refined jet fuel for navy applications that is compatible for use in marine diesel and gas turbine engines and thus could meet all U.S. Navy at-sea liquid-fuel requirements. It differs from commercial jet fuel primarily in its lower flammability—an added requirement because of the risk of fire aboard ship. A nuclear tanker appears feasible given current directions in research and development. The largest gap is associated with the systems to remove carbon dioxide from the atmosphere where there is a need to reduce the system volume by about a factor of four. The system in the flow sheets is a system optimized for siting on land. No effort was made to develop a conceptual system with a smaller footprint.

- *Special commercial markets*. Most of the cost for liquid fuels production is associated with electricity to produce hydrogen via low-temperature or high-temperature water electrolysis. For parts of the world with very low hydro electricity costs and high liquid fuel costs (such as Iceland), this process could become commercially viable (particularly if there were restrictions on greenhouse gas releases). This option is available for any other stranded non-fossil energy source.

- *Normal commercial markets* If there are severe restrictions on greenhouse gas releases to the atmosphere, liquid fuels from air are one of the few options available and could become the economically preferred option. If atmospheric releases of carbon dioxide are limited[1], there are two feedstocks to produce liquid hydrocarbon fuels: biomass and atmospheric carbon dioxide. In both cases carbon dioxide from burning fuel is recycled through the air back to liquid fuels production.

  Biomass would be the preferred feedstock because biomass is a more reduced form of carbon ($\sim COH_2$) compared to carbon dioxide (CO_2). That implies less energy to convert it into a hydrocarbon fuel. However, total biomass resources are limited and may not be able to meet world liquid fuel demands [15].

  As discussed earlier, HTE is potentially a key technology for integration of nuclear-wind or nuclear-solar electricity hydrogen systems [11]. In these applications, the HTE system uses steam (heat) from nuclear plants and electricity from the grid to produce storable hydrogen for industrial markets at times of excess electricity production. Large-scale hydrogen storage is cheap and thus the system has the capability to match electricity and hydrogen production with demand when the production is determined by the energy source (wind or solar), not the customer demand. Production of liquid fuels from air could be done using that hydrogen with small modifications of the flowsheets.

VI. Conclusions

Because the supply of air and ocean water is effectively unlimited, this process when coupled to nuclear power plants provide an upper limit for the cost of liquid fuel relative to the cost of electricity—between 2 and 3 times the cost of electricity on a per unit heat basis because costs are dominated by the energy costs. On a per unit energy basis, liquid fuel prices today in the U.S. are about equal to the price of electricity ($4/gallon ~ $0.10/kWh) . The critical technologies are either commercial technologies or in the pilot plant stage of development. For each of the technologies in the pilot plant stage of development, more conventional technologies with somewhat lower efficiencies and higher capital costs are available.

---

[1] There is the option of producing liquid fuels from fossil fuels with removal and sequestration of carbon dioxide from the air. This option depends upon the economic and social viability of carbon dioxide sequestration. This option does not address the national security challenge of dependence on foreign fuels.

VII. References

1. J. M. Galle-Bishop, C. W. Forsberg, and M. J. Driscoll, *Nuclear Tanker Producing Liquid Fuels from Air and Water: Applicable Technology for Land-Based Future Production of Commercial Liquid Fuels*, MIT-NES-TR-13, Center for Advanced Nuclear Energy Systems, Massachusetts Institute of Technology (June 2011).
2. J. M. Galle-Bishop, *Nuclear Tanker Producing Liquid Fuels from Air and Water*, Thesis, Department of Nuclear Science and Engineering, Massachusetts Institute of Technology, Cambridge, Massachusetts (2011)
3. M. Kazimi, E. J. Moniz, C. W. Forsberg, et. al., *The Future of the Nuclear Fuel Cycle—An Interdisciplinary MIT Study*, Massachusetts Institute of Technology, Cambridge, Massachusetts (2011).
4. M. Steinberg, "Nuclear Power for the Production of Synthetic Fuels and Feedstocks," *Energy Policy*, **5**(1) pp. 12-24 (1977).
5. C. Graves, S. D. Ebbesen, M. Mogensen, and K. S. Lackner, "Sustainable Hydrocarbon Fuels by Recycling CO_2 and H_2O with Renewable or Nuclear Energy', *Renewable and Sustainable Energy Reviews*, **15**, 1-23 (2011)
6. K. S. Lackner, "Capture of Carbon Dioxide from Ambient Air," *The European Physical Journal (Special Topics),* **176,** pp. 93-106 (2009).
7. M. Ranjan, "M.S. Thesis: Feasibility of Air Capture," Massachusetts Institute of Technology, http://hdl.handle.net/1721.1/59782, Cambridge, MA. (2010)
8. S. Werner and J. Morgan, *Aquatic Chemistry: An Introduction Emphasizing Chemical Equilibrium in Natural Waters*, Wiley-Interscience, New York, pp. 780 (1970).
9. J. O'Brien, M. McKellar, and E. Harvego, *High-Temperature Electrolysis for Large- Scale Hydrogen and Syngas Production from Nuclear Energy – System Simulation and Economics*, Idaho National Laboratory, INL/CON-09-15563, Idaho Falls, ID.(2009)
10. J. O'Brien, C. Stoots, and J. Herring, J., *High Temperature Electrolysis for Hydrogen Production from Nuclear Energy – Technology Summary*, Idaho National Laboratory, INL/EXT-09-16140, Idaho Falls, ID. (2010)
11. C. W. Forsberg and G. Haratyk, "Nuclear Wind Hydrogen Systems for Variable Electricity and Hydrogen Production, " *Proc. American Institute of Chemical Engineers Meeting,* Minneapolis, Minnesota, October 16-21, 2011.
12. J. McDaniels, "Unlocking Biofuels with Microchannel Technology," Velocys, Plain City, OH., http://velocys.com/ (2010).
13. A. Tonkovich, D. Kuhlmann, and A. Rogers, "Microchannel Technology Scale Up to Commercial Capacity," *Chemical Engineering Research and Design*, **83**(A6) pp. 634-639 (2005).
14. A. De Klerk, A., *Fischer-Tropsch Jet Fuel Process*, US 2010/018568 A1pp. 1-14 (2010).
15. C. W. Forsberg, "Nuclear Energy for a Low-CO2-Emission Transportation System with Liquid Fuels," *Nuclear Technology*, **164**(3) pp. 348-367 (2008).

Energy Production From Municipal Waste Plastics using by Thermal Degradation

Moinuddin Sarker, PhD, MCIC

Vice President, Research & Development

Natural State Research, Inc.

37 Brown House Road (2nd Floor)

Stamford, CT-06902

E-mail: msarker@naturaltateresearch.com

www.naturalstateresearch.com

Natural State Research

AIChE Annual Meeting, 18th October, 2011

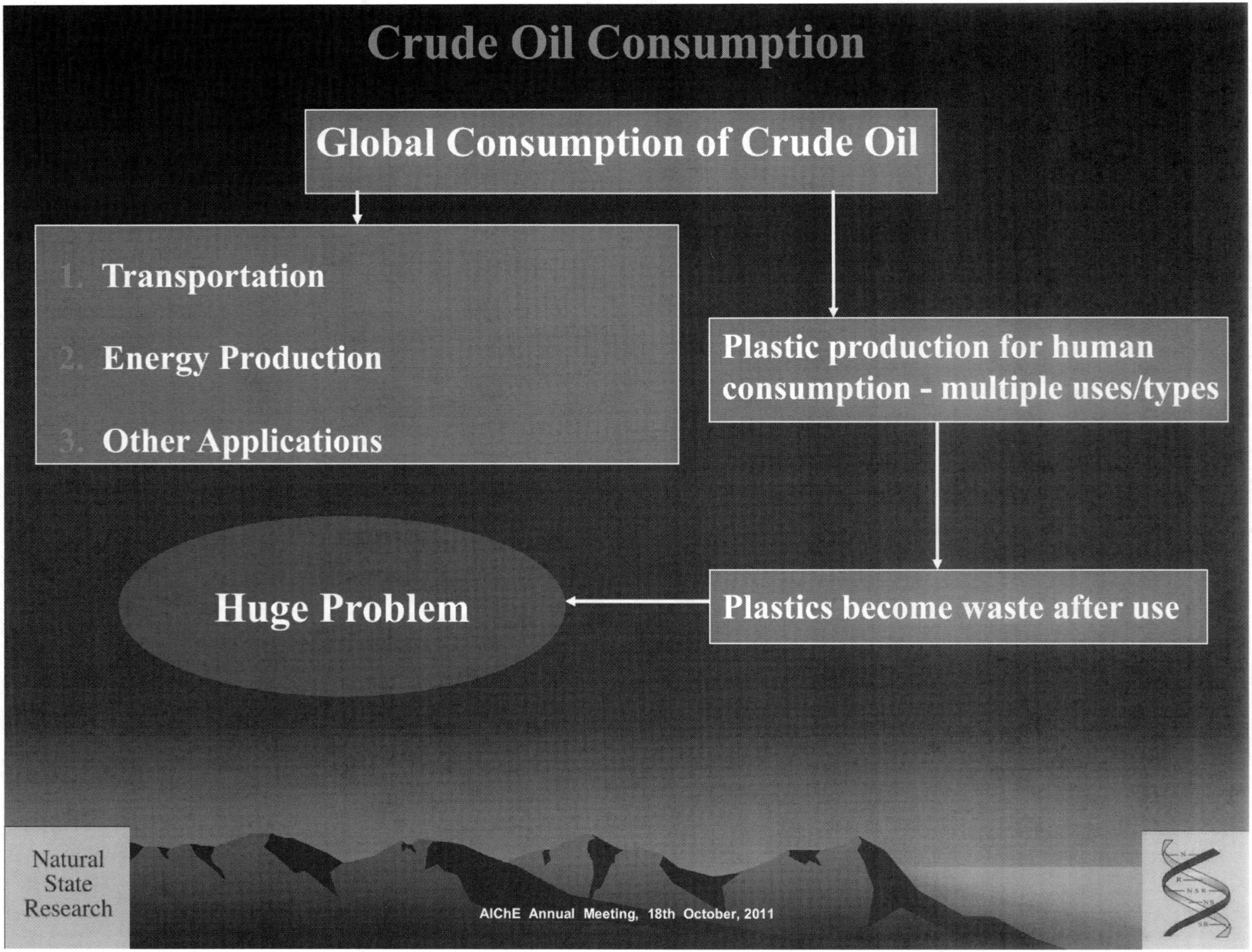
Crude Oil Consumption
Global Consumption of Crude Oil
1. Transportation
2. Energy Production
3. Other Applications
Plastic production for human consumption - multiple uses/types
Plastics become waste after use
Huge Problem
Natural State Research
AIChE Annual Meeting, 18th October, 2011

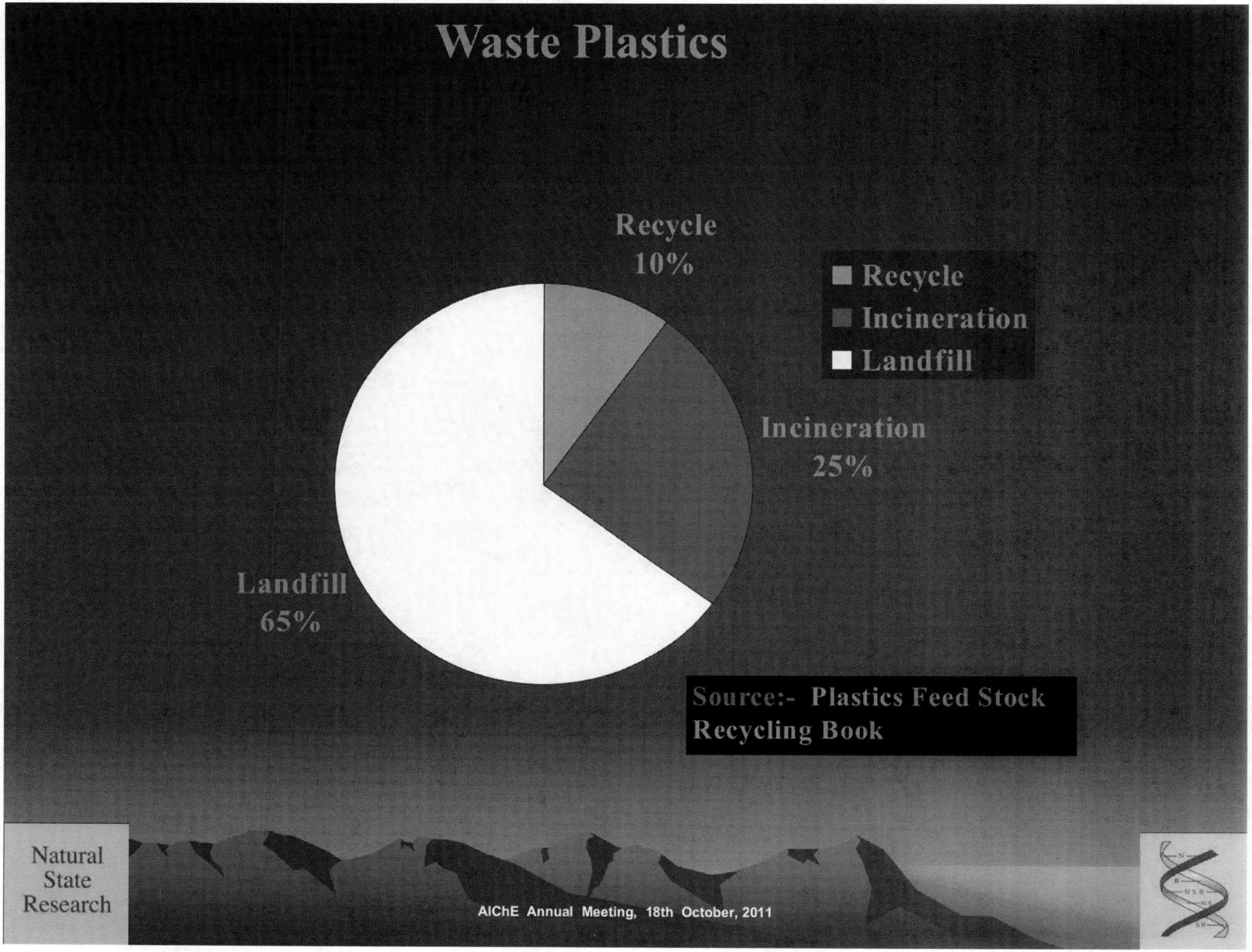
Waste Plastics
Recycle
10%
Incineration
25%
Landfill
65%
Recycle
Incineration
Landfill
Source:- Plastics Feed Stock
Recycling Book
Natural
State
Research
AIChE Annual Meeting, 18th October, 2011

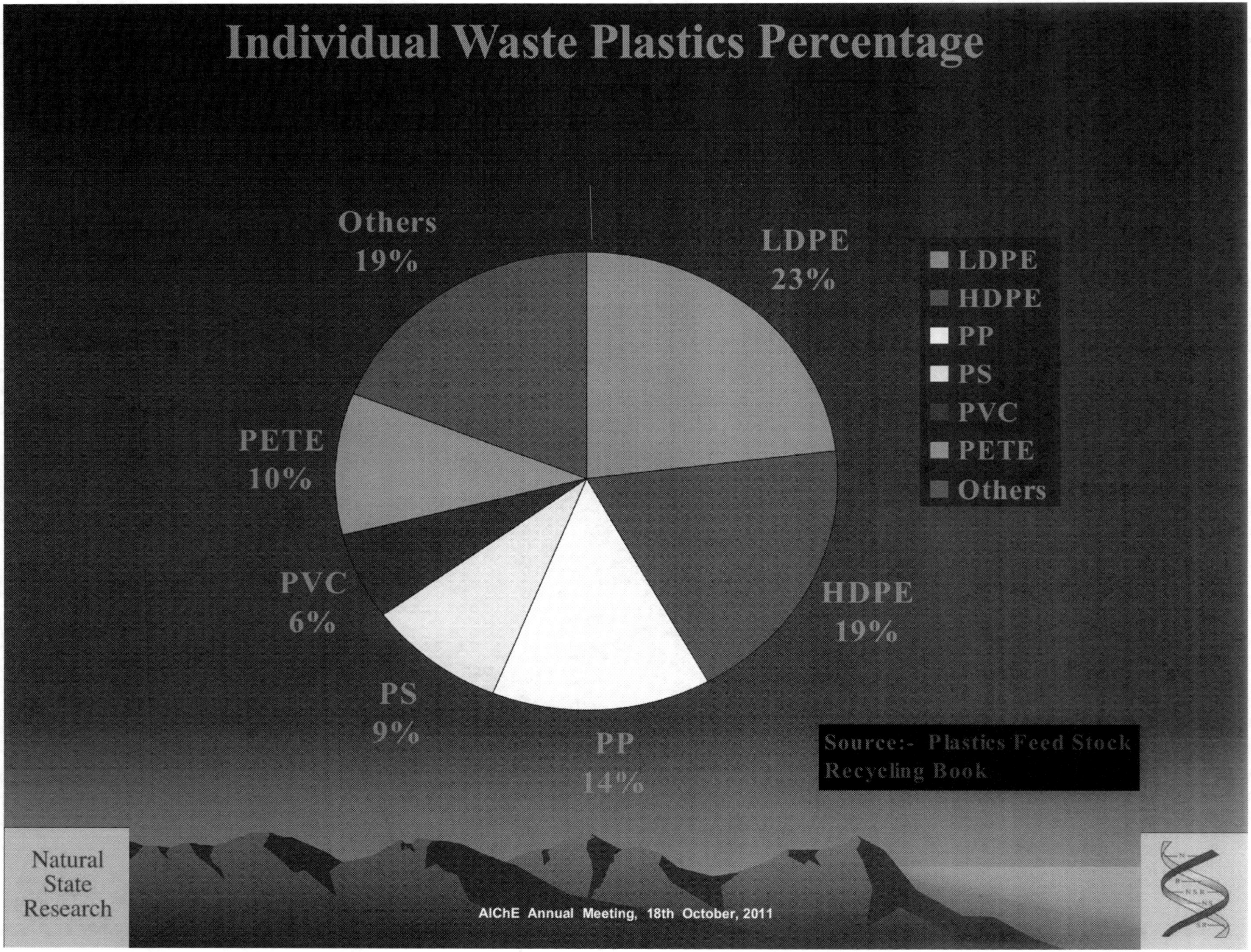
Individual Waste Plastics Percentage
LDPE 23%
HDPE 19%
PP 14%
PS 9%
PVC 6%
PETE 10%
Others 19%
LDPE
HDPE
PP
PS
PVC
PETE
Others
Source:- Plastics Feed Stock Recycling Book
Natural State Research
AIChE Annual Meeting, 18th October, 2011

Plastics Name & Code

Plastic Identification Code	Type of Plastic Polymer	Common Packaging Applications
01 PET	Polyethylene Terephthalate (PET, PETE)	Soft drink, water and salad dressing bottles; peanut butter and jam jars
02 PE-HD	High Density Polyethylene (HDPE)	Milk, juice and water bottles; trash and retail bags.
03 PVC	Polyvinyl Chloride (PVC)	Juice bottles; cling films; PVC piping
04 PE-LD	Low Density Polyethylene (LDPE)	Frozen food bags; squeezable bottles, e.g. honey, mustard; cling films; flexible container lids.
05 PP	Polypropylene (PP)	Reusable microwaveable ware; kitchenware; yogurt containers; margarine tubs; microwaveable disposable take-away containers; disposable cups and plates.
06 PS	Polystyrene (PS)	Egg cartons; packing peanuts; disposable cups, plates, trays and cutlery; disposable take-away containers;
07 O	Other (often Polycarbonate or ABS)	Beverage bottles; baby milk bottles; electronic casing.

Natural State Research

AIChE Annual Meeting, 18th October, 2011

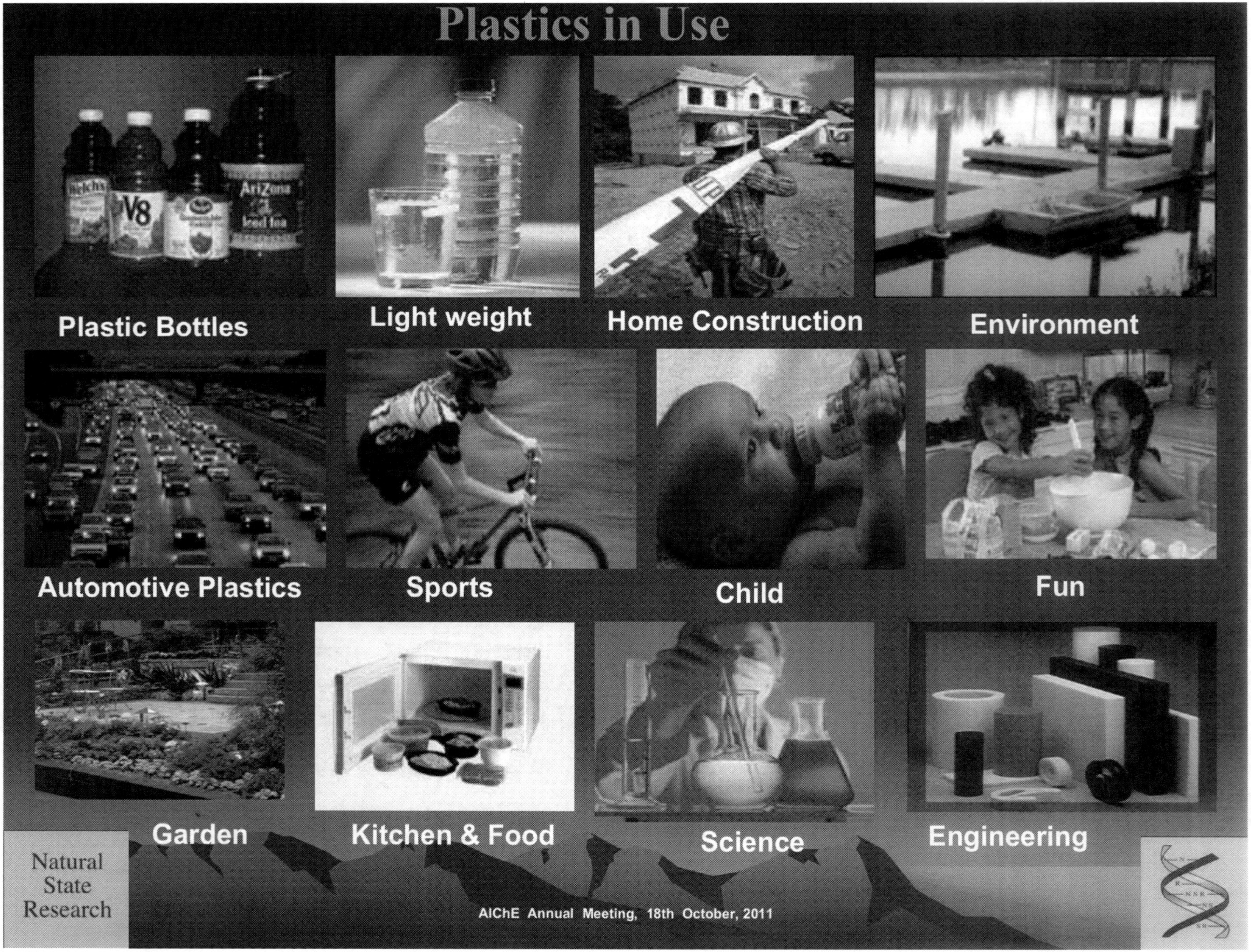
Plastics in Use
Plastic Bottles
Light weight
Home Construction
Environment
Automotive Plastics
Sports
Child
Fun
Garden
Kitchen & Food
Science
Engineering
Natural State Research
AIChE Annual Meeting, 18th October, 2011

Waste Plastics
Baled Polythene Film
Mixed Plastic Films
Plastic Sheet
Mixed Plastic Bottles
Garbage Bags
Plastic Rope /String
Coat Hangers
Film on Rolls
CD's
Shredded Plastics
Granulated Plastics
EPS Briquettes
Car Bumpers
Natural State Research
AIChE Annual Meeting, 18th October, 2011

Alternate Sources of Energy

1. Solar
2. Wind
3. Hydrogen Fuel
4. Bio-mass Fuel
5. Bio-Diesel
6. Green Diesel
7. Bio-ethanol
8. Geo-thermal
9. Ocean Current
10. **NSR???**

A New Alternate/Renewable Source of Energy:

- **Waste Plastics to Liquid Fuel (NSR Patent Pending Technology)**

Natural State Research

AIChE Annual Meeting, 18th October, 2011

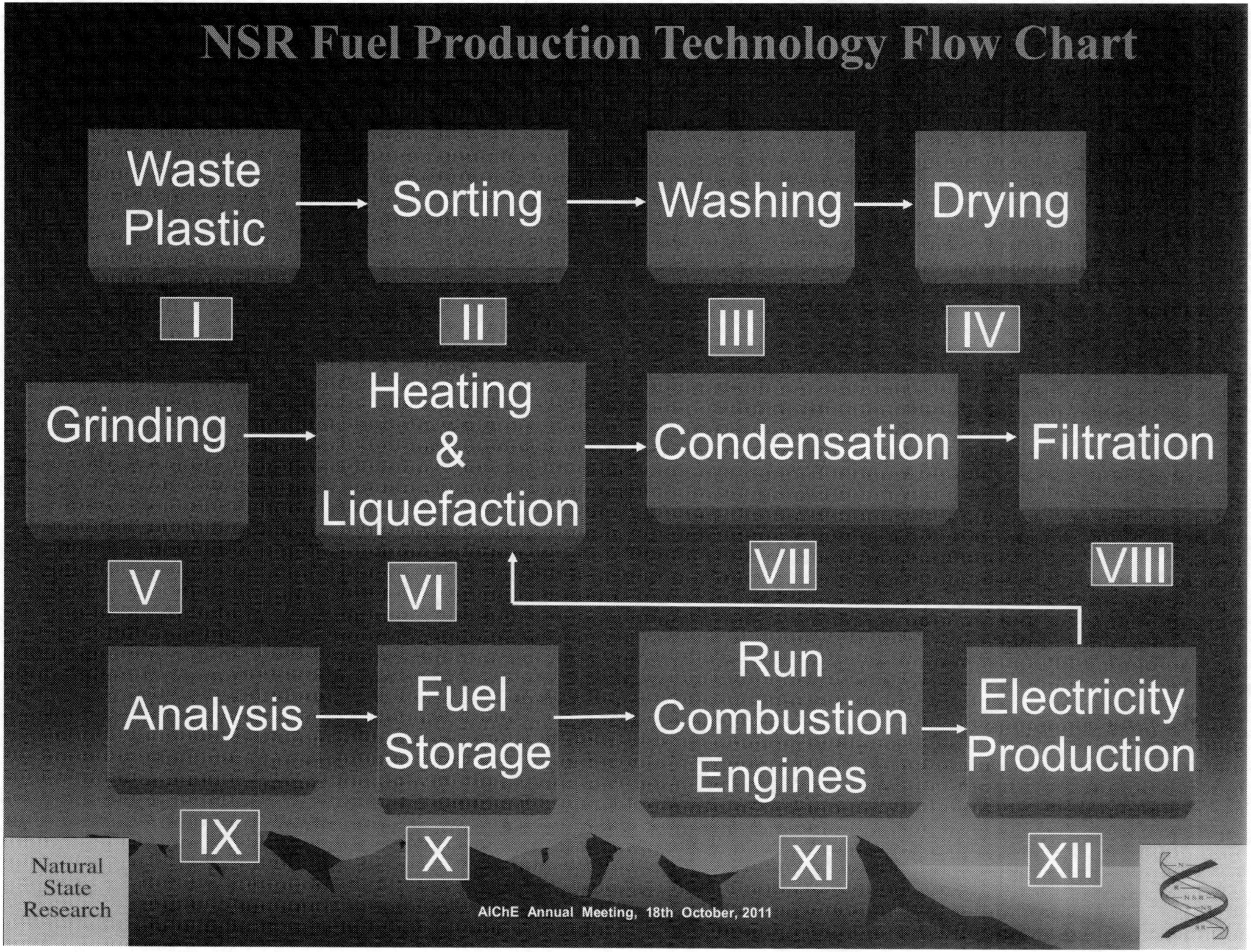
NSR Fuel Production Technology Flow Chart
Waste Plastic
Sorting
Washing
Drying
I
II
III
IV
Grinding
Heating & Liquefaction
Condensation
Filtration
V
VI
VII
VIII
Analysis
Fuel Storage
Run Combustion Engines
Electricity Production
IX
X
XI
XII
Natural State Research
AIChE Annual Meeting, 18th October, 2011

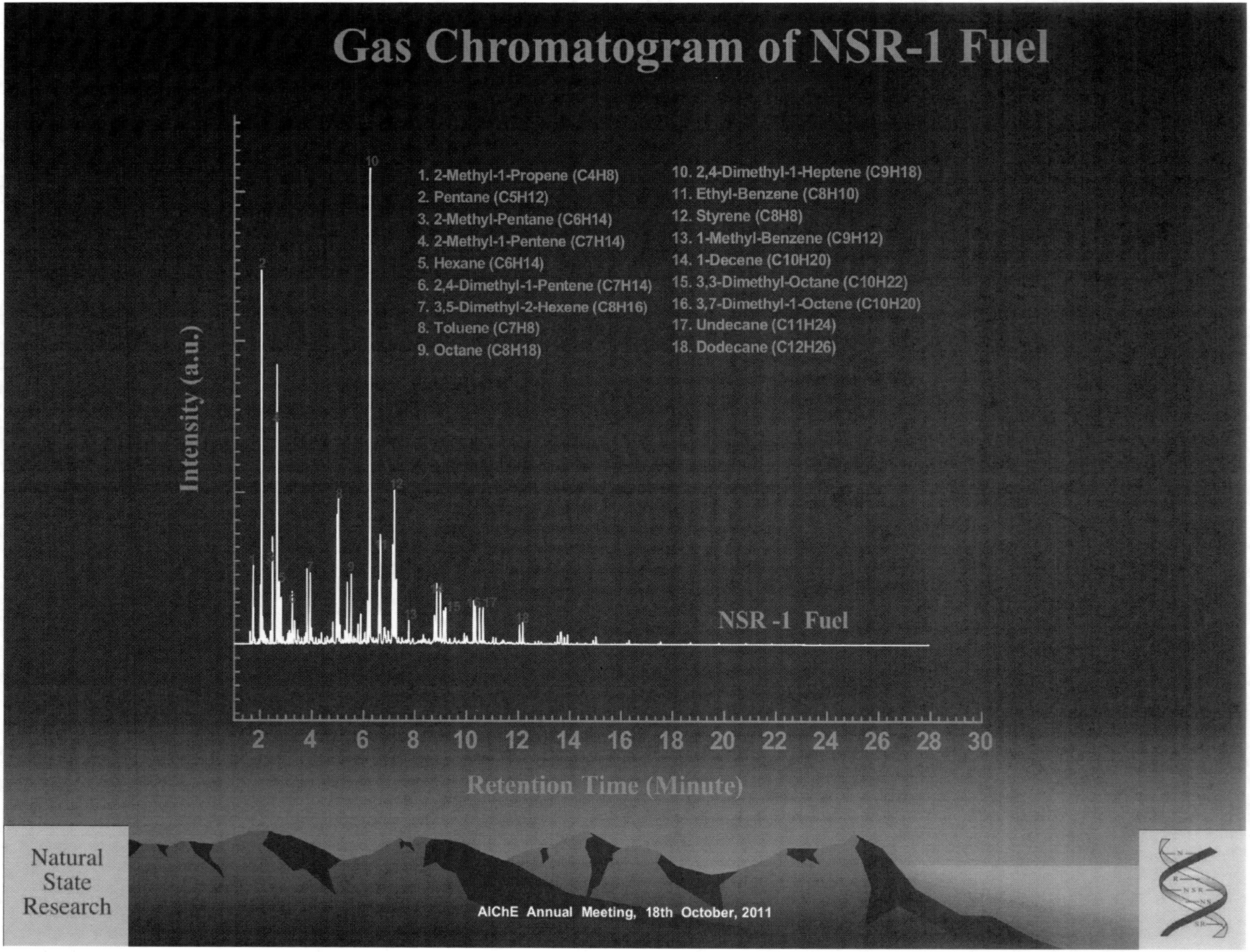
Gas Chromatogram of NSR-1 Fuel
Intensity (a.u.)
1. 2-Methyl-1-Propene (C4H8)
2. Pentane (C5H12)
3. 2-Methyl-Pentane (C6H14)
4. 2-Methyl-1-Pentene (C7H14)
5. Hexane (C6H14)
6. 2,4-Dimethyl-1-Pentene (C7H14)
7. 3,5-Dimethyl-2-Hexene (C8H16)
8. Toluene (C7H8)
9. Octane (C8H18)
10. 2,4-Dimethyl-1-Heptene (C9H18)
11. Ethyl-Benzene (C8H10)
12. Styrene (C8H8)
13. 1-Methyl-Benzene (C9H12)
14. 1-Decene (C10H20)
15. 3,3-Dimethyl-Octane (C10H22)
16. 3,7-Dimethyl-1-Octene (C10H20)
17. Undecane (C11H24)
18. Dodecane (C12H26)
NSR -1 Fuel
2 4 6 8 10 12 14 16 18 20 22 24 26 28 30
Retention Time (Minute)
Natural State Research
AIChE Annual Meeting, 18th October, 2011

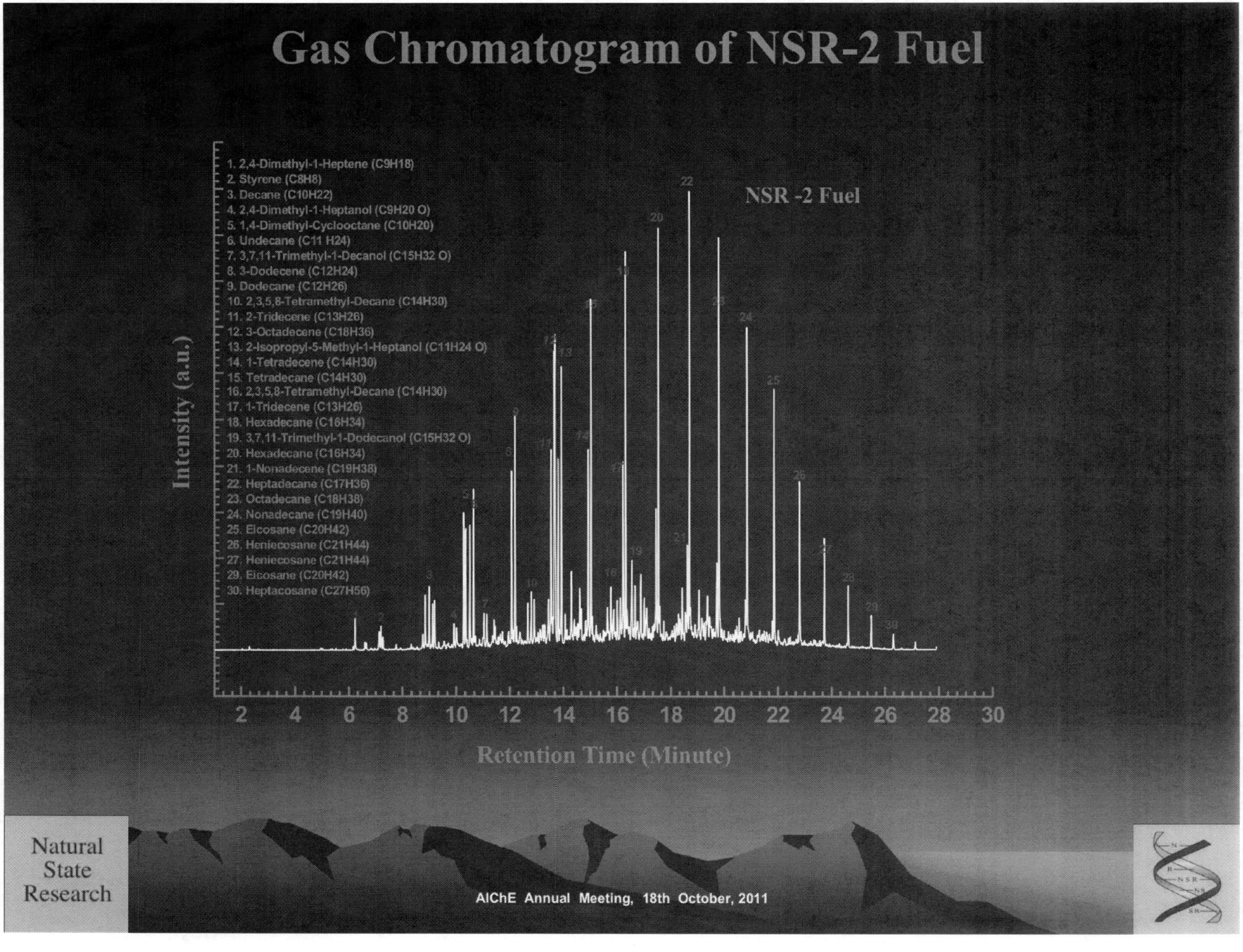
Gas Chromatogram of NSR-2 Fuel
NSR -2 Fuel
1. 2,4-Dimethyl-1-Heptene (C9H18)
2. Styrene (C8H8)
3. Decane (C10H22)
4. 2,4-Dimethyl-1-Heptanol (C9H20 O)
5. 1,4-Dimethyl-Cyclooctane (C10H20)
6. Undecane (C11 H24)
7. 3,7,11-Trimethyl-1-Decanol (C15H32 O)
8. 3-Dodecene (C12H24)
9. Dodecane (C12H26)
10. 2,3,5,8-Tetramethyl-Decane (C14H30)
11. 2-Tridecene (C13H26)
12. 3-Octadecene (C18H36)
13. 2-Isopropyl-5-Methyl-1-Heptanol (C11H24 O)
14. 1-Tetradecene (C14H30)
15. Tetradecane (C14H30)
16. 2,3,5,8-Tetramethyl-Decane (C14H30)
17. 1-Tridecene (C13H26)
18. Hexadecane (C16H34)
19. 3,7,11-Trimethyl-1-Dodecanol (C15H32 O)
20. Hexadecane (C16H34)
21. 1-Nonadecene (C19H38)
22. Heptadecane (C17H36)
23. Octadecane (C18H38)
24. Nonadecane (C19H40)
25. Eicosane (C20H42)
26. Heniecosane (C21H44)
27. Heniecosane (C21H44)
29. Eicosane (C20H42)
30. Heptacosane (C27H56)
Intensity (a.u.)
2 4 6 8 10 12 14 16 18 20 22 24 26 28 30
Retention Time (Minute)
Natural State Research
AIChE Annual Meeting, 18th October, 2011

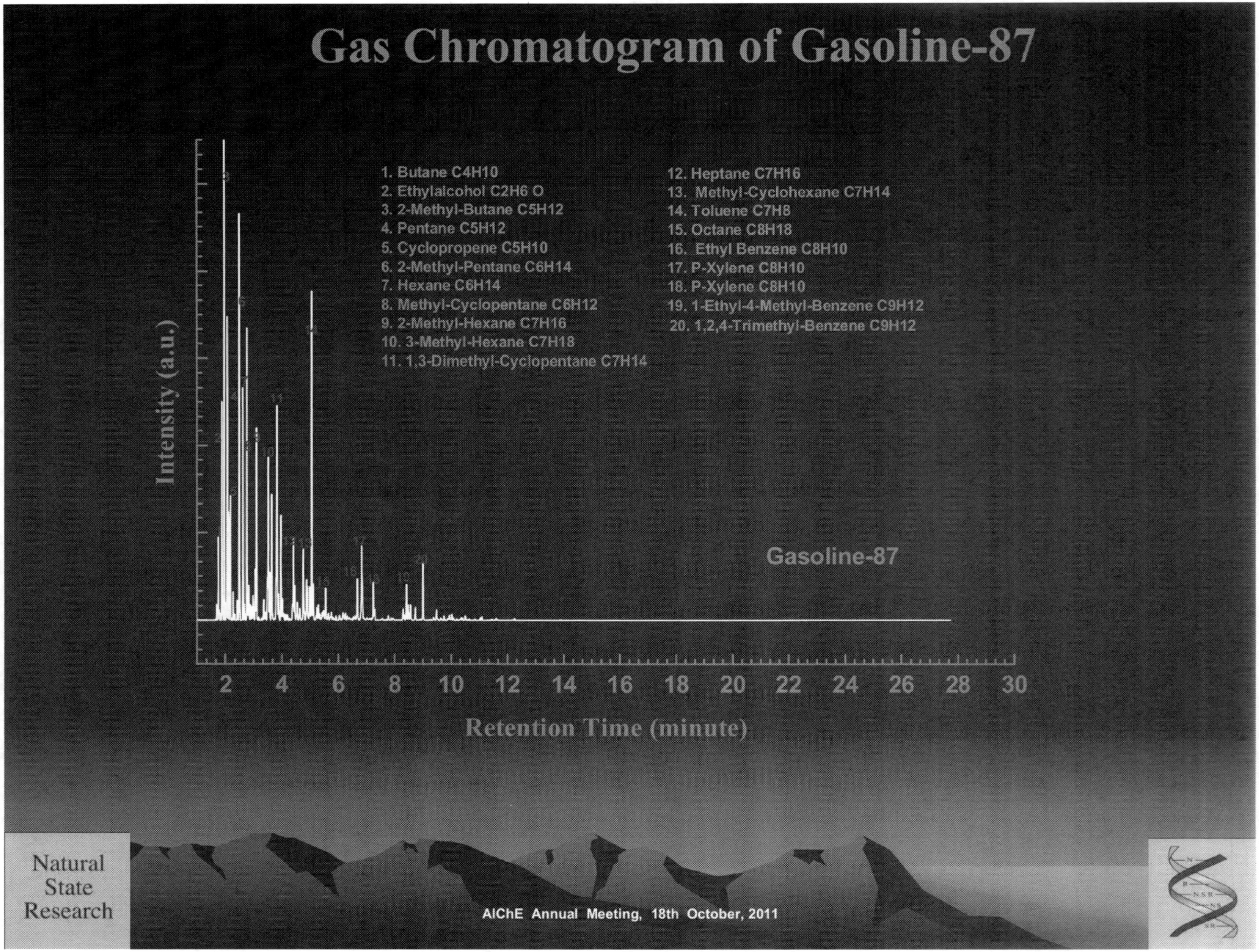
Gas Chromatogram of Gasoline-87
Intensity (a.u.)
1. Butane C4H10
2. Ethylalcohol C2H6 O
3. 2-Methyl-Butane C5H12
4. Pentane C5H12
5. Cyclopropene C5H10
6. 2-Methyl-Pentane C6H14
7. Hexane C6H14
8. Methyl-Cyclopentane C6H12
9. 2-Methyl-Hexane C7H16
10. 3-Methyl-Hexane C7H18
11. 1,3-Dimethyl-Cyclopentane C7H14
12. Heptane C7H16
13. Methyl-Cyclohexane C7H14
14. Toluene C7H8
15. Octane C8H18
16. Ethyl Benzene C8H10
17. P-Xylene C8H10
18. P-Xylene C8H10
19. 1-Ethyl-4-Methyl-Benzene C9H12
20. 1,2,4-Trimethyl-Benzene C9H12
Gasoline-87
2 4 6 8 10 12 14 16 18 20 22 24 26 28 30
Retention Time (minute)
Natural State Research
AIChE Annual Meeting, 18th October, 2011

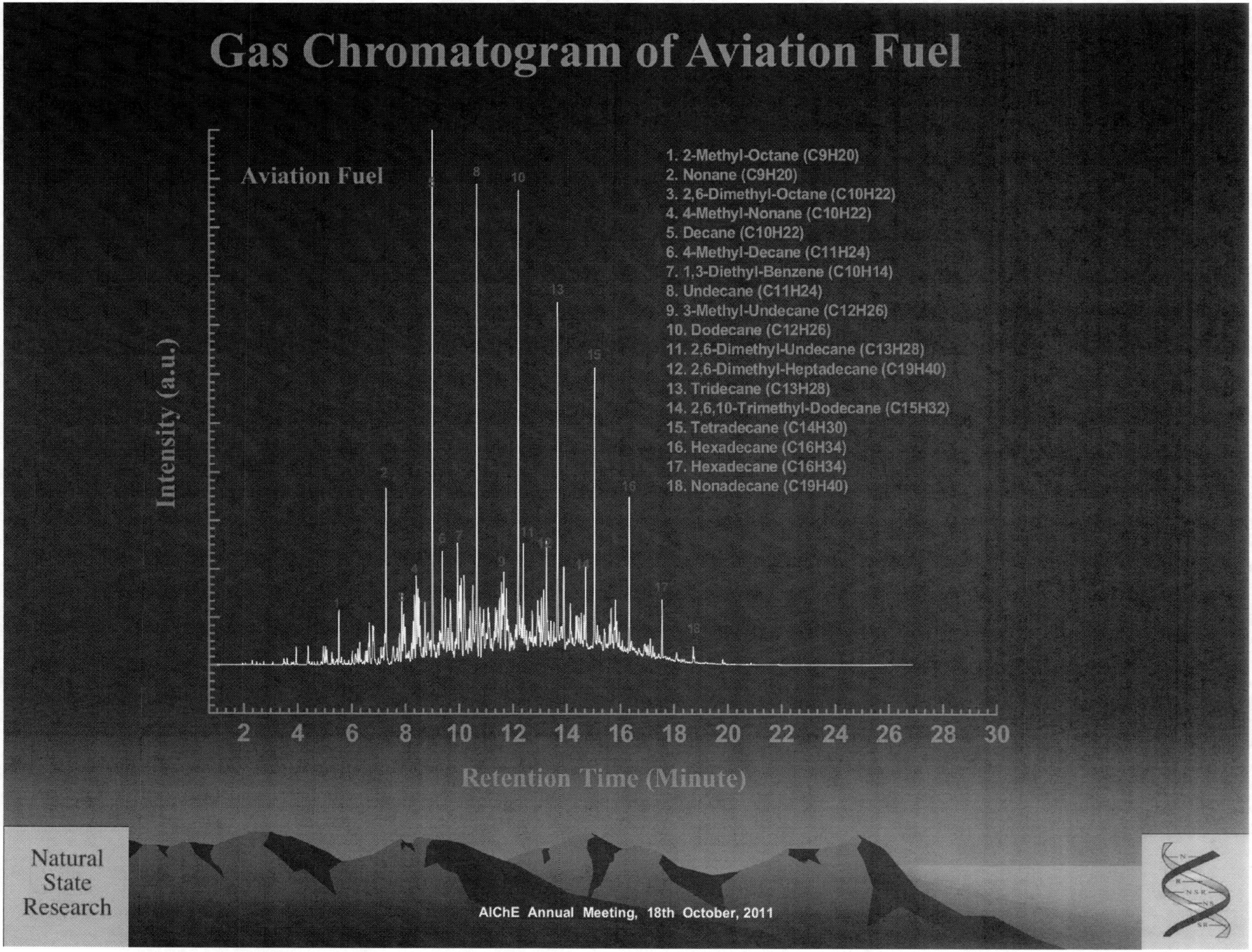
Gas Chromatogram of Aviation Fuel
Aviation Fuel
Intensity (a.u.)
Retention Time (Minute)
2 4 6 8 10 12 14 16 18 20 22 24 26 28 30
1. 2-Methyl-Octane (C9H20)
2. Nonane (C9H20)
3. 2,6-Dimethyl-Octane (C10H22)
4. 4-Methyl-Nonane (C10H22)
5. Decane (C10H22)
6. 4-Methyl-Decane (C11H24)
7. 1,3-Diethyl-Benzene (C10H14)
8. Undecane (C11H24)
9. 3-Methyl-Undecane (C12H26)
10. Dodecane (C12H26)
11. 2,6-Dimethyl-Undecane (C13H28)
12. 2,6-Dimethyl-Heptadecane (C19H40)
13. Tridecane (C13H28)
14. 2,6,10-Trimethyl-Dodecane (C15H32)
15. Tetradecane (C14H30)
16. Hexadecane (C16H34)
17. Hexadecane (C16H34)
18. Nonadecane (C19H40)
Natural State Research
AIChE Annual Meeting, 18th October, 2011

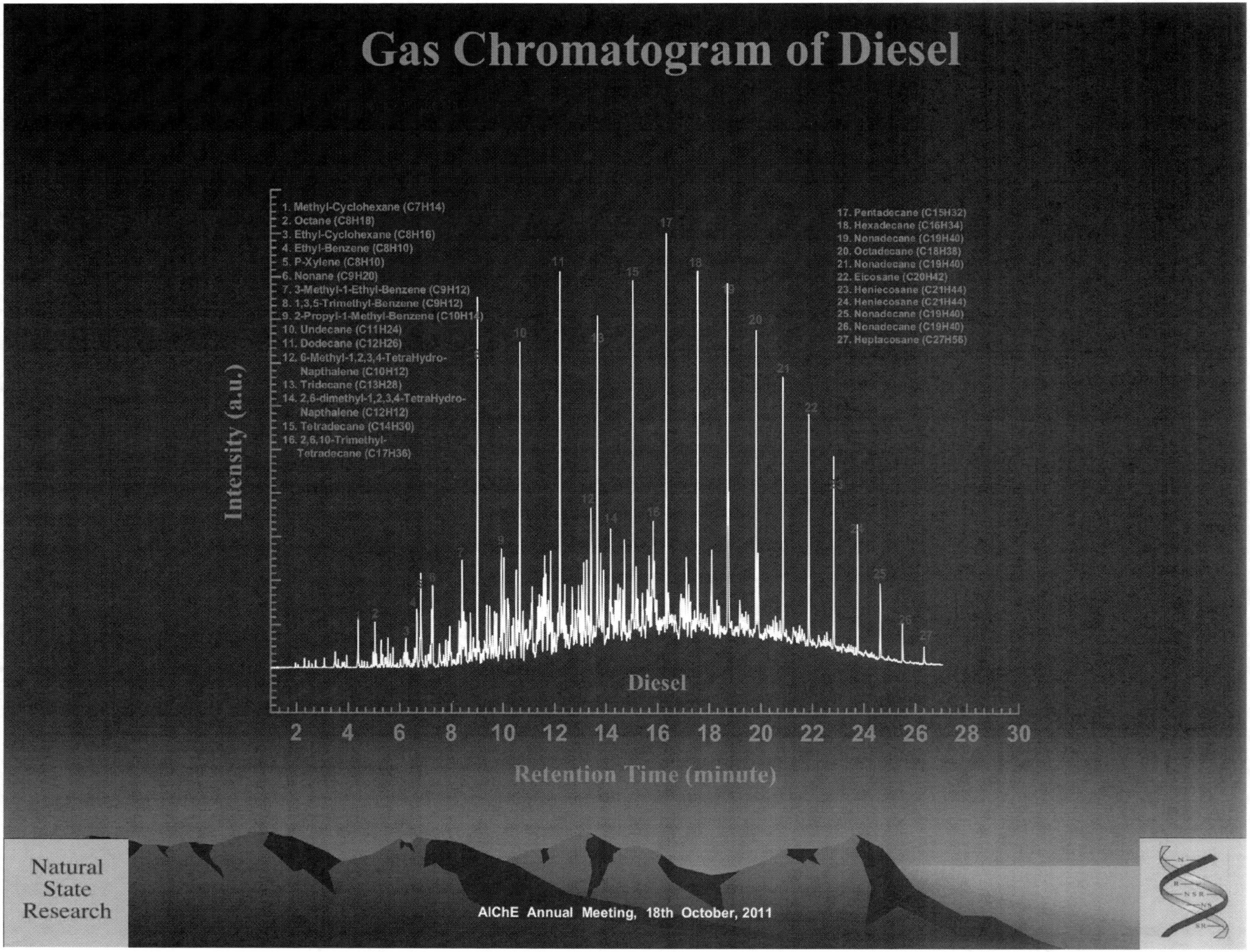
Gas Chromatogram of Diesel
1. Methyl-Cyclohexane (C7H14)
2. Octane (C8H18)
3. Ethyl-Cyclohexane (C8H16)
4. Ethyl-Benzene (C8H10)
5. P-Xylene (C8H10)
6. Nonane (C9H20)
7. 3-Methyl-1-Ethyl-Benzene (C9H12)
8. 1,3,5-Trimethyl-Benzene (C9H12)
9. 2-Propyl-1-Methyl-Benzene (C10H14)
10. Undecane (C11H24)
11. Dodecane (C12H26)
12. 6-Methyl-1,2,3,4-TetraHydro-Napthalene (C10H12)
13. Tridecane (C13H28)
14. 2,6-dimethyl-1,2,3,4-TetraHydro-Napthalene (C12H12)
15. Tetradecane (C14H30)
16. 2,6,10-Trimethyl-Tetradecane (C17H36)
17. Pentadecane (C15H32)
18. Hexadecane (C16H34)
19. Nonadecane (C19H40)
20. Octadecane (C18H38)
21. Nonadecane (C19H40)
22. Eicosane (C20H42)
23. Heniecosane (C21H44)
24. Heniecosane (C21H44)
25. Nonadecane (C19H40)
26. Nonadecane (C19H40)
27. Heptacosane (C27H56)
Intensity (a.u.)
Diesel
2 4 6 8 10 12 14 16 18 20 22 24 26 28 30
Retention Time (minute)
Natural State Research
AIChE Annual Meeting, 18th October, 2011

Carbon Content of Various Fuels

Fuel Name	Carbon Content
NSR -1 Fuel	C_4-C_{12}
NSR -2 Fuel	C_9 - C_{27}
Gasoline (87)	C_4 - C_{11}
Aviation Fuel (Jet Fuel)	C_7- C_{19}
Diesel	C_6 - C_{27}

Natural State Research

AIChE Annual Meeting, 18th October, 2011

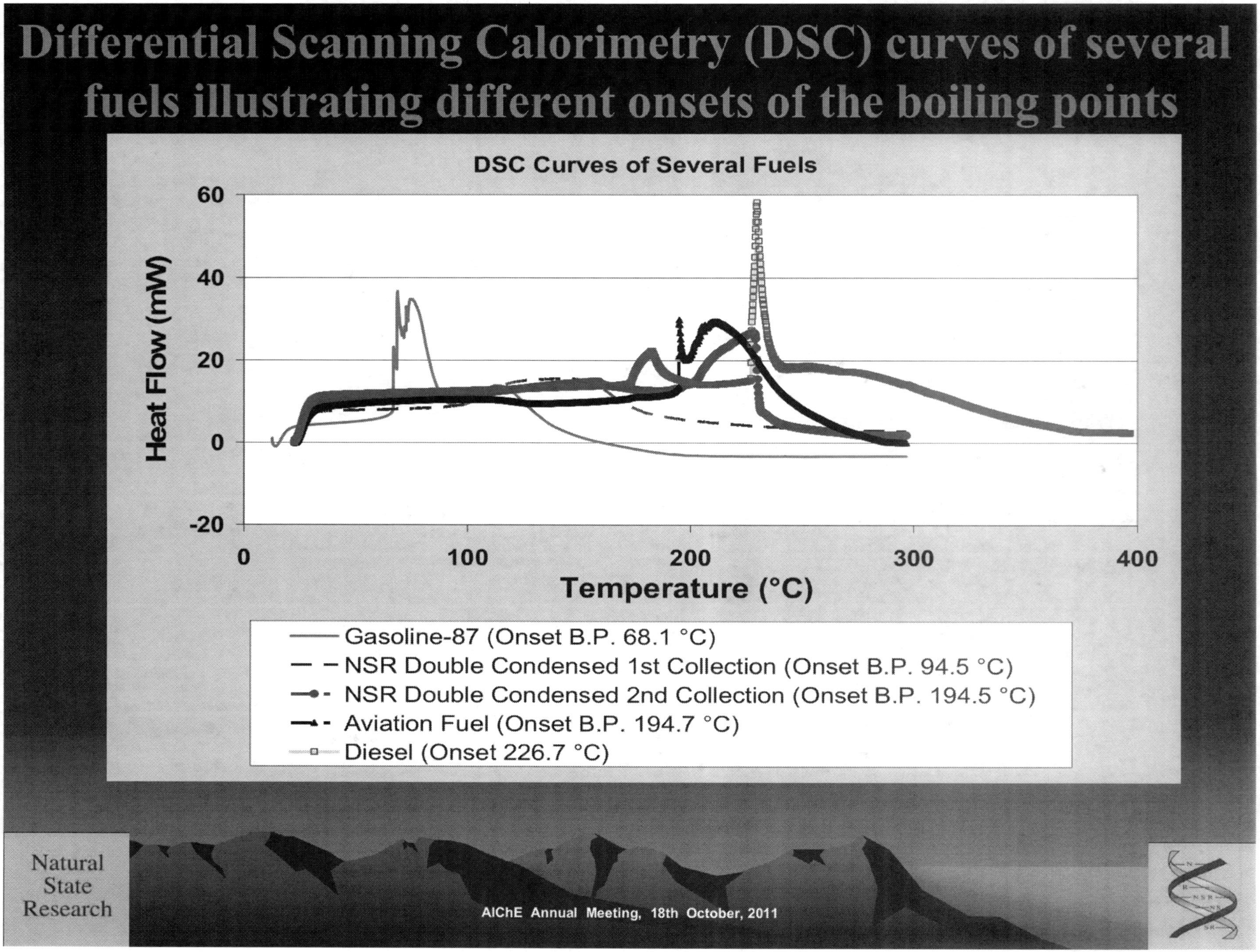
Differential Scanning Calorimetry (DSC) curves of several fuels illustrating different onsets of the boiling points
DSC Curves of Several Fuels
Heat Flow (mW)
60
40
20
0
-20
0
100
200
300
400
Temperature (°C)
Gasoline-87 (Onset B.P. 68.1 °C)
NSR Double Condensed 1st Collection (Onset B.P. 94.5 °C)
NSR Double Condensed 2nd Collection (Onset B.P. 194.5 °C)
Aviation Fuel (Onset B.P. 194.7 °C)
Diesel (Onset 226.7 °C)
Natural State Research
AIChE Annual Meeting, 18th October, 2011

Energy Content of Various Fuels

Type of Fuel	Total BTU /Gallon	Mega Joules /Gallon
NSR -1 Fuel	113,934	120
NSR-2 Fuel	133,366	141
Gasoline - 87	125,000	132
Diesel	138,700	146
Aviation Fuel	120,200	127
Jet Fuel	135,000	142
Compressed Natural Gas	138,700	146
Distillate Fuel	138,700	146
Residual Fuel	149,700	158

1 kWh = 3,412 Btu (British Thermal Unit)
1 kWh = 3.6 MJ (Mega Joules)

Natural State Research

AIChE Annual Meeting, 18th October, 2011

Properties of Various Fuels

Sample	Boiling Point (^0C) (Initial)	Freezing Point (^0C)	Cloud Point (^0C)	Cetane Index	Octane Number	Density (g/ml)	Color / Appearance
NSR- 1	41.5	- 50.0	- 24.0	42.3	**76**	0.74	Pass – clear and bright
NSR -2	171.1	N/A	- 12.0	58.3	N/A	0.80	Pass – clear and bright
Gasoline-87	68.7-204.0	- 72.7	N/A	N/A	87,89,93	0.71-0.75	Pass – clear and bright
Jet Fuel	N/A	- 47.0	N/A	N/A	N/A	0.79-0.84	Clear straw and bright
Diesel	215.5	-10.0 (Gel Point)	- 5.0	42.5	N/A	0.85	Clear greenish and bright

All Preliminary test of NSR Fuels have done by www.intertek.com

Natural State Research

AIChE Annual Meeting, 18th October, 2011

Preliminary Emission Test of NSR Fuels and Comparison with Outside Fuels

DIPE = Di-isopropyl Ether
TAME = Tetra Amyl-Methyl Ether
MTBE = Methyl Tetra-Butyl Ether
DVP = Dry Vapor Pressure (Psi)
E150 = % volume evaporated at 1500C)

Reference Websites:

http://www.med.govt.nz/templates/page 10338.aspx

Sample	DVP (Psi)	E150	Sulfur (mg/kg)	Benzene (Vol. %)	Methanol (Vol. %)	Ethanol (Vol. %)	Oxygen (Wt %)	Olefins	DIPE	TAME	MTBE
NSR 1	4.54	5.6	6.2	0.15	0.00	0.00	0.00	>55	0.00	0.00	0.00
NSR 2	N/A	N/A	4.7	N/A	N/A	N/A	N/A	N/A	N/A	N/A	N/A
Gasoline	6.52	75	150	1	N/A	10	1	18	N/A	N/A	N/A
Diesel	N/A	N/A	50	N/A	N/A	N/A	N/A	N/A	N/A	N/A	N/A

All Preliminary test of NSR Fuels were done by www.intertek.com

Natural State Research

AIChE Annual Meeting, 18th October, 2011

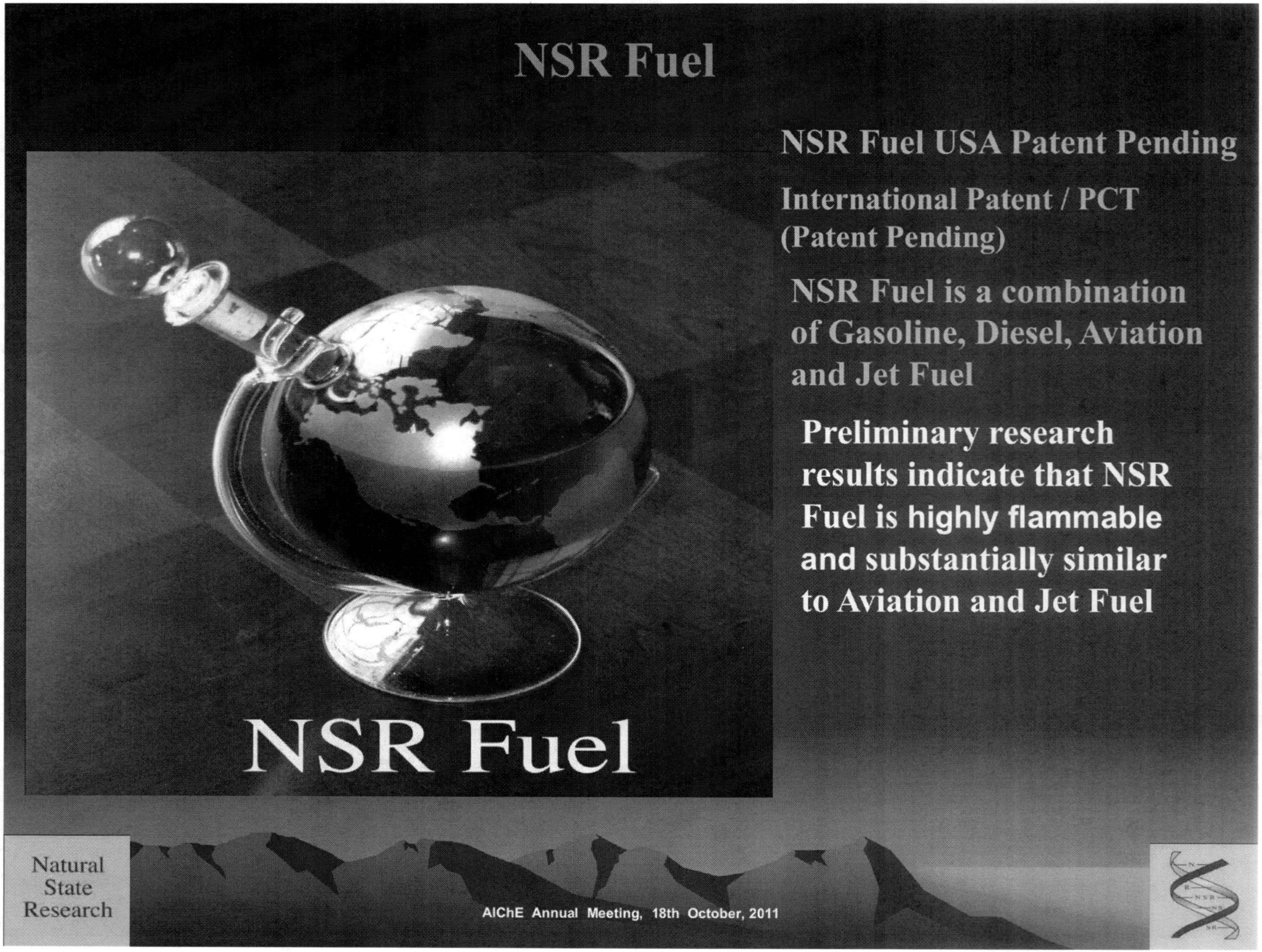
NSR Fuel
NSR Fuel USA Patent Pending
International Patent / PCT (Patent Pending)
NSR Fuel is a combination of Gasoline, Diesel, Aviation and Jet Fuel
Preliminary research results indicate that NSR Fuel is highly flammable and substantially similar to Aviation and Jet Fuel
NSR Fuel
Natural State Research
AIChE Annual Meeting, 18th October, 2011

Driving an Automobile with NSR Fuel

NSR Car Test Experiment

Specification for test vehicle:

1984 Oldsmobile (V-8, 5.0 liters powered engine) and 4 doors Sedan

Experimental Conditions

Driver: Eugene Sheveron (180 lbs)
Along with 3 other researchers (Weight = 550 lbs)
5 Gallon Fuel Container (35 lbs)
Spare Tire (50 lbs)
Total Load : 815 lbs

Mileage obtained:

NSR Fuel (76) = 18 Miles/ Gallon

Gasoline (87) = 15 Miles/ Gallon

Natural State Research

gonne National
boratory, IL, USA.
d June, 2009

AIChE Annual Meeting, 18th October, 2011

Cost Analysis

- Thermal Process at ambient pressure and 370-420 (Degrees Celsius)
- NSR Fuel Energy input = 15-20 kWh

  US$ 1.65 - 2.20

  (excluding overhead)
- Energy output = 55-60 kWh

  US$ 6.05 – 6.60
- 2.90-3.00 Kg of Plastics = 3.8 litter = 1 gallon in Lab Scale (Commercial scale is would be less than US$ 1.00 per gallon)

Present Price:

- Gasoline = US$ 2.50-3.00 per gallon, Energy output: 35-40 kWh
- Diesel: US$ 3.00-3.50 per gallon
- Aviation Fuel: US$ 8.00-10.00 per gallon

Natural State Research

AIChE Annual Meeting, 18th October, 2011

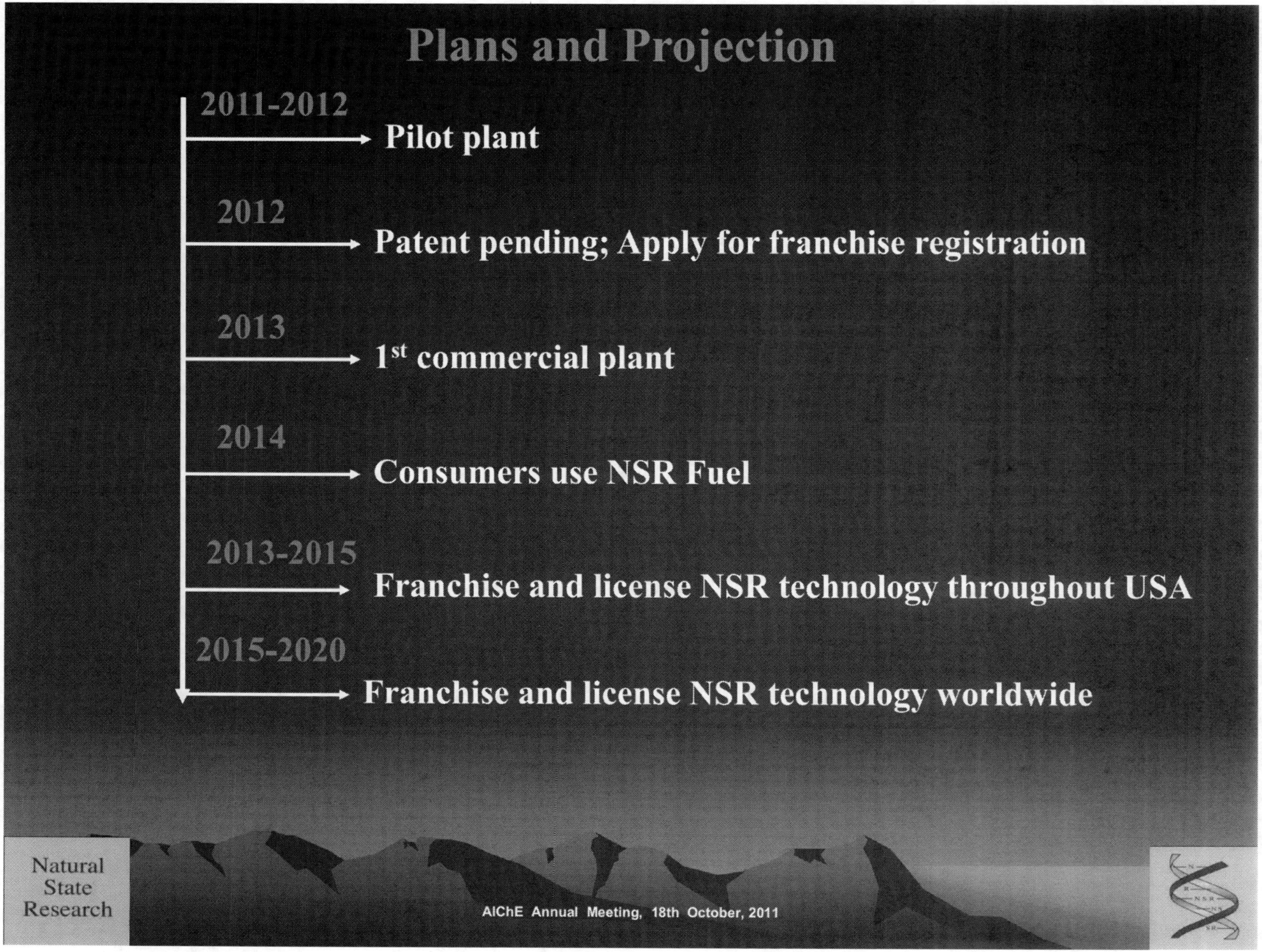
Plans and Projection
2011-2012
Pilot plant
2012
Patent pending; Apply for franchise registration
2013
1st commercial plant
2014
Consumers use NSR Fuel
2013-2015
Franchise and license NSR technology throughout USA
2015-2020
Franchise and license NSR technology worldwide
Natural State Research
AIChE Annual Meeting, 18th October, 2011

Advantage of NSR Technology and Products

1. Efficient
2. Economic (cost effective)
3. Environment friendly
4. 95-98% yield with minimum solid waste/ residue
5. No need to sort out waste plastics (except PETE and PVC)
6. Easy to setup (modular and mobile unit)
7. NSR fuels work with existing internal combustion engine
8. Creates jobs
9. Free/ little cost for raw materials
10. Easy to work with existing recycling and city waste facilities
11. Low investment with high Return

Natural State Research

AIChE Annual Meeting, 18th October, 2011

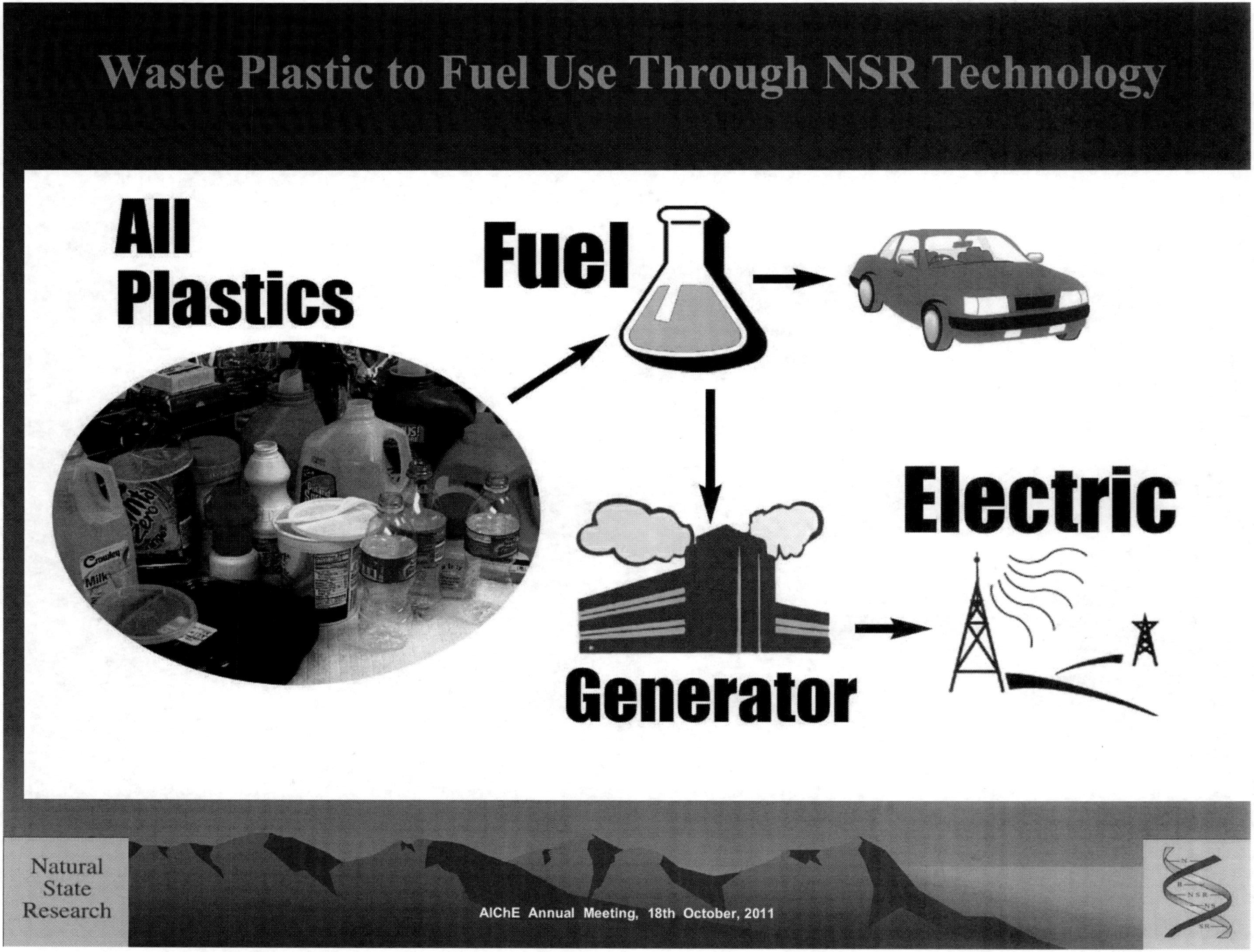
Waste Plastic to Fuel Use Through NSR Technology
All Plastics
Fuel
Electric
Generator
Natural State Research
AIChE Annual Meeting, 18th October, 2011

NSR Technology Award
As recognition of NSR's innovative technology, NSR received the AEI 2009 Tech Award by the Society of Automobile Engineering International at the SAE World Congess and Expo in Detroit, Michigan held on 20-23 April, 2009.
2009 AEI Tech Awards
SAE 2009 World Congress
Plastics to fuel
Natural State Research (NSR) has developed a formula to create liquid fuel from most types of plastic waste. The process turns approximately 95% of the plastic into liquid fuel, while the remaining 5% can be used as a substitute for coal with a higher thermal unit value. The fuel produced is suitable for use in cars, trucks, buses, trains, and heavy equipment. NSR technology does not create greenhouse gas emissions and reduces plastic at overburdened landfills, thereby achieving a balance between energy, economics, engineering, and environmental efficiencies, according to the company. The formula will enable OEMs to reduce costs while stabilizing the U.S. market by establishing a more permanent and predictable resource for fuel.
For more information, visit booth 960
Natural State Research
AIChE Annual Meeting, 18th October, 2011

Innovation Pipeline Award
For recognition for its outstanding achievement, NSR was awarded the Innovation Pipeline Award 2009 "companies to watch" at the Connecticut Technology Council at OMNI Hotel New Haven, CT, held on September, 17 2009 and 2010.
Connecticut Technology Council
The Catalyst for Innovation & Growth
CONNECTICUT TECHNOLOGY COUNCIL
INNOVATION PIPELINE AWARDS 2009
COMPANIES TO WATCH
Natural State Research, Inc.
2009 Company to Watch
Liddy Karter
Program Director, Innovation Pipeline, The Connecticut Technology Council
September 17, 2009
Matthew Nemerson
President & CEO, The Connecticut Technology Council
September 17, 2009
Natural State Research
AIChE Annual Meeting, 18th October, 2011

Conclusion

- Convert waste plastics to hydrocarbon fuel
- Produce electricity using generator / turbine
- Reduce landfill by converting waste plastic to useful hydrocarbon fuel
- Save marine aquatic life by reducing water pollution
- Create local and sustainable jobs

Natural State Research

AIChE Annual Meeting, 18th October, 2011

NSR Research Team

1. Karin Kaufman, PhD
   CEO and President
2. Moinuddin Sarker, PhD, MCIC.
   Vice President, Research and Development (Inventor)
3. Ellery Cressia
   Operation Manager
4. ASM Din Mohammad, M.Sc.
   Lab. Manager
5. Mohammad Mamunor Rashid, M.Sc.
   Plant Manager and Lab Coordinator
6. Nicholas Schiralli
   Come Back Manager
7. Tanya Stewart
   Office Coordinator
8. June Metz
   Business Manager

Natural State Research

AIChE Annual Meeting, 18th October, 2011

9. Mairim Vant
   Network Engineer
10. Aminul Kabir, M.Sc.
   Lab Chemist
11. Ashiquz Zaman
   Research Assistant
12. Mohammad Molla
   Intern
13. Kamrul Islam, M.Sc. MBA
   Lab Technician
14. Sadikur Rahman, M.Sc
   Lab Chemist (Junior)

Consultants:

1. Prof. Richard Parnas, University of Connecticut, USA
2. Sabir Majumder, PhD
3. Dennis Keehn
   Marketing Consultant
4. Fuad Rahman, PhD

Natural State Research

AIChE Annual Meeting, 18th October, 2011

For more information regarding NSR Products and Technology;

Please contact Dr. Moinuddin Sarker at NSR:

+1 (203) 406-0675

OR

E-mails: msarker@naturalstateresearch.com

OR

www.naturalstateresearch.com

Natural State Research

AIChE Annual Meeting, 18th October, 2011

Questions??

Natural State Research

AIChE Annual Meeting, 18th October, 2011

Thank You Very Much

Natural State Research

AIChE Annual Meeting, 18th October, 2011

Volatile Sulfur Compounds Removal in Biogas Using Thermo-Oxidative Pretreatment

Bipro Ranjan Dhar[1], Elsayed Elbeshbishy[2], Hisham Hafez[2], George Nakhla[1,2], Madhumita B. Ray[1,*]

[1]Department of Chemical and Biochemical Engineering, [2]Department of Civil and Environmental Engineering, University of Western Ontario, London, Ontario, Canada, N6A 5B9

Abstract

Thermo-oxidative pretreatment of municipal waste activated sludge was conducted using thermal pretreatment at 60°C in presence of oxidants (0.6 mg H_2O_2+1.5 mg $FeCl_2$/mg S^{2-}) prior to a continuous anaerobic digester operated at 10 days solid residence time (SRT). For the pretreated feed digester, the hydrogen sulphide (H_2S), dimethyl sulphide (DMS) and methanethiol (MT) concentrations in biogas decreased by 75%, 40%, and 10% compared to the control digester, respectively. Compared to the control digester, overall total suspended solids (TSS) and volatile suspended solids (VSS) removal efficiency were 10% and 11% higher, respectively, while methane production rate (L CH_4/day) increased by 20%.

Keywords: Anaerobic digestion, Biogas, Volatile sulfur compounds

* To whom all correspondence should be addressed: Department of Chemical and Biochemical Engineering, University of Western Ontario, London, Ontario, Canada N6A 5B9
Tel.: +1 519 661 2111x81273; Fax: +1 519-661-3498; E-mail: mray@eng.uwo.ca

1. Introduction

Biological treatment of wastewater produces a significant amount of Sludge. Although anaerobic (AD) digestion is an established and widely accepted sludge stabilization process, conventional anaerobic digestion has some limitations and operating challenges. Anaerobic digestibility of municipal waste activated sludge (WAS) is very poor due to the rate-limiting hydrolysis step. In addition, presence of volatile sulfur compounds (VSCs) such as hydrogen sulfide (H_2S), mercaptans in biogas diminishes the utility value of the biogas as they cause corrosion in combustion engines. Pretreatment of sludge prior to AD is a common approach to increase the solid reduction and biogas production in anaerobic digestion.

Various pretreatment techniques have been reported in literature including thermal, chemical, and mechanical pretreatments. Sometimes, different pretreatment technologies are combined together to improve the pretreatment performance. Although the effects of thermal-oxidative pretreatment of sludge have been reported in literature [1, 2], very limited information is available on the reduction of volatile sulfur compounds (VSCs) in biogas. This study investigated the impact of thermo-oxidative pretreatment on VSCs removal during anaerobic digestion. Thermo-oxidative pretreatment of municipal waste activated sludge was conducted using thermal pretreatment at 60°C in presence of 0.6 mg H_2O_2+1.5 mg $FeCl_2$/mg S^{2-} as oxidants with an objective to disintegrate sludge for enhancing anaerobic digestion as well as to remove volatile sulfur compounds generation potential in biogas.

2. Methodology

Thermo-oxidative pretreatment of sludge was conducted using conventional heating at 60°C in presence of oxidants (0.6 mg H_2O_2+1.5 mg $FeCl_2$/mg S^{2-}). The dosages used were based on the theoretical requirement of the chemical dosages for 100% dissolved sulfide (S^{2-}) removal in raw untreated sludge [3]. Thermal pretreatment of 750 ml sludge was done in a flask closed with a rubber cap fitted with a temperature probe and placed on a hot stirred plate (Corning Stirrer/Hot plate, Model PC-420, Corning Incorporated, USA), and heated at the set temperature for 30 minutes. Continuous anaerobic digestion of sludge (750 ml/d) was carried out using two identical completely mixed anaerobic bioreactors (10 L), with a working volume of 7.5 L and a solid retention time (SRT) of 10 days, and maintained at constant mesophilic temperature of 37±1 °C. During start-up the digesters were initially seeded with 7.5 L of anaerobically digested sludge collected from the Guelph wastewater treatment plant (Guelph, Ontario, Canada). Then, the reactors were separately fed with raw and pretreated WAS. Detailed analysis of both liquid and gas samples commenced after steady-state conditions were reached, approximately after three turnovers of the mean SRT. The influent and effluent sludge samples were analyzed for total suspended solids (TSS), volatile suspended solids (VSS), total chemical oxygen demand (TCOD), soluble oxygen demand (SCOD), ammonia nitrogen (NH_4-N), volatile fatty acids (VFAs), different protein and carbohydrate fractions, pH, and alkalinity. The biogas samples were analyzed for methane (CH_4), hydrogen sulfide (H_2S), dimethyl sulfide (DMS), and methanethiol (MT) concentrations.

3. Results & Discussions

3.1 Impact of Pretreatment

The impact of thermo-oxidative pretreatment on various WAS characteristics are shown in Table 1. After pretreatment TCOD remained almost constant, while SCOD significantly increased from 1448±109 mg/L to 3615±164 mg/L due to the release of various extra-cellular polymeric substances (EPSs) such as carbohydrates, proteins etc. Compared to raw WAS, the

TSS and VSS in pretreated sludge decreased by an average of 10% and 15%, respectively. The total volatile fatty acid (TVFA) in pretreated sludge was 17% higher than the raw WAS.

Table 1: Characteristics of raw and pretreated WAS.

Parameter	Raw WAS	Pretreated WAS
TCOD (mg/L)	21328±897	20881±722
SCOD (mg/L)	1448±109	3615±164
TSS (mg/L)	18253±714	16418±752
VSS (mg/L)	13237±609	11132±381
TVFA (mg COD/L)	360±20	420±25
Ammonia (mg/L)	143±10	134±17
TKN (mg/L)	1130±43	1209±125
STKN (mg/L)	700±15	756±19
Particulate protein (mg/L)	2644±275	1965±216
Bound protein (mg/L)	828±60	516±128
Soluble Protein (mg/L)	453±103	1229±166
Particulate carbohydrate (mg/L)	1733±249	1426±110
Soluble Carbohydrate (mg/L)	108±24	364±58
pH	7.1±0.1	6.9±0.1
Alkalinity (mg as $CaCO_3$/L)	3000±200	4500±250

Protein content in sludge is usually divided into three different fractions: cell, bound and soluble [4]. The cell protein represents the protein fraction inside the microbial cell, and the bound protein is the loosely attached protein to the microbial cell wall, while the soluble protein represents the protein fraction available in the aqueous phase. Particulate protein represents the combination of cell protein and bound protein. Bound protein has been identified as a major precursor for volatile organic sulphur compounds generation during digestion [5]. As shown in Table 1, due to thermo-oxidative pretreatment the particulate protein and bound protein decreased by an average of 26% and 38%, respectively, and the average soluble protein concentration in pretreated sludge was 2.7 times higher than the raw sludge. After pretreatment, particulate carbohydrate in pretreated sludge decreased by an average of 18%; while soluble carbohydrate increased by almost 3 times compared to the raw untreated sludge. These results suggest that the thermo-oxidative pretreatment has significantly released the EPSs through the solubilization of sludge. The reduction in total protein and particulate carbohydrate agrees well with the increase in soluble protein and soluble carbohydrate, respectively. No ammonia (NH_4-N) solubilization was observed after pretreatment.

3.2 Anaerobic Digestion

Table 2 summarizes the steady-state digested sludge quality and methane production for the control and pretreated feed digesters. Compared to the control digester, a significant increase of 20% in methane production rate to 3.1 L CH_4/day was observed for the digester fed with pretreated sludge (Table 2). The average TCOD concentrations in influent was almost the same for both digesters, while the SCOD was ~2.5 times higher in pretreated sludge compared to the raw untreated WAS (Table 1). Compared to the control digester, a slight enhancement in TCOD removal was observed for the digester processing the pretreated sludge, while a significant increase in SCOD removal was observed for the pretreated feed digester. The SCOD removal efficiency for control and pretreated feed digesters were 56% and 74%, respectively.

Although prior to the anaerobic digestion the thermo-oxidative pretreatment has significantly reduced the TSS and VSS concentrations compared to the raw WAS, steady-state TSS and VSS removal efficiency during anaerobic digestion for both reactors were almost same. At 10-days SRT, the VSS removal efficiencies for the control and test digester were 40% and 42%, respectively. However, compared to the digestion of raw WAS, the overall (digestion +

pretreatment) TSS and VSS removal in the test digester (including pretreatment) increased by 10% and 11%, respectively. The lack of improvement in suspended solids reduction during digestion due to the pretreatment suggests that the thermo-oxidative pretreatment did not improve the biodegradability of particulate matters.

Table 2: Digested sludge characteristics and methane production.

Parameter	Control Digester	Pretreated Feed Digester
TCOD (mg/L)	13134±320	10800±200
SCOD (mg/L)	630±27	926±84
TSS (mg/L)	12800±350	11300±630
VSS (mg/L)	8000±620	6500±350
TVFA (mg COD/L)	175±23	170±11
Ammonia (mg/L)	635±33	670±47
Particulate protein (mg/L)	1965±216	651±115
Bound protein (mg/L)	516±128	279±103
Soluble Protein (mg/L)	181±58	396±76
Particulate carbohydrate (mg/L)	1053±129	832±80
Soluble Carbohydrate (mg/L)	55±9	80±26
pH	7.5±0.15	7.6±0.15
Methane Production Rate (L CH_4/day)	2.6±0.2	3.1±0.15
CH_4 in biogas (vol. %)	63	66

For the control digester, the particulate protein, bound protein, and particulate carbohydrate removal efficiencies due to digestion were 68%, 62%, and 39%, respectively; while overall removal efficiencies achieved with pretreatment and digestion were 75%, 66%, and 52%, respectively. The results suggest that there was no enhancement in final degradation of bound protein after digestion of both raw and pretreated sludge. For both digesters, the effluent ammonia concentrations increased with solids reduction. As shown in Table 2, the TVFA removal efficiencies for the control and the pretreated feed digester were 51% and 55%, respectively.

3.3 Volatile Sulphur Compounds Removal

The dissolved sulfate and sulfide were analyzed in the raw and pretreated sludge to assess the impact of pretreatment on dissolved sulphide (Table 3). Dissolved sulfide can partition from sludge into the biogas as H_2S. Therefore, a reduction in dissolved sulfide prior to the anaerobic digestion may reduce the H_2S generation potential of sludge during anaerobic digestion. H_2O_2 and $FeCl_2$ can react with dissolved sulfide via a number of different pathways. The following reactions are possible [3, 6]:

$$H_2S + FeCl_2 \rightarrow FeS\downarrow + 2HCl \qquad (1)$$

$$H_2S+H_2O_2 \rightarrow S+2H_2O \qquad (2)$$

$$H_2S+4H_2O_2 \rightarrow 4\ H_2SO_4 +4\ H_2O \qquad (3)$$

$$2FeS + 3H_2O_2 \rightarrow 2S+2\ Fe\ (OH)_3 \qquad (4)$$

However, the reaction pathways at elevated temperature due to thermal pretreatment are still unknown. After thermo-oxidative pretreatment, dissolved sulfide concentration decreased from 80 mg/L to 65 mg/L, while dissolved sulfate concentration remained almost constant. Dissolved sulfide might be converted to ferrous sulfide (FeS) or colloidal sulfur according to reactions (1) and (4). Compared to the raw WAS, 19% removal in dissolved sulfide was observed due to thermo-oxidative preteratment, while the oxidant might react with other compounds in sludge during pretreatment resulting in lower than expected degree of removal of sulfide. The dissolved sulfate and sulfide removal efficiencies were same for both digesters. The very low sulfate removal (<10% for both digesters) during anaerobic digestion indicates the low numbers of sulfate reducing bacteria (SRB) in the anaerobic digesters.

In this study, concentrations of three major VSCs (H_2S, MT and DMS) were regularly monitored in the biogas for both the digesters. The average H_2S, MT and DMS concentrations for control digester were 15, 4 and 0.16 ppm, respectively. H_2S, DMS and MT concentrations in pretreated digester biogas decreased by an average of 75%, 40% and 10% compared to the control digester, respectively (Figure 1). The statistical paired student's t-test indicates H_2S and DMS removal efficiencies in biogas were statistically significant at the 95% confidence level, while MT removal efficiency was insignificant at 95% confidence level. The observed decrease in VSCs concentration in pretreated feed digester might be due to the reduction of two major precursors of volatile sulphur compounds (dissolved sulphide and bound protein) removal through pretreatment.

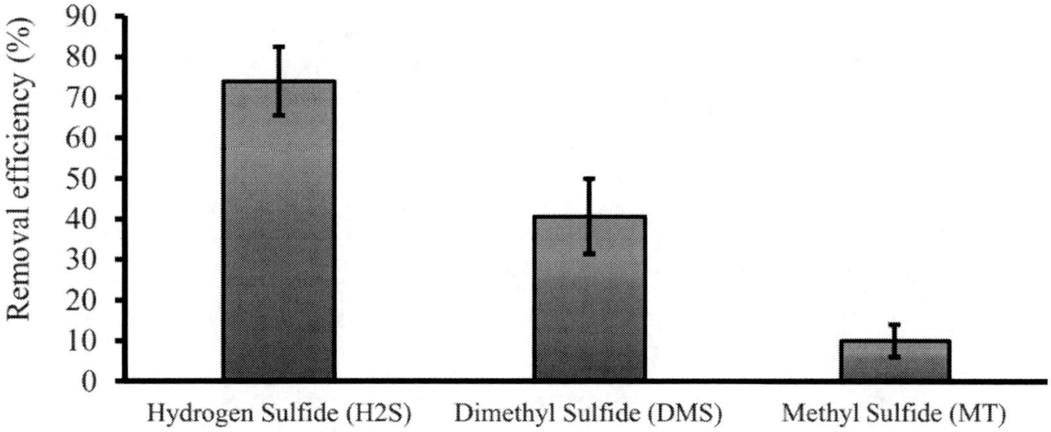

Figure 1: Average volatile sulfur compounds (VSCs) removal in biogas compared to control.

4. Conclusions

The thermo-oxidative pretreatment has shown significant improvement on sludge solubilization along with a significant release of EPSs (protein and carbohydrate). However, solids removal efficiencies during digestion for both digesters were almost the same, although overall (pretreatment and subsequent digestion) TSS and VSS destruction efficiencies with pretreatment were 10% and 11% higher than the control, respectively. Methane production rate for pretreated feed digester increased by 20% compared to the control digester. For the pretreated feed digester, the H_2S and DMS concentrations in biogas decreased by 75% and 40% compared to the control digester, respectively.

References

[1] J. A. Cacho, N. Madhavan, M.T. Suidan, P. Ginestet, P., J.M. Audic, Oxidative co-treatment using hydrogen peroxide with anaerobic digestion of excess municipal sludge, Water Environ. Res. 78 (2006) 691-700.

[2] C. Eskicioglu, A. Prorot, J. Marin, R.L. Droste, K.J. Kennedy, Synergetic pretreatment of sewage sludge by microwave irradiation in presence of H_2O_2 for enhanced anaerobic digestion, Water Res. 42 (2008) 4674-4682.

[3] J.R. Walton, M.S. Velasco, E. Ratledge, Peroxide Regenerated Iron-Sulfide Control (PRI-SC)™: Integrating Collection System Sulfide Control with Enhanced Primary Clarification by Adding Iron Salts and Hydrogen Peroxide, Water Environment Federation Technical Exhibition and Conference (WEFTEC), Los Angles, California, 2003.

[4] R. Dimock, E. Morgenroth, The influence of particle size on microbial hydrolysis of protein particles in activated sludge, Water Res. 40 (2006) 2064-2074.

[5] M. Higgins, D. Glindemann, J.T. Novak, S.N. Murthy, S. Gerwin, R. Forbes, Standardized biosolids incubation, headspace odor measurement and odor production consumption cycles, in: Proceedings Water Environment. Federation and AWWA Odors and Air Emissions Conference, Bellevue, Washington, 2004.

[6] ASCE, Odor control in wastewater treatment plants, American Society of Civil Engineers Publications, USA, 1995.

[7] B.R. Dhar, E. Youssef, G. Nakhla, M.B. Ray, Pretreatment of waste activated sludge for volatile sulfur compounds control in anaerobic digestion, Bioresour. Technol. 102 (2011) 3776-3782.

(413d) Multidisciplinary Approach In Developing Region Specific Models for Sustainable Biorefining

Sumesh Sukumara[1], Jeffrey Seay[1] and Fazleena Badurdeen[2], (1)Chemical and Materials Engineering, University of Kentucky, Lexington, KY, (2)Mechanical Engineering, University of Kentucky, Lexington, KY

Use of foreign fossil fuel for transportation and producing energy has led to energy dependence and growing environmental concerns. Among the various choices of renewable energy resources, biomass intrigues many researchers due to its widespread availability, cost effectiveness and its applicability as a sustainable energy source. Large scale biomass production and improved manufacturing techniques are some of the major areas where researchers are working on. However, because of high variations in the availability of biomass from one region to another there is a necessity to develop region specific economic models for biorefining. Developing economic models requires a multidisciplinary approach as the variation in biomass results in significant change in collection, storage techniques, supply and processing. This presentation aims to show how different disciplines of engineering and sciences can be combined to develop a region specific economic model for sustainable biorefinery based on Geographic Information Systems (GIS), supply chain optimization and process systems engineering. The Jackson Purchase in western Kentucky is the region where this model will be applied first. The Jackson Purchase region is not only rich in coal and biomass but also has a large number of poultry farms. Hence poultry litter can be a potential feedstock for integrated biorefinery in this region. Forest residue, corn stover and chicken litter are some of the feedstocks that are considered in this model for producing liquid transportation fuels. However, the distribution of lignocellulosic biomass in the country is very uneven and the first challenge is to locate the biomass. GIS is a technique by which biomass in a region can be located and by coupling it with supply chain optimization, location of potential biorefinery can be suggested. ILOG OPL is the tool used for optimizing the transportation cost of biomass from supply points to the potential biorefineries. Process systems engineering can be used to develop and optimize the processes. Aspen plus V-7.2 process engineering suite is the process simulation software used for this purpose. Gasification is a proven thermochemical process that is optimized in this model to produce synthesis gas and liquid hydrocarbons from forest residue and chicken litter. Also, a biochemical process is optimized in which corn stover is used to produce ethanol which can be marketed as transportation fuel. This model also looks into costs associated with storage and preprocessing of biomass and chicken litter. Detailed ash analysis will also be done in order to suggest most environmentally friendly conversion techniques among the various available choices. Finally, in future all the above optimized models will be combined to estimate the total production cost of various marketable products from different locally available feedstocks. The results, thus, obtained from these models will not only determine the location and number of biorefineries required to sustain a given region, but will also estimate the adequate monetary investment on sustainable biorefining required for that specific region.

(413e) Environmental Releases In the Fuel Ethanol Industry

Debalina Sengupta, Chemical Engineering, US Environmental Protection Agency, Cincinnati, OH and Raymond L. Smith, US EPA, Cincinnati, OH

Corn ethanol is the largest produced alternate biofuel in the United States. More than 13 billion gallons of ethanol were produced in 2010. The projected corn ethanol production is 15 billion gallons by 2015. With increased production of ethanol, the environmental releases from ethanol industry are going to increase.

The common studies with fuel ethanol plants typically report criteria air pollutants and greenhouse gas emissions. This prompted the need to evaluate the hazardous air pollutants released from ethanol plants and an overview of these studies will be presented in this paper.

This research shows a relative comparison of toxic releases in the United States, and particular to the corn-ethanol industry. Data is collected from the TRI reporting database and compiled for emissions for the toxic chemicals. Ethanol is produced industrially by two processes, wet milling, and dry grinding. In the wet milling process, there is a range of products pertaining to the food and feed industry other than ethanol. In the dry grind process, the products are ethanol and distillers grains with solubles. The processes produce ammonia, acetaldehyde, formaldehyde, acrolein, etc. as toxic releases. These releases are reported to the TRI database, and the releases for the ethanol industry are worth studying for the environmental effects and trends for future plants.

The present work attempts to capture these trends and present them for further evaluation.

See more of this Session: Sustainable Fuel From Renewable Resources II
See more of this Group/Topical: Environmental Division

(413f) Membrane Separation of Ethanol From Bio-Ethanol Containing Gasoline

Akihiro Yamasaki[1], Yohei Ueda[2], Naoki Takahashi[2], Yuka Sakai[2], Shigeo Satokawa[2] and Atsushi Iizuka[3], (1)Department of Materials and Life Science, Seikei University, Musashino, Tokyo, Japan, (2) Seikei University, Musashino, Tokyo, Japan, (3)Research Center for Sustainable Materials & Engineering, Institute of Multidisciplinary Research for Advanced Materials, Tohoku University, Sendai, Miyagi, Japan

Polymeric membranes for enriching ethanol from bio-ethanol containing gasoline (BEG) have been developed for the on-board separation process. Poly (vinyl alcohol) (PVA)-base membranes with controlled hydrophilicity were prepared and their separation performances were examined based on the pervaporation experimental studies. The hydrophilicity of the PVA-based membranes was controlled by changing the degree of saponification (the fraction of hydroxyl groups). The PVA-based membranes showed high selectivity for ethanol, of which the concentration in the permeate side was higher than 90% for a wide range of the feed composition studied. Increasing the hydrophilicity of the membranes increased the ethanol selectivity over a model hydrocarbon (n-heptane), and the flux decreased. The highest flux observed was as high as 100 $g/m^2/h$. The results suggest that the membranes can be applicable for the separation processes of ethanol from BEG.

See more of this Session: Sustainable Fuel From Renewable Resources II
See more of this Group/Topical: Environmental Division

(413g) Life Cycle Analysis of Geothermal Energy Extraction

Victor R. Vasquez and O. Hanbury, Chemical Engineering, University of Nevada, Reno, Reno, NV

As petroleum resources dwindle, many alternative transportation fuels have been proposed as replacements. One of the most versatile of these is electricity, which is source agnostic from the perspective of the vehicle, allowing any power generating process, either traditional or renewable to be used. It is then important to assess the externalities of these sources to meet modern emissions standards. In this work, we use a cradle-to-grave LCA approach to evaluate the life cycle issues of geothermal energy extraction (hydrothermal reservoirs) and compare it to the environmental impact of competing energy and fuel sources.
For comparisons purposes, we use the results of modeling frameworks such as GREET to estimate the baseline environmental impact of more traditional fuel pathways (i.e., petroleum or biomass-based). These are used to evaluate other fuel candidates or validate the potential renewable energy sources from an environmental impact perspective.
This work follows ISO 14040 standards for life cycle analysis applied to geothermal energy. This breaks down energy production into the discrete stages of drilling, production, and site reclamation and allows for further extensive modularization. The environmental impacts from geothermal energy depend heavily on the chemistry of the reservoir, the power facility equipment, and the manner in which the spent fluids are handled.
These factors are summed over the life of a facility using Monte Carlo simulation, to give an estimation of the total impact that a geothermal facility will have in its lifetime. The results show that while the impacts of even renewable energy sources are non-negligible, their use can reduce the influence transportation has on the environment.

See more of this Session: Sustainable Fuel From Renewable Resources II
See more of this Group/Topical: Environmental Division

(433a) Developing a Novel Preparation Method for Residual Biomass Polymer Blends

Jinwen Zhang, Composite Materials and Engineering Center, college of Engineering and Architecture, Washington State University, Pullman, WA

There is a growing interest in polymeric materials from renewable resources. Nevertheless, in facing a society where petroleum-based plastic products are almost ubiquitous, the challenge for the emerging biobased polymer materials is twofold: being cost effective and performance competitive while lowering the environmental footprint.

Soy protein meal, a residue after oil crush, has being explored as a new resource for plastics. However, owing to some obstacles such as water/moisture sensitivity, high melt viscosity, narrow process window and brittleness, neat soy protein (SP) plastics have little practical values. To overcome these disadvantages, blending SP with other biodegradable polymers has received increasing attention. Sugar beet pulp (SBP) is another agricultural residue which has received similar attention for plastic applications as SP. In the current study of SP or SBP polymer blends, SP and SBP are merely used as organic fillers in most cases. Although certain processing and end use properties can be improved with respect to that of neat SP and SBP plastics, the mechanical properties of the blends are often not satisfactory and even lower than that of both the neat polymer and the neat SP (or SBP) plastics.

In this work, we introduced a novel method to prepare polymer blends of residual biomass and biodegradable thermoplastic polymer. SP and SBP were used as plastic components during the compounding of the blends. The polymers used were poly(lactic acid) (PLA) and poly(butylene adipate-*co*-terephthalate) (PBAT), respectively. The mixing of the blends was carried out in a twin screw extruder and the test specimens were prepared by injection molding. Plasticization of SP and SBP by water during compounding played a critical role in determining SP (or SBP) to be a plastic component or filler in the mixing and hence the morphological structure of the resulting blends. When extra water was added, SP (or SBP) behaved like a melt during the compounding, consequently the resulting blends demonstrated significant different properties and morphological structures from those made of dry SP (or SBP). Because water evaporated during mixing and after drying, the resulting blends actually became in situ formed thermoplastic composites. Percolation network structure of SP (or SBP) was noted in some of these blends. Different effects of SP (or SBP) as a plastic component and filler on phase structure and mechanical properties of the resulting blends were investigated. Molecular weight changes of the thermoplastic PLA and PBAT after processing were examined. The change of SP molecular structure caused by melt processing was also evaluated by protein solubility test. Effects of various compatibilizers on the properties of the blends were also studied. The results suggest that when SP and SBP were used as plastic components in mixing, the resulting blends demonstrated an overall superior performance than when they are used as fillers.

(433b) Air-Templated Nanocellulose Reinforced Macroporous Foams

Alexander Bismarck, ICL, London, United Kingdom

In progres

See more of this Session: Biobased Materials I: General
See more of this Group/Topical: Forest and Plant Bioproducts Division

(433c) Polymerization of Lactide to Polylactic Acid and Co-Polymers of Polylactic Acid Using High Viscosity Kneader Reactors

Boyd T. Safrit and George E. Schlager, LIST USA, Inc., Charlotte, NC

Polymerization of lactide to polylactic acid (PLA) can be performed using conventional reactor technology such as stirred tank reactors, but the conversion and/or final molecular weight must be controlled to a lower level. At higher conversion and/or molecular weight, the reaction mass will become very viscous , which limits the ability of conventional reactor technology to provide adequate mixing, minimize mass transfer effects on reaction kinetics, remove exothermic heat of reaction and ensure proper heat transfer in order to eliminate hotspots/thermal degradation.

Kneader reactor technology has been used over 60 years in many high viscosity applications such as reactions and polymerization, devolatilization, and drying. This technology can handle the higher conversion and molecular weight polymerizations of lactide and other copolymers of lactide, while also providing the heat transfer required for proper temperature control. Using model kinetics and rheology data, a study was performed that shows the capability of kneader reactor technology for lactide polymerizations as well as other copolymers. Kneader reactor technology can also be used to remove the uncovered monomers from the polymer and expected results from the continuous operation of a polymerizer and finisher will be shown.

See more of this Session: Biobased Materials I: General
See more of this Group/Topical: Forest and Plant Bioproducts Division

(433d) Biodegradable Poly(hydroxyalkanoate) Foams

Amy Tsui, Qi Liao and Curtis W. Frank, Chemical Engineering, Stanford University, Stanford, CA

Poly(3-hydroxybutyrate-*co*-3-hydroxyvalerate) (PHBV) is a member of the bacterial polyester family of poly(3-hydroxyalkanoates) (PHAs) known for its ability for enzymatic degradation and good mechanical properties. The use of biodegradable polymers for producing sustainable materials has received increasing attention, and PHBV is an excellent candidate, shown to have similar properties to polypropylene (PP), known to be a tough and robust material. Specifically, PHBV could be a useful material for construction materials such as insulation foam. Current insulation foams are made primarily of polyurethane, polystyrene and polyethylene which are derived from non-renewable (petroleum) feedstock. Conventionally, foam structures in nonaqueous systems, such as polymers are stabilized simply by solidification due to vitrification, cross-linking, or crystallization (1). This requires that the polymer matrix have a high enough melt strength to retain cell structures until solidification occurs. However, processing of the PHBV into foam is limited by its low melt elongational viscosity and narrow thermal processing window. Low melt elongational viscosity contributes to cell coalescence and collapse, resulting in undesirable high density foam. The poor foamability of PHBV can be addressed via the addition of a nucleating agent, orotic acid.

Orotic acid (OA), a biodegradable and nontoxic chemical, was shown to be a very effective nucleating agent (NA) for crystallization in poly(3-hydroxybutyrate) (PHB) (2,3). Nucleating agents can also be used to promote bubble formation (4). With NAs, more nucleation sites are created for gas to diffuse into so bubbles are limited to a smaller, more uniform cell size (5), while also increasing cell density (6). Additional benefit to using nucleating agents is that they often act as plasticizers which reduce the processing temperature needed for melt extrusion (4). As NAs also promote crystallization, the addition of NAs may also reduce the time required for stabilization of polymeric foam structure.

To first evaluate the nucleating efficiency of OA in PHBV, films of PHBV and OA were made using the solvent casting method. The compositions ranged from 0 to 10 wt% OA. Polarized light optical microscopy (PLOM) showed that the OA particles were well-dispersed throughout the films, which is optimal for homogeneity of nucleation sites. Differential scanning calorimetry (DSC) indicated that 1 or 2 wt% OA resulted in the highest percent crystallinity. It was also seen that presence of OA resulted in more uniform crystal size, as indicated by the narrower full-width-half-maximum (FWHM) values of the melting endotherm. Further studies are being performed to evaluate the effect of OA content on crystallization kinetics.

Foaming of poly(3-hydroxybutyrate-co-3-hydroxyvalerate) (PHBV) with 5% hydroxyvalerate content (Tianan Biologic Materials Company, China) was performed using an activated azodicarbonamide (AZ) (Actafoam 780, Chemtura) chemical blowing agent that decomposed at 140-150°C, generating approximately 150 cc of nitrogen per gram of solid. After blending and drying the PHBV and blowing agent, materials were extruded through a ¾-inch single-screw extruder (L/D ratio 25:1, compression ratio 3:1, C.W. Brabender) equipped with a 2-inch horizontal flex-lip ribbon die. The extruder temperature profile was chosen based on the temperature requirements for melting the PHBV and decomposing the blowing agent, while preventing PHBV decomposition (7-9). Foaming occurred upon leaving the die. SEM images obtained on cryo-fractured surfaces of extruded foam samples showed varying cellular morphologies with changes in AZ content. Cell coalescence and collapse occurred when the amount of blowing agent AZ exceeded 1.5 parts per hundred (phr), suggesting that the PHBV melt strength does not withstand more increase in pressure posed by gas generation and expansion (10). Using the same procedure as for the PHBV foams, results from the addition of 1 or 2 wt% OA will be shown.

The biodegradable polyester PHBV is a promising material for replacing conventional petroleum-based polymers. However, the melt strength and thermal processing window of PHBV must be improved in order to develop an alternative insulation foam. Adding orotic acid as a nucleating agent for both crystallization and bubble formation could be a promising method for making such enhancements to the foamability of PHBV.

References

1. Thareja P; Ising BP; Kingston SJ; Velankar SS. Macromol. Rap. Comm. 2008, 29, 1329

2. Jacquel N; Tajima K; Nakamura N; Miyagawa T; Pan P; Inoue Y J. App. Polym. Sci. 2009, 114, 1287

3. Jacquel N; Tajima K; Nakamura N; Kawachi H; Pan P; Inoue Y J. App. Polym. Sci. 2010, 115, 709

4. Kim SG; Lee JWS; Park CB; Sain M J. App Polym. Sci. 118, 1691

5. Mills N Polymer Foams Handbook 2007, Chapter 3, 40

6. Rodrigue D; Gosselin R. The Effect of Nucleating Agents on PP Foam Morphology 2003, 169

7. G. Schmack; D. Jehnichen; R. Vogel; B. Tandler J. Polym. Sci. 2000, 38, 2841

8. M. Yamaguchi; K. Arakawa Euro. Polym. J. 2006, 42, 1479

9. R. Vogel; B. Tandler; D. Voigt; D. Jehnichen; L. Hausler; L. Peizsch; H. Brunig Macrom. Biosci. 2007, 7, 820

10. Q. Liao 2010 Ph.D. Thesis, Stanford University, Stanford, CA.

(433e) Carbon Fiber Reinforced BioPlastics: Present Status and Future Directions

Manju Misra, School of Engineering and the Department of Plant Agriculture, University of Guelph, Guelph, ON, Canada and **Amar K. Mohanty**, Department of Plant Agriculture & School of Engineering, Guelph, ON, Canada

In Progress

See more of this Session: Biobased Materials I: General
See more of this Group/Topical: Forest and Plant Bioproducts Division

(433f) Modification of Biomass Rheology: The Influence of Physical Chemistry

Joseph R. Samaniuk[1], C. Tim Scott[2], Thatcher W. Root[1] and Daniel J. Klingenberg[1], (1)Chemical and Biological Engineering, University of Wisconsin-Madison, Madison, WI, (2)USDA Forest Products Laboratory, Madison, WI

Chemicals derived from lignocellulosic biomass, including liquid fuels, are a sustainable alternative to petroleum derivatives. Many different processes have been proposed to convert lignocellulosic materials to liquid fuels but a common operation required in all of them is heating to high temperatures (110-230 ^{o}C). Reducing the concentration of water during this step decreases utility costs and decreases overall capital costs by reducing equipment size. Increasing the solids concentration however, results in rheological changes like a rapidly increasing yield stress. Mixing, pumping and pouring biomass slurries becomes difficult, if not impossible, above 20-30 wt% solids.

We have shown previously that the rheology of biomass can be controlled with the use of certain water soluble polymers (WSPs). AFM experiments have revealed that WSPs alter stresses in cellulose fiber suspensions by lubricating the surface of individual fibers. Using these additives, we have been able to mix and convey untreated biomass at solids concentrations up to 50 wt%. However, the efficacy of WSPs is reduced after certain treatments such as acid hydrolysis. How changes in the physical chemistry of biomass during treatment affect rheology modification with WSPs is unknown.

In this presentation our focus will be on the influence of the physical chemistry of both biomass and WSPs on rheological modifications. The physical chemistry of the biomass is altered by the treatment type, e.g. acid hydrolysis, and the treatment severity. Specific changes to the physical chemistry of the biomass are probed with fiber suspensions containing known chemistries, e.g. cellulose fibers vs. hemicellulose-rich fibers. The physical chemistry of the WSPs are altered by varying the type of polymer, the molecular weight, the degree of substitution, and by the addition of chemical additives. We will show how these variables influence not only the instantaneous rheological modification but also its behavior over time.

See more of this Session: Biobased Materials I: General
See more of this Group/Topical: Forest and Plant Bioproducts Division

Torrefied wood pellets production using natural biomass compounds

Venkata K Penmetsa and Philip H Steele*
Forest Products Department, Mississippi State University,
100 Blackjack Rd. Starkville, MS 39759. Tel. no: 662-325-8083
kpenmetsa@cfr.msstate.edu

The U.S. coal industry is highly interested in the potential for combusting torrefied wood as a co-feed with coal. Unlike biomass torrefied wood can be cofed with coal at a high percentage. Biomass combustion facilities cannot use the existing infrastructure and new biomass fed plants cost from $140 to $200 million to construct.

Torrefaction of biomass has attracted considerable attention as a means to reduce biomass weight by 70% through removal of water and the hemicellulosic fractions while retaining most of the original high-energy cellulose and lignin content. A further advantage of torrefied wood is that it has anti-microbial properties as well as limited water resistance such that microbial degradation is eliminated.

Unlike biomass torrefied wood can be co-fed with coal at a high percentage and requires no major infrastructure investment. One issue is the current inability to ship torrefied wood as coal is currently shipped. A waterproof pelletized product is required to allow bulk shipping via open railcar or by barge. Production of a waterproof torrefied wood pellets or briquettes has not been achieved to date.

Mississippi State University has developed a low-cost binder to produce a high-density pellet with high water resistance. The binder is produced from biomass and has a neutral pH of 6.2 such that the pellets do not produce corrosive gases during their combustion. No chemical cross linking agents, such as formaldehyde, are required making this binder very environmentally friendly. The binder is suitable to produce densified water repellent pellets with many different biomass feed stocks. The pellets have been subjected to crushing strength and water repellant test.

Pellet hardness was determined for each test specimen as described in ASTM standard D143 for small clear specimens. Each specimen was weighed immediately before and after the test and the density values were computed. All test results are given in Table 1.

The pellets were also tested in a water absorption test. The pellets were weighed before immersing in water and were fully immersed in water for 14 days. The pellets were weighed every 24 hours to monitor any increase in the weight of each pellet due to water absorption. The weight difference quantifies the amount of water absorbed inside the pellet pores. Figure 1 demonstrates the behavior of the pellets immersed in water over 14 days at various binder loadings as the weight percent of the total pellet weight. The 30%, 20% and 15% binder proportions showed very minor weight gains of water ranging from 0.05% to 0.6%.

These results indicate that at these high binder ratios the pellets were so densified that limited pore spaces are available for water to penetrate over time. These results on water weight gain over time indicate that binder loading magnitude was directly linked to water absorbance. The pellets with the lowest binder proportion had the highest water absorption values. Over 14 days the pellet produced with 2% binder had maximum weight gain of about 11%; the 5% binder loaded pellet had a water weight gain of 9.6%; that with 10% binder addition had a water weight gain of only 1.2%.

The study results showed that the tested binder produces a densified and water repellent bond for torrefied wood that is helpful in producing high initial strength with this strength substantially maintained following severe water exposure is more helpful in pelletizing torrefied wood. Both initial pellet compressive strength and strength following long-term water soaking the pellet strength is a relatively linear function of proportion of binder applied. However, even pellets with only 2% binder loading produced an initial strength of 764 kg/m^2 and 68% of this initial strength was maintained following the 14 days of water soaking followed by high-temperature drying to zero moisture content. While binder loadings of 30, 20 and 15% were tested and showed superior performance it is unlikely that these loadings will be economical due to the high amount of binder required. In addition, for production of a waterproof pellet at 10% loading appears to be more than adequate. Over the 14 days of soaking only 1.2% of water absorption was observed for this loading. In addition, nearly 80% of the initial compressive strength was maintained after water exposure and oven drying.

Table 1. Pellet weight (g) and compression strength of test pellets immediately after production compared to pellets water soaked for 14 days and oven dried to oven dry water content.

Spec. no.	Weight (g)	Binder composition (%)	Compression strength (kg/m^2)	Compression strength (kg/m^2) after 14 days in water
1	16.80	30	1095	960
2	16.92	20	987	846
3	16.76	15	934	800
4	16.90	10	905	718
5	16.39	5	841	630
6	16.42	2	764	519

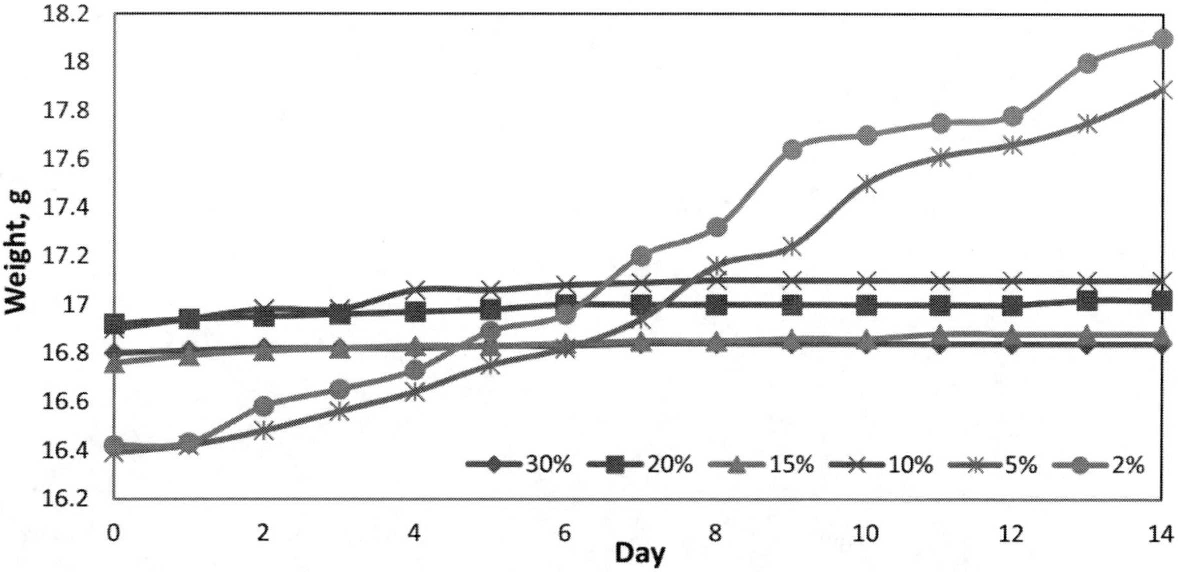

Figure 1: Daily water weight absorption values by pellet binder proportion for 14 days of soaking

For most applications it appears that a 2% loading will produce adequate initial compression strength at 764 kg/m^2. In addition, following water soaking 68% of the initial strength (519 kg/m^2) were maintained. It is not clear that the extra expense of producing a 10%, or higher, binder loading pellet would be cost effective versus the added cost of the additional binder.

References:

1. Fonseca Felfli, F., Luengo, C.A., Bezzon G. and Beaton Soler, P., 1998. *Bench unit for biomass residues torrefaction*, Biomass for Energy and Industry, Proceeding of Int. Conf., Würzburg, Germany, 8-11June, Ed. by Kopetz, Weber, Palz, Chartier and Ferrero, C.A.R.M.E.N., Rimpar Germany, 1593-1595.

2. Arcate, J.R., 2000. *New process for torrefied wood manufacturing,* Bioenergy update, vol 2, No. 4. (www.techtp.com/Bioenergy%20Update%20.pdf).

3. Wescott, J.M. and C.R. Frihart. 2004. *In* Proceedings of the 38th International Wood Composite Materials Symposium, Washington State University, Pullman, Washington. pp 199-206.

(435a) Hot Water Extraction with pH Control and Subsequent Kraft Pulping of Pine Wood Chips

Allen J. Smith, Gopal Krishnagopalan and Harry T. Cullinan, Chemical Engineering, Auburn University, Auburn, AL

This research investigated the effects of pre-extraction time, temperature, and pH on the rate and quantity of sugars recovered from Loblolly pine wood chips and the impact this treatment has on pulp kappa number and yield. Composition analysis of raw chips, extracted chips, pulp, and hydrolyzate were performed to assess the hemicellulose extraction efficiency and subsequent yield loss during kraft pulping of Loblolly pine.

A pseudo kinetic model of prehydrolysis extraction was developed from the hydrolyzate sugar concentration data. This modified H-factor model described the sugar extraction data except when significant degradation of sugars was observed. The data for chip weight loss and pulp yield also fit a smooth curve when plotted against the modified H-factor.

See more of this Session: Biomass Fractionation
See more of this Group/Topical: International Congress on Energy 2011

Which Fractionation Process can Overcome the Techno-Economic Hurdles of a Lignocellulosic Biorefinery?

Adriaan van Heiningen[1,2], Mikhail Iakovlev[2], Minna Yamamoto[2], Evangelos Sklavounos[2], Kristian Melin[2] and Tom Granstrom[2]

[1]University of Maine, Dept Chem. and Biol. Eng., Orono, ME, USA (avanheiningen@umche.maine.edu)
[2]Aalto University School of Chemical Technology, Espoo, Finland.

Introduction

Two pressing global issues, climate change and limited accessible petroleum reserves, are driving the interest to use lignocellulosic biomass for production of transportation fuels. Unfortunately, two techno-economic barriers hinder commercial implementation of renewable transportation fuel production from wood and other lignocellulosic biomass. First, capital cost is high because the present largest lignocellulosic processing facilities, as exemplified by pulp mills, are more than one order of magnitude smaller in feed stock energy feed rate than a modern oil refinery. The second barrier is high operating cost to convert lignocellulosics into biofuels. This is caused by the natural resistance of woody biomass to decomposition, i.e. the "recalcitrance" to convert lignocellulosics into monosugars and lignin at high yield (Kumar et al., 2009).

Integrated Forest Biorefinery

The capital cost disadvantage may be reduced by integrating biofuels production into existing wood products/pulp/paper facilities. In such an Integrated Forest Biorefinery (IFBR) the collection of lignocellulosic biomass is combined with existing wood procurement and delivery infrastructure, and the biomass conversion processes use available steam/power and water effluent treatment facilities (van Heiningen, 2006). Shown in Figure 1 are several different forest biorefinery pathways which are integrated with an existing forest products mill. Different feedstocks are listed on the left of the diagram, including (pulp/lumber) extracts and pulping liquors created as intermediary streams in a forest products mill. The other four (4) feedstocks are forest biomass, MSW, agro-lignocellulosics (straw/corn stover, etc.) and sludges. The diagram shows five pathways to fuels and chemicals listed on the right. There are two pathways which volatilize the entire feedstock (thermochemical conversion) to syngas ($CO+H_2$) or pyrolysis-oil. The three other pathways generate intermediates by liquid phase processing. The highest severity treatment (in terms of temperature and acidity), leads to dehydration of C6 and C5 sugars to levulinic acid and furfural respectively and transforms lignin in a char. The lowest severity pathway called consolidated fermentation is slow anaerobic fermentation by a natural mixture of microbes which produce organic acids, methane and a sludge at a near neutral pH maintained by adding a base such as $Ca(OH)_2$. The intermediate pathway called pretreatment/fractionation + hydrolysis generates C6 and C5 mono sugars, lignin and acetic acid. An important difference between the pathways is that the bottom three pathways do not utilize lignin as a product contrary to the top two pathways. On the other hand, the thermochemical pathways require energy-consuming size reduction and drying of the feedstocks.

The goal of the fractionation pathway is to separate the lignocellulosics biomass into its principal constituents; cellulose, hemicellulose and lignin (Bozell, 2010; Reith et al., 2011). This is a difficult task because lignin and hemicelluloses are bound to each other by covalent bonds (Fengel and Wegener, 1989), and the three components are arranged inside the fiber wall in a complex manner. Fractionation of lignocellulosics leads to release of cellulosic fibers (for pulp-based products) and opens the cell wall structure by dissolution of lignin and hemicellulose between the cellulose microfibrils. The fibers are now better accessible for hydrolysis by enzymes. When the sugars in lignocellulosics are used as feed stock for fermentation, the process to open-up the cell wall structure is called pretreatment. Pretreatment has been intensively studied because it is the most important step affecting the production cost of lignocellulosic ethanol (Elander et al., 2009). While access to cellulose by enzymes is the primary goal of pretreatment, clean separation of the major lignocellulosic polymers is the key objective of fractionation (Bozell, 2010). There is however a continuum between the two processes; from steam explosion – dilute acid hydrolysis – SPORL ($NaHSO_3$ and H_2SO_4 treatment plus mechanical refining, Zhu et al., 2010b) – Lignol (Ethanol-water with H_2SO_4, Mabee et al., 2006) to AVAP (Ethanol-water with SO_2, Retsina and Pylkkanen, 2007). Of all lignocellulosics, enzymatic hydrolysis of pretreated softwood biomass is most difficult (Galbe and Zacchi, 2007). Despite many pretreatment studies, it is still not possible to predict the enzymatic hydrolysis behavior of pretreated biomass based on their chemical and physical composition (Kumar et al., 2009).

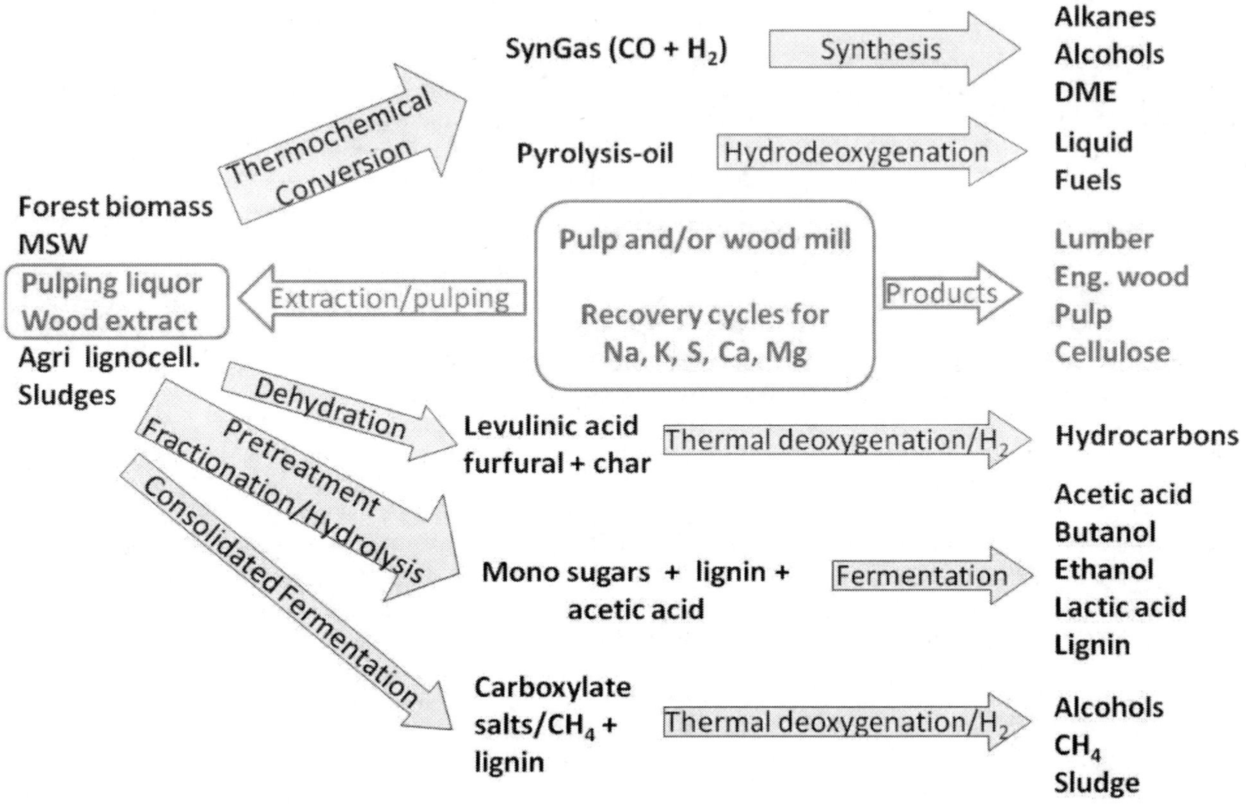

Figure 1. Forest Biorefinery Pathways.

SO_2-ethanol-water fractionation

A highly efficient fractionation method for lignocellulosic biomass is treatment by an ethanol-water mixture with dissolved SO_2 at modest temperatures (130 – 160°C) (Iakovlev et al., 2009 and 2011). The absence of a base (Mg, Na, etc) simplifies the recovery of SO_2 and ethanol to simple absorption, distillation and washing unit operations. The following characteristics make the SO_2-ethanol-water fractionation process uniquely suitable for lignocellulosics. It is the only process which is able to digest different biomass species – softwoods, hardwoods and annual plants – at close delignification rates (Primakov, 1961, Iakovlev et al., 2011, Yamamoto et al., 2011). This increases the potential feedstock supply and size of operation, and thus lowers capital cost. The presence of SO_2 eliminates troublesome lignin-based precipitates and hydrolyzes the hemicelluloses. The presence of ethanol leads to rapid impregnation of the biomass, so that neither a separate impregnation stage nor size reduction smaller than wood chips are needed, thereby avoiding electricity consuming size reduction unit operations and allowing high space velocity in a fractionation reactor. Most importantly the dissolved hemicelluloses are neither dehydrated nor oxidized (Iakovlev, 2011). During recovery of ethanol and SO_2 from the spent fractionation liquid the hemicelluloses are almost fully hydrolyzed to monomers, thereby eliminating costly acid or enzymatic hydrolysis for this wood component (Sklavounos et al., 2011). Evaporation of ethanol from the spent fractionation liquid produced at a practical L/W ratio of about 2-3 L/kg would increase the hemicellulose concentration to about 100 g/L. The fractionation spent liquor recovery to produce monomer sugars is part of the patent pending biorefinery process termed AVAP™ pioneered by American Process Inc. Cellulose is fully retained in the solid phase and can subsequently be hydrolyzed to glucose. The mixture of hemicellulose monosugars and cellulose derived glucose may be used for production of biofuels and chemicals (Rakkolainen et al., 2010).

Qualitative Comparison of Different Pretreatment/Fractionation Processes

Considerable research has been devoted to studying the effect of various pretreatment methods on biomass enzymatic digestibility. The enzymatic digestibility of untreated biomass is less than 20% mostly due to inaccessibility of cellulose being physically protected by lignin and hemicelluloses (Mosier et al., 2005; Zhang and

Lynd, 2004). Furthermore, lignin causes non-productive enzyme adsorption which also contributes significantly to the low efficiency of cellulose hydrolysis (Wyman et al., 2005). More recently, Varnai et al. (2010) found that lignin and/or lignin-carbohydrate complexes (LCCs) were the major component restricting the enzymatic hydrolysis of pretreated (steam plus SO_2) spruce. After chlorite delignification the hydrolysis of cellulose itself was the major limiting factor. Xylan also restricted the hydrolysis of cellulose despite their low content in the pretreated and delignified spruce, suggesting that the location of xylan also plays an important role. Examples of pretreatment/fractionation methods include steam explosion, dilute acid (0.1-3.0% sulfuric acid), SPORL (1-8% sulfite and 0.5-2% sulfuric acid), AFEX (ammonia fibre explosion), ARP (ammonia recycled percolation, 10-15% ammonia), and lime treatments. These methods increase the accessibility of cellulose by removing the protective layers of either of hemicelluloses (acidic processes) or lignin (alkaline processes) (Mosier et al., 2005). Alkaline processes are considered less attractive due to the difficulty to recover alkali/ammonia (Mosier et al., 2005; Wyman et al., 2005). Lime as an alkali source is cheap when used in moderate amounts, and does not require recovery. However, lime pretreatment suffers from extensive scaling (Zhu and Pan, 2010). Although in acidic pretreatments lignin is mostly preserved in the solid residue, it is said that it is considerably altered (melted, condensed, agglomerated) and this is seen as a reason for the enhanced digestibility. Many of the acidic pretreatments are energy intensive as they operate at temperatures higher than 150°C and sometimes even over 200°C with associated high pressures (Zhu et al., 2010a). A significant water requirement related to the high required liquid-to-solids ratios, the necessity to neutralise the acids prior to fermentation and associated salt (generally gypsum) disposal problem, difficulty to treat softwoods, and corrosion problems are other important drawbacks of acidic pretreatment processes. Finally, an operating problem which has mostly been overlooked for acidic pretreatment is formation and precipitation of sticky lignin on reactor walls and piping (Leschinsky, 2009). The explanation of the lack of attention to this problem is that it only becomes apparent in continuous larger scale operation after one to two week operation (Gullichsen, 2007), i.e. it is a real obstacle at laboratory and/or short pilot scale operations. No economical solution to this problem is presently available (Gutsch and Sixta, 2011).

A qualitative comparison of the different pre-treatment/fractionation processes is shown in Table 1.

Table 1. Qualitative comparison of pre-treatment/fractionation processes

Pretreatment or Fractionation	Full utilization of hemicelluloses	Low energy need	No sticky lignin issue	Omnivorous	Simple recovery
Alkaline treatment	-	+	+	-	-
Steam Explosion	-	-	+-	-	+
Autohydrolysis	+-	+	-	-	+
Acid Hydrolysis	+-	+-	-	-	-
Lignol (EtOH-H_2O)	+-	-	+	-	+
SPORL	+-	+-	+	+-	-
SEW (SO_2-EtOH-H_2O)	+	+	+	+	+-

The table shows that the SO2-EtOH-H2O fractionation process used in the AVAP biorefinery process is the only process which does not suffer from significant hemicellulose degradation, has a relatively low energy requirement because of the low temperature (130-150°C) and liquid-to-wood ratio of 2-3 L/kg, does not create sticky lignin precipitates, can simultaneously process softwoods and hardwoods biomass, and only needs simple evaporation of ethanol and SO_2 for recovery of the chemicals. Details will be provided to quantify this comparison.

Conclusions

The SO_2-Ethanol-Water process is highly attractive to serve as the central fractionation method in a future lignocellulosic biorefinery to produce hemicellulose sugars, cellulose fibers and weakly sulfonated lignin. The absence of a base (Mg, Na, etc) reduces the recovery of SO_2 and ethanol to simple absorption, distillation and washing unit operations. Full recovery of unreacted SO_2 is thus achieved. Also, the absence of bisulfite in the fractionation solution avoids degradation of the dissolved sugars, and the total sugar recovery of 91% for spruce compares well with the highest values obtained by SO_2-catalyzed steam explosion, Lignol and SPORL treatments. The presence of SO_2 eliminates troublesome lignin-based precipitates, hydrolyzes the hemicelluloses, and allows fractionation of a mixture of lignocellulosic biomass, including softwoods. At the same time the presence of ethanol leads to rapid impregnation of the biomass, thereby avoiding energy consuming size reduction unit operations and allowing high space velocity in the fractionation reactor. The low L/W ratio of 2-3 L/kg, relatively low fractionation temperature, and lower specific heat of ethanol compared to water lead to a modest energy demand, and allows the use of medium pressure steam. The produced cellulosic fibers may be used as feed stock for glucose production by enzymatic hydrolysis and then fermented to biofuels. Alternatively the fibers may be used for conventional pulp products, either paper, tissue or dissolving pulp. The omnivorous nature of this process increases the feedstock supply, and thus reduces the capital cost by increasing the size of production.

References

Bozell, J.J. (2010) An evolution from pretreatment to fractionation will enable successful development of the integrated biorefinery. BioResources 5(3):1326-1327.

Elander, R.T., Dale, B.E., Holtzapple, M., Ladisch, M.R., Lee, Y.Y., Mitchinson, C., Saddler, J.N., Wyman, C.E. (2009) Summary of findings from the Biomass Refining Consortium for Applied Fundamentals and Innovation (CAFI): corn stover pretreatment. Cellulose (Dordrecht, Netherlands) 16(4):649-659.

Fengel, D., Wegener, G. (1989) Wood – Chemistry, Ultrastructure, Reactions. Walter de Gruyter, Berlin.

Galbe, M., Zacchi, G. (2002) A review of the production of ethanol from softwood. Applied Microbiology and Biotechnology 59(6):618-628.

Gütsch, J.S., Sixta, H. (2011) Purification of Eucalyptus globulus water prehydrolyzates using the HiTAC process (high-temperature adsorption on activated charcoal). Holzforschung 65(4):511-518.

Gullichsen, J. (2007) Personal communication, Espoo, Finland.

van Heiningen, A. (2006) Converting a Kraft pulp mill into an integrated forest biorefinery. Pulp Paper Canada 107:38-43.

Iakovlev, M. (2011) SO_2-ethanol-water fractionation of lignocellulosics. PhD Thesis, Aalto Univ., Espoo, Finland.

Iakovlev, M., Pääkkönen, T., van Heiningen, A. (2009) Kinetics of SO_2-ethanol-water pulping of spruce. Holzforschung 63(6):779-784.

Iakovlev, M., Sixta, H., van Heiningen, A. (2011) SO_2-ethanol-water (SEW) pulping: II. Kinetics for spruce, beech and wheat straw. Journal of Wood Chemistry and Technology 31(3):250-266.

Kumar, P., Barrett, D.M., Delwiche, M.J., Stroeve. P. (2009) Methods for Pretreatment of Lignocellulosic Biomass for Efficient Hydrolysis and Biofuel Production. Industrial and Engineering Chemistry Research 48:3713-3729.

Leschinsky, M. (2009) Water prehydrolysis of Eucalyptus globulus: Formation of lignin-derived precipitates that impair the extraction of hemicelluloses, PhD thesis, University of Hamburg, Hamburg, Germany.

Mabee, W.E., Gregg, D.J., Arato, C., Berlin, A., Bura, R., Gilkes, N., Mirochnik, O., Pan, X., Pye, E.K., Saddler, J.N. (2006) Updates on softwood-to-ethanol process development. Applied Biochem. and Biotechn., 129-132:55-70.

Mosier, N., Wyman, C., Dale B., Elander, R., Lee, Y.Y., Holtzapple, M., Ladisch, M. (2005) Features of promising technologies for pretreatment of lignocellulosic biomass. Bioresource technology 96(6):673-686.

Primakov, S.F. (1961) Delignification of various wood species with aqueous-alcoholic solutions. Nauchn. Tr. Vses. Nauchn.-Issled. Inst. Tsellyulozn.-Bumazhn. Prom. 47:69-75.

Rakkolainen, M., Iakovlev, M., Teräsvuori, A.-L., Sklavounos, E., Jurgens, G., Granström, T.B., van Heiningen, A. (2010) SO_2-Ethanol-Water fractionation of forest biomass and implications for biofuel production by ABE fermentation. Cellulose Chemistry and Technology 44(4-6):139-145.

Reith, H., de Uil, H., van der Linden, R., Huijgen, W., Wildschut, J. (2011) A step towards the development of a biorefinery. NPT Procestechnologie 18(1):26-28.

Retsina, T., Pylkkänen, V. (2007) Back to the biorefinery: a novel approach to boost pulp mill profits. Paper 360°, February issue, 18-19.

Sklavounos, E., Iakovlev, M., Yamamoto, M., Teräsvuori, L., Jurgens, G., Granström, T., van Heiningen, A. (2011) Conditioning of SO_2-Ethanol-Water spent liquor from spruce for the production of chemicals by ABE fermentation. Holzforschung 65(4):551-558.

Varnai, A., Siika-aho, M., Viikari, L. (2010) Restriction of the enzymatic hydrolysis of steam-pretreated spruce by lignin and hemicellulose. Enzyme and Microbial Technology 46(3-4):185-193.

Wyman, C.E., Dale, B.E., Elander, R.T., Holtzapple, M., Ladisch, M.R., Lee, Y.Y. (2005) Coordinated development of leading biomass pretreatment technologies. Bioresource Technology 96(18):1959-1966.

Yamamoto, M., Iakovlev, M., van Heiningen, A. (2011) Total mass balances of SO_2-Ethanol-Water (SEW) fractionation of forest biomass. Holzforschung 65(4):559-565.

Zhang, Y.-H.P., Lynd, L.R. (2004) Toward an aggregated understanding of enzymatic hydrolysis of cellulose: Noncomplexed cellulase systems. Biotechnology and Bioengineering 88(7):797-824.

Zhu, J.Y., Pan, X.J. (2010) Woody biomass pretreatment for cellulosic ethanol production: Technology and energy consumption evaluation. Bioresource Technology 101(13):4992-5002.

Zhu, J.Y., Pan, X., Zalesny, R.S.Jr. (2010a) Pretreatment of woody biomass for biofuel production: energy efficiency, technologies, and recalcitrance. Applied Microbiology and Biotechnology 87(3):847-857.

Zhu, J.Y., Zhu, W., O'Bryan, P., Dien, B.S., Tian, S., Gleisner, R., Pan, X. J. (2010b) Ethanol production from SPORL-pretreated lodgepole pine: preliminary evaluation of mass balance and process energy efficiency. Applied Microbiology and Biotechnology 86(5):1355-1365.

CELLOBIOSE HYDROLYSIS USING ACID FUNCTIONALIZED NANOPARTICLES

*Leidy Peña[a], Brenton R. Ware[a], Myles Ikenberry[b], Keith Hohn[b], Daniel L. Boyle[c], Xiuzhi Susan Sun[d], and Donghai Wang[a],**

[a] *Department of Biological and Agricultural Engineering,* [b] *Department of Chemical Engineering,* [c] *Division of Biology,* [d] *Department of Grain Science and Industry,*

Kansas State University, Manhattan, KS 66506, USA

Introduction

Uncertainties in the oil supply and environmental problems associated with fossil fuels have motivated the search for another energy sources, such ethanol and other renewable biofuels. The global ethanol production increased 400% in the past decade. In 2009, 19 billion gallons of ethanol were produced worldwide [1]. The U.S. Department of Energy has predicted that the global ethanol production would increase up to 68 billion gallons by 2030 [2]. Ethanol is regarded as renewable energy source because it can be obtained from the alcoholic fermentation of monosaccharides. Sugar and starchy crops are the main source of sugars for the process. However, these materials are also staple food and their use for fuel could strain the food supply. Cellulose is the most abundant polysaccharide on the earth and can be used for ethanol production after the hydrolysis of the polymer to glucose monomers. The catalytic conversion of cellulose to monosaccharides demands very harsh conditions. The enzymatic hydrolysis of lignocellulosic biomass requires pretreatment of the biomass to increase the susceptibility of the material to the enzymatic action. The most common pretreatment methods employ mineral acids and high temperatures to remove the lignin component. Therefore, substantial capital investment is required because corrosion resistant materials need to be used; neutralization, detoxification, and waste disposal are also additional costs. Acid functionalized nanoparticles could, besides catalyze hydrolysis of β-1,4 glycosidic bonds, be easily recovered from products using a strong magnetic field.

Methods

Acid-functionalized metallic nanoparticles were synthesized for the hydrolysis of lignocellulosic biomass by previously published procedures [3,4]. Cobalt spinel ferrite magnetic nanoparticles were synthesized using a microemulsion method and then covered with a layer of silica to protect them from oxidation. The silanol groups of the silica served as the support of the sulfonic acid groups perfluoroalkyl-sulfonic (PFS), alkyl-sulfonic (AS), and butyl-carboxylic (BCOOH) that were later

*To whom correspondence should be addressed. E-mail: dwang@ksu.edu. Phone: (785) 532-2919. Fax: (785) 532-5825.

attached to the surface of the nanoparticles. FTIR analysis was done to identify the chemical groups on the surface of the nanoparticles. TEM images were taken to verify the presence of particles within the nanoscale. Dynamic particle analysis was carried out to measure the size of the aggregates in aqueous solutions. Cellobiose solutions were subjected to hydrolysis in a reactor at 175°C for 60 min in the presence of AS, BCOOH and PFS nanoparticles. Thermogravimetric analysis (TGA) was employed to account for the organic load upon the nanoparticles. TGA experiments were carried out before and after acid functionalization to account for the organic mass gained during the functionalization procedure. TGA experiments were also done after hydrolysis of cellobiose to evaluate the retention of the catalytic sites.

Results and Discussion

FTIR spectra for silica-coated cobalt iron oxide nanoparticles are shown in Figure 1. The peaks at 1620 cm^{-1} and 3390cm^{-1} observed in the spectrum have been attributed to the O-H vibration and stretching vibrations of physisorbed water and surface hydroxyls [4]. The bands at 802 cm^{-1} and 940 cm^{-1} have been attributed to the stretching vibrations of Si-O-Si and Si-O-H groups [5].

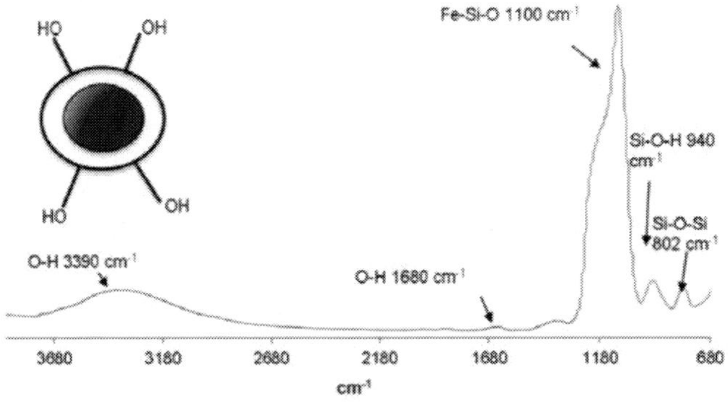

Figure 1 FTIR spectra of silica-coated cobalt iron oxide nanoparticles.

TEM images confirmed the presence of monodispersed nanoparticles (Figure 2). However, some of the nanoparticles aggregated and formed larger clumps. Dynamic particle analysis showed that silica-coated cobalt iron oxide aggregates had a narrow distribution with mean 2.16 μm (±0.27). The material functionalized with PFS acid groups showed a bimodal distribution, the smaller distribution could be attributed to silica-coated iron oxide which escaped the functionalization. The smaller material had a mean equal to 2.31 μm (±0.04 μm) which is the same as the one obtained for silica-coated iron oxide aggregates. The functionalization procedure may have promoted the formation of larger aggregates which had a mean of 74.00 μm (±4.07 μm). AS nanoparticles also showed a bimodal distribution; the larger distribution had a mean size of 13.08 μm (±2.48), and the mean size of the small distribution was 0.82 μm (±0.02 μm). BCOOH functionalized particles showed a very broad distribution with mean of 88.00 μm (±29.18 μm). In this case, growth of the nanoparticles by deposition of cyanopropyltriethoxysilane could have been promoted by the acidic conditions of the reaction media.

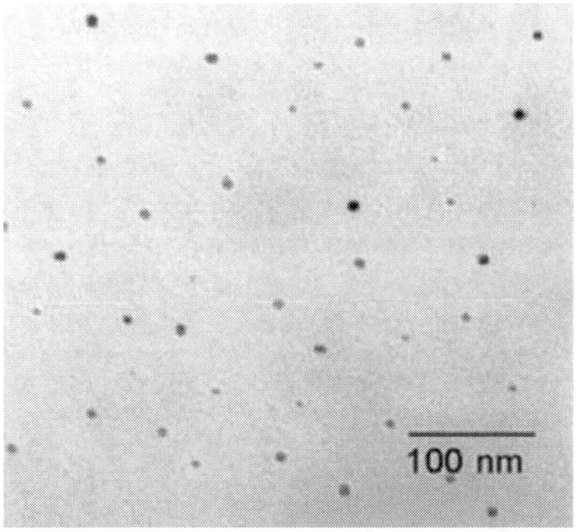

Figure 2 Cobalt iron oxide nanoparticles

Moisture content and organic load were calculated from TGA experiments. AS, BCOOH, and PFS nanoparticles have low moisture contents with values of 2%, approximately. AS nanoparticles show larger rates of dehydration than the observed rates for PFS and BCOOH nanoparticles (Figure 3). Low water contents and high dehydration rates are indicators of the poor water affinity of the nanoparticles. The hydrophobicity of the nanoparticles can cause aggregation problems for reactions carried out in aqueous media. On the other hand, PFS and AS nanoparticles are less thermally stable than BCOOH nanoparticles; when heated up to 450°C, PFS nanoparticles lost 22% (wt.) of their weight which is about 60% (wt.) of their original organic content. Although, AS only lost 10% of their weight, it is also equivalent to about 55% (wt.) of their original organic content. BCOOH nanoparticles showed higher thermal stability when heated up to 450°C because they only lost 9% (wt.); 70% (wt.) of their original organic remains on the nanoparticles.

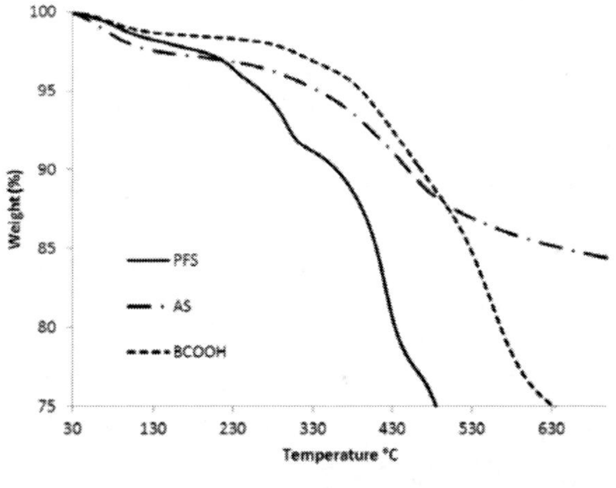

Figure 3 TGA scans for PFS, AS, and BCOOH acid functionalized nanoparticles.

Factors other than hydronium ion concentration such as hydrophobicity and water affinity may also affect the acid hydrolysis of cellulose. When sulfonic acid functionalized carbon was used to break β-(1→4) glycosidic bonds, it was found that the presence of COOH and phenolic OH as surface functional groups help the hydrolytic activity of the sulfonic acid groups [6]. According to Suganuma et al. (2008), COOH and phenolic OH groups help the absorption of β-1,4-glucan that is later hydrolyzed by SO_3H groups. Therefore, it is recommended to functionalize the nanoparticles with both COOH and acidic OH groups in order to improve the catalytic activity of sulfonic acid-functionalized nanoparticles.

Cellobiose hydrolysis was used as a model reaction to evaluate the performance of acid-functionalized magnetic nanoparticles for breaking β-(1→4) glycosidic bonds. A cellobiose conversion of 78% (wt.) was achieved when using AS nanoparticles as the catalyst at 175°C for 60 min, which was significantly higher than the cellobiose conversion for the control experiment 52% (wt.). Cellobiose conversion of 65% (wt.) and 60% (wt.) were obtained for the second and third runs, respectively. After the first run, AS nanoparticles retained more than 60% (wt.) of their original sulfonic acid load.

Conclusions

The alkyl-sulfonic acid functional groups from the magnetic nanoparticles gave a significantly greater cellobiose conversion than the perfluroalkyl-sulfonic and butylcarboxilic acid functional groups did. However, the perfluoroalkyl-sulfonic acid functionalized nanoparticles leached most of their acid groups.

Acknowledgements

This project is supported by the Center for Sustainable Energy at Kansas State University and the NSF EPSCoR Kansas Center for Solar Energy Research.

References

1. Renewable Fuels Association (2010) 2010 ethanol industry outlook.

2. Office of the Biomass Program (2010) Biomass: Multi-year Program Plan, U.S. DOE.

3. Gill, C. S., B. A. Price, and C. W. Jones (2007) Sulfonic acid-functionalized silica-coated magnetic nanoparticle catalysts. *J. Catal.* 251:145-152.

4. Phan, N. T. S. and C. W. Jones (2006) Highly accessible catalytic sites on recyclable organosilane-functionalized magnetic nanoparticles: An alternative to functionalized porous silica catalysts. *J. Mol. Catal. A-Chem.* 253:123.

5. Zhao, X. S., G. Q. Lu, and X. Hu (2000) Characterization of the structural and surface properties of chemically modified MCM-41 material. *Micropor. Mesopor. Mater.* 41:37-37-47.

6. Suganuma, S., K. Nakajima, M. Kitano, D. Yamaguchi, H. Kato, S. Hayashi, and M. Hara (2008) Hydrolysis of cellulose by amorphous carbon bearing SO3H, COOH, and OH groups. *J. Am. Chem. Soc.* 130:12787.

(435d) Dilute Acid Hydrolysis of Xylose Oligomers Fractionated by Centrifugal Partition Chromatography

Ching-Shuan Lau[1], Kris Bunnell[1], Edgar Clausen[2] and Danielle Julie Carrier[1], (1)Department of Biological and Agricultural Engineering, University of Arkansas, Fayetteville, AR, (2)Ralph E. Martin Department of Chemical Engineering, University of Arkansas, Fayetteville, AR

Xylose oligomers, which are chains of xylose molecules linked with β-1,4-bonds with a degree of polymerization greater than two, are the intermediate products of xylan depolymerization into xylose monomer. Studying the kinetics of xylose oligomers is critical to describe hemicellulose depolymerization because these results will help researchers understand how the formation of inhibitory by-products occurs. The objective of this research is to determine the rate of depolymerization of xylose oligomers under commonly used dilute acid hydrolysis conditions. The understanding of the depolymerization rate as a function of the hydrolysis conditions will help us determine the optimum range of dilute acid pretreatment conditions, which will hopefully result in the maximum conversion of hemicellulose into xylose monomer.

Xylose oligomers, fractionated by centrifugal partition chromatography (CPC) using a solvent system of dimethyl sulfoxide-tetra hydrofuran-water, were hydrolyzed using sulfuric acid concentrations ranging from 0 to 4%, at 121 to 240 °C for 0 to 120 minutes. The hydrolysates were analyzed for the xylose monomer, oligomers and by-products concentrations to determine the individual rate of degradation and the formation rate of the by-products. Mass balance closure of xylose oligomers before and after the hydrolysis was performed to account for the majority of the oligomer's mass during the hydrolysis process. The results indicated that the use of National Renewable Energy Laboratory's total sugar analysis method, which hydrolyzes the biomass sample in 4% sulfuric acid, at 121 °C for 60 minutes resulted in a 10% loss of xylose, which can be mostly accounted for by the formation of furfural. Moreover, the xylose loss rates were 26% and 5% for pure xylobiose and xylotriose, respectively, when used in the total sugar analysis.

See more of this Session: Biomass Fractionation
See more of this Group/Topical: International Congress on Energy 2011

(435e) Comparison of Ultrasound-Assisted Alkaline Pretreatments for Enzymatic Hydrolysis of *Jatropha Curcas* Fruit Hulls

Duong TT. Nguyen, Chemical and Biomolecular Engineering, National University of Singapore, Singapore, Singapore

Comparison of ultrasound-assisted alkaline pretreatments for enzymatic hydrolysis of *Jatropha curcas* fruit hulls

Jatropha curcas has gained growing interest in tropical and sub-tropical countries due to its hardness, easy propagation, drought tolerance, rapid growth, adaptation to wide agro-climatic conditions, and most importantly due to its high oil content (40% of the seed weight). *J. curcas* fruit hulls accounting for 37.5% fruit weight obtained as a low-value by-product of *J. curcas* biodiesel facilities, represents a collected source of lignocellulosic biomass that could be utilized on-site for fuel bioethanol production.

Lignocellulose is difficult to hydrolyze using only cellulolytic enzymes due to its recalcitrant and heterogenous structure, which primiarily consist of cellulose, hemicellulose and lignin. Therefore pretreatment that alter the lignocellulosic structure to provide the enzymes with greater access to the cellulose is a crucial step in the cellulose convesion. Alkaline pretreatment methods using sodium hydroxide, lime, or ammonia have been applied to some crops such as straw, corn stover to effectively enhance digestablility of lignocellulosic biomass.

On the other hand, the ultrasound in the field of biorenewables is relatively new method. Ultrasound generates acoustic cavitation with extremely high local temperature, high pressures and heating/cooling rates. Furthermore, local turbulences and liquid micro-circulations are formed that might be utilized for the enhancement of transport processes and the elimination of mass transfer resistance in heterogeneous system. Sonication has been applied in enhancing the pretreatment and saccharification of cellulose, however limited studies address the application of sonication together with alkaline method to modify the biomass structure thus enhancing the sugar release and subsequently ethanol production.

The study aims to evaluate the pretreatment of *J. curcas* fruit hulls by sonication in combination with alkaline pretreatment methods using sodium hydroxide, lime and ammonium hydroxide. The effectiveness of the pretreatment was then determined based on the sugar released in saccharification.

See more of this Session: Biomass Fractionation
See more of this Group/Topical: International Congress on Energy 2011

(435f) Cellulose Extraction From Rice Hulls Using Ionic Liquids

Joan G. Lynam, Chemical and Materials Engineering, University of Nevada, Reno, Reno, NV and
Charles J. Coronella, Chemical Engineering, University of Nevada, Reno, Reno, NV

Overone fifth of calories consumed by humanity are from rice. Consequently, vastquantities of rice hulls are disposed of as waste, since the hull constitutes20% of rice mass. Standard practice for disposal of rice hulls is burning,generating pollution and CO_2, or thrown away in vast quantities. Analternative would be to extract cellulose from rice hulls them using a lowtemperature, environmentally safe process. Removing cellulose for other uses,such as conversion to glucose and then fermentation to bio-ethanol or use inpolymers, would be a better solution to this problem. In addition, thispretreatment process could render the rice hull residue into a biomass readilyaccessible for enzymatic hydrolysis, so even more cellulose could be convertedto glucose for biofuels use. Hemicellulose or pretreated biomass high inhemicellulose can also be used for enzymatic hydrolysis and fermentation tobiofuels.

Ionicliquids are basically molten salts with a melting point at least below 100 Cand generally below room temperature. Ionic liquids have been found whichdissolve isolated lignin and cellulose, and which dissolve these components inwood powder. However, the dissolution capability of these solvents has not beendetermined for rice hulls. Also, when ionic liquids are applied to biomass, thebehavior of hemicellulose has not been well characterized.

Inthis study, the ionic liquids 1-butyl-2,3-dimethylimidazolium chloride,1-hexyl-3- methylimidazolium chloride, 1-ethyl-3- methylimidazolium acetate,and 1-allyl-3- methylimidazolium chloride were used to dissolve rice hulls. A10% (w/w) loading of rice hulls was used. The room temperature ionic liquid1-ethyl-3- methylimidazolium acetate (EMIM Ac) was found to completelydissolve the rice hulls when a temperature of 110C was used with a holding timeof 24 hours. When EMIM Ac was used on rice hulls for 4 hours at 90C, nohemicellulose was removed. For EMIM Ac, the conditions of 4 hours at 110Cremoved one third of the hemicellulose and 8 hours at 90C removed one half ofthe hemicellulose. The reaction conditions of 110C for 8 hours removed 86% ofthe hemicellulose when EMIM Ac was used. Cellulose and lignin were alsoextracted from the rice hull biomass by EMIM Ac treatment and precipitation bywater addition.

See more of this Session: Biomass Fractionation
See more of this Group/Topical: International Congress on Energy 2011

BIO-REFINERY BASED ON INDIAN PAPER INDUSTRY WASTES

Sanjay Tyagi*, Narayan Chandra Mishra, and A.K.Ray
Indian Institute of Technology, Saharanpur Campus, Saharanpur, India
*Central Pulp and paper Research Institute, Saharanpur, India

ABSTRACT

Paper Industry is one of the major industries which discharge some wastes that can produce many valuable by-products. These are black liquor, steam and electrical energy, wood/ bark wastes, sawdust/ bamboo dust, solid wastes like lime mud, grits, dregs, ETP sludges and silica. These wastes can be converted to many value added chemicals, such as activated carbon from black liquor, sawdust, bamboo dust, broken chips and partly pulverized bark.

Black liquor /sulphite liquor produces ligno sulphones from rice straw, wheat straw, bagasse, bamboo, sarkanda and hardwood. It has diversified uses e.g. as a base for industrial detergents(washing powder and liquid soaps), oil-well drilling mud, cement and concrete additives, dispersants for a variety of chemicals, emulsifiers and stabilizers, brightening agents for electrolytic refining during electroplating, binders and adhesives for pellets and briquettes chemicals such as clay, calcium carbonate and titanium di-oxide resin ingredients, lignin-phenol-formaldehyde mixtures as adhesives for ply wood and other wood based panels, rubber additives, adsorbent/desorbent, micronutrient for agricultural applications, asphalt emulsions. Besides black liquor can also be converted to lignin-poly isocyanate foam, lignin reinforced polymers, lignin adducts, chlorobrominated lignins as effective fire retarding agents, composites of alkali lignin and biologically active agents such as fungicides, insecticides, nematocides, herbicides etc. ,activated carbon by coking or gasification of black liquor, hydrolytic degradation of ligno-cellulosic waste to sugars such as xylose and xylitol used as food stuff as sweeteners, synthetic fuel using solar furnace, silica from straw black liquor, hydrogen gas by electrolysis, vanillin, DMSO, methanol, acetate, formate, actate, saccharinic acids, lactones, turpentine and tall oil, alkali lignin, guaiacol, pyrocatechol, acetoguaiacone, biogas(methane) etc.

Cellulosic wastes can also produce cellulosic bio-ethanol using thermophilic bacterial species like Clostridium thermocellum. In the similar manner, the sludge can produce glucose by treatment with cellulase enzymes. Syngas from gasification of pith of some nonwood based plants can also be produced.

From straw soda black liquor, silica can be precipitated for industrial applications such as amorphous silica (for high quality synthetic pigments as paper fillers which can replace soap stone and TiO_2), filler, retention aid and as flocculating agent, for production of ceramics. The later can be employed for household china, fine earthen ware, concrete block, electric porcelain, silica light weight brick, clay brick, calcium silicate thermal insulating materials, reinforcement in rubber, in paint manufacture, adsorbent, grease thickener, insecticide and cosmetic industry.

An attempt has been made in this paper to review the existing status of Indian industry to examine up to what extent they are able to exploit the potential of aforesaid cellulose and lignin by-products. Further exploratory investigations were made to look into the feasibility of their manufacture in plant practice in India for its sustainability.

INTRODUCTION:

Pulp and paper Industry is the World's largest non-food biomass collection system. India is the 15th largest paper producer in the World. The installed capacity of Indian paper industry is 12 million tonnes and 11 million tonnes is in operation, however annual production is around 10 million tonnes with per capita consumption of 10.5 million tonnes. Now the Indian paper industry is growing at 9.5%CAGR per annum along with booming economy with GDP growth rate 8.6 % .It is projected a demand of paper and paper board in the country will be 15 million tonnes by the year 2015 and 22 million tonnes by the year 2020[1]. On the utilization of recovered paper side, India imports around 4.5-5 million tonnes of waste paper as the recovery rate at India rests at meager value of 20-22% [1]. In the coming decade Asia will be major hub for paper production and trade, particularly in a time when drastic paper production cut is taking place in Europe (minus 0.3%),North America (minus 0.5%) and other parts of the World**[2].**As a major economy of Asia, compounded with the booming economy of India and increasing demand of paper and paper board, India will provide consistent growth for decades to come. Together India and China are estimated to make up 59% of global incremental paper and paper board demand during 2009-2020. At the same time , North America and Western Europe is forecast to decline from 46% to 38% of global production by 2020.However, Indian manufacturers have to be more conscious on the emerging opportunities and growing market size as stringent competitions are ahead for them which have to be met by adopting clean technologies and quality products [2].

Pulp and paper industry is capital-intensive industry, intensive in terms of energy, chemical, water, labor and raw materials concomitant with high pollution load generation. Besides it demands various large expensive pieces of equipments for its manufacture to produce quality products. Because of its capital intensiveness, investment in the paper industry is always a long-term proposition. The technology of operations is also very much intricate. Today's plant personnel need the concept of production integrated with energy-environmental-productivity-quality issues. The discharges from paper industry include solid waste, liquid effluent and gaseous emissions. There are regulatory limits enforced on the discharge of these pollutants by the regulatory authorities and the industry is expected to treat its effluents to meet the standards. In some of the states zero effluent concepts are being implemented in waste paper based mills. The new process must induct lower energy, water consumption along with little or no impact on environment. In order to follow this strategy the technology has been continuously passing through many revolutionary changes over the last decades to achieve low manufacturing costs, to earn profits derived from the ability of the producer and to face international competition for quality standards in the global market.

PAPER MILL BASED BIOREFINERY:

The pulp and paper industry is at the forefront of the concept of bio-refinery since it is already a biorefinery -1st generation of Biorefinery –enabling sustainable production of pulp/ paper enhanced by the coproduction of bioenergy and biofuels and bioproducts, through matured technological processes that maximizes the value chain of pulpwood **[3].** However, the decreasing profitability of standard pulp and paper mills and the world's demand for a non-fossil Carbon Economy resulted a new biorefinery platform based upon complex technology concepts with higher revenue potential [3]. Biorefinery is an overall concept of an integrated and diversified processing plant where biomass feedstocks are converted into a wide range of valuable products, much likewise to petroleum refineries. Integrated birefinery is a processing facility that extracts carbohydrates, oils, lignin, and other materials from biomass, converts them into fuels, high value chemicals and other materials, with a near zero waste draw [4]. Paper Industry is one of the major industries which discharges some wastes as given below that can produce many valuable by-products [5-26].These are black liquor, steam and electrical energy, tall oil, wood/ bark wastes, sawdust/ bamboo dust, solid wastes like lime mud, grits, dregs, ETP sludge and silica. These wastes can be converted to many value added chemicals, such as activated carbon from black liquor, sawdust, bamboo dust, broken chips and partly pulverized bark, turpentine and tall oil. Important chemicals like methanol, acetate, formate, lactate, saccharinic acids, lactones, alkali lignin, guaiacol, pyrocatechol, acetoguaiacone can also be prepared. These are discussed as under:

Solid waste from big paper mills[12]

(Avg. data based on data collected from 8 paper mills)

Organic Waste: (T/T of Paper)

1.	Raw Material Preparation (Wood /bamboo dust)	0.1
2.	Pulp mill rejects	0.01
3.	Paper machine cleaner rejects	0.1
4.	ETP Sludge	0.2
	Sub-Total	0.41

Inorganic Waste: (T/T of Paper)

1.	Ash (Cinder + Fly ash) from boilers	0.80
2.	Lime sludge and grits (Recovery + Bleach liquor preparation)	0.70
	Sub-Total	1.50
	Total	1.91

Utilization of wastes from a paper industry can have enough potential for a perfect biorefinery in India. Production of many value added products can be explored. There are many Indian industries engaged in establishing biorefineries for producing multiple products. Some are mentioned below:

UTILIZATION OF WASTES FOR BIOREFINERY IN INDIA

1. Utilization of wood/nonwood wastes:

Large amounts of ligno cellulosic waste are generated through forestry and agricultural practices , pulp and paper industries, timber and agro-industries.

1.1 Production of bioethanol:

Cellulosic wastes, solid wastes like wood dust, bagasse pith, waste paper and even sludge can produce cellulosic bioethanol [8]. Alcohol can be produced from cellulosic wastes using thermophilic anaerobic bacteria species like Clostridium thermocellum at 60°C and pH of 7.5 under control anaerobic conditions [14]. Bioethanol from bagasse pith has been produced in the laboratory using thermophilic ethanologen, yeast strain at 50°C through sachharification both chemically and enzymetically (using hydrolytic enzymes like celluloses, laccases and xylanase) followed by fermentation of the hydrolysate. It is reported that almost 80% glucose (30 g/l)of enzyme hydrolysate was fermented to ethanol(9.5gm/l) within 40 h and thereafter some of the oligosachharides also converted to ethanol bringing the final ethanol concentration 11.5 gm/l in the broth [6,7]. Naithani et al. [9] describes utilization of Lantana Camara and pine needle as lignocellulosic wastes for ethanol production with a yeast culture of Saccharomyces Cerevisiae NCIM-3570 with final pH to 4.5 at a temperature of 30°C. The acid hydrolysate of hemicelluloses extracted bagasse with the yeast S.Cerevisiae for fermentation can produce ethanol after concentration and supplementation can produce 8.3% alcohol with fermentation efficiency of 83.3% compared to 7.8% for molasses medium of 28°Bx with FE of the order of 83.3%[25].

2. Utilization of fiber rich Sludge:

In the similar manner, the sludge can produce glucose by treatment with cellulase enzymes [10]. Treatment of cellulosic sludge with cellulase enzymes rapidly shortens the fiber length and then dissolves the fibers in to glucose. The glucose can then be sent to an ethanol plant for fermentation to ethanol. The economics are based on sludge containing 50 tonnes per day of cellulose. The enzyme cost is projected at $6,000 per day. Water treatment cost is $1,000 per day (including capital and O &M) assuming reverse osmosis will be needed. Thus the total costs are $7,100. The value of the glucose produced (assuming 70% conversion from fiber) is $13,000 per day. The value of sludge disposal avoidance is taken at $3,000 per day for a total benefit of $16,000 per day. Addition of c-PAM accelerates the rate of fiber hydrolysis. The other costs considered for investment are due to some equipment to handle the residual solids,acid and power. The cost-benefits have been found favorable [10]

3. Biomass Gassification:

Syngas from gasification of pith of some nonwood based plants such as bagasse can be produced [11]. Biomass thermo-conversion for energy production from bagasse has been attempted by Umrao et.al. [11] for pyrolysis in a free fall reactor followed by gasification in a quartz reactor to obtain a more efficient and complete conversion of the bagasse as well as production of a gas with high heating value or with a high H_2 + CO (Syngas) content to be used as a fuel for chemical industry. Pyrolysis of bagasse is done on a fluidized bed at high temperature due to which charcoal is formed and the volatile components are released. The charcoal is sent to the two stage gasifier. As a result solid ashes, and a gas mixture which is composed of CO_2,CO,H_2, water vapour, pyrolysis product , tar and hydrocarbons. This process saves one liter of liquid fuel by using 4-5 kg of bagasse.

4. **Bio-gas production [21, 24]**

Primary and secondary (biological) sludges generated in pulp and paper industry can be digested anaerobically to produce bio-gas through acid fermentation followed by conversion to methane and carbon di oxide by methane forming bacteria. Benefits of anaerobic digestion are:

- Production of Bio-gas which can be utilized as a fuel
- Reduction of organic solids (and stabilization)
- Sludge can be used as a soil conditioner or fertilizer.
- Sludge can be treated without pre-dewatering.

Typical values of bio-gas production and organic solid waste reduction using anaerobic digestion of pulp and paper mill sludges are:

Reduction of organic solids : 45-75%)

Bio-gas produced : 0.5-0.7 m^3/kg organic solids

The process details for a typical 30TPD bagasse based soda pulp mill in India producing 0.45 - 0.50 m^3/kg COD gas (75% CH_4 + 25% CO_2) with calorific value of 5400 kcals/m^3 is available. The process removes 90% BOD, and 70% COD at temperature range 35-37°C with saving of 4000kWh /day power and 5 tons per month nutrients [24].

5. **Particle board**: Pith + Lignin can make particle board.
6. **Other important value added products:**

Many valuable chemicals can be manufactured from paper industry wastes. Majority of the processes for chemical manufacture from paper industry wastes are patented [22].These are excellently reviewed by a number of investigators [19-23]. Some of the patented processes are highlighted below:

6.1 Activated carbon:

Ligno-cellulosic material such as saw dust, bamboo dust, broken chips, and partly pulverized bark is reported to have been converted into activated carbon.

Method I: The material was first treated with inorganic acid to remove sugars at low temperature range. Acid extracted material was then heated at **450°C for nine hours and** for one hour at 800°C. While steam was being injected, the tar was removed as condensate in this high temperature range. About 22 % yield of activated carbon was reported.

Method II:

Continuous production of activated carbon can be possible by carbonization of cellulosic material impregnated with activating agent such as zinc chloride. The process involved continuous supply of activating agent-impregnated material into carbonizing zone containing previously carbonized cellulosic material at such constant flow rate as to allow the impregnated material to reside for fifteen minutes in the carbonizing zone. The impregnated material was carbonized with the carbonized cellulosic material heated to 350°C by counter currently supplied hot gases and which formed an agitated fluidized bed by being fluidized with hot gas stream and being stirred by a mechanical means. The total weight of the product was two third of the original weight of the wood material supplied.

6.2 Chemicals from Bark:

Bark along with other refuse creates a disposal problem for paper mills using hard wood as raw material. Utilization of bark can be developed on following lines.

I. As a source of fuel for steam generation
II. As a source of chemicals, particularly tannins and tannin like compounds, which would be recovered by water/acid/alkali extraction of bark, prior to burning.
III. Marginal utilization as soil conditioner, extender for thermosetting resins' oil well mud additive, reinforcing and stabilizing agent for polypropylene moulding compounds.

6.3 Hydrolytic Degradation of Lignocellulosic waste to sugars, Xylose and Xylitol:

During acid hydrolysis of lignocellulosic biomass a number of chemicals are produced. These are sugars (Xylose), aliphatic acids (acetic, formic and levulinic), furan derivatives, furfural and 5-hexamethyl furfural(HMF) and phenolic compounds. Hydrolytic degradation of lignocellulosic waste to sugars such as xylose and xylitol used as food stuff as sweeteners can be potential chemicals. Hardwood and agricultural residues are rich in pentosans which mainly contain xylan. Waste and refuse generated from hard woods/straw/bagasse etc. while preparing these materials for pulping, could be degraded hydrolytically to D (+) Xylose and Xylitol, which are used in food stuff as sweeteners.

Hydrolysis of waste was carried out in first stage with sodium hydroxide of 4% concentration which removes acetic acid chemically bonded in the wood as sodium acetate. The liquid extract of this stage can be re-used after it has been brought to the alkali concentration to be used.

In the second stage of processing the residue from the first stage containing pentosan and cellolignin was further hydrolyzed with mineral acid e.g. sulfuric acid. When working without pressure, 1.5 to 6 % strength of H_2SO_4, and a ratio of liquid to solid of 3 to 6 parts by volume was preferred. Temperature was preferably 50 to 125°C and time was 2-4 hours. After completion of the second stage, the batch was filtered; the filtrate could be worked to Xylose or directly to Xylitol.The insoluble residue from the second stage could next be used to obtain organosolve lignin which was thermoplastic and very reactive; the material, this lignin preparation could be used as a base material for dyestuffs and pesticides. The residue remained after removal of lignin, on treatment with sodium chlorite followed by sodium hydroxide could be converted to practically white cellulose which could be used for industrial applications. Thus all wood waste could be converted to useful chemicals.

6.4 Synthetic Fuel using Solar Furnace:

A process of producing gases such as CO_2, CO, CH_4 and H_2 from pyrolysis and gasification of solid organic waste contained within a reactor has been patented [22]. The process comprises solar heating of working fluid (steam, CO_2 or a mixture of gases) to 600°C to 700°C by means of solar top furnace. The working fluid was contained in the lower section of the reactor situated at the focus of the solar top furnace. The heated working fluid was injected under pressure into porous reactant bed on which the catalyst ($NaHCO_3$) - mixed organic waste was situated. The catalyst-mixed organic waste was pyrolyzed causing the formation of CO_2, CO, CH_4 and H_2 gases and char residue.The hydrogen produced could be utilized to replace natural gas in a hydrogen economy, to make fertilizer or as a chemical feed stock. It is claimed that process did not require a source of pure oxygen, and provided an attractive method for using solar energy.

7.0 Basic chemicals from wastes

Wood waste has a complex carbohydrate lignin structure requiring severe conditions to reduce it to such basic chemicals as Methanol, Ethanol, Furfural and Phenols. Chemical production costs are directly related to scale and complexity of operation **[22].** Utilization of pine needles and bagasse as bed material in solid state fermentation for production of lactic acid by Lactobacillus strains has been an active research in IITRoorkee[26].

8.0 Cellulose nano crystals and nanocomposites:

Cellulose nanocrystals have been largely applied as reinforcing fillers in the preparation of nano composites materials with improved mechanical and barrier properties.

9.0 Utilization of black liquor and sulphite waste liquor

Pulp and Paper industry has a very huge lignin potential. According to rough estimate for every ton of bleached pulp produced, about 0.5 ton of lignin is either wasted to sewer or used as a low value fuel. If the means could be found to isolate it from spent pulping liquor and to convert it to usable chemicals, the financial returns might be sufficient to support separation and processing lignin preparations from spent liquor, thus helping abatement of stream-pollution. In many ways Kraft lignin is a versatile raw material. It can be easily isolated by acidification, preferably carried out with carbon dioxide (stack gas), whereby phenolic group are liberated and lignin precipitates. The precipitated lignin can be removed by filtration at elevated temperature and purified further by re-dissolution in water and re-precipitation by sulphuric acid.

9.1 Lignin Isolation and separation:

Lignin compounds from sulphate black liquor could be precipitated by bubbling CO_2 into it at about 60°C. The precipitated lignin was separated by filtration and purified by dissolving in dry hexane and re-precipitating in ether. Alkali lignin could be isolated, from soda process black liquor by neutralizing it with CO_2 from stack gases to pH 8.3 to 8.5; about 40% of the lignin was reported[9] to have been precipitated. The mix was heated to 85°C – 90°C to coagulate the lignin, was filtered. Residual lignin could be further precipitated by addition of sulfuric/hydrochloric acid. Isolated lignin was washed first with cold water and then with dilute sulfuric acid and dried on double drum drier.

The slimy, gelatinous and hard to filter form of the precipitated lignin from black liquor could be changed to a crystalline, easily filterable and quick settling precipitate by mixing small amount of water-immiscible solvent such as chloroform, methelyene bromide, l-nitropropane or the like, with lignin containing liquid such as Kraft/soda black liquor. The chloroform, or other agent, may then be recovered from the filtered liquid by vacuum- or steam distillation.

Chemical modification of Kraft lignin drastically changes its properties and product obtained are of commercial value. Common modifications can be summarized as follows:

1. Formation of ethers by reaction with alkyl/aryl halides
2. Formation of esters by, reaction with carboxylic acid
3. Formation of polymeric condensation product with phenols, aldehydes and amines
4. Conversion to sparingly soluble phenol derivatives in reaction with various salts of heavy metals
5. Sulphonation to Kraft lignosulphonates
6. Nitration carried out with nitric acid
7. Chlorination with aqueous chlorine
8. Oxidation, with air or oxygen produces derivatives with changed solubility and melting point
9. Demethylation of lignin can be accomplished by acidic and alkaline treatment and results in the formation of catechol group increasing the reactivity in condensate e.g. with formaldehyde.

9.2 Lignin derivatives:

Lignin thus isolated could be converted to large surface area. Lignin is useful in drug formulations for adsorbing acids in the digestive system, due to their extremely high surface area and their acid-alkali resistance. Other suggested uses include carriers for adhesives, catalyst substrate, reinforcement for plastics and carriers for insecticides and herbicides.

A method to produce biodegradable uncolored lignin solutions was described wherein aqueous lignin solutions such as pulp mill effluent were treated with polyvalent metal cations. The amount of cations should be sufficient to react with lignin, and the aromatics to produce a water insoluble precipitate which was separated from remaining liquor. The precipitate was subsequently acidified with acid solution and was allowed to stand for sufficient time to permit digestion. The precipitate thus formed was separated by centrifugation; and was treated with alkali to produce aqueous basic solution. This solution was found to be effective in controlling Scale formation in boiler water systems, when added to these systems in amount of 5-50 ppm. Thus method not only recovers valuable product from effluent but also reduces the amount of non-biodegradable and toxic materials prior to discharge. In addition color of the final discharge is also reduced.

Sulpho methylated alkali lignins mixed with aqueous ferrous salt and dichromate and spray dried to a powdered product was reported to be useful as drilling mud thinner.

9.3 Lignin Reinforced Polymers:

A method for forming lignin reinforced polymers was developed by co-precipitating lignin/latex slurry and gas drying the slurry to leave a lignin reinforced polymer free of inorganic salts. In a process, alkali lignin was dissolved in ammonium hydroxides; and the ammonium lignate thus formed was then modified by adding hexamethylenetetramine to the lignate and heating to 90°-110°C for two: hours. Thus treated lignin was mixed with latex to give emulsion, which was precipitated by CO_2. When the precipitate was dried at 300°F or below, inorganic salts formed in precipitation, sublimed leaving lignin/latex polymer.

9.4 Lignin- polyisocyanate/polyurethane foam:

These foams are useful as plastics and adhesives. Lignin-polyisocyanate foam can be prepared from polyethylene glycol solution of lignin when the later is treated with toluene di-isocyanate at 115°C which afforded foam having good compressive strength and 74% recovery. Polyurethane type foam from lignin preparations and organic isocyanates are also reported.

9.5 Lignin as dispersants:

Method: preparation of ozone treated lignin from Kraft black liquor by treating lignin with oxygen containing 3% ozone at 50°C for two hours; and spray drying the resulting solution has been reported. Oxidized lignin was found to be useful as dispersant for a variety of inorganic materials such as Clay, Cement, Calcium Carbonate, and Titanium Dioxide.

9.6 Lignin Adducts as surfactants:

Lignin adducts with 3-chloro-2-hydroxypropane-l-sulphonate, chloromethane sulphonate, acrolein and the like compounds are also reported to be useful as surfactants in dyestuff compositions.

9.7 Adhesive from lignin: Lignin-phenol-formaldehyde mixtures as adhesives are used for ply wood and other wood based panels:

Method: An adhesive from Kraft or Soda black liquor was prepared by reacting the black liquor (dry matter more than 25%) with formaldehyde and combining the product with separately prepared phenol-formaldehyde adhesive.

9.8 Lignin as coagulants and flocculants:

Lignin preparations from pulping liquor chemically modified by reaction with compounds such as formaldehyde, triazines, alkylamines were found to be useful as coagulants in treatment of sewage and other wastes. Cationic water soluble lignin amines prepared by reacting Kraft/Soda lignins with the reaction product of epichlorohydrin and a tertiary amine were suitable for a variety of uses such as flocculating agents, coagulants, retention aids, dispersing agents, asphalt emulsifiers and emulsion stabilizers.

9.9 Fire retarding agents:

Chlorobrominated lignins were highly effective as fire retarding agents. In one method, the inflammable material was coated or soaked with an aqueous slurry or alkaline solution of chlorobrominated lignin followed by drying. In addition chlorobrominated lignins could be combined with bonding compositions used for the production of bonded articles from particulate material, such as fibers, chips, granules, saw dust etc. eg. in the formation of chip board, particle board and the like. Chlorobromination was affected by adding chlorine and bromine concurrently or in two step operation. After completion the reaction, the chlorobrominated lignin was filtered, washed and dried.

9.10 Composites

Composites of alkali lignin and biologically active agents such as fungicides, insecticides, nematocides, herbicides and the like are useful as controlled release pesticides. These composites were prepared by intermixing ammonium salt of lignin with active agent under conditions of high shear mixing to form stable emulsion; acidifying the emulsion with mineral acid at pH 4-5 to precipitate the active agent/lignin emulsion with the active agent being physically absorbed within the lignin and drying the precipitate if desired.

9.11 Coke and activated carbon from black liquor

Process for activated carbon by coking or gasification of black liquor: In a process black liquor was heated in absence of free oxygen to an elevated temperature (450°-625°F) and pressure to prevent vaporization of the water, for a period sufficient to produce granular coke. The coke was separated from the aqueous phase to yield an aqueous effluent of low COD. The coke produced might be burned to supply heat, or used as a soil conditioner or carbonized to produce high grade coke for metallurgical purposes or used to produce activated carbon.

Organic values of black liquor have been converted into heat energy and activated carbon by a method which eliminates the recovery boiler from pulping liquor regeneration systems. Black liquor containing 16% solids or even lower or at any solids content up to 65% was hydropyrolyzed under very high pressure at 550°-650°F. The organic matter separates as a water-insoluble precipitate along with inorganic constituents such as Na_2CO_3, and minor amounts of Na_2SO_4, Na_2S. The organic matter could be utilized to produce heat energy or activated carbon which itself could be used for purifying effluents from various section of paper mill.

10.0 By-products from lignin

Generally by-products from lignin can be categorized as follows:

1. Low molecular weight compounds
2. Polymeric products

Table 1 gives a detailed picture of by-products (both commercial and non-commercial) obtained from Kraft cooking of wood using different methods of isolation. It can be seen from the table that the main by-products of Kraft cook are the degradation products of carbohydrates and lignin. The most consistent efforts on dry distillation of the alkaline black liquor powdered together with excess lime or alkali were made to obtain methanol, acetone, methylethylketone, higher ketones and other components of light and heavy oils in appreciable quantity (60-70 kg/ton of pulp each of the low molecular and high molecular compounds).

Low molecular weight compounds

The most practical approach to synthesis low molecular weight compounds from lignin is to heat the black liquor for 10-20 minutes at 250-290°C together with small amount of alkali and sodium sulphide. Table 2 covers the chemicals obtained during this process with their corresponding yield in kg/PTP. Under optimum conditions (291°C, 20.4 % of organic substance, 3.2 % Na_2S and 1.6 % of the NaOH based on total liquor) a maximum yield of 33 % of the organic products can be obtained as ether soluble phenols.

Of all chemicals listed in Table 2 the following chemicals are most important.

Use of lignin and lignin derivatives: Lignin Sulphonates and Kraft Lignin, Vanillin, Dimethyl sulphonates, Dimethyl Sulphoxide, Furfural, Methyl Mercaptan, Phenolics,thermosetting resins and graft polymers including polyurethanes, polyesters, polyamines and epoxies[13].

1. Vanillin
2. Furfural
3. Organosulphur compounds
    a) Dimethyl sulphide (DMS)
    b) Methyl mercaptans (MM)
    c) Dimethyl sulphoxide (DMSO)
    d) Dimethyl sulphone ($DMSO_2$)
4. Low molecular weight Phenolic compounds

10.1 Vanillin

Vanillin has a wide range of industrial application. Vanillin is the most widely flavoring agent,, commonly known as vanilla, due to its pleasant aroma and its property of enhancing other flavors.. It is universally used as flavoring agent with a property to mask other odors. Vanillin also finds use as a reductant and as material absorbing ultraviolet light. It can be incorporated into various plastics such as regenerated cellulose, polyvinyl ethers and polyamides to protect them against ultraviolet radiation. Ointments for protection against sunburn on the skin can be prepared with vanillin as active agent. Polymeric materials prepared from vanillin and vanillic acid have potential as fibres and films.

Generally vanillin is synthesized from Guaiacol or Eugenol.

It can be produced from lignin through oxidation of spent liquor with suitable oxidising agent or by extraction by a suitable solvent. There are a few methods for manufacturing Vanillin.

Method I: Catalytic oxidation under pressure in pressure vessel by prior raising pH of black liquor from 9.5 to 13 by sodium hydroxide and then heating for about 2 hours followed by extraction with chloroform. The aqueous layer is acidified with sulphuric acid to bring down the pH to 3.0 and the filtrate is again extracted with chloroform.

Commercial process based on alkaline oxidation of lignin with oxidising agent as cupric oxide (2% wt/wt.), various nitrobenzene compounds, mercuric oxide etc. have reported a better yield of vanillin from lignosulphonates. For example cupric hydroxide oxidation of lignin gives 22 % of vanillin at 160°C and 12.5 % at 102°C. In comparison a maximum yield of 23 % is obtained from nitrobenzene oxidation under optimum conditions of 1 hr at 180°C. Table No.2 gives the product obtained by alkaline oxidation of lignin. While vanillin is the most commonly reported simple compound obtained by alkaline oxidation of lignin, there are a number of other products that are formed which may eventually prove to be of commercial interest.

Method II: The black liquor is dissolved in chloroform and then extracted with saturated sodium bisulphate solution.

Purification of vanillin is an essential step when synthesized from lignin. Most practical way to purify it is to distillate under reduced pressure or co-distillation with saturated hydrocarbon.

10.2 Organo sulphur compounds from lignin

The most important products are DMS and DMSO.

DMSO is used as a catalyst and organic solvent for dyes and other chemical reactions. Since DMSO is an excellent solvent for many carbohydrate materials, it has found application in preparation of sugar derivatives. DMSO also has found utility as a material to protect cellulose tissue from freeze damage. In this case, it has been shown to penetrate-the cell walls better than glycerin and thus prevents excessive concentration of electrolytes in the cell during freezing. It has also been used successfully to protect red blood carpusals against freezing. Widespread interest in DMCO as a solvent for carrying drugs across cell membrane and as a drug itself have been developed.

Depending upon the source of lignin whether softwood or hardwood, the methoxyl contents of Kraft lignin varies from 14-20 %. Since methyl group represents about half of the weight of the methoxy group, it is theoretically possible to recover 7-10 % of lignin calculated as methyl radicals.

Synthesis of DMS

Method of production of DMS and MM is by heating Kraft spent liquor with sodium sulphide at 200°C and to purify the product obtained. The yields of DMS are about 3 % and MM 0.2-3 per cent on lignin solids.

Commercial process consist of heating concentrated spent liquor with elemental sulphur to a temperature of 200-250°C. The DMS formed is flashed from the reactor along with some water, MM and H_2S and collected by means of a condenser. The crude DMS is scrubbed by passing it through Sodium hydroxide and distilled.

DMS itself does not find any application but is a precursor for its halogenated and oxidized derivatives.

Synthesis of DMSO

This is a compound obtained from lignin with a variety of applications in chemical and pharmaceutical industries. This can be commercially obtained by oxidizing DMS by nitrogen-dioxide under optimum conditions. The reaction can be summarized as

$$3Me_2S + 2HNO_3 \rightarrow 3Me_2SO + 2NO + H_2O$$

The application of DMSO has extended in many fields. It has been used as a outstanding medium for promoting organic reactions and as a reactant solvent itself. The following points outline some of the most potential application of DMSO.

10.3 Polymeric compounds from lignin:

Lignosulphonates:

Bulk of the materials are imported from Russia and China **[19]**

This class of lignin polymers has attracted a great deal of interest because of their diversified uses. They can also be utilized as the base for-industrial detergents such as washing powders and liquid soaps.

10.3.1 Synthesis of lignosulphonates

Lignosulphonates can easily be prepared from purified Kraft lignin by sulphonation with solutions of alkali sulphites at elevated temperature and pressure. Sulphite spent liquor contains lignosulphonates which are formed during pulping reactions. Practical approach to isolate and purify this compound is to remove most sulphite and sulphate by lime addition upto pH of 10.5, more lime is added to the filtrate to give basic calcium lignosulphonate which precipitates in the pH interval of 10.5 - 12.2. More than half the lignosulphonate is thereby removed and separated by filtration. By the addition of sulphuric acid and magnesium and sodium lignosulphonates can be obtained.

10.3.2 Application of Lignosulphonates:

Black liquor /sulphite liquor produces lignosulphones from rice straw, wheat straw, bagasse, bamboo, sarkanda etc. and also from hardwood. Fundamental studies on the colloidal properties of purified lignosulphonates demonstrated their specific value for emulsification, defloccuation, chelation, adsorption and adhesion uses. It has diversified industrial applications; some major ones are as under:

i. Oil well drilling muds: In the drilling of oil wells, water and dispersants are continuously fed into the hole, to form, together with suspended fine clay, called as 'drilling mud'. Oil in water emulsions in a drilling fluid has been in use since the 1940's. Lignin sulphonates are valuable for stabilizing the oil in water emulsions which give the mud improved properties in control of water loss, reduction of torque on the drill stem, increase of bit life and general improvement of bore hole condition.

ii. Cement and concrete additives: Lignosulphonates are used as grinding aids in Portland cement and concrete. Their function is to reduce the agglomeration of the-ground particles, to reduce surface-tension at the solid-air interference of the particles-and to keep the surface of the grinding media clean & free. It reduces settling time; water reduction and these can also be used as air entraining agents in concretes. The entrained air gives increased resistance to spalling and cracking of the finished concrete. The dispersive action of ligno-sulphonates is useful in mixing, and placing concrete. Incorporation of 0.07 - 0.15 kg of lignosulphonates-concrete admixture per sack of cement will allow water reduction up to 20%. This results in better compressive strength and durable concrete. In oil well cementing it thus controls settling time and strength.

iii. Dispersants for variety of Chemicals: Lignin sulphonate dispersants usually have a molecular weight in the range of 10,000 - 40,000. The colloidal lignin coats the surface imparting a negative charge to the particle. The similarly charged particles now repel one another and a good dispersion is obtained. This is used as dispersants in refractories, tiles, absorbants, cosmetics, crayons, medicines, paint, paper coatings, tooth powder and dyes. It can also be base for industrial detergents (washing powder and liquid soaps

iv. Ore floatation: Grinding aids, refining of ores as floatation agent, improves filtering properties in Pb-Zn system.

v. Emulsifiers and stabilizers: Lignin emulsifiers are quite competitive with other commercial emulsifiers stabilizing oil-in-water emulsions. Emulsions formed are very stable even under frozen or boiling stage. The stabilization of fire fighting foams that results from the interaction of sodium bicarbonate solution and alum is achieved by an addition of sodium salt of Kraft lignin. Better results can be obtained by using lignosulphonates.

vi. Electrolytic Refining: Lignosulphonates have been successfully used as 'Brightening' agent for the plated surface preventing pitting and deposition of rough surface in electroplating.The electrolytic refining of copper , zinc etc. are some examples.

vii. Binders and adhesives: Lignosulphonate enhances the properties of the specific products e.g., pellets, briquettes. Large tonnage of spent liquor solids are used as binders for pellets and briquettes. 2% calcium lignin sulphonate in iron ore gives a pellet ten times as hard as that made with bentonite binder. The most familiar adhesive product is linoleum cement containing principally lignin sulphonates and clay. These cements are also used as adhesive for ceiling tiles, road surfacing, hard durable surface and laying of dust, chemicals such as clay, calcium carbonate and titanium di-oxide resin ingredients,

viii. Resin ingredients: Combination of lignin sulphonates, with low molecular weight urea and phenol formaldehyde resins have received attention as adhesive or binding agent for fibrous materials such as plywood, particle and chip board and paper laminates. Many good quality glues can be produced by incorporating lignin sulphonates with polyvinyl acetate or alcohol.

ix. Rubber additives: As extender, modified reinforcing pigment. Many types of lattices such as neutral nitrile and neoprene are compatible with lignin either in the acid or sodium salt form. Mixture of rubber with Kraft lignin is also suitable for use as rubber to glass adhesives.

x. Protein precipitants: Forms insoluble complex with proteins.

xi. Tanning agents: Forms more soluble tanning solution which facilitates recovery procedures.

xii. Sequestering agents: Ca, Ni, Sn, Zn & Al can be sequestered by lignosulphonates above pH 7.

xiii. Miscellaneous uses of lignosulphonates:

Other uses of lignosulphonates are lime plastering medium, storage battery plates-as expander and component of separator., lime plaster-use of less water with improved settling properties, crystal growth inhibitor, Ingot mold wash, and many more.

11.0 Other uses of black liquor: **Hydrogen gas by electrolysis,** Gasification of black liquor, Biogas from black liquor, Preparation of rubber additives, adsorbent/desorbent, micronutrient for agricultural applications, asphalt emulsions,

12. Utilization of lime sludge:

As lime sludge contain high silica content, so its burning is not considered economical. However, the following approach can be considered.

a. Burning to get low grade lime
b. In the cement industry for partial substitution of lime stones
c. As a conditioner for soil especially acidic one.
d. Burning after desilication
e. Lime mud comprising calcium carbonate was found to be useful in removing colour from pulping liquid waste effluents. In a process[32] effluents were treated with lime mud followed by clarification, separation of effluents from classifiers to remove colour bodies.
f. Puzzulanna Cement

13.0 Silica recovery from from straw black liquor:

Amorphous silica (for high quality synthetic pigments as paper fillers which can replace soap stone and TiO_2), filler, retention aid and as flocculating agent, for production of ceramics. The later can be employed for household china, fine earthen ware, concrete block, electric porcelain, silica light weight brick, clay brick, calcium silicate thermal insulating materials, reinforcement in rubber, in paint manufacture, adsorbent, grease thickener, insecticide and cosmetic industry.

Silica from boiler Fly ash, and Sludge are also used for manufacturing light weight bricks.

14. Industry Status in India:

- **Mission New Energy** : Uses agri-residues using enzymatic hydrolysis process produces 0.07Ml/year of cellulosic ethanol[3]
- **Utilization of bamboo dust:** Gasfication, Syn Gas by Hindustan Paper Corporation(HPC), **Nogaon,** Assam
- **Pith + Lignin to particle board**: Kailash Paper Mill , Awagwanpur, Moradabad, India
- **Biomethanation Process**: Satia Paper Mills, Pudumjee Paper Mills
- **Biomass cogeneration:** ABC paper mill

- **Sludge and Boiler Fly Ash to light weight Bricks**: ITC, Bhadrachalam, Rainbow Papers, Ahmedabad.
- **Corrugated roof sheets and Diesel from waste plastics in recycled paper:** Rainbow Papers, Damanganga
- **Producer gas plant using wood chips :** Andhra Pradesh Paper Mill (APPM), now International Paper, USA
- **Biogas, cogeneration and biomass gasification:**TNPL
- **PAFC :** BHEL, India[15].
- **Installation of biomethanation plant in pulp and paper industry with UASB reactor:** United Envirotechier
- **Biomass gassifier :** W2E projects and Systems
- **Lignosulphonates:** ABC paper mill, SailaKhhurd, Punjab

15. Status in Research Institutions

ETP sludge, pith and paper to Bio ethanol: Indian Institute of petroleum (IIP, Dehradun), Central pulp and paper research institute (CPPRI), Cellulose and paper branch,Forest research institute(FRI), National Sugar Institute , Kanpur,Basant Dada Patil Sugar Institute, Pune.

Cellulosic waste to alcohol: College of technology,Osmania University, Hyderabad, National Sugar Institute, Govt. of India, Indian Institute of Petroleum, Dehradun, Vasant Dada Sugar Institute, Pune, IITRoorkee.

Utilization of lime sludge,Fly ash :Central building research Institute, National thermal power corporation(NTPC)

Lignin grafted copolymers:Regional Research Laboratory(RRL at Bhubaneswar,Jammu, Jorhat)

Biogas –fuel cell based co-generation: Tata Energy Research Institute (TERI)

Production of lactic acid from waste sugarcane bagasse derived cellulose,cellulase production with Tricoderma citriviridae on solid bed ,use of acid hydrolysates of pine needles,and bagasse for lactic acid production using various strains such as Lactobacillus delbrueckii or lactobacillus pentosus have been explored in Department of paper technology, IITRoorkee( ), National sugar Institute , Kanpur, H.B.T.I., Kanpur

Ministry of New and Renewable Energy(MNRF) is keen to provide technical and financial assistance for setting up of energy generating plant based on biogas, cogeneration and biomass gasification

The technical parameters of a typical HRAD-PAFC (high rate anaerobic digester with phosphoric acid fuel cell) cogeneration system for a 50 TPD paper mill have been estimated by

TERI and development of a prototype PAFC have been explored by Bharat Heavy Elecricals Ltd.[15].

Table 1: By-products (both commercial & non-commercial) of the kraft cook

Compound	**Origin**	**Quantity kg per ton of pulp**
Methanol	Methoxyl groups of lignin & possibly glucuronoxylan	5
Acetate	Acetyl groups of the xylan as well as degraded carbohydrates	100-200
Formiate	Degradation of carbohydrates	40
Lactate	Degradation of carbohydrates	100
Saccharinic acids & lactones	Degradation of carbohydrates	250
Aliphatic sulfur compounds (methyl mercaptan, sulfide, disulfide)	Methoxyl groups of lignin and possibly glucuronoxylan	1
Turpentine	Terpenes	8-10
Tall oil (fatty & resin acid salts, phytosterols, etc.)	Wood resin components	20-100
Alkali lignin	Lignin	400-600

Table 2: By-products of the Kraft cook and from subsequent pressure heating of the black liquor with alkali and sulfide at 250-285°C (48 atm)

Before Pressure Heating	**kg ptp**	**After Pressure Heating**	**kg ptp**
Tall Oil	50	Acetic acid	60
Turpentine	17	Formic acid	60
Acetic acid	40	Methyl sulfide	40
Formic acid	40	Methyl mercaptan	8

Methanol	5	Pyrocatechol	30
Vanillin	2	Methyl catechol	11
Acetoguaiacone	2	Ethyl catechol	9
Guaiacol	2	Other ether-soluble phenols	150
Pyrocatechol	1	Homoprotocatechuic, protocatechuic and other phenol carboxylic acids, perhaps	150
Other phenols	42	Ether soluble acids of carbohydrate origin	150
Phenol carbonic acids	45	Demethylated lignin of novalac properties	100
Non-volatiles and lactic acids	145	Butanol-soluble lactones	35
TOTAL	**391**	**TOTAL**	**803**

CONCLUSIONS

Paper industry wastes can be properly utilized to produce value added products in order to act as second generation biorefinery. Lignin in Black Liquors has excellent potential for reuse in Fine Chemical Manufacture. The Indian Paper Industry is not yet ready for such a drastic transformation, since it will jeopardize the Chemical Recovery Processes. The Paper Industry is totally dependent upon the lignin black liquors as a source of energy fuel in Recovery Boilers. Serious attempts have to be made for medium and small scale manufacturing process. Some initiatives have been taken by Indian pulp and paper industry in recent years. A few research institutions have started research activities for establishing biorefinery based on paper/ sugar mill wastes.

It is necessary to determine the quantum of the wood waste/ agri-residue facility for the maximum size consistent with availability of raw material. A single product facility appears not to be competitive. In order to be economic utilization of the wood waste, multiple product facility is essential.

REFERENCES

1. Mukundan, P.G., Editorial, Inpaper International, Vol.12, No.2, Jan.-March 2010, Vol.13, No.1, Oct.-No.2010, www.inpaper.com.
2. Inpaper India, Vol.13(3), 14(2), March-April,2011, pp.24, 30.

3. Gabriel D.A., Sousa, Biorefinery development pathways: A survey for the pulp and paper industry, Proceedings of XXI Encontro Nacional da TECNICELPA/VI CIADICYP 2010, 12-15 Outubro 2010-Lisboa , Portugal.
4. Florbela Carvalheiro, Luis C.Duarte and Francisco M Girio, Hemicellulose biorefineries: a review on biomass pretreatments, JSIR, Vol.67, November, 2008, pp.849-864.
5. 56th Chemical weekly, July 05, 2011.
6. Thakur Vasanta V, Diwakar Pandey, R.K.Jain, R.M.Mathur, Second generation biofuels: Bioethanol from lignocellulosic materials, Inpaper India, Vol. 14(2), March-April, 2011, pp.4-12.
7. Jain, R.K., Thakur Vasanta V., Pandey Diwakar, Adhikari, Dixit A.K. and Mathur R.M., Bio ethanol from Bagasse Pith-A lignocellulosic waste biomass from paper / sugar Industry, IPPTA J., Vol.23, No.1, Jan-March, 2011, 169-173.
8. Mark E. Reno, Monique STREFF, Andrew Hird and Anjan Ray, Conversion of biomass to fuel as an environmental impact reduction opportunity for the pulp and paper industry, Inpaper International, Vol.13, No.3, April-June, 2011, pp.18-20.
9. Naithani Sanjay, Ganesh Joshi and Alok Kumar Dubey, Detoxification of Acid Catalyzed Lignocellulosic Hydrolyzate to enhance Fermentation Efficiency for Ethanol Production, INPAPER INDIA, Vol.14(2), March –April 2011, pp.14-22.
10. Banerjee Sujit, Energy from Pulp Mill Sludge,INPAPER INTERNATIONAL, Vol.13, No.1, Oct.-Nov. 2010, pp.10-12.
11. Umrao S., Pankaj Kumar and Sadhana Sachan, Method of producing Syngas from gasification of bagasse, Proceedings on "Advances in Chemical Engineering"-MacMillan, Advanced Research series February 27-28, 2009, Patiala, pp. 445-450.
12. Dhingra H.K. and Mohendru V.K., Status of Solid Waste Pollution From Indian Paper Industry and Future Prospects of its Management, IPPTA Vol.8,No.1, March 1996, pp.49-55.
13. Sahoo N, Rohella R.S.,Choudhury S.,Paul S.C.,and Chakravortty, V., Preparation of Value-Added Products From Black Liquor,IPPTA, Vol.8(3), September, 1996,pp.45-50.
14. Padmaja A, Venkateshwar S,Seenayya G and Bhagavanth Rao M, Evaluation of cellulosic substrates in the conversion of biomass to ethanol,Indian Chemical Engr., Section A . Vol.37 No.3, Jul-Sep(1995),pp.140-144.
15. Kurup ,G. Rajendra Chawla S.K.,Mathur A., Biogas –fuel cell based cogeneration in the pulp and paper Industry, IPPTA convention issue ,1992,pp.25-30.
16. Singh S.P., Recent developments in characterization and utilization of black liquor lignins from small pulp and paper mills, IPPTA, Vol.2 (3), 1990, pp.11-15.
17. Adsul, M.G., Verma A.J. and Ghokle, D.V., Lactic acid production from waste sugarcane bagasse derived cellulose, Green chemistry, 9, 58-62, 2007.
18. Guerra, G.,Casado, M.R.L.G., Arguelles,M.I., Sanchez M.I.,Manzano, A.M., and Guzman,T., Cellulase production with sugarcane straw by Trichoderma citrinoviridae on solid bed, sugar Tech.,8(1),30-35.

19. Managaonkar , N.D. and Kulkarni, Y.S.,Preparation and utilization of lignin by-products and lignosulphonates obtained from wood black liquor,IPPTA convention Issue,1987 pp.40-52
20. Ramaswamy, V. B iotechnology –applications in waste utilization and pollution abatement, ibid, pp.1-13.
21. Jain Sailendra, Environmental protection in Paper Industry, ibid, pp.53-61.
22. Dhake, J.D., Utilization of pulp and paper industry waste for chemicals,ibid, pp.19-29.
23. Trivedi, M.K., Lignin recovery from alkaline pulping of bagasse, ibid, pp.30-35.
24. NIEM (UNEP) Project Under Phase II- Activity III, October, 1990, Institute of Paper Technology (University of Roorkee), Saharanpur-247001 (India), pp.64-65, 66-67.
25. Lehri Madhu, Vartika Saxena and P.K. Agarwal, Preliminary studies on ethanolic fermentation of acid hydrolysates of hemicelluloses-extracted bagasse,pp.G587-G593, , Proceedings of the 57th Annual Convention of STAI, New Delhi, 13th -15th October,1995

26. Manoj Kumar Ghosh, Uttam K. Ghosh, Utilization of pine needles as bed material in solid state fermentation for production of lactic acid by Lactobacillus strains, Bio Resources, Vol.6, No.2,2011, pp.1556-1575.

Lignin – A Renewable Precursor for Carbon Nanofibers

John F. Kadla,[1] Ian Dallmeyer,[1] and Frank Ko[2]
[1]Biomaterials Chemistry Lab, Dept. of Wood Science, Faculty of Forestry
[2]Advanced Fibrous Materials Lab, AMPEL, Dept. of Materials Engineering
[1,2]University of British Columbia, Vancouver, BC

Abstract

Here we report on our research efforts aimed at developing carbon nanofibers through electrospinning of Kraft lignin. The research described herein focuses on understanding the effect of polymeric additives in influencing viscoelastic fluid properties which control the fiber morphology and diameter of fibers electrospun from blend solutions of softwood Kraft lignin (SKL) and poly(ethylene oxide) (PEO). To characterize the relevant fluid properties we report on the solution shear properties as well as the extensional properties of spinning solutions. We demonstrate that viscoelasticity can be varied in order to control the spinning process in our system by changing the concentrations of SKL and PEO as well as the PEO molecular weight. We also report on the mechanical properties of CNF precursor fibers as a function of thermal processing parameters. The results of these studies indicate that electrospinning is a promising platform for the development of micro- and nanofibrous carbon materials from technical lignins.

Introduction

Lignin is a naturally abundant biopolymer which reinforces the carbohydrate matrix, facilitates water and nutrient transport, and protects against invasion by microorganisms in the stems of trees and other terrestrial plants.[1,2] In terms of its chemical structure, lignin is an aromatic heteropolymer consisting of an irregular sequence of phenylpropane units connected by a variety of interunit linkages.[1,2] The renewable nature, abundance, and availability of lignin from the pulp and paper industry[3] are driving forces for the development of value-added products from lignin. Carbon fibres (CFs) are one example of value-added products that can be derived from lignin.[4] CFs have numerous applications including reinforcement in advanced light weight composite materials,[5] catalyst substrates,[6] adsorbents,[7] electrodes,[8] and tissue engineering scaffolds.[9] The most common precursors for commercial CFs are petroleum pitch, synthetic copolymers based on polyacrylonitrile (PAN), and rayon.[5] While PAN-based CFs have superior tensile strength, the high cost of PAN and non-renewable nature of petroleum-derived precursors are disadvantages.

Lignin is a potential renewable precursor for CFs, but the complex, heterogeneous chemical structure, the variability in properties among different types of isolated lignins, and the comparably poorer mechanical properties of lignin-based CF are barriers to their application.[4,10,11] Nevertheless, significant research efforts are being devoted to the production of CFs based on lignin, for example, at Oak Ridge National Laboratory[12]. Researchers in our lab have reported on thermal extrusion of lignin and lignin-polymer blends and conversion of these precursors to CFs by thermal treatment during the last decade.[4,13,14] Polymer blending was employed as a strategy to improve the spinnability of lignin[4], increase the tensile strength[13], or modify the surface porosity[14], depending on the blending polymers and their extent of interactions with lignin.[15-18] While studies conducted to date have reported improvements in fibre properties, a deeper understanding of the relationship between the complex macromolecular structure of lignin, interactions with additives, processibility, and material properties is needed to expand into a wider range of applications for lignin-based CFs. In addition, novel processing strategies must be explored to improve properties for specific applications and overcome some of the difficulties associated with lignin mentioned above.

One strategy to improve the properties of lignin-based CFs is to reduce the fiber diameter, since the tensile strength of CFs is known to increase with decreasing fibre diameter,[4,19,20] and applications such as filtration and catalysis depend on high specific surface area, which increases with decreasing diameter. Electrospinning is a robust strategy to produce fibers with diameters in the range of a few hundred nanometers or less.[21-23] Electrospinning of CF precursors such as PAN,[24-26] phenolic resin,[27] and pitch[28] has been reported. Electrospinning of organosolv lignin[29,30] and blends of sulfonated alkali lignin with PAN[31] have also been reported in the literature. However, lignin in general has been observed to display a high tendency to electrospray into droplets instead of forming fibers. To overcome this limitation, we have reported on the electrospinning of blends of several technical lignins with small amounts of poly(ethylene oxide) (PEO).[32] The advantages of our system compared to other reported systems are applicability to different technical lignins, the ability to spin fibers with greater than 99% lignin content with a single spinneret, and the ability to control the fiber diameter by varying the viscoelasticity of the spinning solution. Our current research in the area of lignin electrospinning is aimed at two main objectives: 1) To understand the role of lignin structure and interactions with additives in determining the viscoelastic fluid properties which govern the spinning process, and 2) To thermally process the electrospun fabrics and characterize the material properties as a function of processing parameters. In the following sections we provide an overview of our research efforts in the areas of rheology and its effect on the electrospinning process and processing and characterization of nonwoven fabrics produced from fractionated softwood Kraft lignin.

Materials and Methods

Lignin and solution preparation:

Softwood Kraft Lignin (SKL, Indulin-AT, Meadwestvaco, Glen Allen, VA, USA) was washed with dilute HCl (pH 2) to remove Na^+ ions and fractionated by sequential extraction with organic solvents according to a procedure based on that of Morck et al.[33] The F_4 fraction (according to the nomenclature of Morck et al[33]) was selected for rheological and electrospinning experiments. Polyethylene oxide (PEO) with M_v of 1000 and 5000 kDa (denoted PEO1M and PEO5M) were obtained from Sigma-Aldrich. N,N'-dimethylformamide (DMF, ACS Reagent grade) was obtained from Fisher Scientific (Ottawa, ON) and used as received.

Solution preparation:

SKL and SKL:PEO solutions in DMF were prepared as follows: SKL solutions were first prepared by adding dry lignin powder to the solvent. In the case of SKL:PEO solutions, PEO was first dissolved by briefly heating the solution at 80°C for 10-15 minutes, then SKL was added to the PEO solution. The mixture was heated at 80°C for 2 hours with intermittent vortexing every 30 minutes for 1-2 minute intervals. Solutions without PEO were prepared at concentrations of 25, 30, 35, 40, 45, and 50% by mass. Solutions with PEO were prepared with 25-45% SKL and 0.1-0.4% PEO (M_v = 1 or 5 x 10^6 g/mol).

Rheology:

The shear viscosity η was obtained from small amplitude oscillatory shear as the limit of $G''(\omega)/\omega$ at low ω or from steady shear measurements increasing shear rate from 1-1000 s^{-1} using an AR2000 torsional rheometer (TA Instruments, Grimsby, ON) with 40 mm diameter cone geometry and cone angle of 2°. A capillary breakup extensional rheometer (Haake CaBER 1, Thermo Fisher Scientific, Waltham, MA) was used to characterize the extensional fluid properties. In CaBER experiments, plate diameters of 4 mm, initial plate separation of 2 mm, final plate separation of 12 mm, and plate separation time (strike time) of 50 ms were used. Where applicable (for elastic solutions), the characteristic time scale of viscoelastic stress growth in uniaxial extensional flow λ (herein referred to as the relaxation time) was obtained by fitting the portion of the thinning profile where exponential thinning was observed as described by Rodd et al[34] using the equation:

(1) $D_{mid}(t) = D_1((G^*D_1)/4\sigma)^{1/3}\exp(-t/3\lambda)$

Where $D_{mid}(t)$ is the midpoint filament diameter, D_1 is the filament midpoint diameter just after the upper plate stopped moving, G is the elastic modulus, and σ is surface tension.

Electrospinning:

Electrospinning experiments were conducted using two different apparatuses. One was a self-built apparatus previously described[32] with a flat-tip needle as spinneret (gauge 18, 21, 22, or 25), operating voltage of 15 kV, flow rate of 0.03 mL/min, and spinneret-to-collector distance of 15-20 cm. The second was a Kato Tech Nanofiber Electrospinning Unit (NEU) (Kato Tech Co., LTD, Kyoto, Japan) using a horizontal configuration, rotating drum collector, and the same operating parameters listed above.

Thermal processing of electrospun fabrics:

Electrospun fabrics were heated under different conditions in a Hewlett Packard 5890 Series II gas chromatography oven under air. The samples were mounted on a Teflon-wrapped glass sample holder and heated from 25°C at a specified heating rate up to a specified final temperature and holding at the maximum temperature for 1 hour. Heating rates of 0.5, 5, and ~40 °C/min and final temperatures of 150, 200, and 250°C were used. Each condition was repeated three times and tensile testing was carried out for each of the three runs using 20 samples per run. The results were reported as the average of the three runs +/- the standard deviation. Following thermostabilization, carbonization was carried out in a GSL 1100X2 tube furnace (MTI Corporation, Richmond, CA) under N_2 heating at a rate of 10°C/min to 1000°C and holding at 1000°C for 1 hour.

Characterization of fibers and fabrics:

Optical and scanning electron microscopy was used to characterize fiber appearance using and Olympus BX41 Microscope (Richmond Hill, ON) and Hitachi S3000N SEM using gold-coated samples and accelerating voltage of 5 kV. Fiber diameters were measured using ImageJ analysis software counting a minimum of N = 100 fiber diameters per sample. The tensile properties of electrospun nonwovens fabrics were measured on a KES-G1 Kawabata microtensile tester as described elsewhere.[35] A density of 1.3 g/mL was obtained from an MSDS for alkali lignin and used to convert tensile data from units of N/tex to units of MPa.

Results and Discussion

Shear rheometry:

Shear rheometry has been used to characterize the viscosity of electrospinning solutions in the literature and correlations between shear viscosity, electrospinnability, and fiber diameter have been reported for a variety of different systems.[36-38] We first reported on the shear properties of lignin solutions with and without PEO in DMF in a published study.[32] In general, the shear properties of lignin solutions was Newtonian, with a viscosity independent of shear rate. However, it was clear in previous work that different lignins have different viscosities.[32] The observed differences underscore the inherent variability in lignins obtained by different isolation procedures or from different botanical origins. To focus our study, we selected to characterize SKL in greater detail.[32] The viscosity of SKL solutions varied over roughly 3 orders of magnitude when increasing from 25-50%. The rather strong dependence of viscosity on SKL concentration $\eta \sim c^7$ indicated that intermolecular interactions between SKL molecules increased the viscosity. Interestingly, solutions with PEO displayed a similar dependency of viscosity with concentration, with slightly higher values of viscosity but overall showing the same trend. However, SKL:PEO solutions were electrospinnable at shear viscosities far lower than those of SKL solutions. This result indicated that shear viscosity was not the key fluid property influencing the electrospinning behavior, since typically higher viscosity promotes fiber formation.

CaBER:

While viscosity affects the electrospinning process, it is also known that the elastic properties of polymer solutions strongly affect the ability to form fibers and the fiber diameter.[39-43] Based on comparison of our results with reports from the literature, we hypothesized that elasticity resulting from PEO addition may be responsible for the improved electrospinnability of lignin:PEO solutions. To test our hypothesis we varied the elasticity by varying the PEO molecular weight and the concentrations of SKL and PEO and investigated the ability to form fibres and the fibre diameter. Furthermore, we characterized the extensional properties of our solutions using a capillary breakup extensional rheometer (CaBER).

Figure 1 shows the thinning profile of the SKL solutions without PEO for concentrations of 40, 45, and 50%. Time is on the abscissa and D_{mid} as recorded by a laser micrometer is shown on the ordinate normalized by the initial filament diameter D_1. In general SKL solutions at or below 35% concentration could not be measured using the CaBER due breakup occurring before the end of the initial step. In Figure 1 it can be observed that the filament thinning profiles were linear in time, consistent with the behavior of a Newtonian fluid, in agreement with shear measurements.[44]

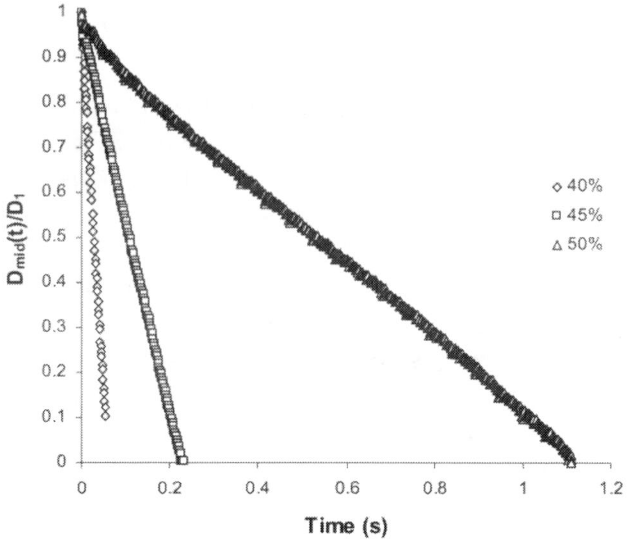

Figure 1: Thinning profiles of SKL solutions without PEO from CaBER experiments

In contrast, solutions with PEO displayed very different thinning behavior. Figure 2 shows the thinning profiles of several solutions with different SKL concentration, PEO concentration, and PEO molecular weight. Solutions with PEO display exponential thinning behavior, which indicates elasticity.[44] Elasticity implies that the extensional viscosity η_{ext} is not constant during thinning but increases with decreasing filament diameter (strain hardening). The relaxation time λ was determined from the CaBER data. Table 1 summarizes the compositions of the fluids and the values of λ obtained from the data. It can be seen in the table that λ increases with SKL concentration, PEO concentration, and PEO molecular weight, in agreement with published results.[45] Taken together these results show that the viscoelasticity of the SKL:PEO spinning solution can be varied by changing the relative amounts of SKL and PEO as well as the PEO molecular weight. The results also suggest that strain hardening occurring in extensional flow (which is absent in solutions without PEO) may provide the stability previously observed for lignin:PEO solutions in electrospinning.[32] We now discuss the effect of viscoelasticity on electrospinning.

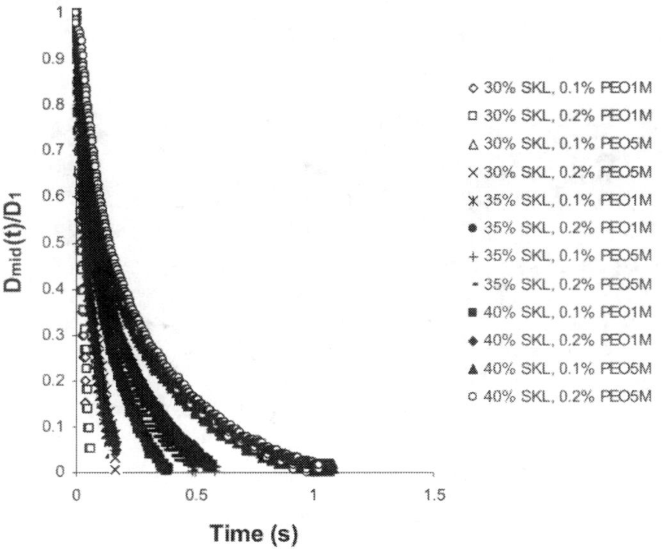

Figure 2: Thinning profiles of SKL:PEO solutions from CaBER experiments

Table 1: Fluid compositions and relaxation times

SKL concentration (%)	PEO concentration (%)	PEO mol. weight (g/mol)	λ (ms)
30	0.1	1×10^6	7.3
30	0.2	1×10^6	11.4
30	0.1	5×10^6	21.7
30	0.2	5×10^6	24.7
35	0.1	1×10^6	19.1
35	0.2	1×10^6	21.1
35	0.1	5×10^6	50.6
35	0.2	5×10^6	49.5
40	0.1	1×10^6	39.9
40	0.2	1×10^6	63.1
40	0.1	5×10^6	92.3
40	0.2	5×10^6	99.3
45	0.1	1×10^6	134.8
45	0.2	1×10^6	150.6
45	0.1	5×10^6	240.8
45	0.2	5×10^6	408.8

Electrospinning:

Example SEM images are shown in Figure 3 to illustrate the results of spinning experiments. Similar to our previous observations, only electrospray was formed without PEO below a concentration of 45-50%, with beaded fibers forming at 45%. In contrast, nearly all of the solutions with PEO were capable of forming fibers even though the amount of PEO in solution was less than 1% relative to SKL, producing fibers with SKL contents greater than 99%. The ability to achieve processibility with low PEO content is an advantage of the SKL:PEO system. Comparing rheology and electrospinning experiments, it can clearly be seen that fiber diameter increases with increasing λ. Figure 4 shows a plot of diameter vs. λ. The mean fiber diameter correlates well with λ, indicating that elasticity is correlated with the fiber diameter. Furthermore, the error bars show the standard deviations of the measured fiber diameters. It is also clear from the plot that the width of the diameter distribution tends to increase with increasing λ. This is partially due to the flattened, twisted morphology which tended to occur in solutions with higher viscoelasticity. The flattened morphology may be due to unevaporated solvent trapped in the core of the fiber followed by collapse of the fiber during the bending instability. The results presented here demonstrate that by controlling the viscoelasticity of the spinning solution, the mean diameter and width of the diameter distribution can be controlled.

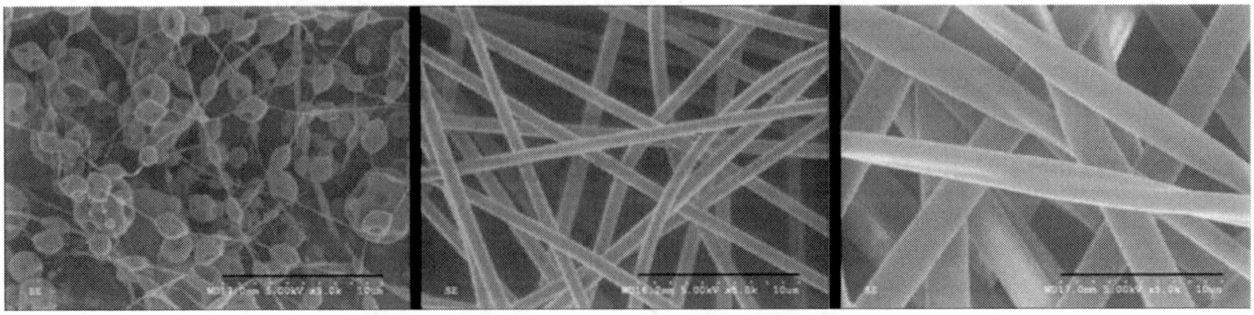

Figure 3: SKL:PEO fibers produced from solutions with different viscoelasticities (Scale bar = 10 μm)

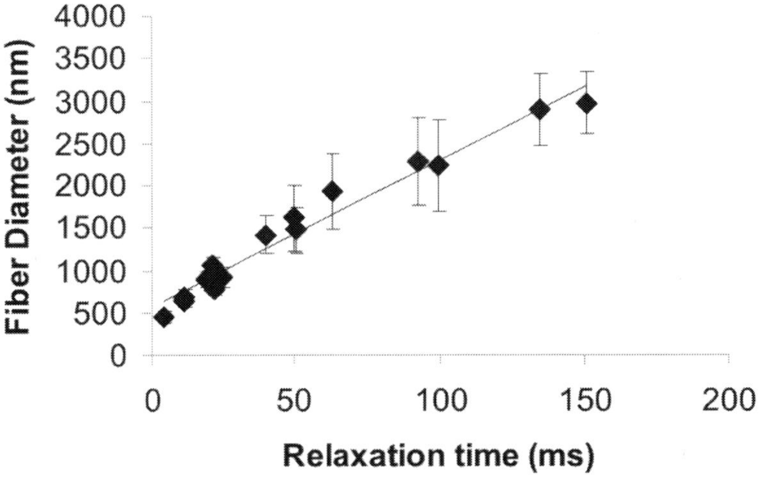

Figure 4: Fiber diameter vs. relaxation time for SKL:PEO solutions

Thermal Processing:

To prepare carbon fibers from lignin, an oxidative thermostabilization stage is necessary to prevent fibers from fusing during carbonization.[4] In previous work a slow heating rate was required to prevent fusion of melt-spun hardwood Kraft lignin (HKL) fibers.[4] The necessity of using a slow heating rate can be considered a disadvantage since it increases the time and energy required to generate carbon fibers. Due to its different chemical structure, SKL is expected to be more thermally stable.[4] However, previous attempts at melt spinning SKL were unsuccessful due to the tendency of SKL to char, so the thermostabilization of SKL fibers has not been reported in the literature. An SEM of a sample carbonized with and without thermostabilization is shown in Figure 5, showing fusion of fibers on the left and non-fused fibers on the right. To investigate thermostabilization of SKL fibers, we subjected our electrospun fabrics to different heating rates and final heating temperatures. It was observed that in contrast to melt-spun fibers or non-thermostabilized samples, none of the samples fused, indicating good stability. These samples were therefore suitable for conversion to carbon nanofibers. An SEM of carbonized 99% SKL fibers is also shown in Figure 5.

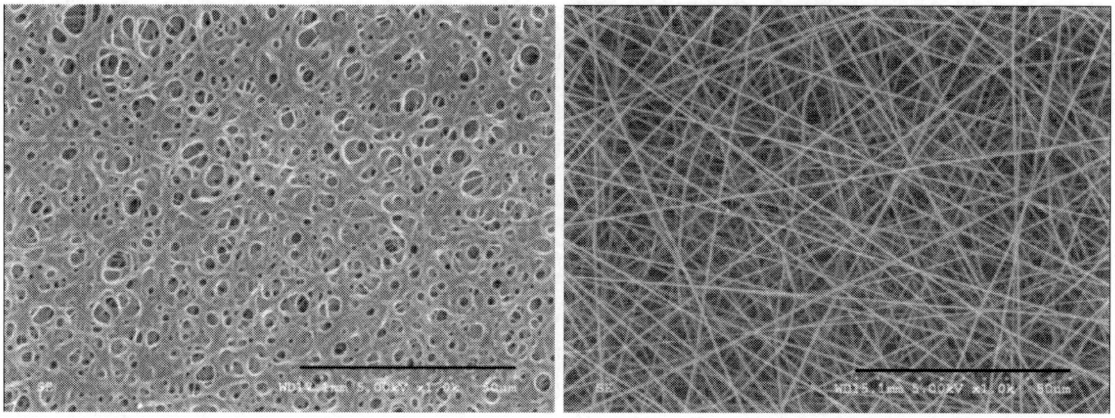

Figure 5: Carbonized 99% SKL fibers without thermostabilization (left), and after thermostabilization at 5°C/min to 250°C (right). Scale bar = 50 μm

Tensile testing:

To investigate the mechanical properties of SKL nonwoven mats, samples thermostabilized under different heating rates and temperatures but with similar diameters on the order of 1 μm were subjected to microtensile testing. The modulus and tensile strength values for different heating rates and thermostabilization temperatures are shown in Table 2. A steady increase in tensile strength was observed as the maximum treatment temperature increased from roughly 2-4 MPa for fibers heated to 150°C to values of more than 20 MPa for samples heated to 250°C. A similar trend was observed for elastic modulus, which increased from roughly 300 MPa to values of 1 GPa for fibers treated at 150 and 250°C, respectively. Interestingly, fibers heated at the lowest heating rate had a lower tensile strength, while comparable tensile strength was observed for samples heated at 5°C/min or faster. This could indicate that longer heating times lead to degradation of the mechanical properties of SKL fibers. The results presented here demonstrate the feasibility of generating precursors for carbon nanofibers from SKL by electrospinning.

Table 2: Mechanical properties of thermostabilized SKL:PEO nonwoven fabrics

Heating rate (°C/min)	T_{max} (°C)	Tensile strength (MPa)	Standard Deviation (MPa)	Elastic Modulus (MPa)	Standard Deviation (MPa)
As-spun	n/a	4.0	1.1	327.6	67.0
0.5	150	2.9	1.0	324.8	147.3
0.5	200	12.4	3.4	938.9	199.0
0.5	250	14.5	2.9	818.2	160.1
5.0	150	1.7	0.6	231.7	117.1
5.0	200	13.3	3.0	966.1	147.0
5.0	250	24.0	5.9	1055.9	269.5
47.0	150	3.7	1.0	322.6	121.6
41.0	200	10.7	1.9	849.4	101.6
39.0	250	22.1	3.7	1131.4	149.9

Summary:

The results from rheological experiments have demonstrated for the first time that while SKL solutions without PEO behave like Newtonian fluids while undergoing extensional flow, SKL:PEO solutions display non-Newtonian elastic behavior. We have demonstrated that SKL:PEO solution viscoelasticity strongly influences the fiber diameter and the width of the diameter distribution using shear rheometry and CaBER coupled with electrospinning experiments. By varying the SKL concentration, PEO concentration, and PEO molecular weight, we have demonstrated the ability to vary the viscoelastic properties over a wide range with less than 1% PEO added to the spinning solution (relative to SKL mass). With a sufficient understanding of SKL:PEO solution viscoelasticity, we then carried out thermostabilization experiments and measured the tensile properties of nonwoven SKL fabrics as a function of temperature and heating rate. In contrast with some previous reports on melt-spun fibers produced from HKL, electrospun SKL fibers did not fuse under any of the thermostabilization conditions tested, whereas carbonization performed without thermostabilization resulted in fiber fusion. In future work we hope to further optimize the process of lignin electrospinning to move toward toward carbon nanofiber precursors with properties suitable for real-world applications.

Acknowledgements

We would like to acknowledge North Carolina State University, the U.S. Department of Agriculture, and the Natural Sciences and Engineering Research Council of Canada (NSERC) for supporting this research.

References

1) Boerjan, W., Ralph, J., Baucher, M. (2003) Lignin Biosynthesis. *Annu. Rev. Plant Biol.* 54; 519-546
2) Ralph, J., Lundquist, K., Brunow, G., Lu, F., Kim, H., Schatz, P.F. Marita, J.M., Hatfield, R.D., Ralph, S.A., Christensen, J.H., Boerjan, W. (2004) Lignins: Natural polymers from oxidative coupling of 4-hydroxyphenylpropanoids. *Phytochemistry Reviews* 3; 29-60

3) Brodin, I., Sjoholm, E., Gellerstedt, G. (2009) Kraft lignin as feedstock for chemical products: The effects of membrane filtration. *Holzforschung*: 63, 290-297.
4) Kadla, J.F., Kubo, S., Venditti, R.A., Gilbert, R.D., Compere, A.L. (2002) Lignin-based carbon fibers for composite fiber applications. *Carbon* 40: 2913–2920
5) Morgan, P (2005). Carbon fibers and their composites Taylor and Francis, Boca Raton
6) Im, J.S., Park, S.J., Kim, T., Lee, Y-S. (2009) Hydrogen storage evaluation based on investigations of the catalytic properties of metal/metal oxides in electrospun carbon fibers International Journal of Hydrogen Energy; 34: 3382-3388
7) Osmond, N.M. (2000) Activated carbon fibre adsorbent materials. *Adsorption science and technology* 18, 6, 529-539
8) Sharma, C.S., Katepalli, H., Sharma, A., Madou, M. (2011) Fabrication and electrical conductivity of suspended carbon nanofiber arrays. *Carbon*; 49: 1727-1732
9) Aoki, K., Usui, Y., Narita, N., Ogiwara, N., Iashigaki, N., Nakamura, K., Kato, H., Sano, K., Ogiwara, N., Kametani, K., Kim, C., Taruta, S., Kim, Y.A., Endo, M., Saito, N. (2009) A thin carbon-fiber web as a scaffold for bone-tissue regeneration. *Small* 5: 13, 1540-1546.
10) Kubo S, Ishikawa N, Uraki Y, Sano Y. (1997) Preparation of lignin fibers from softwood acetic acid lignin—Relationship between fusibility and the chemical structure of lignin. *Mokuzai Gakkaishi* 43, 8, 655–662.
11) Dave,V., Prasad, A., Marand, H., Glasser, W. (1993) Molecular organization of lignin during carbonization. *Polymer* 34, 15, 3144-3154.
12) http://www.ornl.gov/ornlhome/news_items/news_100920.shtml
13) Kubo, S., Kadla, J.F. (2005a) Lignin-based Carbon Fibers: Effect of Synthetic Polymer Blending on Fiber Properties. *Journal of Polymers and the Environment*, Vol. 13, No. 2. 97-105.
14) Kubo, S., Yoshida, T., Kadla, J.F. (2007) Surface Porosity of Lignin/PP Blend Carbon Fibers. *J. Wood Chemistry and Technology* 27; 257-271.
15) Kubo, S., Kadla, J.F. (2005b) Kraft Lignin/Poly(ethylene oxide) Blends: Effect of Lignin Structure on Miscibility and Hydrogen Bonding. *Journal of Applied Polymer Science* 98; 1437–1444
16) Kadla, J.F., Kubo, S. (2003) Miscibility and Hydrogen Bonding in Blends of Poly(ethylene oxide) and Kraft Lignin. *Macromolecules 36:* 7803-7811
17) Kadla, J.F., Kubo, S. (2004) Lignin-based polymer blends: analysis of intermolecular interactions in lignin-synthetic polymer polymer blends. *Composites, Part A: Applied Science and Manufacturing* 35; 395-400
18) Kubo, S., Kadla, J.F. (2006) Effect of poly(ethylene oxide) molecular mass on miscibility and hydrogen bonding with lignin. *Holzforschung*. 60; 245–252.
19) Tagawa, T., Miyata, T. (1997). Size effect on tensile strength of carbon fibers. *Materials Science and Engineering* A238, 336–342
20) Tanaka, T., Nakayama, H., Sakaida, A., Horikawa, N., (1999) Estimation of tensile strength distribution for carbon fiber with diameter variation along fiber. *Materials Research Science International* 5, 2, 90-97.
21) Rutledge, G.C., Fridrikh, S.V. (2007) Formation of fibers by electrospinning. *Advanced Drug Delivery Reviews* 59, 1384-1391.
22) Reneker, D.H., Yarin, A.L. (2008) Electrospinning jets and polymer nanofibers. *Polymer* 49, 2387-2425

23) Greiner, A., Wendorff, J.H. (2007) Electrospinning: A fascinating method for the preparation of ultrathin fibers. *Angew. Chem. Int. Ed.* 46, 5670-5703.
24) Joshi, P., Zhang, L., Chen, Q., Galipeau, D., Fong, H., Qiao, Q. (2010) Electrospun Carbon Nanofibers as Low-Cost Counter Electrode for Dye-Sensitized Solar Cells. *ACS Applied Materials and Interfaces*; 2: 12, 3572-3577.
25) Arshad, S.N., Naraghi, M., Chasiostis, I. (2011) Strong carbon nanofibers from electrospun polyacrylonitrile. *Carbon*; 49: 1710-1719
26) Moon, S., Farris, R. (2009) Strong electrospun nanometer-diameter polyacrylonitrile carbon fiber yarns. *Carbon* 47: 2829-2839
27) Imaizumi, S., Matsumoto, H., Suzuki, K., Minagawa, M., Kimura, M., Tanioka, A. J. (2009) Phenolic Resin-Based Carbon Thin Fibers Prepared by Electrospinning: Additive Effects of Poly(vinyl butyral) and Electrolytes. *Polymer Journal* 41: 12, 1124-1128.
28) Park, S.H., Kim, C., Yang, K.S. (2004) Preparation of carbonized fiber web from electrospinning of isotropic pitch. *Synthetic Metals* 143: 175-179.
29) Lallave, M., Bedia, J., Ruiz-Rosas, R., Rodríguez-Mirasol, J., Cordero, T., Otero, J.C., Marquez, M., Barrero, A., and Loscertales, I.G. (2007) Filled and Hollow Carbon Nanofibers by Coaxial Electrospinning of Alcell Lignin without Binder Polymers. *Advanced Materials* 19; 4292-4296
30) Ruiz-Rosas, R., Bedia, J., Lallave, M., Loscertales, I.G., Barrero, A., Rodrı´guez-Mirasol, J., Cordero, T. The production of submicron diameter carbon fibers by the electrospinning of lignin. *Carbon* 48 (2010) 696–705
31) Seo, D.K., Jeun, J.P., Kim, H.B., Kang, P.H. (2011) Preparation and characterization of the carbon nanofiber mat produced from electrospun PAN/lignin precursors by electron beam irradiation. *Rev. Mater. Sci.*; 28: 31-34.
32) Dallmeyer, I., Ko, F., Kadla, J.F. (2010) Electrospinning of technical lignins for the production of fibrous networks. *Journal of Wood Chemistry and Technology* 30: 315-329.
33) Morck, R., Yoshida, H., Kringstad, K.P., Hatakeyama, H. (1986) Fractionation of Kraft lignin by successive extraction with organic solvents. *Holzforschung* 40(suppl.): 51-60.
34) Rodd, L.E., Scott, T.P. Cooper-White, J.J. McKinley, G.H. (2005) Capillary breakup rheometry of low viscosity elastic liquids. *Applied Rheology* 15: 12-27.
35) Gandhi,M., Yang,H., Shor, L., Ko, F. (2009) Post-spinning modification of electrospun nanofiber nanocomposite from Bombyx mori silk and carbon nanotubes. *Polymer* 50: 1918-1924.
36) Gupta, P., Elkins, C., Long, T.E., Wilkes, G.L. Electrospinning of linear homopolymers of poly(methyl methacrylate): exploring relationships between fiber formation, viscosity, molecular weight and concentration in a good solvent. *Polymer* 46 (2005) 4799–4810
37) McKee, M.G., Wilkes, G.L., Colby, R.H. Long, T.E. (2004) Correlations of solution rheology with electrospun fiber formation of linear and branched polyesters. *Macromolecules* 37, 1760-1767
38) Shenoy, S.L., Bates, W.D., Frisch, H.L., Wnek, G.E. (2005). Role of chain entanglements on fiber formation during electrospinning of polymer solutions: good solvent, non-specific polymer–polymer interaction limit. *Polymer* 46, 3372–3384
39) Feng, J.J. (2002) The stretching of an electrified non-Newtonian jet: A model for electrospinning. *Physics of Fluids* 14, 11, 3912-3926.
40) Feng, J.J. (2003) Stretching of a straight electrically charged viscoelastic jet. *J. Non-Newtonian Fluid Mechanics* 116, 55-70.
41) Carroll, C.P. Joo, Y.L. (2006) Electrospinning of viscoelastic Boger fluids: modeling and experiments. *Physics of Fluids* 18, 053102-1-14.

42) Yu, J.H., Fridrikh, S.V., Rutledge, G.C. (2006). The role of elasticity in the formation of electrospun fibers. *Polymer* 47, 4789–4797
43) Helgeson, M.E., Grammatikos, K.N., Deitzel, J.M. Wagner, N.J. (2008) Theory and kinematic measurements of the mechanics of stable electrospun polymer jets. *Polymer* 49, 2924-2936
44) Entov, V.M., Hinch, E.J. (1997) Effect of a spectrum of relaxation times on the capillary thinning of a filament of elastic liquid. *J. Non-Newtonian Fluid Mechanics* 72, 31-53.
45) Clasen, C., Plog, J.P., Kulicke, W.-M., Owens, M., Macosko, C., Scriven, L.E., Verani, M., McKinley, G.H. (2006) How dilute are dilute solutions in extensional flows? *Rheology* 50: 6, 849-881.

(499c) Effect of Lignin Obtained From Ionic Liquid Pretreated Poplar Wood On the Thermomechanical Properties of Polypropylene

Srikanth Pilla, Wisconsin Institute for Discovery, University of Wisconsin-Madison, Milwaukee, WI, Anantharam P. Dadi, SuGanit Systems, Toledo, OH, Balakrishna Maddi, Chemical Engineering, University of Toledo, Toledo, OH, Craig Clemons, USDA-Forest Products Laboratory and Joseph Lawrence, Bioengineering, University of Toledo, Toledo, OH

PP-Lignin Composites

See more of this Session: Biobased Materials II: Lignin-Based Materials
See more of this Group/Topical: Forest and Plant Bioproducts Division

(499d) Development of Nylon Biocomposites Through the Torrefaction of Waste Stream Agricultural by-Products

Jessica L. Lattimer and Chad A. Ulven, Mechanical Engineering, North Dakota State University, Fargo, ND

In the last several years the use of biobased materials for fillers in thermoplastics has seen a remarkable increase. The desire to produce lighter, stronger, and more ecologically friendly materials being key driving forces behind the movement. While the focus has thus far been on producing biobased composites out of commodity polyolefins or bio-derived resins, little work has been done in the realm of engineering thermoplastics. For engineering thermoplastics, the increased processing temperatures are unfavorable for biocomposite production, causing significant damage to the biobased fillers. As the automotive industry is one of the largest consumers of nylon and the ever increasing push to make cars green, there is a natural fit for research into producing biobased nylon composites. Finding a way to drive off the constituents within the biomass that degrade above the traditional processing temperatures of polyolefins would provide a new biomass that could withstand the processing of engineering thermoplastics.

From previous studies into the use of biobased materials for energy production it is known that torrefaction, conducted between 220-300 °C, will drive off the hemicellulose, fats, waxes, and other constituents yielding a biomass consisting of mostly cellulose, degrading between 300-375 °C, and lignin, degrading slowly over 250-500 °C. As cellulose is the constituent in biomass that acts as a reinforcing agent in the biocomposite, this is a potential method to prepare biobased fillers for introduction in to engineering thermoplastics. This study describes the development of nylon based biocomposites using torrefied sunflower hulls (TSFH), oat hulls (TOH), and flax shive (TFS).

To prepare the biocomposites, nylon was blended with each of the torrefied biomasses analyzed based on weight percentage. This study looks at three grades of the biocomposite; neat nylon-6, 16 wt% filler in nylon-6, and 20 wt% filler in nylon-6. The biocomposite samples prepared were tested for static and dynamic properties in addition to characterizing their resistance to moisture uptake as a function of torrefied biomass loading. Correlations between biomass input and loading are discussed.

See more of this Session: Biobased Materials II: Lignin-Based Materials
See more of this Group/Topical: Forest and Plant Bioproducts Division

(499e) Lignin Based BioFoam Composites From Functionalized Soy Oil Based Biopolyurethane

Manju Misra, School of Engineering and the Department of Plant Agriculture, University of Guelph, Guelph, ON, Canada

uploading in progress

See more of this Session: Biobased Materials II: Lignin-Based Materials
See more of this Group/Topical: Forest and Plant Bioproducts Division

(499f) Lignin-Based Carbonaceous Nano-Fibrous Felts

Chuilin Lai[1], Lifeng Zhang[1], Hao Fong[1] and **Lew P. Christopher**[2], (1)Chemistry, South Dakota School of Mines & Technology, Rapid City, SD, (2)Center for Bioprocessing Research and Development and Department of Chemical and Biological Engineering, South Dakota School of Mines & Technology, Rapid City, SD

Lignin is a naturally existing biopolymer widely found in lignocellulosic plant biomass at an average of 20-30%. It is also a high-volume byproduct of wood pulping which exceeds 15 million tons per year in the US. Nonetheless, lignin has been utilized mainly as a low-grade fuel for production of heat and power to date. There is however a growing demand to identify applications with high economic value for lignin. Although production of carbon fibers from lignin has been reported in literature, the low mechanical properties obtained so far have prevented their commercial use in reinforcement and/or composites applications. Nano-fibrous felts consisting of porous carbon nanofibers derived from lignin, however, have potential applications that do not rely on mechanical properties but on large specific surface area. The use of low-cost lignin as the precursor to develop carbonaceous nano-fibrous felts is anticipated to have a great impact on these applications. The nanomaterials processing technique of electrospinning provides a straightforward approach for convenient preparation of fibers with diameters ranging from tens to hundreds of nanometers. In this research, innovative carbonaceous nano-fibrous felts with high specific surface area were developed by electrospinning aqueous mixtures of alkali lignin and poly(vinyl alcohol) with varied mass ratios followed by heat treatments for stabilization and carbonization. The developed carbonaceous nano-fibrous felts had a large pore volume and specific surface area and are therefore expected to find important applications as catalyst support, supercapacitor electrode, and hydrogen storage materials.

See more of this Session: Biobased Materials II: Lignin-Based Materials
See more of this Group/Topical: Forest and Plant Bioproducts Division

(554a) Integration of a Membrane Based Extraction Process Into a Lignocellulosic Biorefinery

David Grzenia, Chemical and Biological Engineering, Colorado State University, Fort Collins, CO, Dan Schell, National Renewable Energy Laboratory, Golden, CO and **S. Ranil Wickramasinghe**, Ralph E Martin Department of Chemical Engineering, University of Arkansas, Fayetteville, AR

In the future, it is anticipated that multi-product biorefineries will be used to convert lignocellulosic biomass to fuels and chemicals, in which, it may be possible to use membrane-based separations to recover fuels and various products and to improve energy efficiency. To biochemically convert biomass a pretreatment step is required to increase the susceptibility of the biomass to enzymatic cellulose hydrolysis. However, many pretreatment processes release not only fermentable sugars but other compounds that are toxic to the fermenting microorganism. In this study, we investigated the feasibility of using membrane extraction for removal of acetic acid, sulfuric acid, furfural, HMF and other toxic compounds from dilute acid pretreated corn stover to reduce its toxicity. Acetic acid in its protonated form was extracted from the liquor portion of a corn stover hydrolysate across a membrane into an organic phase consisting of oleyl alcohol and Alamine 336, the latter a tertiary amine containing aliphatic chains of 8-10 carbon atoms. Other toxic compounds such as HMF and furfural are also co-extracted into the organic phase.

The sugars in hydrolysates that were detoxified using our membrane extraction process as well as conventional ammonium hydroxide detoxification, were fermented to ethanol by glucose-xylose fermenting *Zymomonas mobilis* 8b. The best ethanol yields and rates were achieved for hydrolysates using membrane extraction with oleyl alcohol as the organic phase followed by ammonium hydroxide conditioned hydrolysates. Hydrolysates that were detoxified using membrane extraction with octanol as the organic phase were unfermentable because highly toxic octanol was found in the hydrolysate. Our results indicate that membrane extraction may be an effective membrane detoxification technology.

See more of this Session: Separation Processes In Biorefineries I: Membrane and Chromatographic Separations
See more of this Group/Topical: International Congress on Energy 2011

Title: Modified Anopore^{TM} membranes for the microfiltration of cellulose hydrolysis mixtures

Yong-Hong Zhao and Daniel F. Shantz

Artie McFerrin Department of Chemical Engineering, Texas A&M University, College Station, TX 77843-3122, United States

ABSTRACT:

Three classes of nonfouling substances, i.e., poly[poly(ethylene glycol) methyl ether methacrylate] (POEM), poly(sulfobetaine methacrylate) (PSBMA), and perfluoroalkyl tagged poly(ethylene glycol) surfactants (fPEG), were covalently tethered onto the surface of Anopore^{TM} membranes to form a series of organic-inorganic hybrid membranes. These membranes were then used for the microfiltration of cellulose hydrolysis mixtures. It was found that the fluxes of the three synthesized hybrid membranes could be largely recovered by washing with methanol and water after the microfiltration of cellulose hydrolysis mixtures, showing much better antifouling properties than the bare Anopore^{TM} membrane. Furthermore, the stability of the membranes in acidic environment was enhanced significantly as compared to the unmodified membranes when the membrane surface was grafted with a thick enough polymer layer. The synthesized membranes were also characterized by scanning electron microscopy (SEM), Fourier transform infrared spectroscopy (FTIR) and X-ray photoelectron spectroscopy (XPS), which were used to understand the morphology and chemical compositions of the membrane surfaces, and the results of these methods correlate with the microfiltration data presented. Mathematical modeling of the microfiltration data indicates that the pore constriction model best describes the fouling behavior of the bare Anopore^{TM} membrane, while the cake filtration model fits the data most satisfactorily for the three groups of hybrid membranes. The current work presents an alternative approach to membrane design for biomass related separations.

Keywords: hybrid membranes; biomass separation; antifouling; surface modification

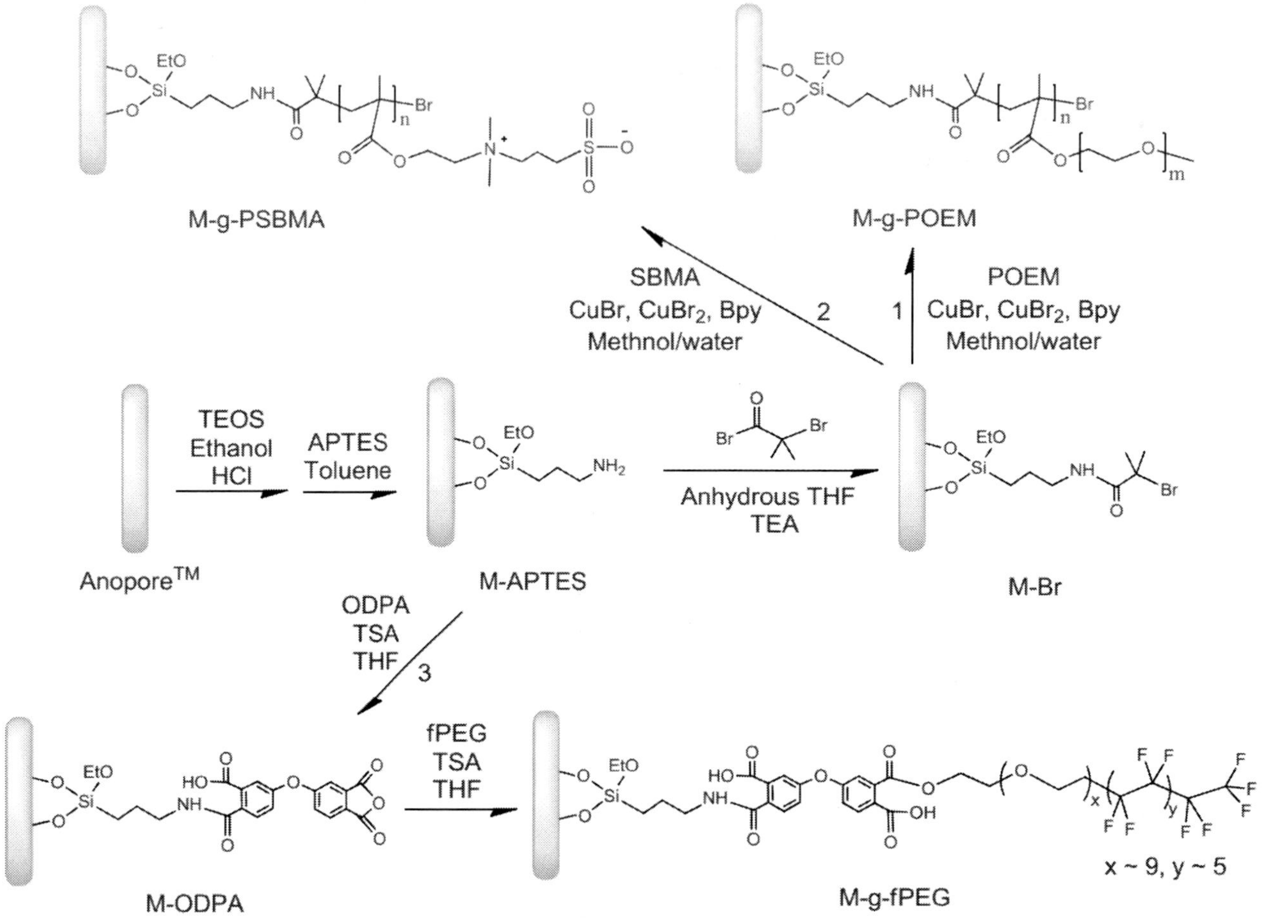

Surface modification protocols for forming hybrid membranes

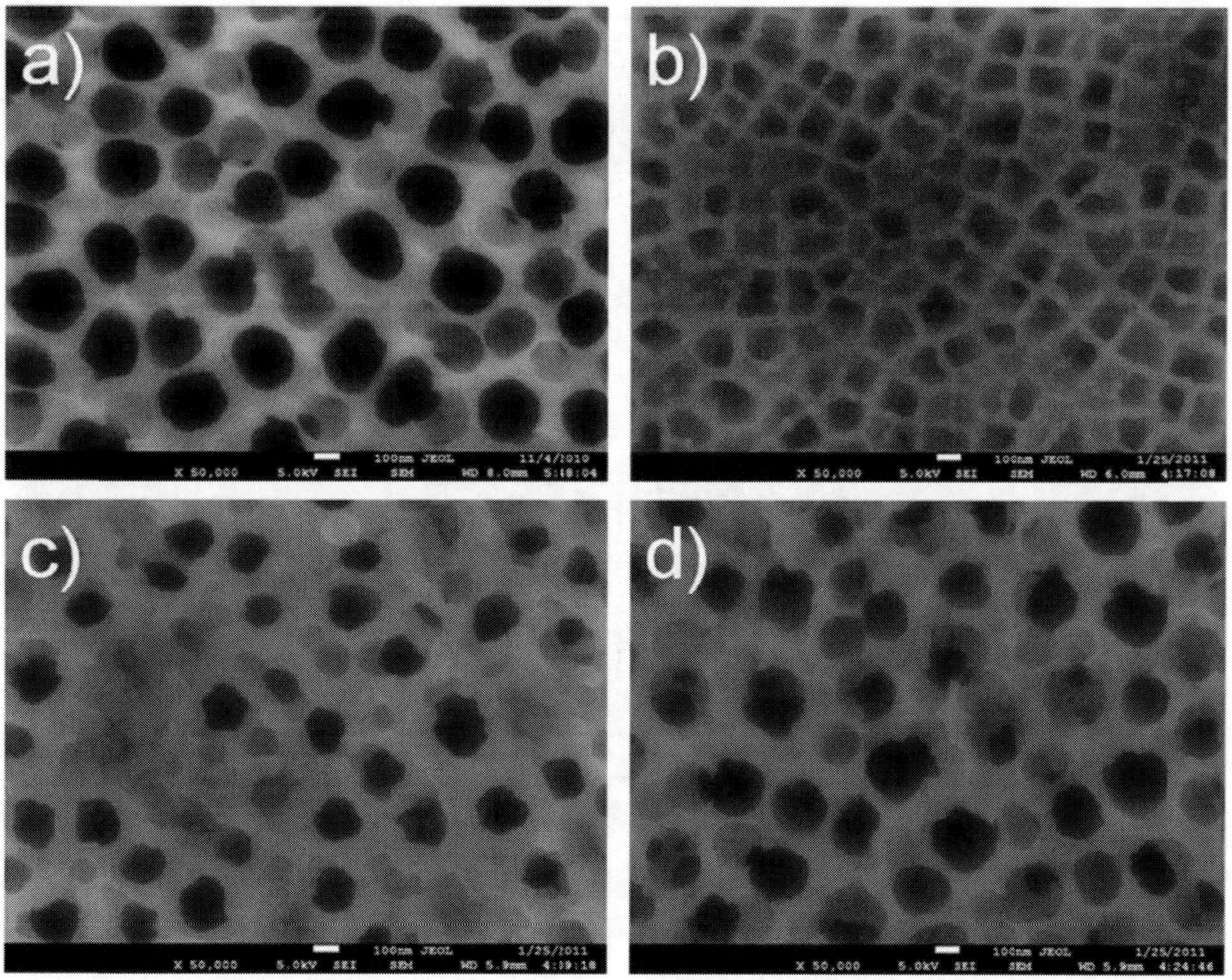

SEM images for a) bare Anopore^{TM} membrane; b) M-g-POEM-24h;

c) M-g-PSBMA-24h and d) M-g-fPEG.

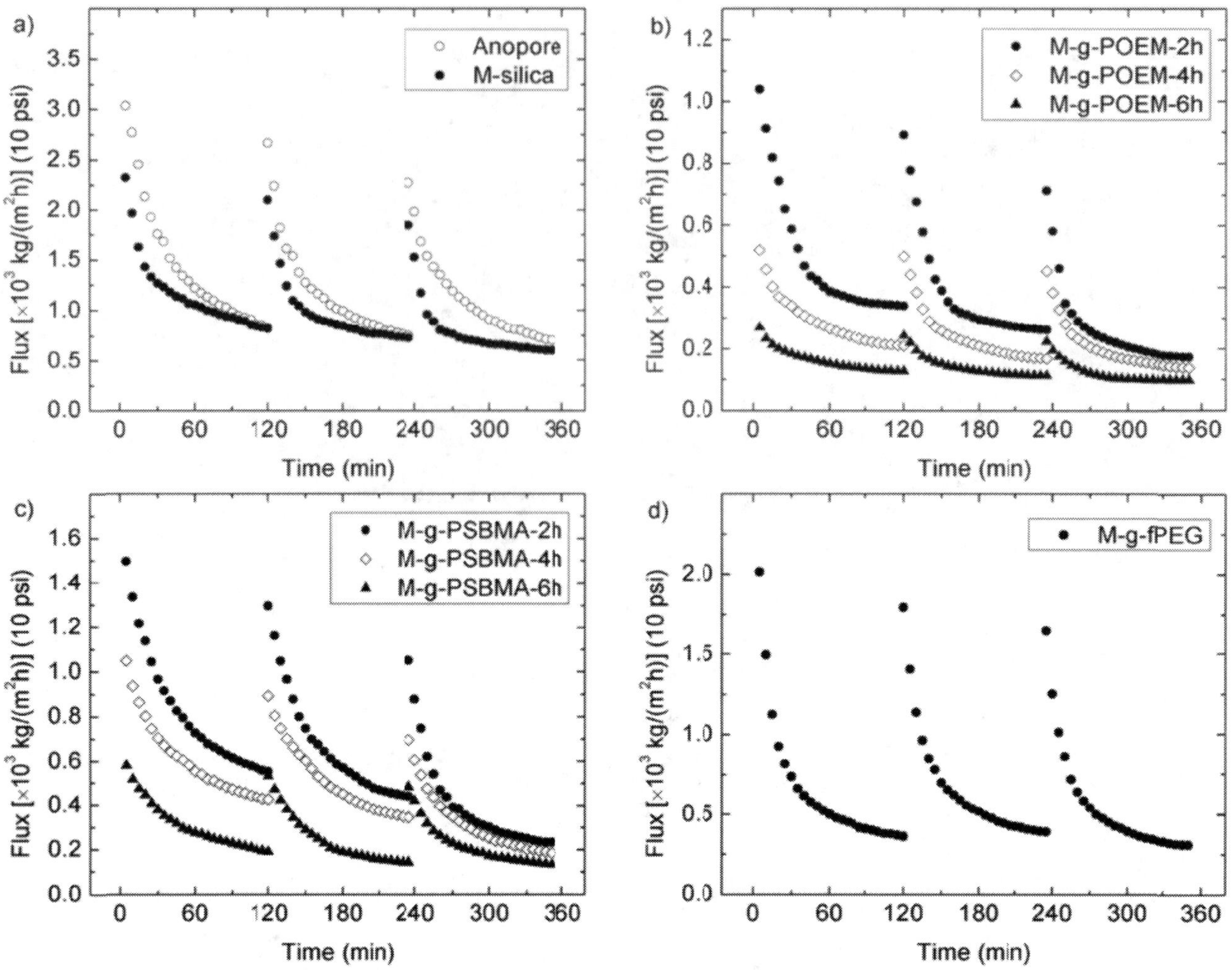

Flux measurement for a) Anopore™ and M-silica; b) M-g-POEM; c) M-g-PSBMA and d) M-g-fPEG. The solution is cellulose hydrolysis mixtures at pH 7.

(554c) Polymer Enhanced Microfiltration of Lignocellulosic Hydrolyzates

Rakesh Yasarla, State University of New York, Syracuse, NY and **Bandaru V. Ramarao**, Paper and Bioprocess Engineering, Empire State Paper Research Institute, Syracuse, NY

Lignocellulosic hydrolyzates produced by hydrothermal treatment of wood chips contain lignin and carbohydrates in colloidal forms that can interfere with downstream processing. Separation of such materials is necessary to improve the fermentability of sugars into biofuels and other bioproducts. Flocculation of such hydrolyzates prior to microfiltration improves their filterability. The effect of prior flocculation on filtration resistance, steady state fluxes and the influence of solid concentration and temperature are discussed.

See more of this Session: Separation Processes In Biorefineries I: Membrane and Chromatographic Separations
See more of this Group/Topical: International Congress on Energy 2011

(554d) Acetic Acid Removal From the Hydrolysate Using Reactive Extraction In a Membrane Extractor/Reactor

Rahul Patil[1], Glenn Lipscomb[1] and Frederick F. Stewart[2], (1)Department of Chemical and Environmental Engineering, University of Toledo, Toledo, OH, (2)Interfacial Chemistry, Idaho National Laboratory, Idaho Falls, ID

Production of bioethanol from lignocellulosic biomass requires pretreatment of the biomass to hydrolyze hemicellulose, separate lignin and reduce cellulose crystallinity. During the pretreatment by dilute acid hydrolysis, several grams of acetic acid per liter of hydrolysate is generated. Acetic acid needs to be removed from the hydrolysate since it can dramatically reduce the efficiency of the subsequent fermentation step by inhibiting enzyme activity. Lime treatment, the current method of hydrolysate detoxification, consumes significant quantities of calcium hydroxide, involves solids handling and does not produce any valuable products.

We propose to remove the acetic acid with an alternative process whereby acetic acid is extracted into 1-octanol in a membrane extractor and reacted with 1-octanol *in situ* using a membrane containing an embedded catalyst to yield 1-octyl acetate. Acetic acid concentration in the hydrolysate will be reduced from 12.5 g/L to < 1 g/L. The product ester will be removed periodically from the organic phase. The product has very good solvent properties and can be sold to improve process economics.

In our study, we prepare membranes with a suitable catalyst (initial efforts utilize strong acid ion exchange resins) embedded in the membranes to achieve extraction and reaction in a single step. The effects of catalyst concentration, operating temperature and flow rates are investigated to optimize the performance of the process. Thus, the membrane extractor/reactor removes acetic acid from the hydrolysate and produces a valuable by-product in an environmentally friendly process.

See more of this Session: Separation Processes In Biorefineries I: Membrane and Chromatographic Separations
See more of this Group/Topical: International Congress on Energy 2011

(554e) Evaluation of Membrane Separations Opportunities within Lignocellulosic Biorefining Processes: Tailoring Membrane Properties

Jennifer M. Leberknight[1], Amitkumar Gautam[2] and Todd J. Menkhaus[2], (1)Chemical and Biological Engineering, South Dakota School of Mines and Technology, Rapid CIty, SD, (2)Chemical and Biological Engineering, South Dakota School of Mines and Technology, Rapid City, SD

Evaluation of Membrane Separations Opportunitieswithin Lignocellulosic BiorefiningProcesses: Tailoring Membrane Properties

Jennifer Leberknight, Amitkumar Gautam, and Todd J. Menkhaus

South Dakota School of Mines and Technology

Department of Chemical and Biological Engineering

501 East Saint Joseph Street

Rapid City, SD 57701

Tel: (605)-394-2422

Email: Todd.Menkhaus@sdsmt.edu

Membraneseparations can be integrated into a biorefinery at many locations and offertremendous potential to reduce water usage and energy consumption, decreaseproduction costs, simplify purification operations, and offer opportunities forrecovery of valuable co-products. Here weexplore five locations for integration of membrane technology into the basic biochemicalproduction process for fuels and/or chemicals.The biomass is initially pretreated and enzymatically hydrolyzed toconvert the cellulose and hemicellulose into fermentable sugars, after which asolid-liquid clarification operation to remove solids (primarily lignin) may beemployed (Membrane Separation 1). Theliquid stream can then be further separated to recycle the hydrolysis enzymes (MembraneSeparation 2) and/or to concentrate the sugars by reverse osmosis (Membrane Separation3). Increased fermentation efficiency couldbe achieved by integrating a membrane step after fermentation to separate andrecycle the fermentation organism (Membrane Separation 4), while membraneextraction or pervaporation can be used for *in-situ*product recovery (Membrane Separation 5).Within the various membrane processes, fouling is a major challenge toovercome. We have correlated membranefouling tendencies within the different membrane separation locations tomeasureable properties of the membrane and foulants such as ionic charge,hydrophilicity, size, and surface roughness to determine why membranesfoul. Through a combination of analytical andmodeling approaches we postulate fouling mechanisms and identify methods tomitigate fouling. Finally, simple empiricalmodels using variable concentrations of fouling components and quantitativemembrane properties have been developed to help establish predictive performanceabilities. As detailed examples, bothmicrofiltration (Separations 1 and 4) along with pervaporation of ethanol froma fermentation broth (Separation 5), were evaluated. Both selectivity and fluxmust be considered in new fouling-resistant membrane designs. The unique soluble compounds (phenolics,furans, and organic acids) in the lignocellulosic derived feeds greatly complicatethe fouling and separation, as will be discussed.

See more of this Session: Separation Processes In Biorefineries I: Membrane and Chromatographic Separations
See more of this Group/Topical: International Congress on Energy 2011

(554f) Membrane Filtration of Lignocellulosic Hydrolyzates: Comparison of Crossflow and Dead-End Microfiltration

Bhavin Bhayani, Empire State Paper Research Institute, Syracuse, NY and Bandaru V. Ramarao, Paper and Bioprocess Engineering, Empire State Paper Research Institute, Syracuse, NY

Depleting oil reserves have encouraged the development of technologies to harness the natural resources to sustain the energy needs. Renewable lignocellulosic biomasses have been focus of many studies but advances in the purification and recovering of the products from the product medium has been very limited. Amongst many separation processes membrane separation offers better prospects to simplify purification operations, offers valuable by-products, decrease energy consumption and reduce water requirements. A membrane separation system can be either crossflow or dead end and both have unique advantages and disadvantages. While Cross flow filtration offers higher feed velocity i.e. cross flow velocity and better control over cake layer development, dead end filtration can result in energy savings. The focus of this study is to characterize the flow and the effect of cake layer development under different pressures and flow rate. A detailed comparison of flow characteristics has been studied. It is also found that the temperature of the hydrolyzate influences separation in a very significant manner.

See more of this Session: Separation Processes In Biorefineries I: Membrane and Chromatographic Separations
See more of this Group/Topical: International Congress on Energy 2011

(554g) Effect of Column Temperature On HPLC Separation of Acetic Acid and Levulinic Acid

Rui Xie and Maobing Tu, Auburn University, Auburn, AL

High performance liquid chromatography (HPLC) is a routine method in the analysis of degradation products in the prehydrolysate or the hydrolysis liquor of lignocellulosic biomass. Cation exchange column chromatography with a metal counter ion has been used for the quantitative analysis of carbohydrates with water as the eluent. A cation exchange column with exchangeable hydrogen ion (H^+) has been employed in quantification of organic acids and furan derivatives with acidic water as the mobile phase. In particular, the Aminex cation-exchange HPX-87H column has been developed for determining organic acids, furan derivatives and alcohols. Although this column has been broadly used in biomass analysis, potential issues with co-elution of target analytes have been reported by several groups. Scarlata and Hyman (2010) suggested that Aminex HPX-87H column was not suitable for biomass sugar analysis because xylose, galactose and fructose are co-eluted together, and glucose and mannose are co-eluted together. The co-elution of acetic acid and levulinic acid has not been addressed indicated the value of acetic acid in biomass hydrolysate was not reliable at high severity due to continuous degradation of furfural and HMF to levulinic acid.

As one of the principle components in hemicelluloses, acetic acid always has a noticeable amount in the hydrolysate of lignocellulosic biomass. Characterization of this organic acid becomes even more important during fermentation due to its inhibition to the microorganisms. Levulinic acid, one of the normal degradation compounds of HMF, is present in the hydrolysate at low concentrations. The separation of these two acids could be easily ignored due to their co-elution in the hydrolysate liquor. Moreover, the separation performance of HPLC column gradually deteriorates during its lifetime worsening co-elution of these two acids. Solutions to improve the separation of these organic acids included the use of a dual column system, employment of a capillary electrophoresis to separate and quantify the concentration of acetic acid and levulinic acid from softwood hydrolysate. However, no method has been developed to resolve co-eluting acetic acid and levulinic acid by optimizing the HPLC chromatographic conditions. In the present study, the improvement of HPLC separation of acetic acid and levulinic acid on the Aminex HPX-87H column was investigated by varying column temperature, flow rate, and concentration of sulfuric acid in the mobile phase. Resolution and capacity factors were used as indicators for measuring the effectiveness of separation using pure standard compounds as well as real biomass hydrolysate from woody biomass.

See more of this Session: Separation Processes In Biorefineries I: Membrane and Chromatographic Separations
See more of this Group/Topical: International Congress on Energy 2011

(566a) Fast Responsible Water Stimulated Shape Memory Aerogel From Native Cellulose Nanofibers

Wei Zhang, Ipst, Georgia Institute of Technology, Atlanta, GA, Yaan Zhang, Georgia Institute of Technology, Atlanta, GA and **Yulin Deng**, School of chemical and biomolecular engineering, Georgia Institute of Technology, Atlanta, GA

Recently, shape-memory materials have attracted significant attention from both industrial and academic researchers due to their useful and fascinating functionality. Shape memory materials have great potential for application in sensors, actuators, smart devices, and media recorders. For example, shape-memory materials have been instrumental in significantly improving the quality of diagnostics, treatments, and surgeries.

Water-driven shape-memory material is a relatively new type of materials that was reported recently by Huang et al. The effects of moisture on the polymer glass transition temperature can be utilized for water triggered shape recovery of polyurethane. Moisture controlled shape-memory hydrogels made from crosslinked poly(vinyl alcohol) were also reported. Chitosan based self-expandable biodegradable stents, crosslinked with ethylene glycol diglycidyl ether and blended with glycerol and PEO, were prepared by Chen et al.

In this study, we report a robust aerogel from chemically crosslinked native cellulose nanofibers that show superabsorbency of water and unique fast shape recovery behavior under water actuation. The water-driven shape-memory materials could be potentially used as new biomedical engineering materials.

See more of this Session: Biobased Materials III: Value-Added Coproducts
See more of this Group/Topical: Forest and Plant Bioproducts Division

(566b) Dual Photo and Thermal Polymerizable Monomers Derived From Methyl Esters of Vegetable Oils

Brian Dillman, Chemical and Biochemical Engineering, Univerisity of Iowa, Iowa City, IA
See more of this Session: Biobased Materials III: Value-Added Coproducts
See more of this Group/Topical: Forest and Plant Bioproducts Division

UTILIZATION OF SUGAR INDUSTRY BY-PRODUCTS , THE BAGASSE PITH FOR MANUFACTURE OF FURFURAL

A.K. Ray, K. M. Srinivas, S. Gupta, S. Chattopadhyay, A.K.Tiwari and M. Kumar

DEPARTMENT OF PAPER TECHNOLOGY
INDIAN INSTITUTE OF TECHNOLOGY ROORKEE
SAHARANPUR CAMPUS, SAHARANPUR-247001

ABSTRACT

Most of Indian chemical industries have concentrated so far on their production of chemicals mainly on the petroleum based refinery, while other alternatives like those offered by sugar cane have received little attention over the decades. Time is now ripen to pay more attention to explore the possibilities of manufacturing both low volume valued added chemicals and high volume low cost energy from sugarcane economically through both chemical and biochemical routes in most economic ways. Sugarcane offers number of wastes, besides deteriorated sucrose and cane juice, cane leaves and tops, molasses, filter mud, fusel, starch, molasses, bagasse along with pith and bagacillo. The pith, mainly composed of parenchyma cells, has low fiber content which needs lower activation energy compared to any other pentosan bearing materials. This pith is detrimental, not only for production of paper but also it creates health and fire hazardous problems.

In this present study, utilization of pith is first reviewed both from research and industrial production point of view for manufacture of furfural and bioethanol in India. It is evident that while the production of the former is in matured state in the industry, the later is in preliminary stage, confined only in the research laboratory. Presently it is recommended by many global industries that furfural and methyl –tetrahydrofuran-based biorefinery. A biorefinery that focuses on converting biomass pentosans to furfural, as a precursor to liquid fuels, appears in many ways to be superior to the ethanol-based concept.

Furfural itself is widely used as a solvent in the petroleum refinery, for purification of lubricating oils, plastics and resins, for manufacture of furane resins, for making nylon 6 and nylon 66 (used in the manufacture of parachutes), pharmaceuticals and many others. It is also an important industrial chemical intermediate and can produce wide range of products. Furfural and its derivatives can be manufactured by acid hydrolysis from various raw materials containing cellulose like woods, saw/wood dust, paper dusts, agricultural residues like bagasse, bagasse pith, bagacillo, rice straw, wheat straw, jute stick etc. Even agricultural resources like corn cobs, rice husk, peanut husk, etc. can also been employed. Most of the reported processes use lignocellulosic materials like bagasse, rice straw, barley stalks, zea maize stem pith, rice hull, menthe wastes, oat hulls, cotton seed hull bran, corn stalks, chhestnut etc. However, majority of Indian plants employ either bagasse, or rice husk as raw materials.

The process involves hydrolyzing the pentosans of the agricultural resources by digesting them under pressure using mineral acids like sulphuric, hydrochloric, phosphoric and formic acids in presence of metal salts as catalysts such as aluminium, calcium and zinc chloride, besides, pyridine in a fixed or fluidized bed batch reactors and the product is steam distilled. This is then concentrated by fractional distillation.

An attempt has been made in the present investigations to explore the possibility of getting furfural using bagasse pith as raw material which needs low thermal energy, low reaction time and operates at comparatively low pressure.
The present study has been undertaken to investigate the possibilities of predicting the reactions at low temperature and at atmospheric pressure under the presence of non-polluting catalyst (calcium Acetate) with low heat-input to utilize waste and to produce furfural having no impact on environment.
Experiments were conducted in the laboratory starting from isolation of pith, pith characteristics and production of pentosan and furfural. The effects of raw material ratios, temperature, concentration of reactants, reaction time, catalyst type and concentration were optimized for maximum yield. An attempt is also made to compute activation energy which helps to develop an appropriate kinetic model.

INTRODUCTION:

Furfural is widely used as a solvent in the petroleum refinery, for purification of lubricating oils, for manufacture of furane resins, for making nylon 6 and nylon 66 (used in the manufacture of parachutes), pharmaceuticals and many others. It is also an important industrial chemical intermediate and can produce wide range of products.
As a matter of fact,Furfural, a sister chemical to the increasingly popular hydroxymethylfurfural or HMF molecule, is regaining attention as a biobased alternative for the production of everything from antacids and fertilizers to plastics and paints.
For the most part,thus far,the production of ethanol has been the goal of biorefinery concepts proposed. A biorefinery that focuses on converting biomass pentosans to furfural, as a precursor to liquid fuels , appears in many ways to be superior to the ethanol -based concept.Hydrogenated furfural called 2-methyl tetrahydrofuran (2-MTHF) has the potential to become a high value fuel additive. It is partially oxygenated , giving it a high octane rating , and is cyclic , giving it a low vapour pressure relative to other CS species. Furthermore , it can act as a co-solvent for alcohols such as ethanol and methanol with gasoline.
It has been indicated that furfural and methyl-tetrahydrofuran –based biorefinery can have good potential compared to ethanol based fuel .In the later case more energy is expended in growing , harvesting, transporting and processing the raw material than is recovered in the form of ethanol. Therefore ethanol can never be competitive because the cost of energy increases, ethanol processing costs increase as well.Integration of 2-MTHF production from the hemicelluloses part of biomass, cellulose to alcohol, and lignin to clean heavy fuel or to convert to lignosulphonates can be a promising biorefinery in near future.
It is normally manufactured from various pentosans bearing raw materials of vegetable origin including agricultural resources like rice straw, corn cobs, rice husk, peanut husk, bagasse etc. Higher the amount of pentosans, higher is the production of furfural. Bagasse is one of the major resources in India which can help produce furfural as it contains substantial amount of pentosans. The importance of furfural and its profitable utilization of lignocellulosic wastes has been extensively studied by various investigators(1-23).
Therefore these agricultural wastes can be profitably utilized for development of product as well as abatement of pollution. The basic steps are as follows:

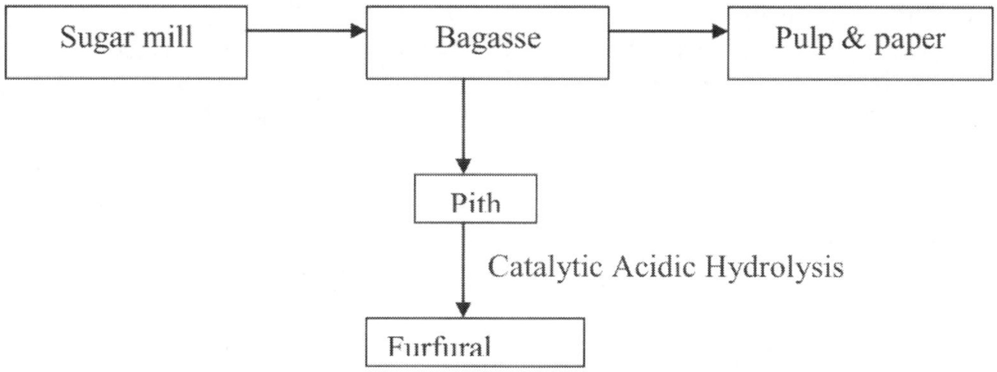

The production of furfural seems to be very much promising in comparison to others. The production of this chemical is advantageously in two folds. Firstly, to produce the useful chemicals which are at present imported from foreign countries by spending enormous amount of foreign exchange. These can be saved if proper economical exploitation is made. Secondly, the paper industry condition at this present contest is extremely precarious due to shortage of raw materials, high cost of energy and unavailability of machinery and various kinds of utilities, restricted labour relations, change of attitude of operating personnels, ecological restriction & problem of pollution control. A complex of paper, sugar and chemicals can easily be established if one follows the above guide lines:

Though bagasse is one of the raw materials for the production of furfural, the portion of bagasse which is more suitable for the production of furfural is the pith. Pith is undesirable constituents of bagasse,typically 20%,with a calorific value of 1500-2000 kCal/kg with a bulk density of 100-150 kg/ m^3 bulk density.Thus, the pith which is separated in the sugar factories and bagasse based paper factories can be advantageously used for the production of furfural and many derivatives of furfural as indicated. No literature or information is available regarding production of furfural for pith obtained from Indian bagasse.

Various chemicals are obtained using the above raw material by hydrolyzing the pentosans present in the presence of catalyst at high temperatures (413 K to 453 K) and at moderately high pressures (344.6 kN/m^2 to 1033.67 kN/m^2) with reaction time 3 to 10 hours followed by steam distillation. The catalysts employed are acids like sulphuric, hydrochloric, phosphoric,HF,CH_3COOH and formic and salts like aluminium, calcium and zinc chloride besides, pyridine in a fixed or fluidized bed batch reactors. This method suffers from disadvantages in that they are mostly batched and requires high heating energy and high consumption of steam to recover furfural from the reactor liquor is abnormally higher, on the order of (20 kg steam/kg of furfural).

An attempt has been made in the present investigation to explore the possibility of producing furfural in bench scale experiment using a suitable agricultural raw material which needs low thermal energy, low residence time and at comparatively low pressure. The raw material used was bagasse pith mainly composed of parenchyma cells, has low fibre content which needs lower activation energy compared to any other pentosan bearing materials. This bagasse pith is detrimental, not only for paper making point of view but also it creates health and fire hazardous problems in both the industries. Besides, this furfural draws a considerable amount of foreign exchange by import. This can be saved if proper economic exploitation is made. The previous investigation is normally directed towards reaction at high temperature & pressure and use of catalyst which creates pollution. The reaction also needs high level of thermal energy. No attempt has been made to produce furfural from bagasse pith at catalytic reaction conditions occurring at low temperature and pressure. The present study has been undertaken to investigate the possibilities of predicting the reactions at low temperature and at atmospheric pressure under the presence of non-polluting catalyst with low heat-input to utilize waste and to produce furfural having no impact on environment.

Current status and Economics:

Every year , more than 200 million tonnes of bagasse (based on 600 million tonnes of world production of sugar cane) are obtained are obtained together with sugar, in all cane producing countries , 95 percent of which is used as fuel in the mills , which represent a saving of about 40 million tonnes of oil.Generally it is assumed that 1 tonne of mill wet bagasse(of 50% moisture) can be replaced by 0.173 tonne of fuel oil or 0.263 tonne of bituminous coal. In practice, about 25 tons of mill-wet bagasse (12.5 tons of bone dry bagasse) are required to produce 1 ton of furfural of 99.0% with 35 tons steam requirements.Global production is reported of the order of 450,000 ton at a stable price of 1,000 $ per ton. The US consumption of furfural is only about 40 million lb/year. China is the biggest supplier of this product with about half of global capacity. In India there is only one plant known as Southern Agri-furane Ltd. in Tamil Nadu to produce about 3000T of furfural from bagasse. Capital cost for a 5000 tonnes /yearly plant , generally considered as the minimum economic capacity, would be about US$ 9 million and the production cost about US$ 450 per tonne of furfural. Furfuryl alcohol is produced by the catalytic hydrogenation of furfural.Starting from bagasse , a plant to produce 4500 tonnes yearly of furfuryl alcohol would cost US$ 12 million to US$ 13 million and would require some 150,000 tonnes of mill-wet bagasse. The production cost would be about US$ 1250 per tonne of furfuryl alcohol.

PREVIOUS WORK:

Profitable utilization of waste food grains and ligno-cellulose waste including agri-residues, to produce value added chemicals like furfural has been extensively studied by a number of investigators, using various processes and operational conditions(1-23). Rao (1) emphasized the use of pith for production of furfural but no information is available regarding optimal operational conditions. Bains etal. (2) had investigated experimentally using rice straw, corn cobs and mentha waste under various pressures, temperatures and, HCl /H_2SO_4 concentrations. Maximum yield of 6.5% is obtained from rice straw at a temperature of 150°C, 2.5 hrs with 1.6% H_2SO_4 and 1:4 bath ratio. Further, increase of temperature up to 180°C with 2 hrs time yielded only 1% more furfural. Corncobs produced almost the same level of yields, 6.8 percent and 7.2 percent respectively. Azeotropic distillation of agro-waste with material: salt ratio 1:1.2 gave expt. yield as 34% of O.D. material whereas actual yield were obtained for rice straw, corncobs and mentha waste of the order of 8.7, 12.8 and 6.2 percent respectively. Mittal and Biswas (3) used Zea Mays stem pith, sugarcane bagasse and saw dusts. Studies at lower temperatures (293K,303K) are also reported by Trivedi and Bhattacharya (4) for removal of pentosans using both HCl and H_2SO_4 with varying concentrations ranging 1% -30%. Maximum 50% removal of pentosans at 25% HCl has been made possible.Further it is found that H_2SO_4 is more effective than HCl in removing pentosans. Jacob et al. (5) adopted corn cobs, oat hulls, & bagasse under digester pressure at temperature of 403 K-453K with acid and found maximum theoretical yield 20% of oat hull. Ganasekharan and Gananasekaran (6) have investigated the formation of furfural from rice hull in a fixed bed reactor under influence of catalyst (calcium chloride) concentration 1-10% in the temperature range of 433 K-493 K using N_2 gas as a carrier. The results of furfural yield at pressure range 345-1034 kN/m^2 and reaction time 3-10 hours are compared with those obtained from theoretically developed model. The reaction rate constant and gas phase mass transfer coefficient are evaluated for each catalyst concentration and temperature using data and model. The acetic acid formation, a major co-product, is also observed. Though the study draws interesting observations, it has been directed only for the process control purpose. In order to develop kinetics Aguilar et al.(17), and Lavarak etal.(18), carried out extensive experiments with bagasse and bagacillo for the production of xylose, arabinose, glucose, , furfural, acid soluble lignin etc. through various assumed schemes.

Das (22) presented an excellent review of production of furfural in recent times.

From literature survey it is evident that hydrolysis reaction of lignocellulosic materials and grains at higher temperature and pressure except the work of Trivedi etal.(4),Friese(24) at 50°C-120°C and some experiments performed by Lavarak etal. and Aguilar etal.at 80°C-100°C. In order to verify the production of furfural at low temperature,the present investigation were carried out at lower temperature but at presence of metal catalyst to compensate the higher temperature effect.

Mechanism of Formation of Furfural:

Jacob et al. (5) Blatt (7), Clark, (8) and many other investigators (7, 10-16) have widely discussed about mechanism of the reactions involved.The main mechanism (7,10-18) involved in the reaction is considered to be the cleavage of glycosidic bonds in the polysaccharides (mainly cellulose & Hemicellulose) with the formation of simple sugars such as D-xylose,L-arabinose,D-glucose,D-mannose,and D-galactose and acids(acetic and glucuronic) polymer. The hexose sugars on hydrolysis and dehydration gives hydroxymethyl furfural, levulinic acid while the pentoses yield furfural depending upon the variety, age, drying condition and the nature of the polysaccharides. Though the conversion of pentosan to furfural is considered to be accomplished as one step procedure, the mechanism of formation of furfural from pentosans is not so simple(2,12 - 18). It is recognized that there are several intermediate steps in the overall reaction. Broadly speaking the pentosans content in the raw material are hydrolysed with acids to aldo-pentose, chiefly xylose (88-90%, because the characteristic sugar unit of the cell-wall is xylose though some amount of arabinose might be present varying between 9%-11% and ribose and lyxose are completely absent) followed by the elimination of 3 molecules of water to affect the five membered furan ring with the –CHO group in the 2nd position. The reactions are represented as follows:

1. Hydrolysis of pentosans to aldopentose(Xylose/ Arabinose)

$$(C_5H_8O_4)_n + nH_2O \longrightarrow nC_5H_{10}O_5$$

2. Dehydration of aldo-pentose to furfural

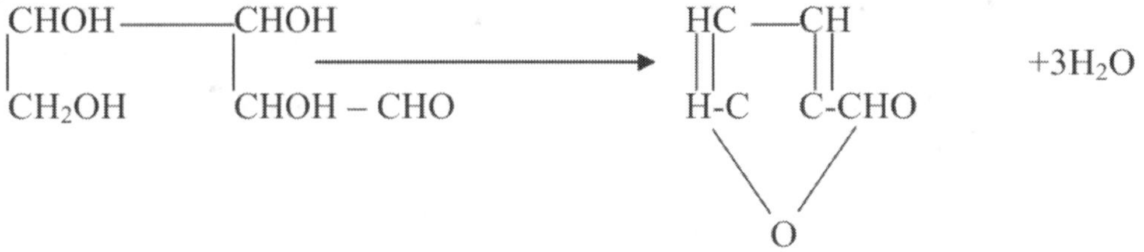

Furfural

The latter is not a simple reaction of dehydration. It has been observed that at a given acid concentration and temperature, the hydrolysis reaction takes place at a greater rate than does the succeeding step viz., conversion of pentose to furfural(11-13). Accordingly, under selected conditions, pentosans can be hydrolysed largely to pentose without formation of furfural. In all probability the second step of the over-all process, conversion of pentose to furfural proceeds by way of one or more intermediate products. In addition, indirect evidence was obtained for a reaction between furfural and an intermediate to give resinous products. A summary of the over-all process is given below:

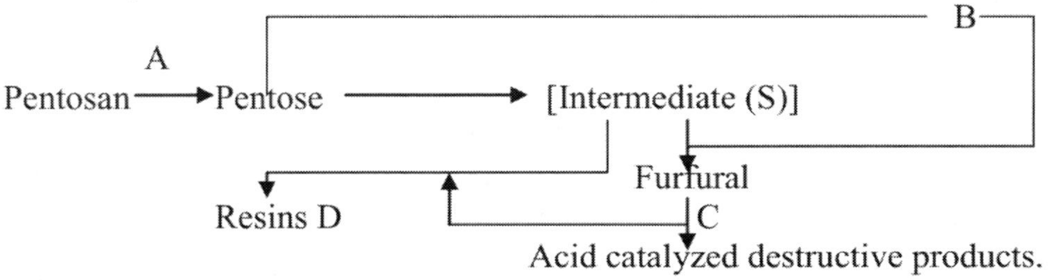

According to Shilling (9) though the above steps are involved to produce final product furfural but no information regarding kinetics pertaining to complex consecutive reactions are expressed explicitly. A general kinetic equation for depolymerization is derived by Emsley and Stevens(19) for the degradation of cellulose over a wide range of material types and environmental conditions. The activation energy is 111 ± 6 kJ/mol and is independent of reaction conditions. In oder to develop kinetics and to get the optimum production, Aguilar et al.(17), Lavarak etal.(18) and others assumed various schemes and mechanisms as summarized below:

Aguilar etal(17)developed the following model by solving the differential equation based on the following scheme and applied for hydrolysis of sugar cane bagasse.

$$\text{Polymers(P)} \xrightarrow{k_1} \text{Monomers(M)} \xrightarrow{k_2} \text{Decomposition products}$$

$$M = M_0 e^{-k_2 t} + \alpha P_0 [k_1/(k_2 - k_1)](e^{-k_1 t} - e^{-k_2 t})$$

where k_1, the rate of the generation reaction and k_2 is the rate of the decomposition reaction, M, the concentration of monomer(Xylose,Glucose,Galactose and Mannose) ,and P(Glucan, Xylan,Galactan and Mannan) the concentration of polymer, t, the time and the subscript 0 indicates initial conditions,α, the ratio between of susceptible xylan/total xylan(usually between 0.55-1.0).They found that kinetic coefficients for formation of xylose is 80 fold higher than that due to degradation reactions. The average values activation energy obtained was of the order of 109 kJ/mol, and Arrhenius frequency factor,$5.293x10^8$.The equation has been applied for xylose, glucose, and arabinose. Modified Arrhenius equation of the following form has been used

$k_1= aC^n e^{-E/RT}$ where a and n are regression parameters and C is in %(w/w).

In the hydrolysis of sugar cane bagasse , furfural (a decomposition product of pentoses)and phenolic compounds (decomposition product of lignin) are generated.

Furfural concentration, F as a function of time,t

$F=F_0(1-e^{-k_1 t})$.

Lavarak etal.(18) also worked extensively on the kinetics of hydrolysis of bagasse and bagacillo and developed seven schemes with seven kinetic models based on various assumptions. For all the schemes , the reaction rate has been assumed to be first order with respect to the reactants in reaction step. the Arrhenius frequency factor, activation energy In one of the scheme , it was mentioned that the activation energy for the hydrolysis of xylan to xylose is of the order of 82.8-107.2 kJ/mol . They further found that the mole ratio of arabinose to xylose and xylose to glucoronic acid range between 0.019-0.247 and 7.4-100 respectively.The kinetic models under isothermal conditions for xylose, arabinose or furfural formation in the aqueous solution has the same form for the scheme 1 as under:

$C_j^L=C_{j0}^S[k_1/(\varphi k_2-k_1)](e^{-k_1 t}-e^{-\varphi k_2 t})$

where C_j^L= concentration of the component j(xylose,arabinose, glucose or furfural),g/g of bagasse ,C_{j0}^S , the equivalent concentration of component j if all bound xylose,arabinose,bound glucuronic acid or pentosans were converted to xylose,arabinose,glucose and furfural respectively (g/g),φ, the ratio of soild bagasse material to liquid(g/g).

$ki= Ai\ C_{acid}^n\ e^{-E/RT}$

n, being the order of reaction with respect to acid concentration. More complicated expressions are available for schemes 2 and 3 taking care of γ, the fraction of xylan that is 'easy-to-hydrolyse'.However, for furfural formation , scheme 6 is used as under assuming the formation is much faster than the later: Pentosans → Furfural (aq.) → Decomposition products

3.0 EXPERIMENTAL:

Preparation of raw material: Mill wet bagasse obtained from Sri Lord Krishna Sugar Mills Limited, Saharanpur, India containing about 50% moisture were dried in open air upto the moisture content of about 10%. The 50-60 % of total pith was removed from the bagasse by first manually and then on drying depithing process further continued using hammers and finally separated through a 30 mesh screen. The residual material was tested and examined under microscope to study the reliability of complete elimination of pith. The proximate analysis of pith is shown in Table-1.

Experimental procedure:

Initially, 2 gm air-dry pith taken in a round bottomed flask was hydrolyzed by hydrolyzing agent (H_2SO_4) with a 2.5% catalyst (Zinc acetate or Ca-acetate) concentration at constant temperature. The reactant ratio was maintained constant throughout the experimentation. After the completion of hydrolysis (digestion) period, the residual material was washed with distilled water till the acid and other components were freed (tested by $BaCi_2$ solution). Then the residual material was dried in oven and weighed after 24 hours and was tested for its pentosan content according to TAPPI standard. The experiments were carried out with the aim to see: the effect of change in concentration of acid (1%-10%), temperature (313-373K), time (1-2.5hrs), type of catalyst (Zinc acetate or Ca acetate), and catalyst concentration (2.5%). The bath ratio in all the experimental runs was kept constant. The concentration of H_2SO_4 was changed keeping catalyst concentration constant. The hydrolysis experiments were carried out for different concentrations, temperatures, times, & catalyst and the residual pentosans are calculated in each case & the results are tabulated in Tables 2.

The use of Catalysts, namely $CaCl_2$, Ca $(OAC)_2$ Zn acetate and the type of acids (HCl, H_2SO_4) do not present any remarkable difference. Therefore, Ca $(OAC)_2$ as catalyst and H_2SO_4 as acid, both are non polluting ingredients may be considered compared to the pollution prone nature of $CaCl_2$ & HCl. The experimental conditions are shown in the following table:

Experimental Conditions:

Parameters	Values
Bath Ratio	1:10
Acid : Liquor	Varying
Conc. of hydrolytic agent, H_2SO_4	1,3,5,10
Temperature,K	313-373
Time, hrs.	1-2.5
Catalyst Conc.,%	2.5,Zn Acetate
Catalyst type	$Zn(OAC)_2$,$CaCl_2$,$Ca(OAC)_2$

The formation of furfural is confirmed by the presence of precitpitation of phloroglucinal & methylaniline. The solution is also tested by Tollen's reagent, soln. of freshly distilled aniline & glacial acetic acid in distilled water. The above estimation of furfural has been done using distillate from the flask.

4.0 RESULTS AND BISCUSSIONS

The experimental results based on different time, temperature, catalyst and the acid concentrations for production of furfural or removal of pentosans from baggase pith are now interpreted in the following paragraphs:

Proximate Analysis of Pith:

Table 1 indicates ash, alcohol-benzene solubility, pentosans & cellulose. It is indicated that pentosans content varies between 24.87 % to 26.58%. No attempt has been made to determine the polyuranides, xylan,arabinan, glucan, Klason lignin etc. The data of pentosans is very closely tallying with the data of bagasse for pentosan 26.16%(combined xylan and arabinan) presented by Aguilar etal.(17) for bagasse ,26.8% due to Lavarak etal. (18), and 24.5 % due to Biswas et al.(3), Grant(9), Kawamura(16), and 24.4-24.9 for rice straw due to Casey(23), Bains etal.(2) and Mittal etal.(3).It indicates the pentosans content of Indian bagasse pith is of the same order of magnitude with that of Indian rice straw.

Parametric effects:

Effect of concentration on furfural yield as a function of time and temperature is shown in Fig.1 The parametric effects such as time on pentosan residue is drawn in Fig.2,the influence of sulphuric acid concentration on pentosan yield at 40^0C in Fig.3,and the same at 50^0C is plotted in fig.4.To examine the effect of higher temperature and concentration of sulphuric acid on furfural yield at 100°C and time 1 hr in Fig.5.

Effect of concentration of acid:

The formation of furfural with the acid concentration is shown in Fig.1 for two different values keeping bath ratio, catalyst concentration constant. It indicates that the yield of furfural increases with increased acid concentration at lower temperature level. At higher temperature level the yield of furfural is slightly increased.

Effect of Temperature:

With increase in temperature, at same concentration of acid and at constant bath ratio & catalyst concentration, there is a tendency of slight increase of furfural yield.
Experimental data also shows that with the increase of temperature there is an increase of furfural yield at a particular concentration of acid, bath ratio, and catalyst concentration.
It has been observed that around 3% increase of yield for every 10^0C rise of temp at lower level of temperature but for higher acid concentration there is an average of 10% increase in the furfural yield for 10^0C rise of temperature.

Effect of Time:

All the experiments were carried out for four different times (1hr, 1.5 hr, 2 hr, 2.5 hr) at two different temperatures (40°C and 50 °C) and two different acid conc.(1% and 3%). It is observed that with the increase of time, there is an initial decrease of furfural yield, goes to a minimum and again increases. In most of the plots it has been noticed that there is a slight increase of yield with the increase of time, goes to a minimum & then again increases. 1.5 hr reaction time, at 3% concentration of acid has been found to be the optimal time and acid concentration.

Effect of higher temperature:

The experiment has also been conducted at 100^0C and the digestion time is 1 hr. at atmospheric pressure. The concentration of acid is varied form 1-10% whereas bath ratio, catalyst concentration, temperature & time remained constant.
This is depicted in Fig.3.This indicates that the yield of furfural with concentration of acids at first decreases, comes to a minimum, increases further, goes to a maximum and then again decreases. The maximum yield of furfural obtained based on raw material is at 1 ½ hr., 3% acid concentration, 50^0C temperature. 16.27% based on O.D. Pith. The

effects of concentration of sulfuric acid on petosans yield as a function of acid concentration ranging from 1-3% at two different temperatures 40 °C and 50°C are plotted in Figs.3 and 4. Fig.5 depicts the effect of concentration of H_2SO_4 on Furfural Yield at 100°C and time 1 hr. It is evident from the later that the yield remains almost constant.

Steam distillation of the reaction product.

The steam consumption to recover furfural by steam distillation has been calculated on the order of 20 kg/kg of steam. The product is considered entirely furfural which may not be true. Further fractional distillation to purify in order to examine other products in the mass has not been tried.

5.0 Development of Reaction Scheme and Kinetic models

From the literature survey it is observed that pentosan(both xylan and arabinan) is hydrolyzed to both aldopentoses which are converted into two or more steps into furfural. Loss of furfural takes place due to side reactions which leads to condensation and to the formation of resins. Both levulinic acid and furfural can be produced from wood or any biomass & levulinic acid can be converted to succinic acid. The scheme can be depicted as under:

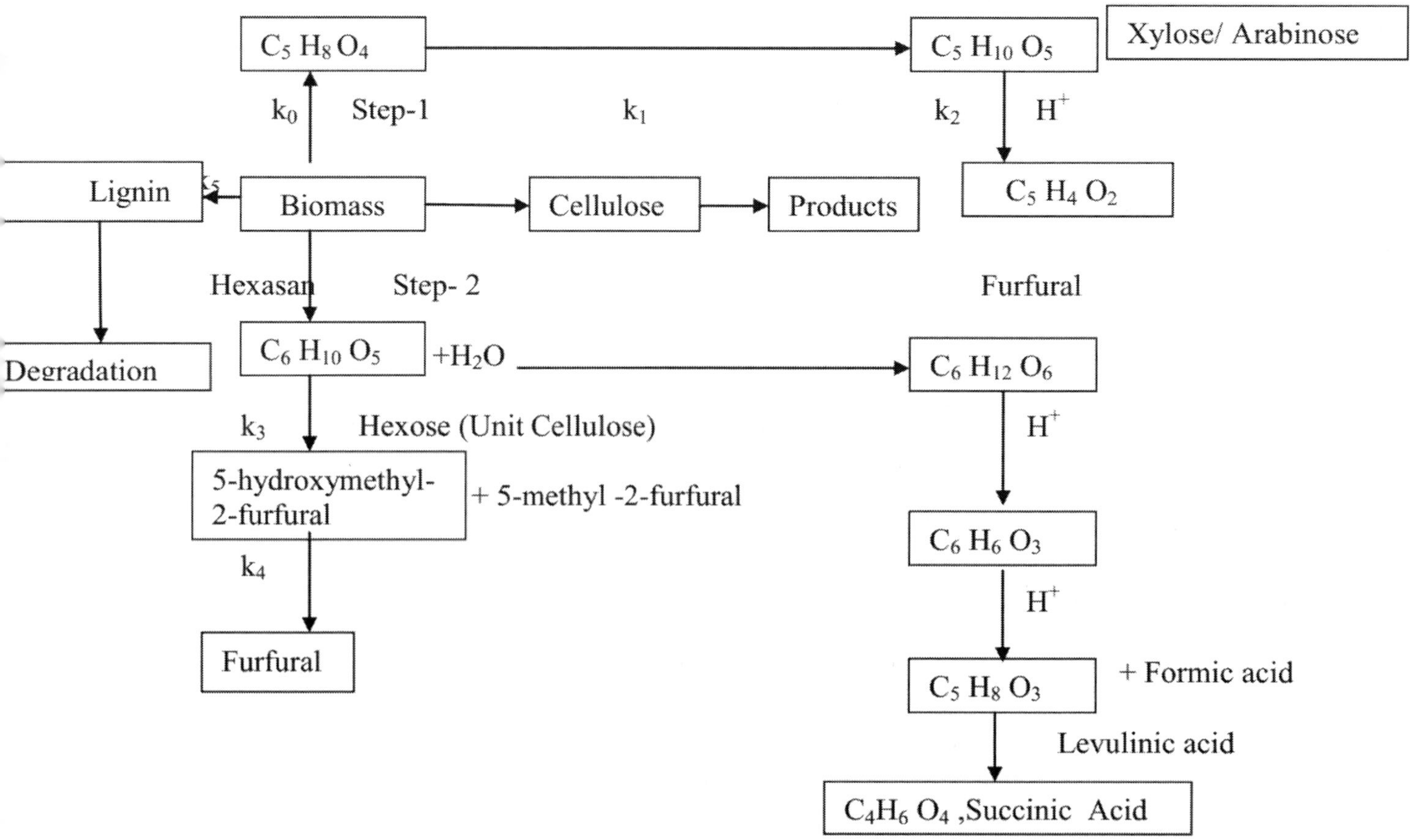

Due to difficulty in finding a strict mechanism for hydrolysis reactions, it is usual to use simplified models to determine the kinetics of hydrolysis of lignocellulosic materials like bagasse, bagacillo, pith, straws, woods etc.The models proposed in the literature use pseudo-homogeneous irreversible first order reactions.
Therefore, an attempt has been made to develop kinetic expression for hydrolysis of bagaase pith. In reality the reaction sequence is very complex involving a consecutive reaction of three steps having rate constant k_1, k_2 and k_3 with many side reactions.The reaction product obviously contain both degradation products from pentoses(Xylose) as well as hexoses. At this stage it is difficult to develop detailed kinetic modeling of furfural production from bagasse pith. Hence the productions of pentoses followed by conversion to furfural assuming formation in equimolar amounts, the following equation are used.
The rate of decomposition of pentoses = the rate of formation of furfural.

$$\text{Pith} \xrightarrow[H^+]{k_1} \text{Pentosan} \xrightarrow[H^+]{k_2} \text{Pentoses} \xrightarrow[H^+]{k_3} \text{Furfural} \xrightarrow{k_4} \boxed{\text{Products}}$$

$$\frac{d\,(\text{Pentosan})}{dt} = -\,k_1\,C_A \quad \ldots\ldots \quad (1)$$

$$r_p = k_1\,C_A - k_2C_P \quad \ldots\ldots \quad (2)$$
$$r_{ps} = k_2\,C_P - k_3C_{PS} \quad \ldots\ldots \quad (3)$$
$$r_f = k_3\,C_F \quad \ldots\ldots \quad (4)$$

The above expressions are equally applicable to both pith, bagasse and rice straw.
It is evident that optimum yield of furfural depends upon various reaction steps.

$$-\,r_A = k_0\,e^{-E/RT}\,(C_A)^n \quad \ldots\ldots (5)$$
$$k = k_0 \exp(-E/RT)$$
$$C_F = C_{A0}[1+1/(k_2-k_3)\{k_3e^{-k_2t} - k_2e^{-k_3t}\}]$$

Following assumptions are made:
At this low temperature , hydrolysis of hexosan(glucan etc.) to form glucose and subsequently furfural, the same from lignin to acid soluble lignin is negligible. Further the furfural is formed from both arabinan and xylan(arabinose /xylose $\approx$0.11) and no other constituents of pith. No effect of acid concentration,type of acid and catalyst type and concentration are considered.
k_1 is slow(needs generally higher pressure and temperature), k_2 is very fast, k_3 is still faster and k_4 is extremely slow.
In these situations the following models are assumed with the change of nomenclature as the concentration has been changed to percentage weight:
Decomposition of pentosans= $P=P_0e^{-k_2t}$
Formation of furfural=$F=P_0(1-e^{-k_3t})$
In order to know the kinetics of the furfural production from bagasse pith in respect to reaction order and activation energy, residual pentosan and furfural from similar kind of raw material, i.e. rice straw data are used. These are shown in Figs. 6-10. If P_0 is the amount of pentosan at t=0, P is the same at t=t, x is the converted amount, then P_0-x=P, By plotting ln P vs t , one gets a straight-line with an intercept ln p_0 and slope=-k=-

0.569. On the other hand ploting ln F_0/F vs. t also gives a straight line where, F_0,F the amount of furfural initially and at time t respectively. Both the kinetics of pentosan decomposition appear to be of pseudo first order with specific rate constants are of the order of 0.569 hr^{-1} and 0.044 hr^{-1} respectively.The activation energy has been found of the order of 31617.6 J/mole. The values of average k,and activation energy for furfural formation from rice straw have been found of the order of 0.02505 hr^{-1} and 23478.26 J/mol Therefore, it is expected that similar expression will be valid for bagasse and bagasse pith also.

Further, attempt has been made to elucidate about the kinetic reaction and level of activation energy for the reaction of pith with acid. Fig. 11-12 indicate the order of the reaction to be of first order. Accordingly, the Arrehenius plot of ln (C_A/C_{A0}) vs. 1/T is made.

Assuming first order reaction rate, n = 1, k_1 and k_2 for two temperatures 313 K and 323 K are calculated as:

$k_2 = 0.04167\ hr^{-1}$

$k_3 = 0.05303\ hr^{-1}$

The activation energy is found out as

$E_1 = 4878.9938$ kCal/kmole ≈ 4879 kCals/kmole=20492 kJ/kmol.

This value is in agreement with those found out Trickett etal.(20).

Further experimentation is required to perform the experiments with varying temperature at higher pressures. It is also important to develop the detailed kinetic expressions for this very complicated multiple series –parallel reaction.

6.0 CONCLUSION

The process development for production of furfural from bagasse pith under digestion with acid at low temperature, atmospheric pressure, in the presence of catalyst has indicated promising results. Most of the previous methods have used high temperature and pressure, high reaction time and require high energy requirements. The maximum yield of furfural obtained based on raw material is at 1 ½ hr., 3% acid concentration, 50^0C temperature,16.27% based on O.D. Pith. This value appears to be high. The steam consumption to recover furfural has been reported to be 20 kg/kg of steam. Besides, most of the works are patented and hence industrially exploited processes are very scanty. It has been found that this reaction has a low level of activation energy of the order of 20492 kJ/mol with specific rate constants 0.053 and 0.0417 hr^{-1} respectively.

Present investigation with preliminary computations of activation energy and specific rate constants also indicate that the production of furfural is possible with low energy inputs. However, effects of catalyst and experiment at higher temperature are required. The attempt to produce furfural has two fold functions for the disposal of waste & abatement of particulate pollution followed by product development. It needs to look afresh to explore the possibility of getting better optimum condition for conversion of pith to furfural with change of catalyst.

Table 1 Proximate Analysis of Bagasse Pith

Ash	:	2.73
Cold Water Solubility	:	7.02

Hot Water Solubility	:	9.84
10% NaOH Solubilty	:	46.23
Alcohol-Benzene Solubility	:	4.37
Pentosans,% (avg.)	:	24.87
Cellulose (Cross & Bevan),%	:	50.0

REFERENCES

1. Manohar Rao, P.J., A waste in bagasse based paper factories,IIPTA, Convention issue, April 9, 1987 ,PP-180.
2. Bains, B.S., Puri, S.C. & J.S. Chawla,Thermal hydrolytic studies on Ligno-cellulogic materials, IPPTA Vol. XIV No.3, July, 1977, PP-209-215
3. Mittal, J.P., & Biswas, B.,Studies in the preparation of furfural from Zea Mays stem pith, chemical examination of Zea Mays stem pith, IPPTA, Vol.XIV No.3, 1977,PP-201-204.
4.Trivedi, M.K. & P.K. Bhattachaya,Upgrading of cellulose pulps , part 4, removal of pentosans and mineral matter,Proc. of IPPTA Souvenier, 1974, PP-62-65.
5.Jacob, S., Balakrishan, E., Kabra, K.D. and Bahety, R.C., Furfural –a by product in prehydrolysed sulphate process, Ibid.
6.Gananasekaran, A. &. C. Sivaprasad Rao,Formation of furfural from rice hull in a fixed bed reactor, Indian Chemical Engineer, Vol. XX,No.2, April- June, 1978, PP-3-8.
7.Blatt, A.H.,Org.Synthesis,Collective Vol.-I, John Wiley & Sons Inc. London, 1959
8.Clerk, H.T."Organic Analysis, Orient Longman Ltd. 1970.
9.Grant ,J., A lab. Hand book of pulp and paper manufacture EdwardArnold Publishers Ltd.,1961, P.40
10. Shilling, W.L. Tappi ,Conversion of Lignocellulosic materials to chemicals and fuels ,October,1965, 48, 105A
11. Colin Cano B.,Cromezrodrigue, J.V.,Experimental determination of the optimal process conditions for the hydrolysis of the pentoses of sugarcane bagasse, A.B.I.P.C.,Vol.53, July-Dec.1982
12. Morozov,E.F., Shkut , Kebich, M.S., Conversion of pentosans into furfural by the action of salts in the solid waste in the solid phase,. A.B.I.P.C., Vol.53, July-Dec.1982
13. Srinivas K.M., Mukesh Kumar, Experimental project report, Kinetic modeling of preparation of furfural from bagasse pith,submitted to University of Roorkee, June,1987
14. KhimDrev. Analysis of plant materials for pentosan contents & potential furfural,(Riga) No.2 :89-93, (March/April 1980),(Russ)

15. Mehlberg, R.L., Purdue Research Foundation, pentose production from hemicelluloses, A.B.I.P.C., Vol.53, Jan-April,1983
16. Kawamura, J., J. Sec. Chem. Ind. Japan, 33, 392, 1930
17. Aguilar R., J.A.Ramirez, G.Garrote, M.Vazquez, Kinetic study of the acid hydrolysis of sugar cane bagasse,Journal of Food Enginnering, 55,309-318,2002
18. Lavarack, G.J.Griffin,D.Rodman,The acid hydrolysis of sugarcane bagasse hemicelluloses to produce xylose, arabinose, glucose and other products, Biomass & Bioenergy 23,367-380,2002
19. Emsley A.M., G.C.Stevens ,Kinetics and mechanisms of the low-temperature degradation of cellulose, Cellulose,1,25-26,1994
20.Trickeett R,Neytzell-de Wilde F., Dilute acid hydrolysis of bagasse hemicelluloses, Chemsa 1982,March,11-5
21.Rodrigues R.C.L., M.G.A. Felipe,J.B. Almeida e Silva, M.Vitolo and P.V.Gomez,The influence of pH, temperature and hydrolyzate concentration on the removal of volatile and nonvolatile compounds from sugarcane bagasse hemicellulosic hydrolyzate treated with activated charcoal before and after vacuum evaporation, Brazilian Journal of Chemical engineering,vol.18(3), sept,201,pp.1-9.
22.Das Majani ,Production of Furfural from Bagasse, Seminar report, submitted to IITRoorkee ,August,2006
23.CaseyJ.P., Pulp and Paper Chemistry and Chemical Technology,Third Edition,Volume I, A wiley Interscience publication, John Wiley &Sons,1981
24.Friese H.,U.S.Patent 3,954,497(1976) through Chem-Abstra;83:62244 e(1976)

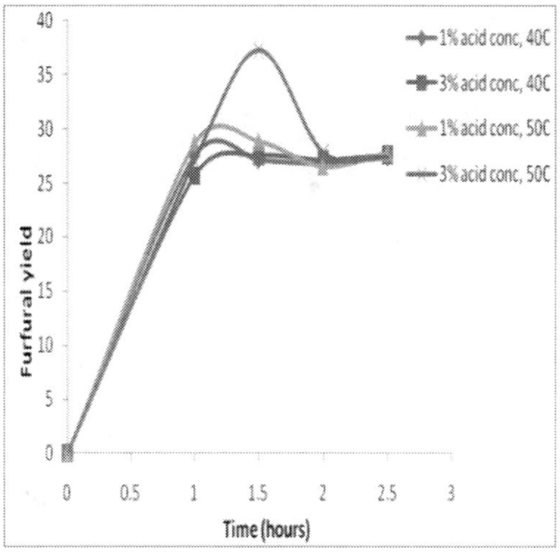

Figure 1. Effect of Time on Furfural Yield

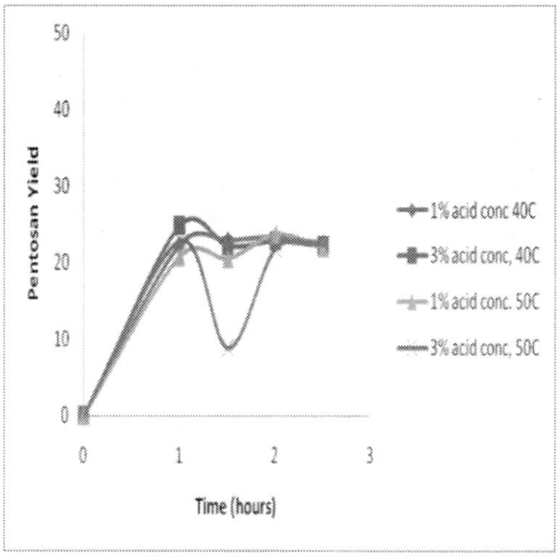

Figure 2. Effect of Time on Pentosan residue

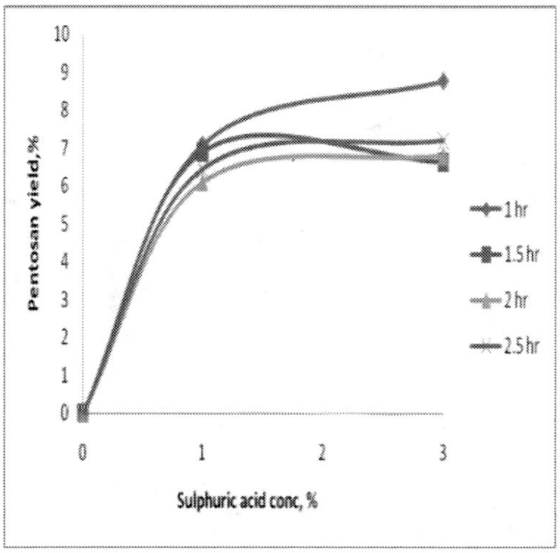

Figure3. Effect of Sulphuric acid concentra- yield at on Pentosan 50^0C

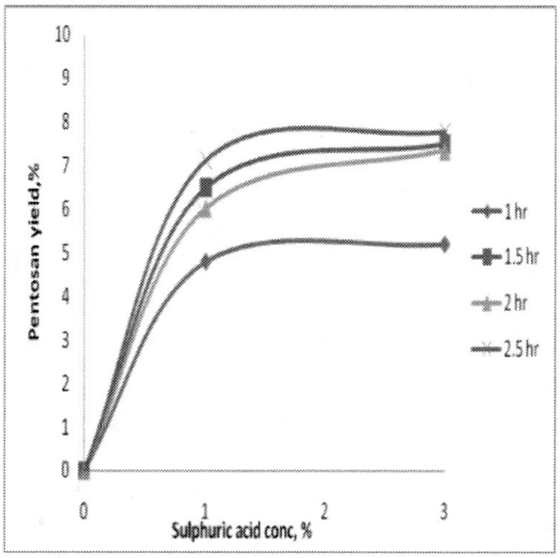

Figure 4. Effect of Sulphuric acid concentration on Pentosan yield at 40^0C

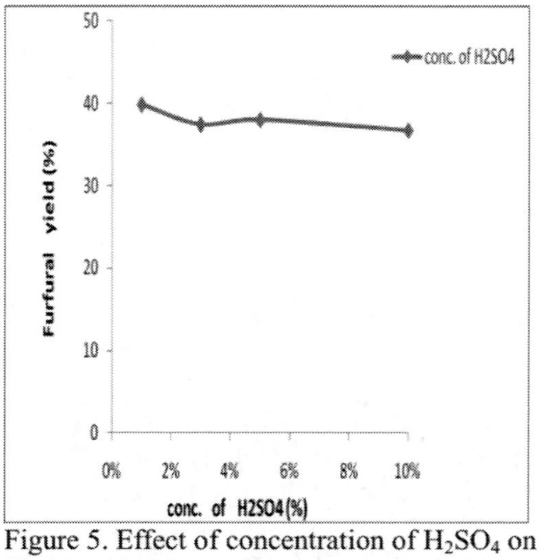

Figure 5. Effect of concentration of H_2SO_4 on Yield at 100°C and time 1 hr

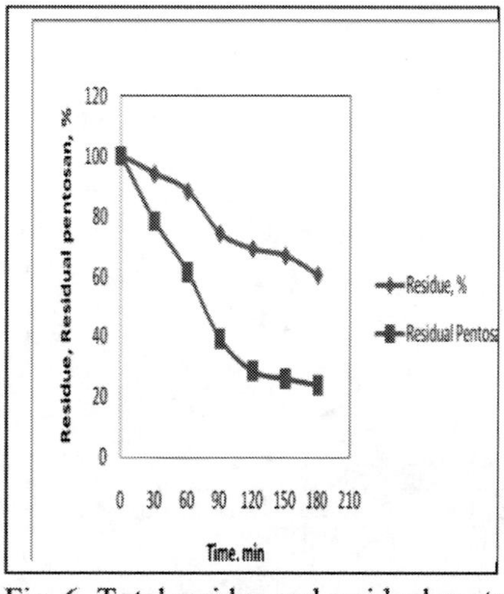

Fig. 6 Total residue and residual pentosan from Furfural rice straw as a function of time

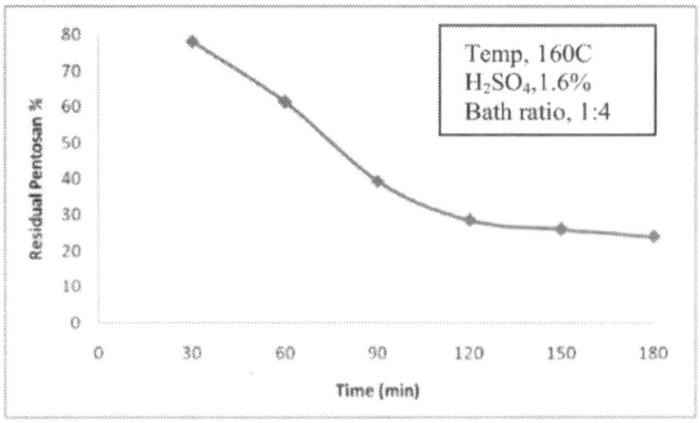

Figure 7. Residual pentosan yield from rice straw as a function of time

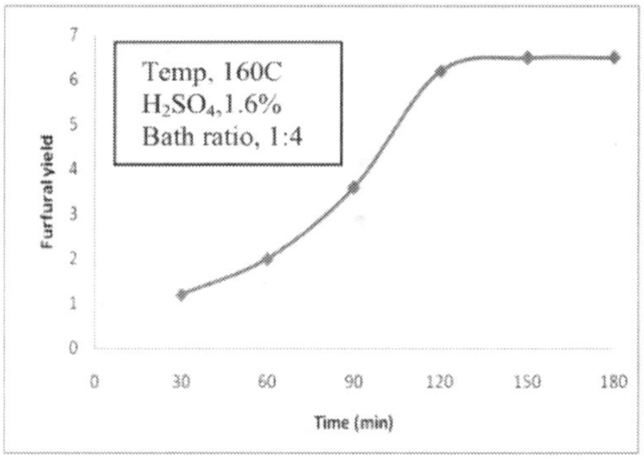

Figure 8. Furfural yield from rice straw as a function of time

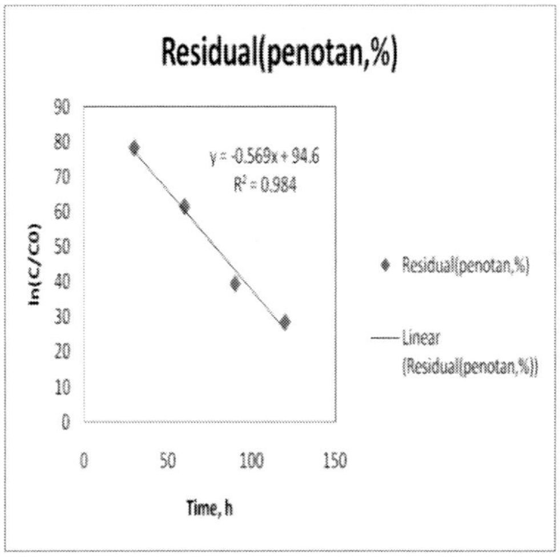

Fig.9. ln (C_A/C_0) , the concentration terms for pentosan as a function of time

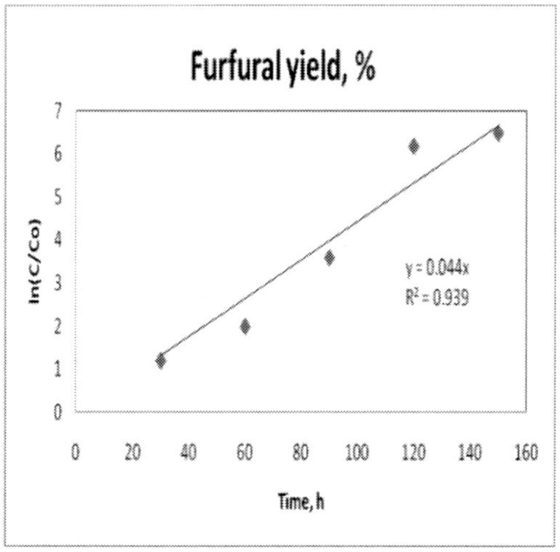

Fig.10 ln (C/C_0) , the concentration terms for Furfural as a function of time
$C=p_0-x$ and $C_0=p_0$,$\ln(C/C_0)=\ln[(p_0-x)/p_0]$

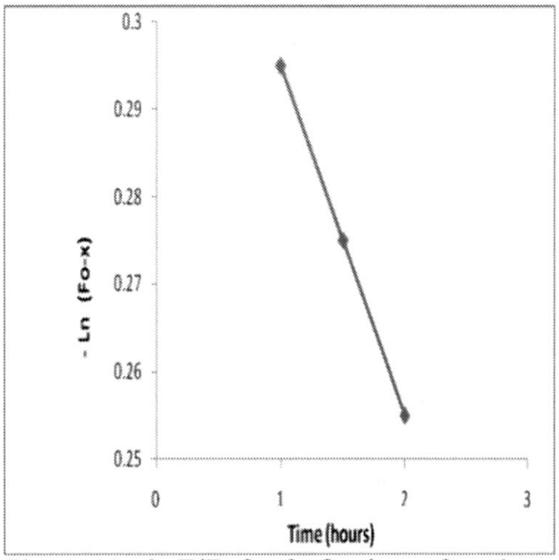

Figure 11: ln F/F_o for furfural as a function of time for bagasse pith.

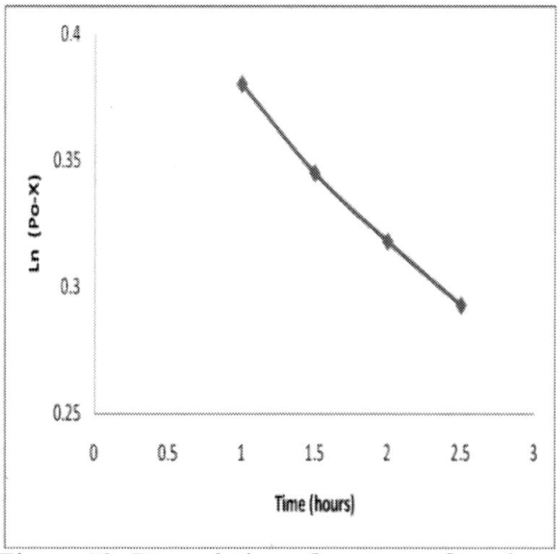

Figure 12: Degradation of pentosan from bagasse pith as function of time (P, P_o expressed in %).

(566d) Production of Biopolyols and Polyurethane Foams From Crude Glycerin and Lignocellulosic Biomass

Shengjun Hu Sr. and Yebo Li, Food, Agricultural, and Biological Engineering, The Ohio State University-OARDC, Wooster, OH

Polyurethane (PU) foams are cellular polymeric materials produced from polyols and isocyanates, both of which are currently largely petroleum derived. In this study, two cheap, abundant, and renewable resources (lignocellulosic materials and crude glycerin, i.e. a biodiesel byproduct) were explored for their potentials in the production of biopolyols and PU foams. Biopolyols were produced by liquefying soybean straw using crude glycerin as liquefaction agent in a one-pot liquefaction process. It was found that the properties of produced biopolyols varied with various liquefaction conditions and that under preferential liquefaction conditions biopolyols showed hydroxyl numbers from 440 to 490 mg KOH/g, acid numbers below 5 mg KOH/g, and viscosities from 16, 000 to 45, 000 mPa.s, which are suitable for the production of rigid PU foams. The produced biopolyols were used directly for the production of PU foams. The produced PU foams showed densities from 0.033 to 0.037g/cm^3 and compressive strength from 148 to 203 kPa and are applicable in a variety of areas. This study suggests that crude glycerin alone can be effectively used to liquefy soybean straw and potentially other lignocellulosic materials as well for the production of biopolyols and PU foams. This technology contributes to the cost-effectiveness of biodiesel industry and agriculture development by converting low value biodiesel byproduct (i.e. crude glycerin) and lignocellulosic biomass to value-added products.

See more of this Session: Biobased Materials III: Value-Added Coproducts
See more of this Group/Topical: Forest and Plant Bioproducts Division

(566e) Thermo-Mechanical Analysis of Thermoplastic Biocomposites

Venkata S. Chevali[1], Jessica L. Lattimer[2], Shanshan Huo[2] and Chad A. Ulven[2], (1)Mechanical Engineering Department, North Dakota State University, Fargo, ND, (2)Mechanical Engineering, North Dakota State University, Fargo, ND

Thermoplastic biocomposites are gaining interest in commercial applications because of their mechanical performance at reduced cost and density as compared to synthetic-fiber-reinforced composites. The long-term durability and heat resistance of thermoplastic biocomposites is dependent on the constituents, applied loads, and environmental influence. The constituent makeup of the lignocellulosic components dictates the effectiveness of the filler in contributing to several functionalities in commercial applications.

This analysis is an investigation of the effects of filler loading level, interfacial modifiers, and the fraction of semi-structural lignocellulosic components on the heat deflection and creep response of the biocomposites. Polypropylene- and ABS-based biocomposite combinations are assessed for their long-term mechanical properties and sustained thermal resistance.

The lignocellulosic fillers used in this study are co-products of different agricultural processes and are fractionated for dispersion-compounding with the thermoplastic matrix using extrusion. The pelletized compound is injection molded as test specimens. The thermomechanical response is critically discussed and correlated with compositional changes brought about by the fillers and hybridization thereof.

See more of this Session: Biobased Materials III: Value-Added Coproducts
See more of this Group/Topical: Forest and Plant Bioproducts Division

(566f) Co-Products: New Applications

Manju Misra, School of Engineering and the Department of Plant Agriculture, University of Guelph, Guelph, ON, Canada

In progress

See more of this Session: Biobased Materials III: Value-Added Coproducts
See more of this Group/Topical: Forest and Plant Bioproducts Division

(566g) Chemical Modification of Polysaccharides Using Reactive Extrusion

Pratik N. Bhandari, Industrial Agricultural Products Center, University of Nebraska-Lincoln, Lincoln, NE and Milford A. Hanna, Biological Systems Engineering, University of Nebraska-Lincoln, Lincoln, NE

A Brabender TSE-20 co-rotating twin screw extruder was used for chemical modification of starch and cellulose. Ether and ester derivatives of the polysaccharides with various degrees of substitutions (DS) were prepared. These derivatives included carboxymethyl starch, carboxymethyl cellulose, starch acetate, starch maleate and cellulose acetate. Carboxymethyl starch, with low DS (0.3-0.4), was prepared for super-disintegrant applications as a pharmaceutical excipient. The physical, chemical, morphological and disintegrant properties of cross-linked carboxymethyl starch were compared with those of a commercial super-disintegrant VIVASTAR®P. High DS carboxymethyl starches (DS 0.68-1.54) also were prepared using the extruder. The effects of the amounts of aqueous ethanol and NaOH and screw configuration on the DS, reaction efficiency and product microstructure were investigated. Carboxymethyl cellulose also was prepared using a twin screw extruder to study the effects of water/ ethanol ratio and amount of NaOH on the product DS, crystallinity, microstructure, viscosity and water absorption properties. Further, starch and cellulose esters also were prepared using reactive extrusion. Reactive extrusion for preparing these modified bio-polymers is more efficient, uses less solvents, is faster and has more complete reaction and fewer byproducts than conventional processes.

See more of this Session: Biobased Materials III: Value-Added Coproducts
See more of this Group/Topical: Forest and Plant Bioproducts Division

(610a) Separation and Purification of Biobutanol During Bioconversion of Biomass

Hua-Jiang Huang and **Shri Ramaswamy**, Bioproducts and Biosystems Engineering, University of Minnesota, St Paul, MN

Biofuels from biomass becomes more and more important, due to the limited availability and increasing costs of fossil fuels, the need for reduction in GHG emissions, and energy independence. ABE (Acetone-Butanol-Ethanol) fermentation is the second largest industrial fermentation process after ethanol. Butanol has several advantages over ethanol as a drop-in bio-fuel such as higher energy content, lower vapor pressure, and lower hygroscopy, and it can be produced from non-food, renewable lignocellulosic feedstocks and can be used in existing transportation fuel distribution infrastructure. Thus, biobutanol has the potential to substitute both bioethanol and biodiesel in the biofuels market estimated to be worth $247 billion by 2020. Today, biobutanol can compete with synthetic butanol in the chemicals market (Green 2011).

"Biotanol" can be produced by ABE fermentation using *Clostridium acetobutylicum* or *C. beijerinckii* under anaerobic conditions (Kraemer et al., 2010). Current commercial biotanol processes are based on fermentation of starch or sugar-based feedstocks. These feedstocks, however, are usually expensive leading to high production cost. In addition, most biobutanol plants use corn, which may compete with food and feed. Thus, there has been a growing research interest in developing non-food, renewable biomass based drop-in biofuels such as biobutanol. However, there are still many challenges to be solved for making biomass based biobutanol economically and commercially viable. The unsolved challenges include (i) high product (especially butanol) inhibition, (ii) low butanol yield and productivity, and (iii) expensive downstream processing (product separation) (Kraemer et al., 2010; Kumar and Gayen, 2011). Among these challenges, the separation and removal of product butanol is a critical issue. In this study, an extensive review of separation and purification of butanol from fermentation broth is provided, including gas stripping, distillation, liquid-liquid extraction, membrane separation (e.g., pervaporation), adsorption (e.g.., molecular sieve adsorption), membrane solvent extraction, and integrated separation-fermentation, etc. This could be helpful in research and development of renewable butanol biofuels.

References

Green E. M., Fermentative production of butanol – the industrial perspective, Current Opinion of Biotechnology, 22: 1-7 (2011).

Kraemer K., Harwardt A., Bronneberg B., and Marquardt W. Separation of butanol from acetone-butanol-ethanol fermentation by a hybrid extraction-distillation process. 20th European Symposium on Computer Aided Process Engineering – ESCAPE20, S. Pierucci and G. Buzzi Ferraris (Ed.), 2010.

(610b) Experimental and Theoretical Study of Lignin Extraction by Ionic Liquids From Straw

Wen-Yan Ji and Hai-Jun Wang, Jiangnan University, Wuxi, China

Lignin is a natural polymer and it can be the source of aromatic compounds. However, efficient means of seperating lignin from straw is still a problem needed to solve. In this work, dissolution experiments were carried out. Three ionic liquids with the same cation [emim]ClA[emim]OAcA[emim]$MeSO_4$ were selected and sythesised , in order to dissolve lignin from straw with the optimal conditions respectively. The lignin was regenerated by precipitation, then the exteraction yields of the above liquids were obtained, so we can get the order of lignin solubility for varying anoins. Microscopically, the interaction between ionic liquids and lignin was also studied by quantum chemical calculations. Thus, the relationship between exteraction yield and interaction energy was revealed. It also indicates that the solubility of lignin in ionic liquid was mainly influenced by the anions part of ionic liquids.

See more of this Session: Separation Processes In Biorefineries II: Ionic Liquids, Reactive Separations and Other Novel Processes
See more of this Group/Topical: International Congress on Energy 2011

(610c) Determination of Significant Cellulase and Hemicellulase Activities Required for the Deconstruction of Ionic-Liquid Pretreated Biomass

Christopher J. Barr[1], Jeffery Mertens[2] and Constance Schall[1], (1)Chemical and Environmental Engineering, University of Toledo, Toledo, OH, (2)NCAUR, USDA-ARS, Peoria, IL

Lignocellulosic biomass, comprised of polysaccharides derived from cellulose and hemicellulose, is a second generation renewable feedstock for carbon-based chemicals and fuels. The recalcitrance of biomass to deconstruction is one of the main obstacles in the conversion of lignocellulosic biomass to monomeric sugars and the production of sugar platform fuels and chemicals. Ionic Liquid (IL) pretreatment creates unique biomass substrates consisting of amorphous cellulose with largely intact hemicellulose and lignin. These substrates differ from many other pretreatments which remove hemicellulose and retain cellulose crystallinity. The unique substrate resulting from IL-pretreatment poses an interesting challenge for enzymatic deconstruction using traditional commercial mixtures.

Mono-component cellulase and hemicellulase enzymes from *Aspergillus nidulans* have been expressed in *Pichia pastoris*, purified and characterized for activity, substrate specificity, and thermal stability. Enzyme combinations were used in series of hydrolysis experiments to determine key enzyme activities and/or interactions that significantly contribute to the conversion of a variety of lignocellulosic biomass substrates to monomeric sugars. Accessory hemicellulases appear to be an essential addition to the mono-component enzyme mixtures. Exploration of these accessory enzymes also yields insight into an understanding of the differences in hemicellulose structure between feedstocks. Some accessory hemicellulases include acetyl xylan esterases (AXE), α-glucuronidases, and arabinofuranosidases. In addition to mono-component enzyme mixtures, next generation commercially-available enzyme mixtures were characterized and formulated using key activities to optimize hydrolysis at low enzyme loadings. These simplified mixtures will be dependent on the substrate. As part of these studies, an analysis of variance (ANOVA) was used to statistically model the main and interaction effects of key enzyme activities.

See more of this Session: Separation Processes In Biorefineries II: Ionic Liquids, Reactive Separations and Other Novel Processes
See more of this Group/Topical: International Congress on Energy 2011

(610d) Catalytic Isomerization/Dehydration of Sugars On Functionalized CNT-Metal Oxide Hybrids At the Liquid-Liquid Interface of Water/ Oil Emulsions

Jimmy A. Faria, Rattiya Saetang and Daniel E. Resasco, School of Chemical, Biological and Materials Engineering, University of Oklahoma, Norman, OK

The conversion of C5-C6 sugars into 2-furfural (FA) and 5-hydroxymethylfurfural (HMF) have been considered as key intermediate step in the production of transportation fuels and chemicals from ligno-cellulosic biomass. Typically, the process involves the dehydration reaction of the sugars catalyze by strong acids, albeit this approach results in loss of selectivity due to the instability of the intermediate species in such acidic conditions. An alternative catalytic strategy is the isomerization of the sugars into the xylofuranose and ribose isomers on a basic catalyst, followed by the dehydration reaction on a weak acid. Recently, we have shown that amphiphilic nanohybrid catalysts based on Carbon Nanotubes (CNT) fused to inorganic oxides can stabilize water/oil emulsions and catalyze reactions at the water/oil interface when properly functionalized [1,2,3,4]. In the present contribution we report a new family of amphiphilic-bifunctional catalysts based on sulfonic- and amine- functionalized CNT fused to metal oxides. These nanohybrid materials are able to perform the isomerization and dehydration of the C6-C5 sugars in a biphasic system with high selectivity. Several advantages are obtained when this unique system is used, such as, enhanced mass transfer between the phases, selective conversion of molecules in each of the phases (phase selectivity), direct partitioning and separation of products, leading to substantial simplification of the purification stages.

References:

1. Crossley, S., Faria, F., Shen, M., and Resasco, D.E. Science 1, 68 (2010).

2. Ruiz, M.P., Faria, J., Shen, M., Drexler, S., Prasomsri, T., and Resasco, D.E. ChemSusChem, In press (2010).

3. Shen, M., and Resasco, D.E. Langmuir 25, 10843 (2009).

4. Faria, J., Ruiz, M.P., and Resasco, D.E. Adv. Synth. Catal. 352, 2359 (2010).

See more of this Session: Separation Processes In Biorefineries II: Ionic Liquids, Reactive Separations and Other Novel Processes
See more of this Group/Topical: International Congress on Energy 2011

(610e) Separation and Purification of Value-Added Co-Products In Biorefinery

Hua-Jiang Huang, Department of Bioproducts and Biosystems Engineering, University of Minnesota, Saint Paul, MN and Shri Ramaswamy, Bioproducts and Biosystems Engineering, University of Minnesota, St Paul, MN

Recently, there has been increasing interests in conversion of biomass to bioproducts including fuels, chemicals, materials, and heat and power, etc. due to the shortage of fossil fuel and its increasing price, the
requirement of national energy independence, as well as the enviornmental concerns about green house gas
emissions. In order to improve the ovreall technical, economic viability and environmental sustainability of biobased
economy, similar to petroleum refineries, it is important to consider producing multiple products – in addition to fuels and energy, value added co-products. Number of plant species contain bioactive compounds
or phytochemicals and they can be extracted as value-added co-products prior to biomass conversion. Extraction and use of phytochemicals have been studied for many years, with more focus on pharmaceuticals
and neutraceuticals from fruits, vegetables, or other food crops/plants, while less research has been done on
saparation of phytochemicals from woody and perennial materials, which can be renewable feedstocks for
biofuel.

Phytochemicals from plants can be used in pharmaceuticals, cosmetics, neutrition, and consumer products.
Carotenoids, polyphenols and flavonols are known phytochemicals that can be used for treatment of a number of disorders. Phytochemicals from plants are usually present in very dilute quantity. The thermo-liable
or heat-senstive properties of phytochemicals and the increased difficulty of solid-liquid extraction over liquidliquid
extraction bring a great challenge for efficient separation of phytochemicals from dilute biomass matrix. Thus, selection of suitable solvents and extraction and purification methods for targeted active components
become critically important. In order to help understand various approaches for phytochemicals extraction,
separation and purification, and aid in the development of appropriate methods for value-added co-products,

here we provide an extensive review on the major separation methods suitbale for co-products in biorefineries. The various separation and purification methods to be considered include solid-liquid extraction (leaching, ultrasound-assisted extraction, microwave-assisted extraction etc.), steam distillation, pressurized liquid extraction or near-critical (subcritical) fluid extraction, supercritical fluid extraction, as well as separation and purification of phytochemicals from dilute solutions or extacts from biorefinery such as liquid-liquid extraction, vacuum distillation, membrane separation, and molecular distillation, etc.

See more of this Session: Separation Processes In Biorefineries II: Ionic Liquids, Reactive Separations and Other Novel Processes
See more of this Group/Topical: International Congress on Energy 2011

(610f) Cellulase Inhibition by Pretreatment-Released Biomass Degradation Products

G. Peter van Walsum and Mahdieh Aghazadeh, Chemical and Biological Engineering, University of Maine, Orono, ME

The influence of pretreatment chemistry and the severity of dilute acid pretreatment on the subsequent enzymatic hydrolysis of soluble oligomer-derived and insoluble crystalline cellulose using a commercial enzyme preparation (Novozymes) was investigated. A dilute acid pretreatment was performed on three feedstocks: corn stover, pine and mixed hardwood, at 0.7 % H_2SO_4 (w/v) and 1% solid concentration with a 100 mL Accelerated Solvent Extraction (ASE, Dionex Corp.). The pretreatment conditions (residence time, temperature) with constant acid concentration were varied over a range of six reaction severities. Different pretreatment chemistries were studied on pine and included dilute acid, liquid hot water, buffered hot water, aqueous ammonia and lime, all carried out at the same severity.

Hydrolysis kinetics were analyzed using a Yellow Springs Instruments (YSI Corp.) glucose analyzer. Enzyme kinetics were measured at different dilutions of biomass hydrolysate, with all three sources of biomass causing inhibition of the enzymes to different degrees. Alkaline pretreatments were observed to cause less enzyme inhibition than acidic pretreatments. It was noted that the presence of soluble oligomers in the hydrolysate resulted in apparent diminution of the inhibitory effect, resulting from an increased release of oligomer-derived glucose at higher concentrations of hydrolysate. Correction for this glucose contribution from the oligomeric sugars rendered the inhibitory effect of the hydrolysate more apparent.

Analytical interrogation of the hydrolysates quantified 40 different compounds present in the hydrolysates. Comparison of analytical to hydrolysis results indicate that enzyme inhibition can be correlated to the presence of select degradation compounds present in the hydrolysate.

See more of this Session: Separation Processes In Biorefineries II: Ionic Liquids, Reactive Separations and Other Novel Processes
See more of this Group/Topical: International Congress on Energy 2011

(610g) Recovery and Characterization of Plant Cell Wall Biopolymers After An Alkaline Pre-Extraction of Hardwoods

Ryan J. Stoklosa, Chemical Engineering & Materials Science, Michigan State University, East Lansing, MI and David Hodge, Department of Chemical Engineering and Material Science, Michigan State University, East Lansing, MI

In an alkaline environment hemicellulose and lignin compounds within plant cell walls have the ability to be extracted into an aqueous form. This is partly due to the amorphous structure of the hemicellulose polysaccharide and its morphological placement in the cell wall. Hardwood trees, a key feedstock for pulp production in North America and northern Europe, contain a hemicellulose polymer of β(1-4) linked xylose residues with side chain substitutions of acetyl and 4-O-methylglucuronic acid units. During an alkaline extraction the hardwood xylan hemicellulose chain is fractured into shorter chain oligomers without significant degradation due to chain stabilization by the 4-O-methylglucuronic acid substituent. After extraction the dissolved material can be removed from the aqueous form through the use of an organic liquid. These steps significantly impact the hemicellulose by providing changes to the chemical and physical structure of the polymer. To understand the changes taking place during extraction and precipitation the recovered material was analyzed for total composition and molecular weight. The hardwood species of trees tested included *Acer saccharum* (sugar maple), *Betula pendula* (silver birch), *Populus nigra X. P. maximowiczii* cv. NM6 (a hybrid poplar), and *Populus tremuloides X. P. tremula* (a hybrid aspen). A sample of switchgrass, *Panicum virgatum* cv. Cave-In-Rock, was used as a comparison feedstock to see differences in the chemical characteristics of HC. Sugar maple and switchgrass performed the best in terms of biopolymer recovery exceeding 70% of the total amount of material dissolved after extraction. Solvents, such as ethanol and methanol, were compared to see if the route of precipitation varies depending on the solvent used. The performance of either solvent in recovery of the dissolved biopolymers will help to understand overall solubility characteristics of the extracted cell wall material.

After precipitation the recovered material was then subjected to chemical and physical characterizations to understand the factors that influence hemicellulose solubilization during extraction and precipitation. Composition studies show how the chemical make-up of the polymeric material can influence recoverability. Physical studies on the material, such as size exclusion chromatography (SEC), allow for analysis on the polymers actual size once extracted and recovered. Degree of polymerization (DP) of the xylan chain was also checked with a novel spectrophotometric method to determine the number of sugar reducing ends by using 2,2'-bicinchoninate acid (BCA). These studies show that changes in the biopolymer influenced by extraction and precipitation affect the overall solubility. The total amount of material recovered is not entirely polysaccharides due to non-covalently and covalently linked lignin fractions being extracted and recovered simultaneously with the hemicellulose forming lignin-carbohydrate complexes (LCC). These compounds, due to the aromatic linkages, have been known to influence polymer aggregation that can inflate molecular weight analysis by SEC. The BCA procedure is sensitive to these compounds as well. A bleaching step with hydrogen peroxide was applied on the precipitated material to remove these interfering compounds for more accurate characterization.

See more of this Session: Separation Processes In Biorefineries II: Ionic Liquids, Reactive Separations and Other Novel Processes
See more of this Group/Topical: International Congress on Energy 2011

(675a) Microbial Oil Production by Oleaginous Fungi and Yeasts Using the Hydrolysate From the Dilute Sulfuric Acid Pretreatment of Wheat Straw

Xiaochen Yu, Jijiao Zeng, Yubin Zheng and Shulin Chen, Department of Biological Systems Engineering, Washington State University, Pullman, WA

Microbial oil production by oleaginous fungi and yeasts using thehydrolysate from the dilute sulfuric acid pretreatment of wheat straw

Xiaochen Yu, JijiaoZeng, Yubin Zheng and Shulin Chen

Department ofBiological Systems Engineering, WashingtonState University,Pullman, WA 99164-6120

Abstract

Takinginto account the fundamental problems around the world: hunger, the lack ofenergy and the deterioration of the environment, biodiesel production fromlignocellulosic biomass has increasingly attracted much attention, since itdecreases the potential to spark conflicts between food and energy issues byusing non-starch, non-food biomass, and additionally it emits less green housegas to the environment than conventional fossil fuels. Generally lignocellulosicbiomass is firstly hydrolyzed either chemically or biologically to fermentablesugars derived from cellulose and hemicellulose, which will be then converted tosingle cell oil by oleaginous microorganisms, as the feedstock for biodieselproduction. Our work comprehensively explores the use of the hydrolysate fromthe dilute sulfuric acid pretreatment of wheat straw for microbial oilproduction. In this study, we investigated the performance of the oleaginousyeast and fungi on both detoxified and non-detoxified hydrolysate. The results showedthat all the strains could use the hydrolysate to produce lipids anddetoxification of the hydrolysate was not necessary for optimal cell growth andlipid production by the fungus and yeast strains. Among all the tested strains,the yeast *Cryptococcus curvatus* andthe fungus *Mortierella isabellina*achieved 34% and 50% lipid content respectively on the non-detoxifiedhydrolysate with up to 20% lipid yield. The overall mass balance shown in Fig.1indicated that lipid yield using the hydrolysatefrom dilute sulfuric acid pretreatment of wheat straw was 4.7 g lipids/100 g initial dry biomass. Our finding is quite significant as (1) itelucidates the feasibility of microbial oil production from lignocellulosicmaterials and suggests an alternative strategy for biodiesel production, (2) itprovides the possibilities for the further scale-up and demonstration, whichwill accelerate the development of biodiesel production towards thecommercialization, (3) it simplifies the hydrolysate fermentation process andcould probably reduce the cost since the detoxification step is not required.

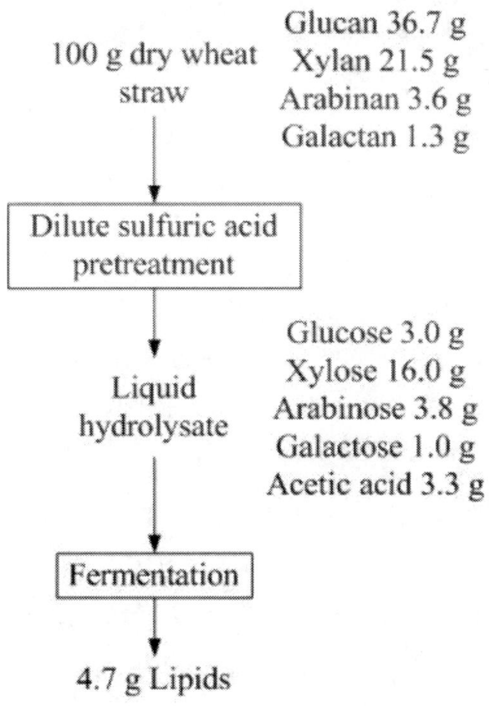

Fig. 1. The overall mass balance ofconverting lignocellulosic biomass into lipids

See more of this Session: Sustainable Forest Product and Process Innovation
See more of this Group/Topical: International Congress on Energy 2011

BIOCORE - On a paradigm for the design and synthesis of real-life biorefineries

Aikaterini D. Mountraki[α], Athanassios Nikolakopoulos[α], Konstantinos A. Pyrgakis[α], Bouchra Benjelloun Mlayah[b], Antonis C. Kokossis[α]

[α] School of Chemical Engineering, National Technical University of Athens, Greece

[b] Compagnie Industrielle de la Matičre Végétale, France

Abstract

As oil is a fast depleting resource (BP, 2010), much is expected by the production of biofuels as valid options to replace fossil-oil transportation fuels. Biorefineries are not an entirely new concept. Many of the traditional biomass conversion technologies such as the sugar and paper industries are aspects of the bio-refining approach. By adapting some processes (such as fermentation of sugars to produce ethanol) these existing industries can develop into biorefineries. Following this skepticism, biomass is trying to act as a replacement for oil and chemicals (P. Nieuwenhuizen, A. D. Little, 2010). Lignocellulosic biomass includes agricultural and forestry residues, municipal wastes and greases. It does not require the cultivation of productive land and uses materials which are normally considered waste, such as wood and crop residues. This paper presents the efforts made through the work of BIOCORE for the transition from an existing paper industry (CIMV), to a sustainable biorefinery, by integrating and assessing diverse transformation processes. Tools used for this purpose, are systemic methods of composition and integration, optimization and process modelling. In design problems, for which there is prior knowledge of chemistry and products available, the synthesis process provides solutions at a conceptual level, defining the optimal portfolio of the products to be produced and enables a systematic partition of options by simultaneous optimization approach. Anticipating the development of new solutions and options, this paper outlines the major challenges for the development effort and the design of new biorefineries. The realization of biorefineries is not far from reality and can emerge from many conventional processes playing an important role in the energy game and the chemical market.

Keywords: Biofuels, Process Design, Flowsheeting, Synthesis, Biorefinery.

(675c) Purification of Lignocellulosic Hydrolyzates by Nonionic Polymer Flocculation and Scale up Studies

Rakesh Yasarla, Paper & Bioprocess Engineering, State University of New York, Syracuse, NY and **Ramarao Bandaru**, Paper and Bioprcess Engineering, State University of New York, Syracuse, NY

Lignocellulosic hydrolyzates produced by hydrothermal treatment of wood chips contain hemicellulose sugars, acetic acid and significant quantities of colloidal material. These particles are mostly composed of lignin and its derivatives, which have a wide range of particle size distribution. Separation of these colloidal materials is necessary to improve fermentability of sugars into biofuels and other bioproducts. Flocculation of wood hydrolyzates prior to microfiltration improves their filterability. In this paper we investigated the separation of sugar maple wood extracts at bench and pilot scales (of 1000 kg) by a non-ionic polymer Polyethylene Oxide (PEO) with two different molecular weights (100 and 1000 kDa).

The rate of flocculation was monitored by sedimentation of the suspensions. The optimal dosage of PEO depended on the temperature and extract concentrations and varied between 20 to 50 ppm. Although the pH of the extracts was varied from 2 through 10, the effect on suspension stability was minimal. The optimal temperature for flocculation at optimal polymer dosage was 21.5°C. The composition of hemicelluloses in the supernatants after flocculation were not altered from the raw extract showing that flocculation does not remove significant amount of fermentable sugars. Hence such pretreatment clarification is not likely to affect downstream bioproduct yields.

Flocculation and clarification with PEO were demonstrated on large pilot scale batches of 1000 kg of extracts. Further the flocculated extract was filtered and the filtrate showed a 99.5% reduction in turbidity, from 12000 NTUs in the raw extract to 50 NTUs in the filtrate.

See more of this Session: Sustainable Forest Product and Process Innovation
See more of this Group/Topical: International Congress on Energy 2011

(675d) Revitalizing the Forest Bioproducts Industry: Steps Required to Overcome Innovation Fatigue

Jeff Lindsay, Asia Pulp and Paper, Shanghai, China, Cheryl Perkins, Innovation Edge, Neenah, WI and Mukund Karanjikar, NA

The challenges of the forest products industries, especially the pulp and paper industries, represent cases of "innovation fatigue" with several well-known "innovation fatigue factors" limiting progress and innovation. We discuss these factors and identify steps needed to revitalize innovation, including steps that can build culture and processes that will support innovation and new product development.

See more of this Session: Sustainable Forest Product and Process Innovation
See more of this Group/Topical: International Congress on Energy 2011

REMOVAL AND RECOVERY OF DYE FROM PAPER MACHINE EFFLUENTS

T. Utgata*, P. K. Mohanty**, S. Gupta, S. Chattopadhyay and A.K.Ray
IIT Roorkee, Saharanpur Campus Department of Paper Technology
Saharanpur, India
*Century pulp & Paper, Lalkua, Uttar Pradesh
**Ballarpur industries ltd., Orissa

ABSTRACT

For production of various kinds of coloured paper, dyes are employed. The un-adsorbed dyes go with the effluents to the water bodies. The dyes used are usually not biodegradable and therefore they must be removed by any means to make the process environment friendly. Adsorption of dye on a suitable adsorbent is thus a plausible alternative of abating colour (dye) pollution from paper machine effluents. Experimental studies have been carried out using Methyl violet and Victoria blue dyes as adsorbents, on adsorbents like wood meal, bagasse pith, saw dust and crushed paper for both batch as well as continuous packed column. Parametric effects of particle size, agitation, initial dye concentration, time, and temperature on the rate of dye removal have been investigated in batch as well as continuous adsorption column.

Results indicate that a rapid initial uptake of dye takes place attributing to surface mass transfer and the rate of dye removal is found to be reduced greatly at the later stage as intra particle diffusion becomes the rate-controlling step. The surface mass transfer coefficients, K_s correlated in terms of dimensionless $Sh/Sc^{0.33}$ term, against the particle variables and the intra particle rate parameter K as a function of agitation, initial dye concentration and particle size are simulated. (98.0-99.5 % of the dye adsorption has been possible within one hour. Regeneration of adsorbent and desorption of adsorbate have been tried. Experiments show promising results.

KEYWORDS

Colour organic Dye, adsorption, wood meal, bagasse pith, sawdust, methyl violet, Victoria blue, rate kinetics, process parameters, batch reactor, models in trap article diffusion, surface mass transfer coefficients

INTRODUCTION:

Effluents from paper machine section are colored due to the use of some colorants, such as colored pigments and dyes,organic direct, basic and sometimes acid dyes, the common dyes presently being employed by the Indian mills are: direct blue, C.I Basic green 4, C.I. acid yellow 36, aquamarine, acid orange, brilliant yellow, methylene blue, methyl violet, rhodamine, Victoria blue pergusol red, nigrosine etc.The commonly used coloured papers are coloured wove, coloured posters, envelope, matchbox etc.Their quantity, however, depends whether the end use is for cultural or decoration purposes.

The dyes are used in the manufacture of paper for a variety of reasons, including eye appeal, color-coding and brand identification. The selection of suitable colorants depends upon a number of factors such as end -use requirements, physical and chemical properties and handling characteristics. Coloring is done: (a) to produce a paper having given color or shade; and (b) to produce a white paper having a desired tint or "off-whiteness". To accomplish these ends the papermaker utilizes a number of different coloring materials that fall into one of five categories. These are non-biodegradable and cause pollution to the water bodies in terms of color, toxicity and loss of aesthetic value; they are stable to sunlight oxidizing and reducing agents and are resistant to aerobic or anaerobic digestion process. In addition to the above characteristics, color dye is one of the costliest items of paper mill inventory. Though various treatment techniques are available for removal of BOD and suspended solids from waste stream, no economically feasible process for dye color removal has been established till today. Of late stringent government legislations have enforced to spur the continuous research activities in order to remove and reuse dye from industrial effluent. Adsorption on solids appears one viable alternative to control the color displayed by dye.

The dye suspension is mixed in beaters/ mixing chests with pulp and sufficient retention is allowed for its stability and optical color pick-up by adsorption or by chemisorption mechanism. The unadsorbed dye goes with white water and gets diluted by fresh water showers. Actual color of the outgoing effluent or the extent of color deposition is rarely measured is the mills .A Major portion of white water is recirculated in well designed mill and a small portion is sent out as effluent.

The important parameters from industry viewpoint are: retention; the components and conditions of furnish (the furnish variables); the paper machine system and the interaction with other chemicals that affect color development obtained from the dye (e. g. Microbiol control agent). In most instances the papermaker desires to retain as much of an added dye as possible. This is possible by studying the equilibrium depending upon the shades-light or deep; heating the stock, adding salts to the dyed pulp and adding alum to the furnish after dye addition and adding a promoter i.e. a dye retention aid. The later is a low molecular weight, high charge density resins. Sources of interfering substances include fresh water and its constituents, pulping residuals, bleaching residuals and the broke.

It has been investigated that there are possibilities of measuring the color removal by adsorption on solid mass from both thick and effluents. Peat, wood meal/wood dust, saw dust, activated carbon, and bagasse pith and crushed pulp etc.; some of which are industrial solid wastes generating particulate pollution. The profitable utilization of the above solid particulate matters will also serve the development of an environmentally compatible process.

Literature Survey:

The data from paper mil survey clearly indicates that the most important paper products use dyes like methyl violet and Victoria blue. Their quantity also varies significantly.

There are innumerous information available about the removal of various pollutants including dye from waste water, either from many chemical process industries ,like sugar mill, paper mill, textile mill , dye industries , distilleries etc. or synthetically prepared solution for removal of acid dye ,indigo dye , Victoria blue,direct dye violet and orange dye, metanil blue,BOD, COD of waste water streams (1-27). The adsorbents used were either from natural sources like saw dust, fly ash, wollastonite, rice husk , waste tea leaves, lignite, peat, wood chip ,babul wood bark, wheat straw, wheat husk,coconut or activated carbon.Space limits to indicate all of them. Some are mentioned here.

Basic dye using natural absorbents (Poots at al., 1976,1978), spruce wood meal for Teflon blue-acid blue 25 (McConvey and McKay, 1977), activated carbon as adsorbents (McKay, 1982) have been tried. The correlation between mass transfer coefficients and intra-particle diffusion rate parameter as function of three variables, namely initial dye concentration, agitation and particle size has been developed (McKay, 1982). The mechanism for preferential adsorption of high and low molecular weight dyes on different preferred sites of granular activated carbon (Paschal, 1976) and drastic reduction of surface area and efficiency due to thermal regeneration were evaluated (Paschal, 1976). The fundamentals of mass transfer in fluid particle and intra-article in slurries has been studied (Furusawa and Smith, 1973) and a model approaching a real system has been proposed. The subsequent researchers adopted the same model for simulation purposes of many important systems.

Freundlichs isotherm, Langmuir isotherm and Lagergren kinetic models are mostly used to explain the phenomena for practical applications. **Deshmukh(2007)** studied on adsorption of basic dye in aqueous solution using activated carbon and coconut shell and corncob as adsorbents. Adsorption kinetics was studied using Lagergren and Helffrichsing Models. **Gautam et al.(2011)**worked for the removal of Victoria blue in simulated waste water using sand as an adsorbent. Equilibrium isotherms were analyzed by Freundlich, Langmuir, Temkin, Dubinin-Radushkevich equations and the goodness of fitting the curve inspected using a linear regression analysis and another two nonlinear error functions; some-of- square –error and Chi square test χ^2. Vrious exponential parameters ,agitation contact times, initial dye concentration, dose of adsorbent, effect of cations and anions, pH, and temperature, of the solution affecting the adsorption behavior of dye on the adsorbent sand has been studied. The thermodynamic parameters, Gibbs free energy (ΔG_o), change in enthalpies (ΔH_o) and change in entropy(ΔS_o) have been calculated

In most of the study the parameter studied were adsorption materials, pH of the solution, adsorbent dose, particle size, initial concentration of dye, agitation , time, temperature and type of activation.

There is no data available regarding the removal of colour (dyes) used by paper mill as a function of input parameters, specially on the adsorbents like hard wood meal, bagasse pith, and saw dust, some of which are industrial solid wastes generating particulate pollution. The purpose of profitable utilization of the above solid particulate matters will also serve the development of an environmentally compatible process. Till now no study has been made on the colour removal kinetic characteristics of two major dyes, namely Victoria blue and methyl violet which are largely used by Indian paper mill. No information is also available on the pretreatment of the adsorbents, which are necessary for accurate design estimates. In the present study the colour removal kinetics of the aforesaid adsorbate- adsorbent combination are attempted in a simple batch reactor as well as continuous reactor. However, results of only batch adsorption experiments are shown in this article.

MATHEMATICAL ANALYSIS:

The wellknown ideal but approximate model of adsorption developed by **Furusawa and Smith (1973)** have been used in the present investigation for estimating removal of colour dye of paper mill with some additional conditions as follows:

The adsorbent must be free from any solute including the adsorbate. Furthermore, the particle size of all adsorbents passing through a particular mesh is assumed to have same specific surface area and pores. Shape factor of all adsorbents are considered equal to 1. This is true in the sense that all fibrous mass is cylindrical in nature. The adsorbent must be sufficiently porous and possesses a very little surface resistance and Knudsen diffusion will not take place.

Based on the above assumptions the following equation relating transient concentration change can be written:

$$\partial C_t/\partial t = -K_s S_s (C_t - C_o);\ \text{at } t = 0,\ C_t = C_o \qquad \text{.......(1)}$$

Where, C_o, C_t are concentrations of dye in bulk liquid initially and at any time t respectively and S_s, the outer surface of the adsorbent particles per unit volume of particle free slurry. The simplified form of the differential mass balance equation can be written in cylindrical coordinates with radius r of the adsorbent particles as:

$$D_v\ (1/r^2\ \partial^2 C_r/\partial r^2 + 2/r.\ \partial C_r/\partial r\ ) - \rho_P\ \partial C_{nr}/\partial t = \varepsilon_P\ \ \partial C_r/\partial t \qquad \text{.....(2)}$$

where C_r, the concentration in the liquid filled pores at radius r, C_{nr}, the concentration of adsorbed dye at radius at radius r of the particle, Dv, the total effective diffusivity, ε_p, the porosity of the adsorbent particle, ρ_P, the particle density. The boundary and initial conditions are:

$$D_v\ (\partial C_r/\partial_r) = K_s\ (C_t - C_o) \qquad \text{.....(3)}$$

and $\partial C_r/\partial r = 0$ when $r = 0$ .....(4)

$C_r = o$ at $t = o$ .....(5)

K_s being external mass transfer coefficient.

Since equilibrium is assumed during adsorption at interior site, n_r, concentration of adsorbed dye at radius r of particle filled pore at radius r are related by the isotherm

$$\partial n_r/\partial t = \partial/\partial C_r\ [K\ C_r/(1+m\ C_r)\ ]\ (\partial C_r/\partial t) \qquad \text{.....(6)}$$

Eq. 6 has been numerically solved to relate C_t vs t curves to the rate coefficients K_s and D_v.

When intraparticle diffusion resistance is negligible and when the isotherm is linear an analytical solution for C_t vs t is possible in the case $C_r = C_s$ and n_r is also uniform throughout. The equations representing the particle mass transfer, Eqs. 2 and 3 can then be replaced by

$$M\ \partial n_r/\partial t = K_s\ S_s\ (C_t - C_s) \qquad \text{.....(7)}$$

and also Eq. 6 becomes

$$dn_r/dt = K dC_s/dt \qquad \text{.....(8)}$$

Analysis of isotherm plotted in the form C_o /n (C_o M/X) against C_0 indicate the Langmuir constant K.

Small cylindrical particle size of saw dust, bagasse pith, and wood meal can be assumed to be spherical particles of diameter D_P and the outer surface area of the particles is obtained from M as

$$S_s = 6M/ D_P \quad \rho_t (1-\varepsilon_P) \qquad .....(9)$$

Eqs. 1, 7 and 8 with initial conditions from eq. 2 and t = 0 can be solved analytically to give

$$C_t/C_o = 1/(1+mK) + mK /(1+mK)[(-1 +mK)/mK]K_s S_s t \qquad .....(10)$$

Where, the constant in Langmuir adsorption isotherm is m, K, being the adsorption equilibrium constant, m, mass of adsorbent particles per unit volume of particle free dye.

From the contact time results, the mass transfer coefficient between bulk liquid and outer surface of particles, K_s can be obtained from initial slope of C_t/C_o vs t curve. Alternatively a plot of In[C_t/ C_o-1/(1+mK] vs ln t will give a straight line at t = o with intercept ln [(-1+mk)/(1+mk)]K_sS_s and slope Validity of the equation at t = o is based on the assumption that the influence of the intra particle diffusion does not affect the results and the isotherm becomes linear at t = o. Therefore from the initial value of (- 1+mK) / mK vs time, K_s can be evaluated.

Diffusion rate processes are usually expressed in terms of $t^{0.5}$. A plot of fraction dye adsorbed against $t^{0.5}$, if shows distinctly different linear regions having different slopes, then the initial part represents the surface adsorption followed by the intra-particle diffusion. The slope of X/M against $t^{0.5}$ is defined as the rate parameter k characteristic of the rate of diffusion in the region where intr-aparticle diffusion is the rate-controlling step.

The film mass transfer coefficients K_s can be presented in the dimensionless mass transfer Numbers $Sh/Sc^{0.33}$ where Sh is the Sherwood Number, Y/D_v.

EXPERIMENTAL

Materials

The adsorbents used in all experiments were Victoria blue (basic blue 26) and Methyl violet supplied by M/s Star paper mill Ltd. India in their original form. The adsorbents were hardwood meal (Eucalyptus Tereticornis) collected from the chipper section of the said mill; bagasse pith obtained after separating the cellulose from bagasse by dry depithing method passing through 30 mesh screen ; saw dust collected from local carpentry shops ,crushed waste paper stock and activated charcoal.

Experimental procedure

The adsorbents were pretreated with hot water and cold-water solubility using Tappi specification (TIS - 59) to remove the water-soluble coloring matter, extractives and resinous materials to ensure that they should not interfere with the colour used during invesigation.

The physico-chemical characteristics of the adsorbents like density, solubility and viscosity were measured. The determined final concentration of dye in the sample was determined spectrophotometrically using an ELICO spectrophotometer ((Model-CIL-125). First standard curves for different dye concentrations at a wavelength 550nm were draw. Size distributions of different adsorbents were determined using Tyler mesh screen sizes. The densities of all the adsorbents were evaluated.

The experiments were conducted in a series of batch reactors to investigate the effect of agitation at atmospheric conditions (303K) with 36 mesh fraction as particle size for both the absorbents, each having 200 ml volume. For bagasse and 100 ml for other adsorbents, the effect of initial dye concentrations with time were studied for all particle sizes, for wood meal, saw dust and bagasse pith for both Victoria blue and methyl violet as enumerated in Table 1.

The characteristics of adsorb ate- adsorbents and the conditions maintained are shown in Table 1.

Table 1 Operating conditions of Batch reactor

Parameters	Details
Adsorbents	Wood meal; 5 gms each Bagasse pith Sawdust
Particle Size, mesh	36, 44, 52,60, 100
Adsorbents	Methyl violet, 100ml / 200ml Victoria blue
Initial dye conc., gpl,	0.5, 1, 2
Time interval, hrs.	0.25-18.75
Temperature, K	303,318,348
Agitation speed, r.p.m.	300,450,550

The concentrations were measured at different time intervals from 3, 5.5, 6, 12 and 18 hours spectrophotometrically. The effect of temperature was studied for the same amount of absorbents for wood meal and bagasse pith only from three levels or temperature, 303,318 and 348K at constant volume of adsorbents and particle size of 36 mesh. Concentrations of dye were measured at three time intervals, namely 0.25, 0.75 and 1.5 hrs.

RESULT AND DISCUSSIONS

The experimental results obtained from the laboratory were analysed and some typical representative figures 1- 13 are shown to indicate the effect of some important parameters: **The above figures are interpreted as under:**

Effect of time:

Fig. 1 is drawn d**imensionless concentration C_t/C_o for bagasse pith of particle size 36 mesh** as a function of time and degree of agitation for both methyl violet and Victoria blue with bagasse pith as adsorbent. The figure also compares the behavior blue and Victoria blue with speed of the agitator as a parameter.

Figure 2 indicates effect of time on concentration profiles for both victoria blue and methylene blue for bagasse pith of 36 mesh at 550 rpm at 2 gpl concentration. It is evident from the plot that up to 15 min's time, nearly 83% for the input dye is removed in case of Victoria blue whereas 50-60% for the case of methyl violet. At the initial stage there is a sharp decrease of C_t/C_o value followed by a much slower dye removal rate in subsequent stage. For 88% dye removal approximately 45 min's are required for Victoria blue in comparison to 70% for methyl violet.

Effect of agitation:

Fig 1 also reflects the influence of agitation on the dye removal kinetics. The results indicate that for the methyl violet dye, the rate of colour removal from the bulk solution is controlled by the degree of agitation. The effect is not so much sensitive in case of victoria blue. However in both the cases it is found that higher the degree of agitation more is the rate of dye removal. The effect of increasing stirring rate is to decrease the resistance to mass transfer in the film surrounding the adsorbent particles. It is evident that there are two distinct regions to each curve. When the solid adsorbent is added to the solution at the start of each adsorption reaction, there is an initial rapid rate of dye removal. The initial region of the plot is attributed to adsorption of dye particles on to the surface of adsorbent particles. At about 15 minutes after start of each run, the rate of dye removal starts to fall off to give a smooth curve. This part of the curve is probably due to intraparticle diffusion. As the entire process is in unsteady state large time period is required to reach 99.9% dye removal.

Effect of initial dye concentration

The results of a series of contact time- concentration experiments for varying initial dye concentration are shown in **Figure 3** as dimensionless concentration, C_t/C_o vs time with initial dye concentration as a parameter.

Figure reveals that the values of C_t/C_o with time for both Victoria blue adsorbate with initial concentrations of 0.5(gpl) kg/m^3 and 2 kg/m^3 respectively. In other words, higher the concentration better is the initial rate of dye removal. But this rate as expected is much lower than those with agitation. Influence of agitations is therefore is very much clear. Further increase of initial concentration beyond 2 kg/m^3 is impractical. As explained earlier, that plot of dimensionless concentration of effluents as a function of reaction time at different rpm, the two regions exist for each curve. The initial region of rapid adsorption on particles lasts for a longer period of time than with lower initial dye concentrations before intraparticle diffusion fully controlling the process. This observation is further explained by the fact that for the same particle size, in each run, the greater the initial dye concentration the greater is the driving force between the bulk solution and particle surface.

Effect of varying particle sizes

Figs. 4-7 indicate the value for concentrations for various particle sizes used in the experiment for Victoria blue and Methyl violet at the same initial concentration, 1 gpl and 2 gpl.for various types of adsorbent.

It is quite clear that with decrease in particle size the value of concentration decreases with time exponentially. This can be explained by the fact that the surface area exposed for adsorptions increased as the particle size is decreased; thereby more mass transfer is inevitable. However the net gain is not much as anticipated for bagasse pith as adsorbent though initially little advantage is obtained for smaller particle sizes. But finally 36 mesh sizes scores over the other. This fact is due to the agglomeration of smaller size particle by prolonged exposure which inhibits mass transfer by increasing the resistance.

Comparison of nature of adsorbents

Figures also compare the relative efficiency of all the adsorbents i.e. bagasse pith, saw dust and wood meal. It is evident that for bagasse the dye removal rate is much faster than those of saw dust and wood meal for both the adsorbents. Wood meal possesses the slowest rate. 90-95% dye removal is accomplished in case of bagasse at about 3 hours' time in comparison to 78-80% for sawdust and maximum 50% for wood meal at the same particle size, and temperature and for methyl violet adsorbent. For victoria blue 95% is achieved within 1.25 hours' time for bagasse whereas at 3 hours 97% for sawdust and maximum 85% in case of wood meal are estimated. A similar behavior is observed for all particle sizes. For cases of sawdust and wood meal, however 36-mesh size gives better performance than other smaller size particles. It may be the fact that the extractives (alcohol- benzene, ethyl ether etc.) or resins present in these adsorbents have the tendency to agglomerate the particles in presence of dye, transforming them to be impervious to mass transfer.

Effect of temperature

The effect of temperature on the dye removal kinetics is shown in **Fig. 8 for** three temperatures, namely, 303K, 318K and 348K for both Victoria blue and methyl violet respectively at three time intervals using bagasse pith of particle size of 36 mesh only.

As it is expected higher the temperature more is the rate kinetics. The increase in rate is due to lowering of viscosity and hence reducing the film resistance.The resistance in the pore diffusion is unaltered which is considered to impose maximum resistance. Therefore not very significant improvement can be achieved. The results at still higher temperature 363K did not yield satisfactory performance because of significant vaporization of solvent at atmospheric pressure and the adsorption equilibrium became disturbed. Hence it becomes unsuitable for prediction or for design use.

The degree of adsorption by the two dyes were more clearly shown in Figs. 9-10.

Evaluation of various parameters

The density, surface area per unit volume and diffusivity of wood meal are 92 kg/m^3, $5.5 cm^{-1}$, 2 $X10^{-5}$ cm^2/s respectively. Whereas the density and viscosity of methyl violet and Victoria blue are 1040,1004.9 kg/m^3 and 0.99 and 1.16 mPas respectively. Using the Furusawa and Smith adsorption model and based on McKay's suggested procedure as already explained in the section of mathematical analysis, the following parameters were estimated graphically: Langmuir Isotherm analysis (from C_o M/X and C_o relationship at various particle diameter; the Langmuir Constant is of the order of 217 cm^3g^{-1}.); the value of mass transfer coefficients between the bulk solution and particle surface from the initial slope of the In (C_t/C_o-1 /(1+mK) against t plot; the effect of diffusion from X/C_o i.e. the fraction of dye adsorbed and $t^{0.5}$ at varying particle sizes; The rate parameters determined from the slopes of the linear parts of graph of X/M against $t^{0.5}$ and its dependence on particle diameter was determined.A plot of log k vs log of (d_{pm}, the mean particle diameter was also made. A straight line is obtained. K_s was determined through the plot of ln [vs t is aC_t/C_o- 1/(1+mK)]. Effect of intra-particle diffusion with varying degree of agitation and the resultant effect on mass transfer coefficient was estimated. It is found that increasing the mean particle diameter results in increase in K_s. The various parameters are shown in the following table.

D_P,mm	K_S	-log K_S	Sh	Log Sh	Sc	$Sh/Sc^{0.33}$	Log of $Sh/Sc^{0.33}$	Log D_P
0.417	0.38	0.42	792	2.8937	57717	21.2566	1.3	-1.37
0.244	0.377	0.4236	459	2.662	57717	12.319	1.09	-1.61
0.208	0.333	0.47753	343	2.535	57717	9.20	0.96	-1.68

This can be explained by the fact that a large surface area is presented to the dye molecules in solution. The opposite effect occurs for large adsorbent particles. The film mass transfer coefficients, K_s were transformed in the dimensionless mass transfer form $Sh/Sc^{0.33}$.

Fig.11-12 indicate the effect of particle size on rate parameter, mass transfer Sherwood No. and $Sh/Sc^{0.33}$ for bagasse pith. All give distinct linear relationship. The dependence of rate parameter K_S on particle diameter shows that the intra-particle diffusion is predominant rate controlling step during a large part of adsorption process.

Simulation study:

The model predicted data were simulated with practical data obtained from a model solution and the conditions of the mill.As there is no effluent available from any Indian mill , the above solution were prepared by pulping rice straw at the conditions of chemical cooking.The conditions maintained during the preparation of solution were as under:°SR:35,Rosin,2%, Alum,4%,Talcom,8-10%, Methyl Violet: 2 gpl which is required for manufacturing coloured poster paper. Trials were made for both methyl violet and Victoria blue with crushed paper as adsorbent. Adsorption of Victoria blue is found to be much higher than that for methyl violet. Approximately 99.9% is adsorbed in case of Victoria blue within 15 minutes time whereas 98.5% is adsorbed in case of methyl violet within the same time for a 5 ch slurry height and 95% is adsorbed at 3 cm slurry height. At most one hour is required for achieving 99% adsorption of the dye on the adsorbent.

RECOVERY OF DYE

Some experimental trials have been carried out to desorb the adsorbed dye from the exhausted adsorbent using several solvents. The solvent chosen were Methanol, Acetone, 1,4 Dioxane and Acetic Acid. Acetone gives best results whereas the acetic acid has poor performance. Almost all the dye was removed by the first three solvent treatment process and very minor amount remained with the adsorbent, which may be chemically reacted (due to chemi-adsorption mechanism) and they could not be removed inspite of use of larger amount of solvent. It may be explained by the fact that the adsorbent microstructure might have formed strong bonds with the dye molecules. The desorption capacity of acetic acid has been found quite low as compared to other solvents.

CONCLUSION

Adsorption rate decreases with an increase of time. Higher initial concentration, temperature, degree of agitation, and particle size improve the dye removal rate. However, the effect of initial concentration and agitation has pronounced effect than temperature and particle size. Without agitation the rate becomes much slower and hence time required will be much more.

Adsorption rate of Victoria blue is found much higher than that of Methyl violet. Approximately 88-90% of Victoria Blue added to the adsorbate is adsorbed within 15 minutes' time with proper agitation, (for the same degree of dye removal without agitation near about 75 minutes are required) using bagasse pith of 36 mesh size whereas 55% is adsorbed in the case of Methyl violet within the same time. Total treatment and regeneration time probably will not cross 2.5 hours' time.

Bagasse pith has been found to be the best adsorbent compared to sawdust and wood meal. Saw dust will be the next preferred one..

Higher particle size than 36 mesh is found to decrease the dye removal rate in case of sawdust and wood meal.

The influence of rate parameter shows that the intraparicle diffusion is predominant rate controlling step during a large part of adsorption process. The equilibrium data follows Langmuir adsorption isotherm.

REFERENCES:

1. Ottamayer, W. (1930). The use of Peat and Wood for the treatment of textile effluent-Gesundhrits- Ingenieur, 53, 185.**(1930).**
2. Poots, V.J.P., McKay, G. and Healy, J.J. (1976). The removal of acid dye from effluents using natural adsorbents-II Wood,Water Research,10,1067(**1976).**
3. **McKay G. and Poots V.J., Kinetics and diffusion in color removal from effluent using wood as adsorbent,J.Chem.Tech. Biotechnol, Vol.30,Pp 279-281,1980**
4. McConvey I.F. and McKay, G. (**1977**) Dimensionless mass transfer correlations for colour removal from effluent using Wood meal as adsorbent. Indian Chemical Engineer,April-June,**1981**, Vol.. XXIII, No.2, PP.33-37
5. McKay, G. **(1982)** Intraparticle diffusion and surface mass transfer processes during colour removal from effluent dye carbon adsorption- Colourage 29,11
6. Paschal, B.D. (**1976**). Proceeding of 31st. Industrial Waste Water Conference, Purdue University, 175.
7. Poots, V.J.P., McKay, G. and Healy, J.J.,Removal of Basic Dye from effluents using Wood as an Adsorbent, J.W.P.C.F.,50,926(**1978**).
8. Furusawa T.,and Smith J.M.(**1973),** Fluid particle and Intraparticle mass transfer in slurries, Industrial Engineering Chem. Fundamentals,12,197
9. Amin Nevine Kamal, Removal of reactive dye from aqueous solutions by adsorption in to activated carbon prepared from sugar cane bagasse pith,Desalination 223(**2008**) 152-161
10. G. McKay, J.F. Porter and G.R. Prasad, The removal of dye colours from aqueous solutions by adsorption on low-cost materials, Water, Air Soil Pollut., 114 (**1998)** 423–438.
11. W.T. Tsai, C.Y. Chang, M.C. Lin, S.F. Chien, H.F. Sun and M.F. Hsieh, Adsorption of acid dye onto activated carbons prepared from agricultural waste bagasse by $ZnCl_2$ activation, Chemosphere,45 **(2001**) 51–58.
12. S.P. Raghuvanshi, R. Singh and C.P. Kaushik, Kinetics study of methylene blue dye bioadsorption on bagasse, Appl. Ecol. Environ. Res., 2 **(2004**) 35–43.
13. G. McKay, G. Chl Wal Hul and G. Buning Chen, Film-pore diffusion modeling and contact time optimization for the adsorption of dyestuffs on bagasse pith, Chem. Eng. J., 84 **(2001**) 77–94.
14. G. McKay, M. El-Geundi and M.M. Nassar, Adsorption model for the removal of acid dyes from effluent by bagasse pith using a simplified isotherm, Adsorp. Sci. Technol., 15 **(1997**) 737–752.
15. G. McKay, M. El-Geundi and M.M. Nassar, Pore diffusion during the adsorption of dyes onto bagasse pith, Process Safety Environ. Protect., 74 **(1996**)277–288.
16. G. McKay, M. El Geundi and M.M. Nassar, Equi- librium studies during the

removal of dyestuffs from aqueous solutions using bagasse pith, Water Res., 21 **(1987)** 1513–1520.

17. N. Kannan and M.M. Sundaram, Kinetics and mechanism of removal of methylene blue by adsorp- tion on various carbons — a comparative study, Dyes Pigments, 51 **(2001)** 25–40.
18. V.K. Garg, R. Gupta, A.B. Yadav and R. Kumar, Dye removal from aqueous solution by adsorption on treated saw dust, Bioresour. Technol., 89 (**2003**)121–124.
19. P. Nigam, G. Armour, I.M. Banat, O. Singh and R. Marchant, Physical removal of textile dyes from effluents and solid-state fermentation of dye- adsorbed agricultural residues, Bioresour. Technol.,72 **(2000)** 219–226.
20. S. Senthilkumaar, P. Kalaamani, K. Porkodi, P.R. Varadarajan and C.V. Subburaam, Adsorp- tion of dissolved reactive red dye from aqueous phase onto activated carbon prepared from agri- cultural waste, Bioresour. Technol., 97 **(2006)**1618–1625.
21. V.J.P. Pools, G. McKay and J.J. Hearly, Removal of basic dye from effluent using wood as an adsorbent, Water Pollut. Control, 50 **(1978)** 926.
22. Vineeta Gautam , Karan Pratap singh, sanjay Kumar and Vijay Laxmi Yadav,Removal of Victoria Blue by using Ganga River Sand as a low cost Adsorbent, Proc. Of 13th Conference of International Academy of physical sciences , held at University of petroleum and energy studies on 14-16th June,**2011** pp.222
23. McKay G,M.J.Bino and A.R.Altamami, The absorption of of various pollutants from aqueous solution on to activated carbon, Water Research,19(**1985**),491-495
24. Pragya Tiwari and N.P.Shukla, Water Pollution Control-Color removal from Textile Effluent, ibid,V 12 –V-14-26.
25. Mall I.D. and K.Mangesh Kishore, Treatment of Metanil Yellow bearing wastewater using coal flyash, Proc. Of the Eleventh National Convention of Chemical engineers, held at Roorkee between Sept.28-29,1995, PP.V-37-V-41,**1995**
26. Khare S.K., Pandey K.K., srivastava R.M. and Singh V.N., Removal of Victoria blue from aqueous solution by flyash, Chem. Tech. Biotechol 38,99-104,**1987**
27. Deshmukh, S.K.,Kinetic Performance and parameters evaluation for decolorization of industrial waste water using low cost adsorbent,Proceedings of 23rd National Convention of Chemical Engineers on recent trends in Chemical Engineering, Oct.,5-7 ,**2007,** pp.340-350

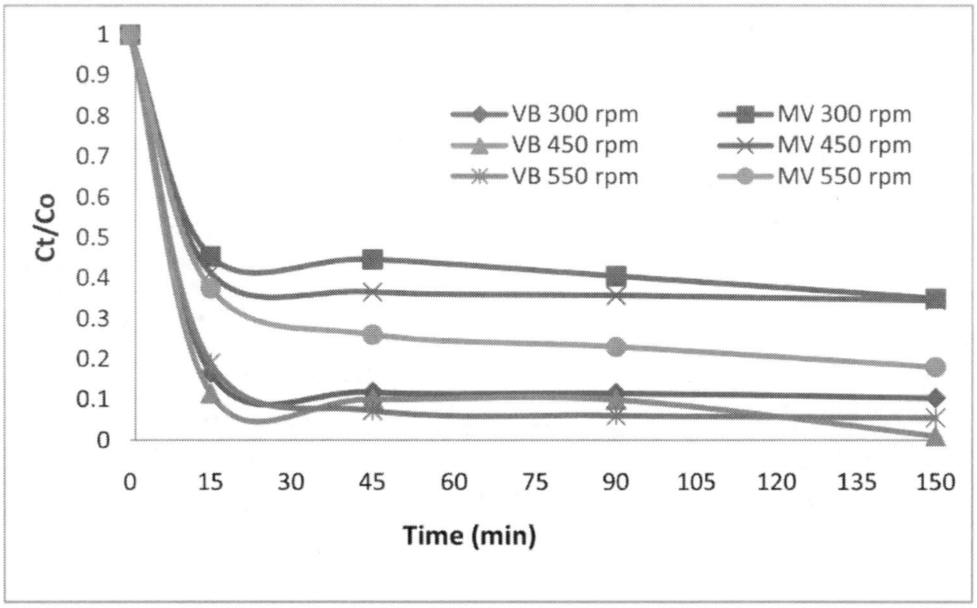

Figure 1. Effect of time and degree of agitation on dimensionless concentration for dyes, Victoria Blue (VB) and Methyl Violet (MV) with bagasse pith as adsorbent.

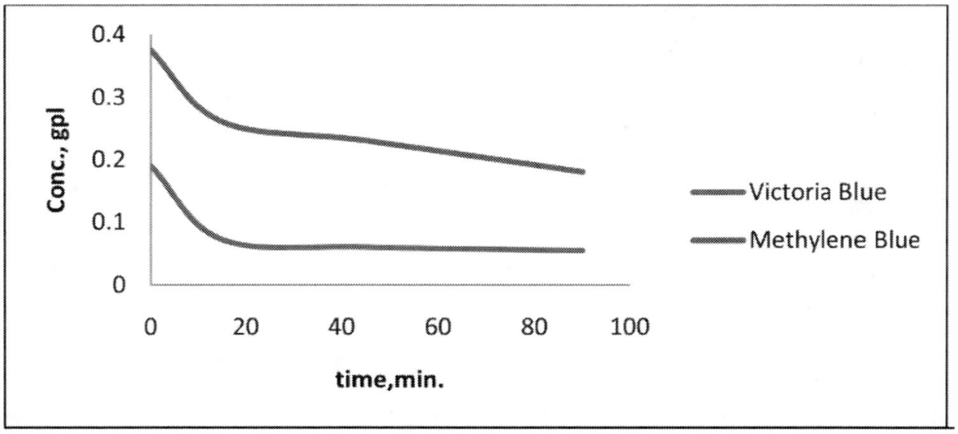

Figure 2 Effect of time on concentration profiles for both victoria blue and methylene blue for bagasse pith at 550 rpm at 2 gpl concentration

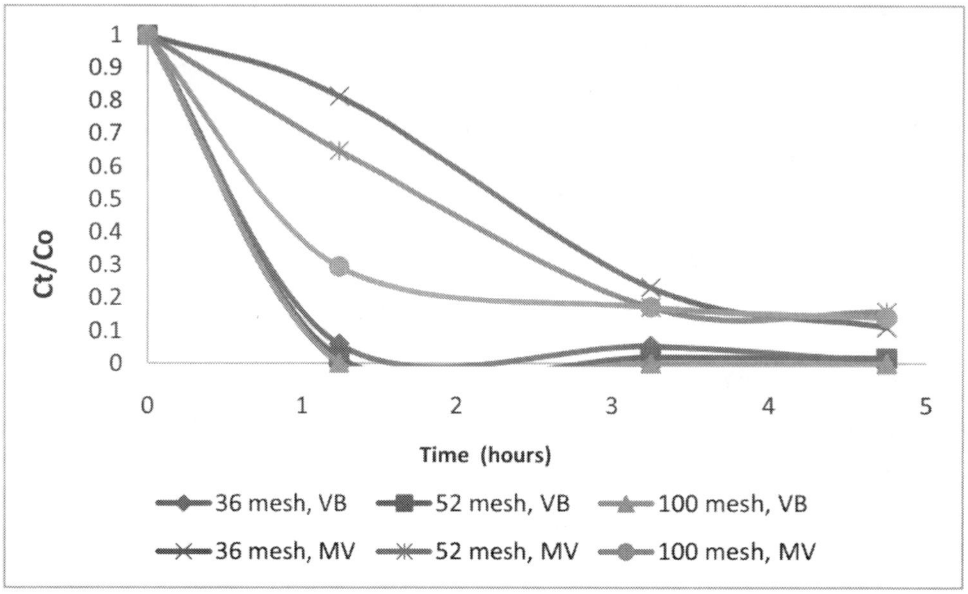

Figure 3. Effect of initial dye concentration at 2gpl on C_t/C_o profiles at various particle sizes

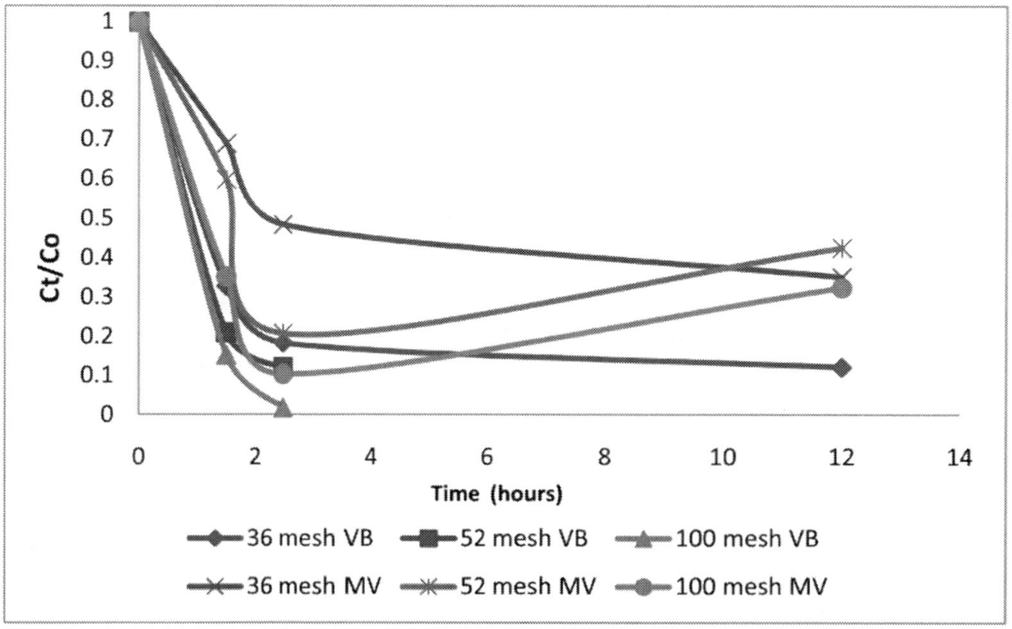

Figure 4. Effect of initial dye concentration at 1gpl for Victoria Blue and Methyl Violet.

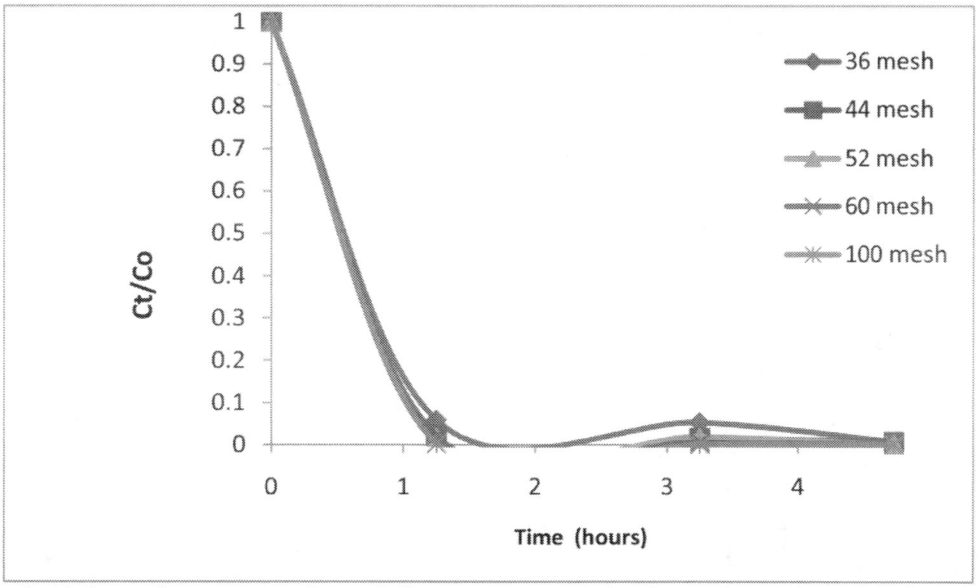

Figure 5. Value of concentration of five range particle size using Victoria Blue (VB)

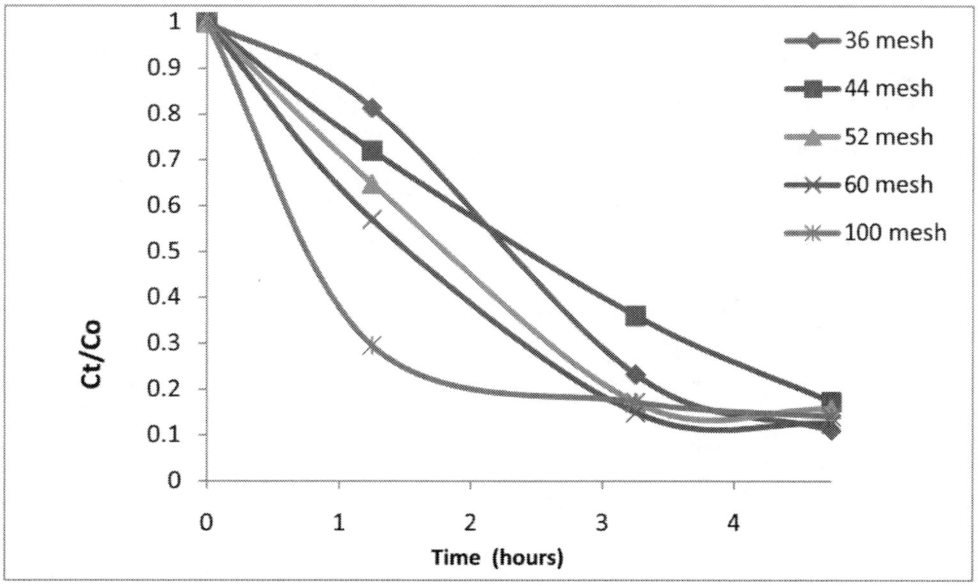

Figure 6. Value of concentration of five particle sizes using Methyl Violet (MV)

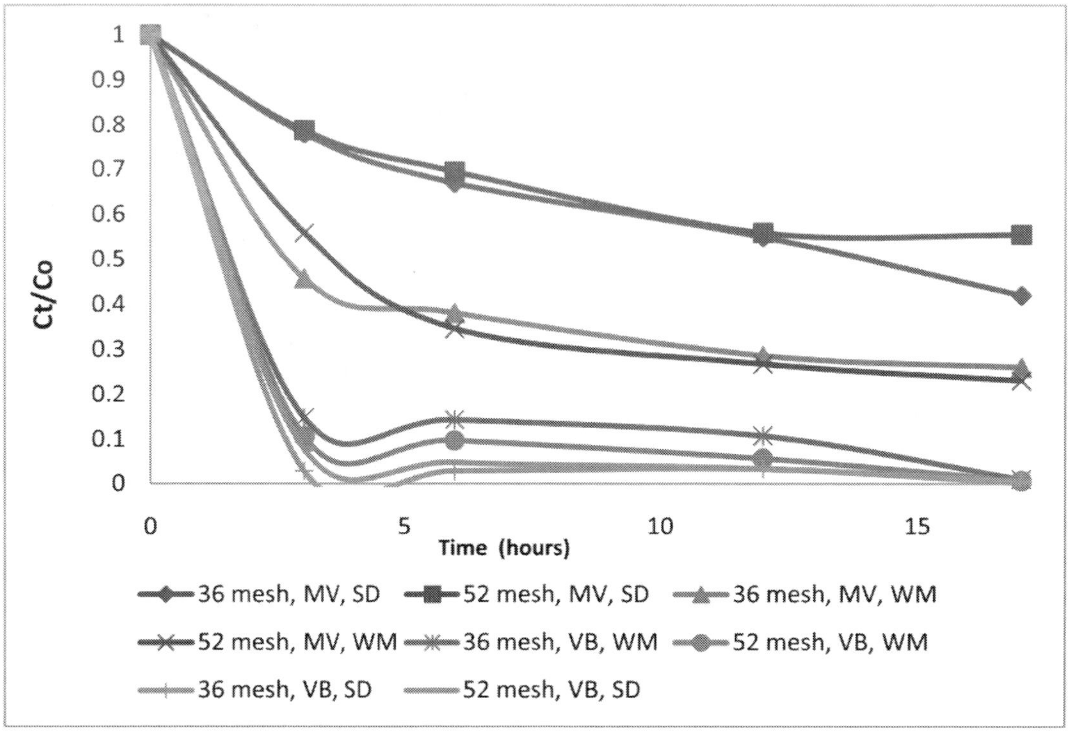

Figure 7. Effect of varying particle size for wood meal and saw dust

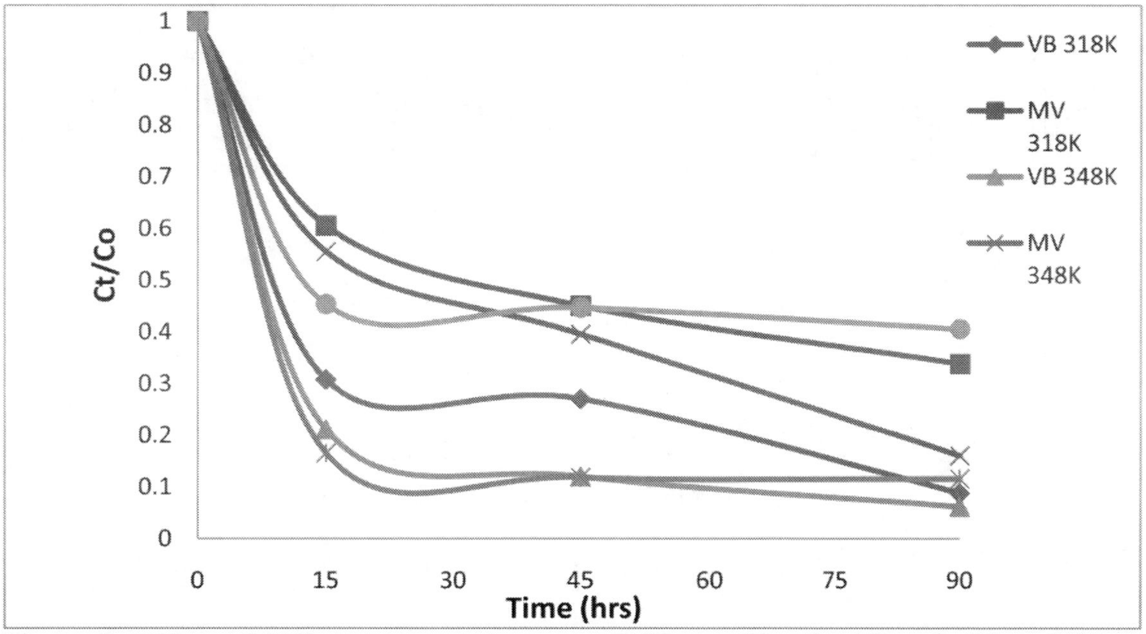

Figure 8. Effect of different temperatures on Ct/Co profiles for Victoria Blue (VB) and Methyl Violet (MV)

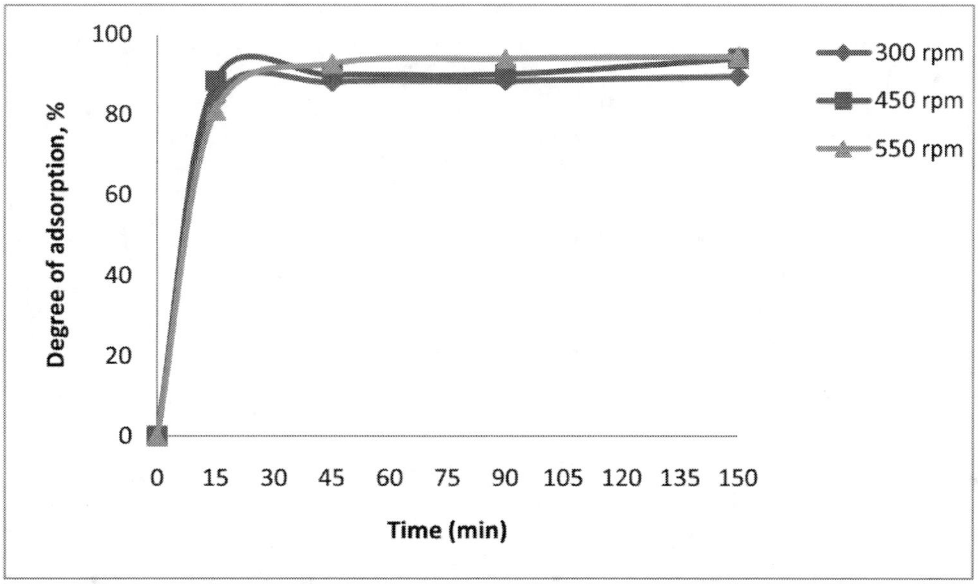

Figure 9.Degree of adsorption as a function of time with RPM as a parameter

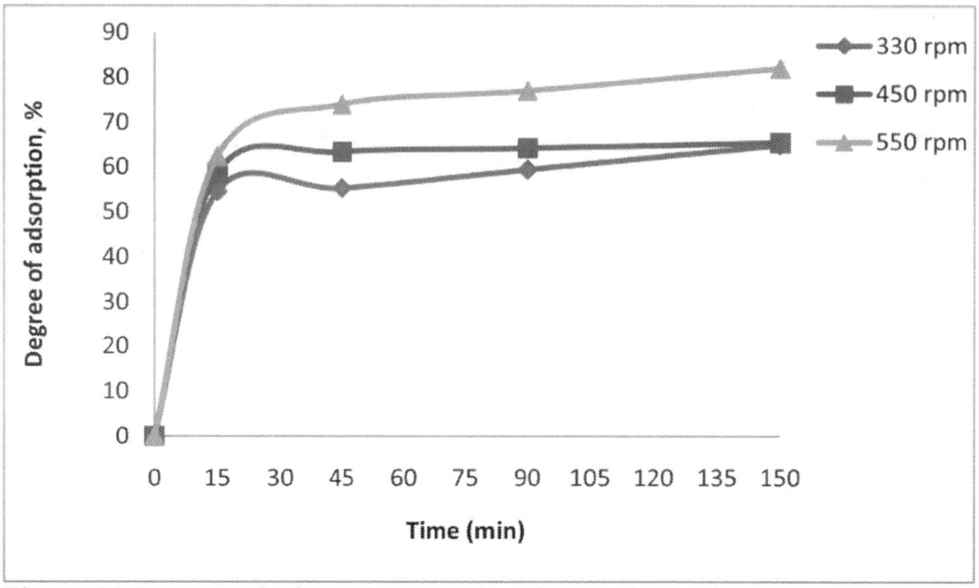

Figure 10.Degree of adsorption as a function of time with RPM as a parameter

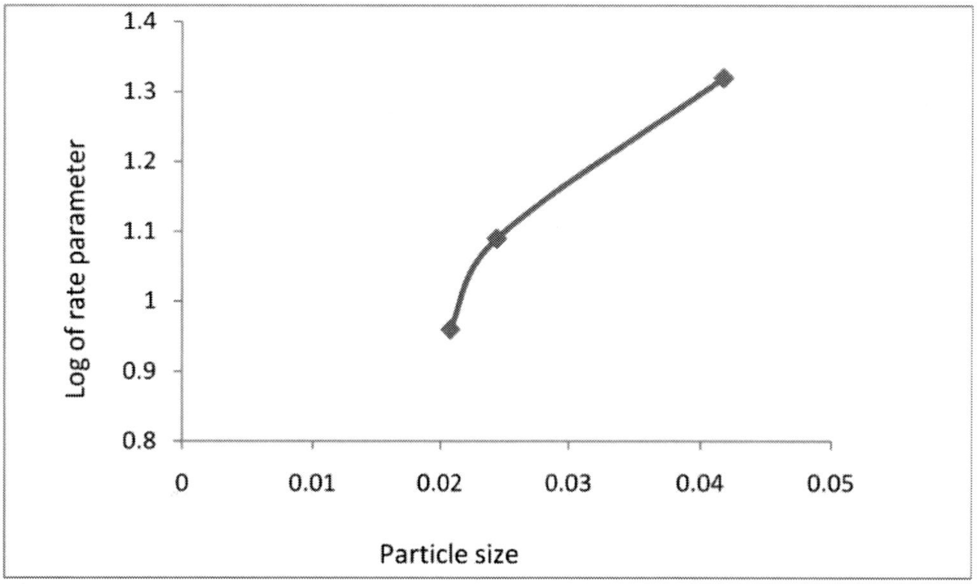

Figure 11. Logarithm of rate parameter versus particle size, Dp, for Victoria blue, for bagasse pith

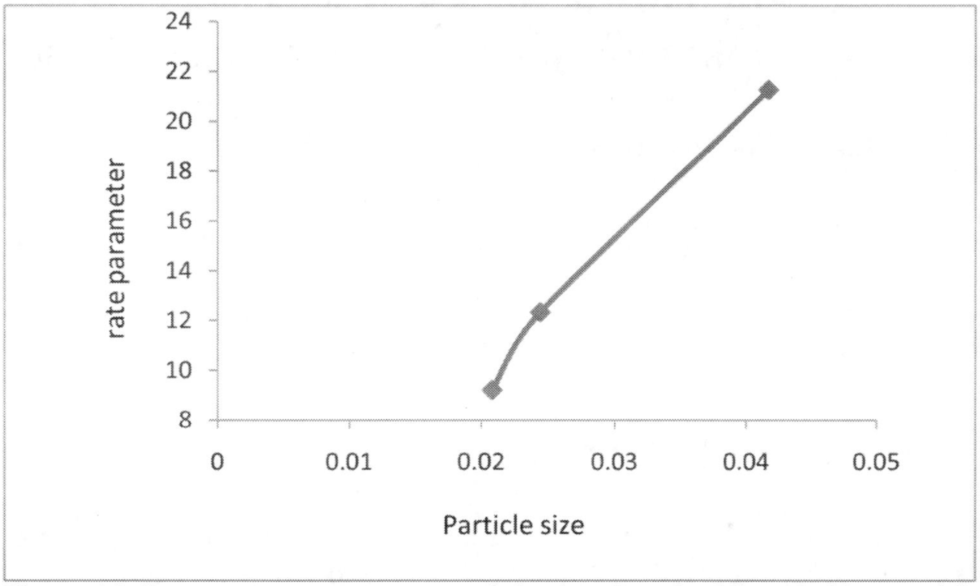

Figure 12. Rate parameter versus particle size, Dp, for Victoria blue, and for bagasse pith

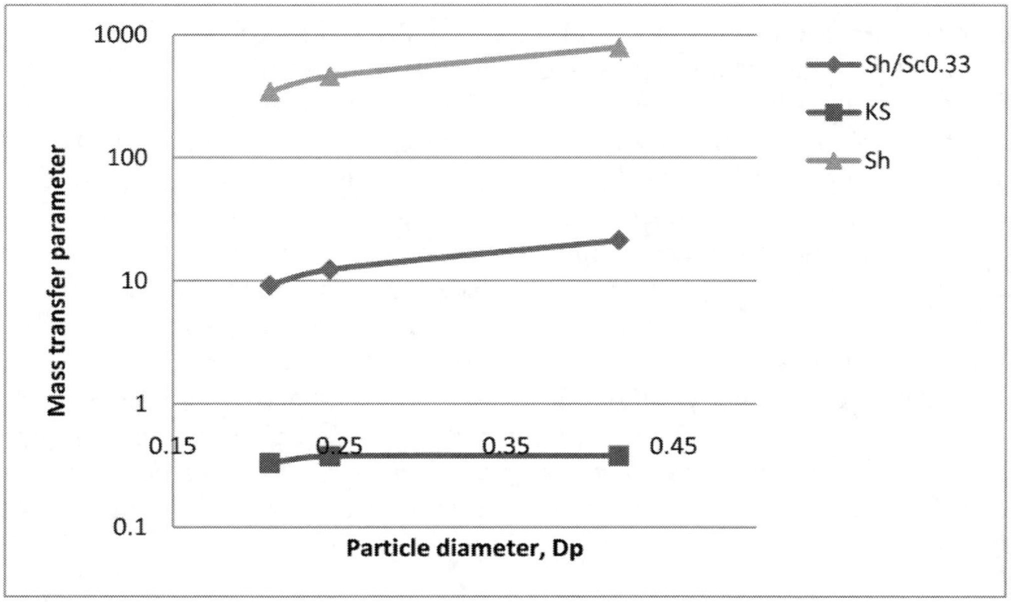

Figure 13. Effect of Particle size on rate parameter, Ks, Sherwood no. and $Sh/Sc^{0.33}$ for bagasse pith.

INFLUENCE OF CALENDERING PARAMETERS ON COATING LAYERS OF AGRI-RESIDUE BASE PAPER FORMED BY DIFFERENT COATING PROCESSES.

Sanjay Tyagi 1 # , ***A K Ray*** [2] *& R M Mathur* [3]

[1] *Scientist C, Paper Testing and Energy Management Division, Central Pulp & Paper Research Institute, Saharanpur, India*
[2] *Professor, Department of Paper Technology, IIT Roorkee, Saharanpur , India*
[3] *Director, Central Pulp and Paper Research Institute, Saharanpur, India*

ABSTRACT

The quality of coated paper is dependent on coating color composition, consisting of binders and pigments, coating process followed by quality of base paper and its source and other external process and design conditions such as type of coating equipments etc. One of the most important parameter next to coating color composition is coating process (such as blade coating, roll coating, air knife coating etc.). This is of particular importance for non wood based pulps which are relatively shorter in size and impart unevenness on the surface, thus causing the problem of fiber rising and fluff generation. It thus becomes very necessary to examine the response of coating process to non wood agro residue based pulp for growing Indian paper industry. Further, effect of calendering variables on the quality of coated paper made from agro residue needs to be examined. Unfortunately, extremely limited literature is available in this regard. In the present work, blade coating and film transfer coating processes were compared on paper made from the blend of agriculture residue pulps and that of softwood through exhaustive experimental trials made on pilot coater. Results indicated that blade coating process reduces the fiber and flock scale unevenness more efficiently than film transfer coating process resulting in improved paper gloss and print gloss mainly after the first pass only. Sheet opacity decreases more gradually in film transfer coating and occurs throughout the calendering nip. Calendering temperature plays more important role for blade coated paper and readily improved the sheet and print gloss. Reduction in PPS roughness of GCC coated paper is more during subsequent calendering trials as compare to china clay and blend of two. The effectiveness of calendering as described by "calenderability" showed increase in gloss and print gloss with increasing number of nips. The value of calenderability also founds to be in good correlation with calendering temperature and with formulations of different coating pigments.

INTRODUCTION

The main aim of coated paper calendering is to make the paper structure more suitable for printing. Coated papers, which are intended mainly for high quality products like magazines, fliers, catalogs and books are often printed in multicolor printing processes where low roughness and uniform pore structure are essential for print result. The quality of colour print is generally related to the uniformity of the coating layer which is affected significantly by the base paper characteristics as well as by the interaction between base and coating, by coating formulation, the coating process and drying and calendering conditions.

The surface of the coated paper is formed during the coating process through consolidation of the coating color and subsequent drying of the coating layer and the base paper. The coated paper surface properties as such, mainly depend on coating color formulation applied on the surface of paper. The coating color properties affects water retention in coating color and its transfer to base paper during coating consolidation at blade pressure and drying of paper [1]. Beside this, it also depends on the type of coating process used (i.e. blade coating, film transfer coating & surface sizing process), coat weight, number of coating layers and intensity of drying. In long dwell time blade coating the time between application and metering normally long enough for structural changes in the base paper to occur before the blade [2]. Because of the base paper roughness, the coating layers thickness after the blade is uneven. The paper becomes rougher during the consolidation process and during drying of the coating layer when water is removed from the coating color. Depending on blade pressure, evenness of coating layer and hence coated surface smoothness decreases after the blade due to spring back effect of under – the – blade compressed base paper areas. It is not very clear how different coating processes influence the formation of coated layer structure on paper and their consequent coated paper properties. In film transfer coating, there are less coating layer thickness variation than in blade coating. This is due to the nature of application process where an even layer of wet coating is applied to the surface of the base paper [3]. Also, the behavior of paper, coated with different coating process during calendering is not very clear. There is also no mention of studies on the mill made paper based on agricultural residues whose surface characteristics are entirely different than those papers made from hardwood and softwood pulps. There is a need to examine these aspects for the base paper indigenously manufactured from non-wood agricultural residue raw materials.

In the present study, an attempt has thus been made to explain how coating process affects the different properties of coated paper made from non-wood agricultural residue raw materials . The influence of calendering variables like number of nips and calendering temperature on coated paper properties were also studied. Printing properties were tested in terms of print gloss. The quantification of the calendering effect of gloss and print gloss of coated using delta (Δ) gloss is meaningful only if the initial uncalendered gloss values of coating are equal. Therefore, effectiveness of calendering process on gloss and print gloss of paper coated with different coating process was studied using a new term called "calenderability" (κ).

EXPERIMENTAL METHODS

All blade coating and film transfer coating experiments in this investigation were performed on pilot scale coater supplied by DT Paper Science, Finland. Pilot coating trials were carried out on the commercial base paper procured from the mill in reel form and made from agricultural residue pulp blended with soft wood pulp in the ratio of 75: 25. The coating pigments and other chemicals were also procured from the mill. The coating formulations used are given in Table I. The base paper was coated on single side with the coat weight of 17 g/m^2 at the speed of 30m/min. The distance between the applicator roll and blade was kept 250mm. The metering blade measuring 84 by 310 mm was used in this trial. The blade angle was kept at 31°. The blade load was adjusted by pneumatic cylinders at the pressure of 1.3 bar. For drying of paper two IRT- Monocassettes of infrared units were used to give a temperature of 60 – 70°C. The temperature of air drying unit was kept at 50°C. For film transfer coating, the coating solution was metered by applicator roll before applying to paper at the pressure of 2.1 bar. The web was then dried by IR dryer and air dryer.

I: Coating formulation used for blade coating and film transfer coating trials on pilot coater

Coating Ingredients	Parts in formulation		
China clay	70 parts	100 parts	-
GCC	30 parts	-	100 parts
Latex	15 parts	15 parts	15 parts
Thickener	0.5 parts	0.5 parts	0.5 parts
Dispersant	0.5 parts	0.5 parts	0.5 parts

All calendering experiments were carried out on laboratory single nip calender (supplied by DT Paper Science, Finland) running at the speed of 22 m/min at the linear load of 95kN/m. For calendering soft nip calendering mode was used. Since, the laboratory calender has only single nip, the paper was passed several times to increase the number of nips. (i.e. for 3 nips we have passed same sheet three times, for 5 nips we have passed paper 5 times and so on). All printing trials were made on laboratory IGT printability tester. (AIC 2 -5). Different paper and printing properties were evaluated using different ISO standard test methods.

RESULTS AND DISCUSSION

(a) Response of calendering nips on coated paper surface properties

The responses of calendering nips are expressed in terms of variation of surface, opacity and printing properties such as sheet roughness (PPS), printing opacity, sheet gloss and print gloss as a function of number of passes in the laboratory calender. The responses of multiple passes in a single nip calender do not exactly emulate the responses from multiple nips in actual calender, the results shown here can only used for comparison purposes between two kinds of coating processes. The results are depicted in Figures 1 to 4. From the detailed study the following note-worthy remarks can be drawn.

- In five pass or nip calendering process approximately 50% of total roughness decrease was occurred in the first nip. The reduction of roughness in calendering was larger for film transfer coated paper compared to those from blade coated paper (Fig.1). In both cases, the main roughness decrease occurred after first pass only. (i.e. about 50% in case of blade coating and 46% in case of film transfer coating). In subsequent 3 and 5 passes the roughness reduction was lower.
- The improvement in paper gloss was more for blade coated paper. Similarly, to roughness, the improvement in gloss was more after first nip (Fig. 2). The improvement in gloss in subsequent nips or passes was less for both cases. Both roughness and gloss results indicated that the deformation of blade coated paper structure occurred more easily in calendering. For blade coated paper, the roughness decrease (and consequently the gloss increase) occurred more during the first three passes and further change in gloss were small.
- The decrease in sheet opacity occurred in both cases, but decrease in opacity was more gradual in case of film transfer coating process and it occurred throughout the calendering sequence, while for blade coated paper maximum decrease in opacity occurred in first pass and there were little change in subsequent stages. (Fig. 3)

- The increase in print gloss of coated paper followed the same trend as the paper gloss. The maximum improvement in print gloss occurred after first pass in both cases. The print gloss improvement in subsequent nips was less in both cases. (Fig. 4)

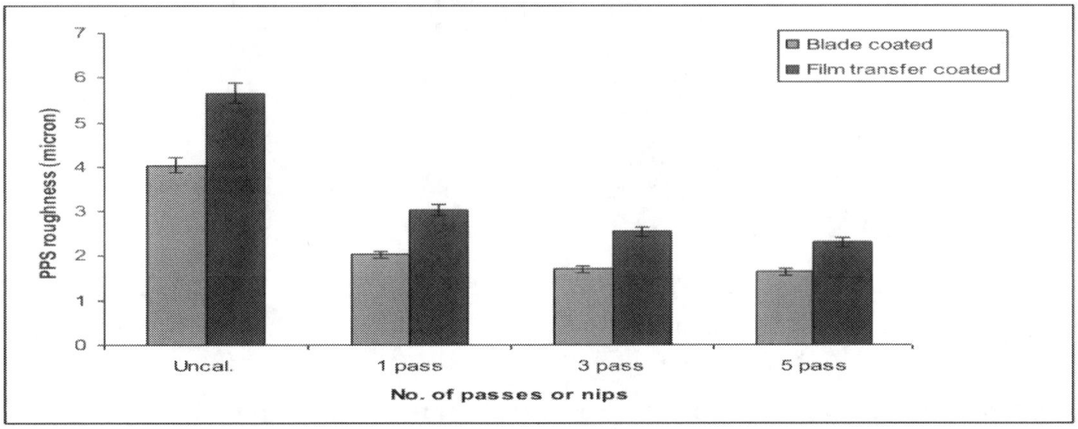

1. Response of calendering nips on PPS roughness of blade coated/ film transfer coated paper

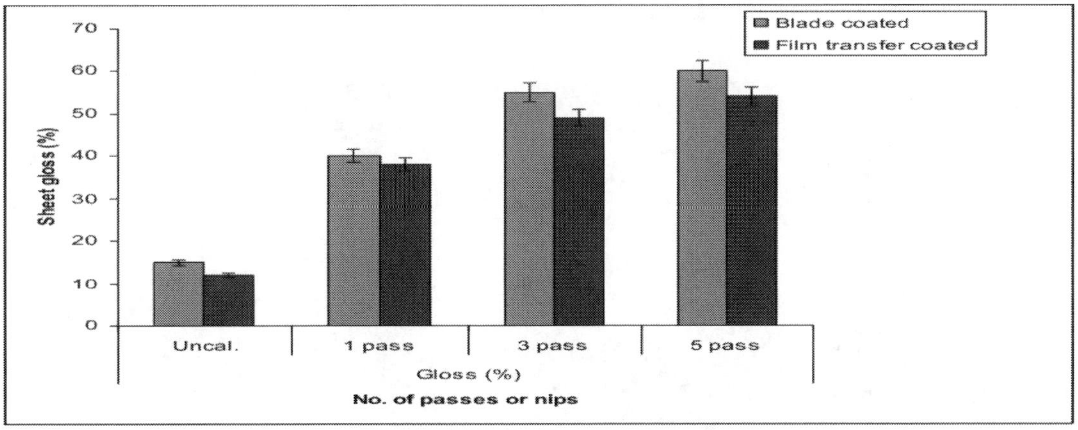

2. Response of calendering nips on sheet gloss of blade coated/ film transfer coated paper

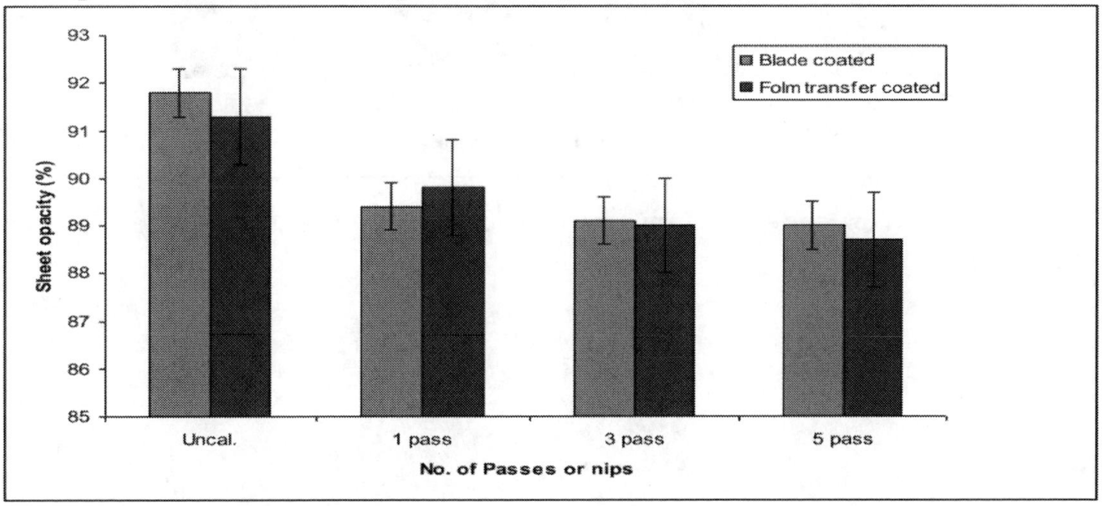

3. Response of calendering nips on sheet opacity of blade coated/ film transfer coated paper

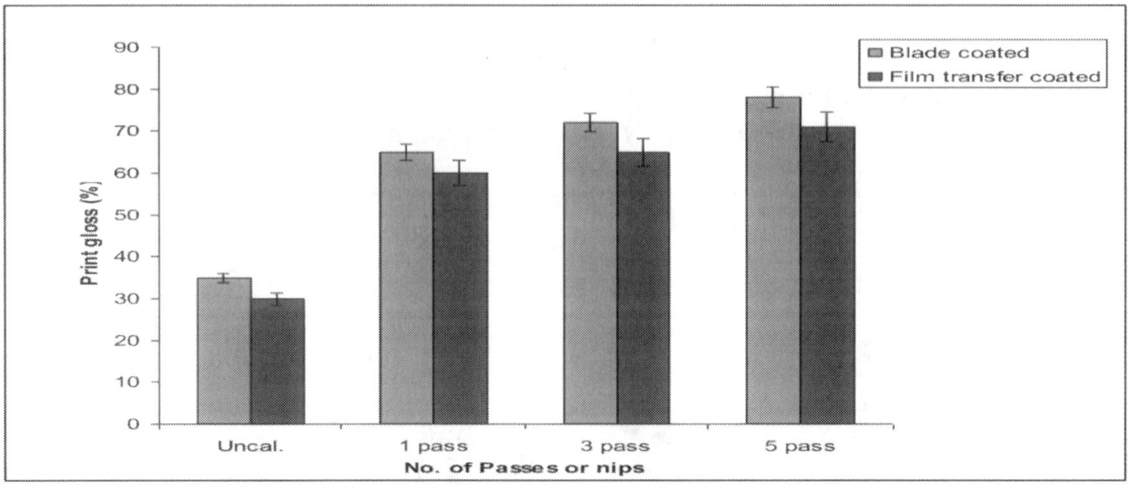

4. Response of calendering nips on print gloss of blade coated/ film transfer coated paper

(b) Effect of calendering temperature on the coated paper surface properties

The effect of calendering temperature on the gloss of paper coated through blade coating process and through applicator roll process is shown in Figures 5 through 8.

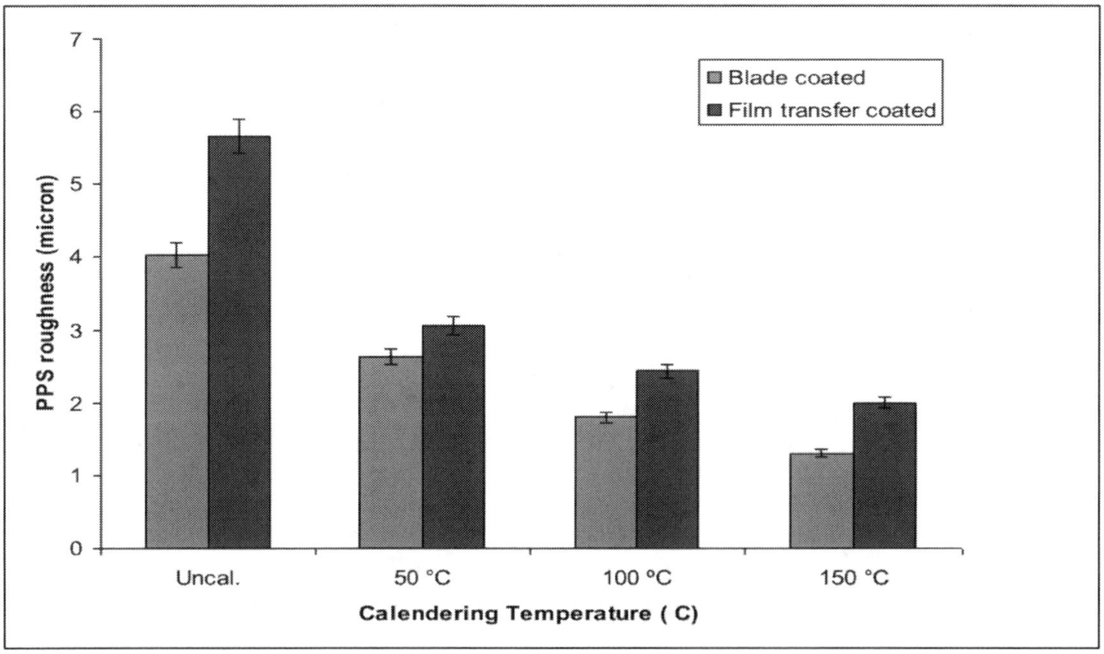

5. Response of calendering temperature on PPS roughness of blade coated/film transfer coated paper

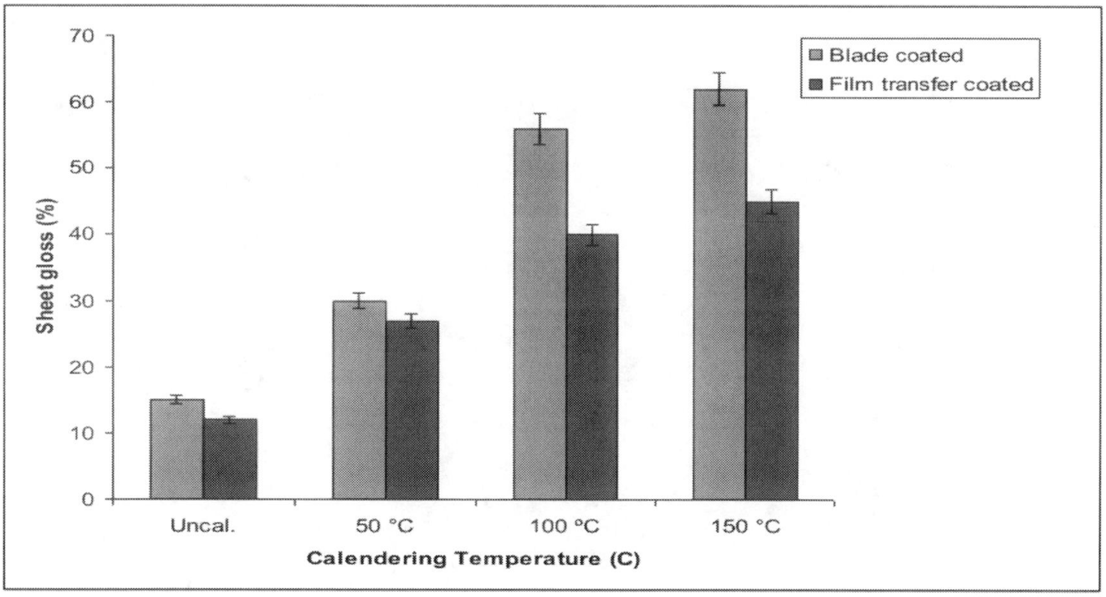

6. Response of calendering temperature on sheet gloss of blade coated/film transfer coated paper

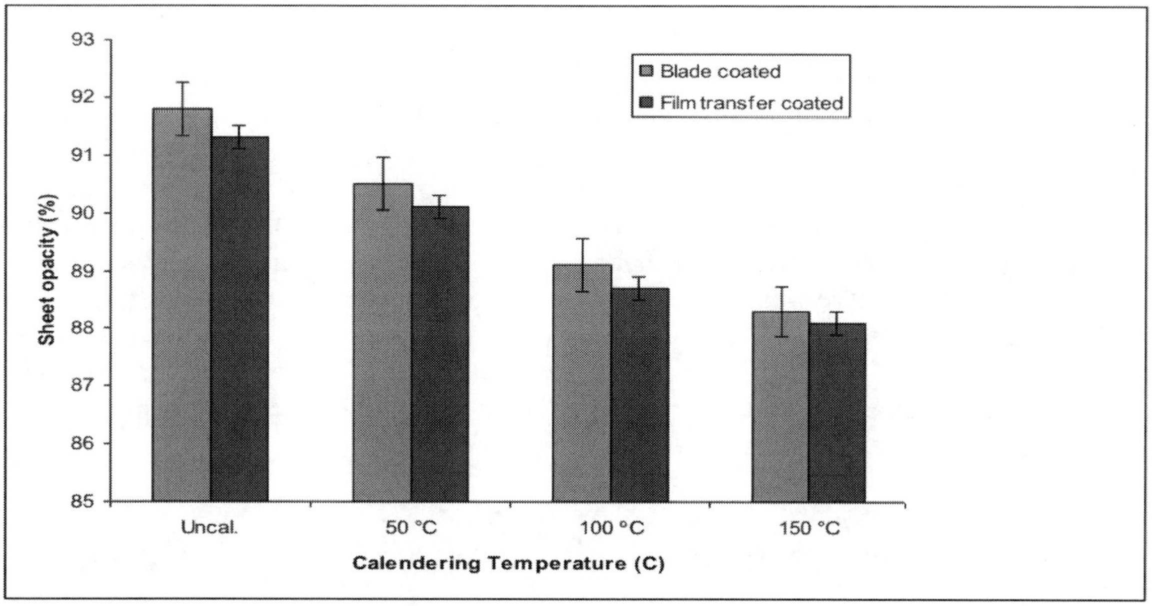

7. Response of calendering temperature on sheet opacity of blade coated/film transfer coated paper

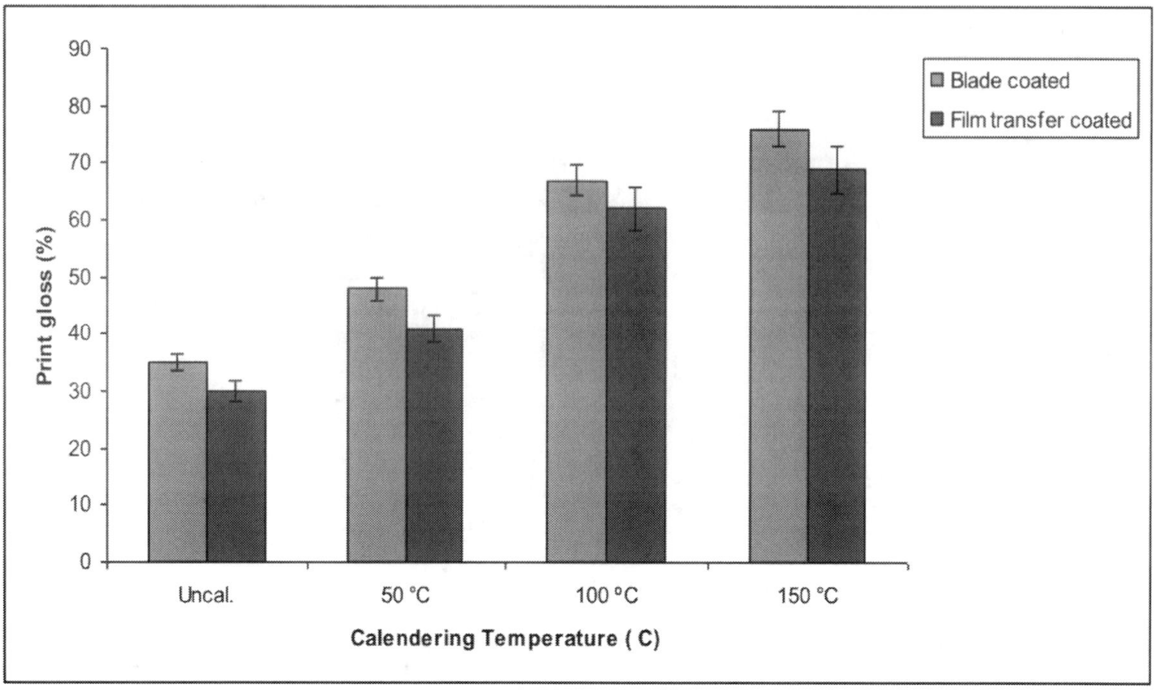

8. Response of calendering temperature on print gloss of blade coated/film transfer coated paper

The analysis of above figures revealed that an increase in calendering temperature increased the gloss and print gloss of coated sheet but reduced the sheet roughness. The main change in paper properties during calendering had occurred at lowest calendering temperature. The subsequent use of higher temperature had only a minor influence on properties of paper. The blade coated paper shows more improvement in gloss and print gloss (Fig. 6 & 8). The improvement in gloss is more upto 100°C in both cases. On further increasing the temperature there was not much improvement in gloss value. There was only a minor influence of temperature on roughness in both cases with film transfer coated paper showing more change as compared to blade coated paper. The effect of calendering temperature on opacity is given in Figure 7. It revealed that the decrease in opacity was nearly same for both blade coated and film transfer coated paper and occurs gradually throughout the calendering process

(c) Effect of base paper roughness on paper coated with different pigments

Roughness of base paper results in higher roughness of coated paper. It increases the coating non uniformity on paper surface and decreases coating coverage. This sometimes lead to print non-uniformity. The effect of calendering conditions of number of passes or nips is represented in figures 9 to 12. The main influence of the roughening of the base paper on the calendering behaviour of the paper was seen in the different roughness reduction of the clay coated and calcium carbonate coated paper. Clearly for uncalendered paper, the roughness of clay coated papers was lower as compared to ground carbonate coated papers. The roughness of paper coated with the blend of fine clay/GCC in the ratio of 70:30 lies in between that of fine GCC and clay. The experiment squeal of five calendering passes and temperature of 50 °C was although taken arbitrary, it is more practically at mill condition. It simulates more practically the temperature condition maintaining at industry of 50°C and number of passes limited to five with six roll calender stack. However, to reduce the reduction of bulk many industries attempted for five passes whereas one or three passes are the common cases of soft nip calenders.

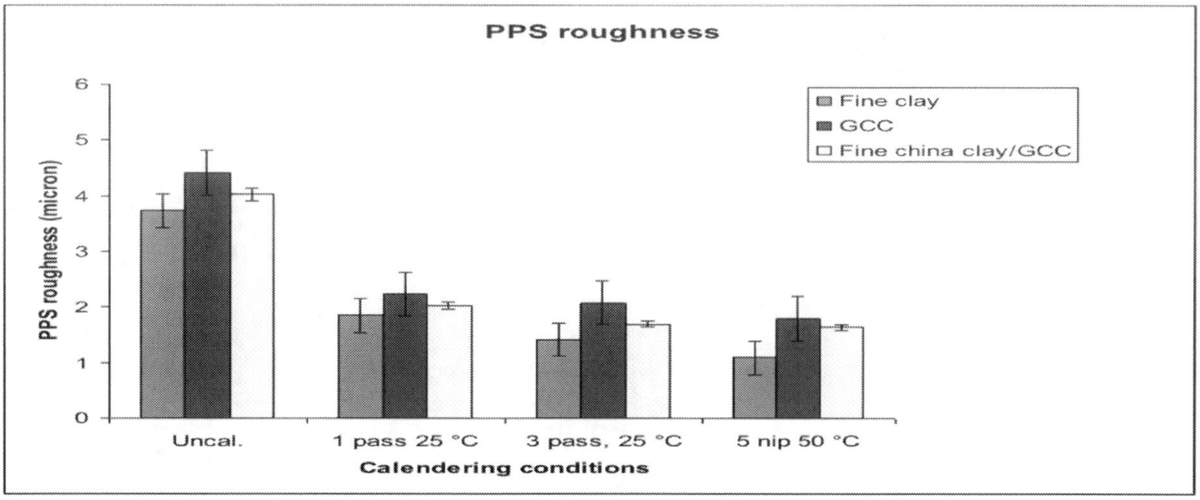

9. PPS roughness of calendered /uncalendered blade coated paper with different coating pigments

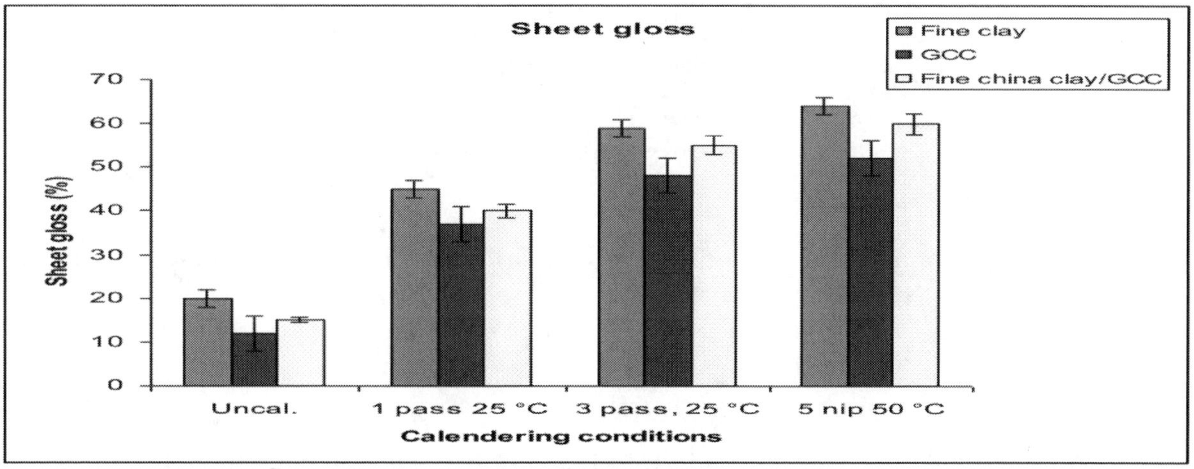

10. Sheet gloss of calendered /uncalendered blade coated paper with different coating pigments

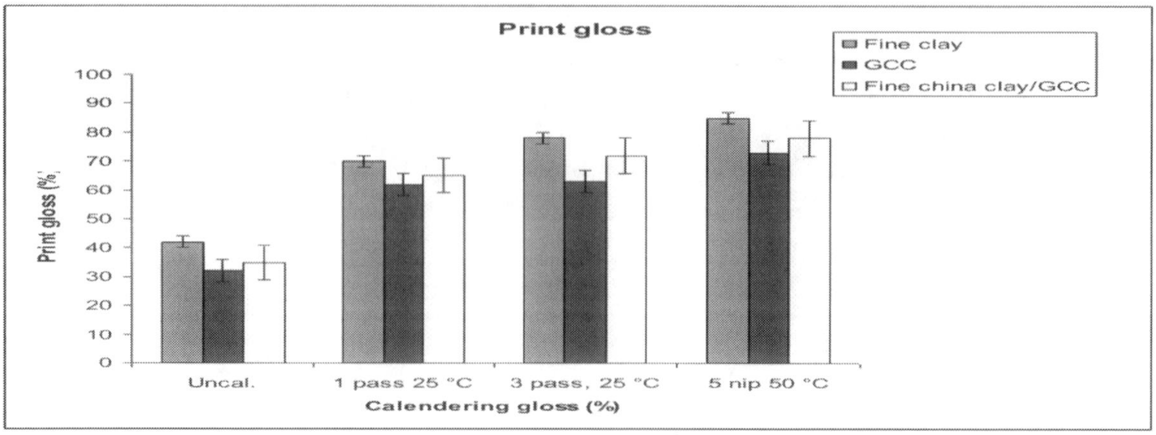

11. Print gloss of calendered /uncalendered blade coated paper with different coating pigments

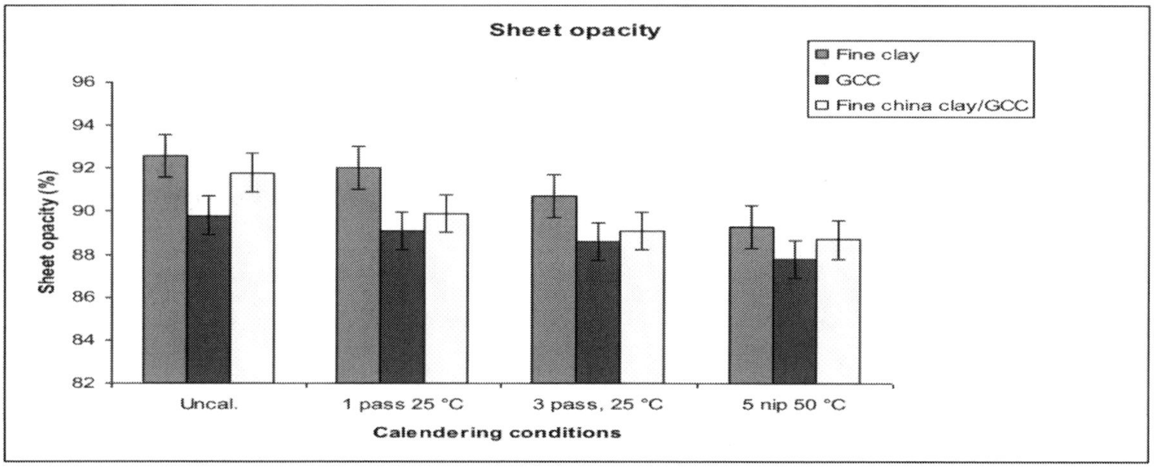

12. Sheet opacity of calendered /uncalendered blade coated paper with different coating pigments

An examination of above plots inferred that at higher temperature, the difference in roughness reduction between clay and carbonate coated papers was small (Fig 9). The papers coated with clay/carbonate blend have values in between fine clay and GCC. The fiber network was bulkier because of roughening and so more prone to deformation during calendering. The difference in paper gloss and print gloss was most likely due to difference in particle size and shape and only partly due to the difference in calendering behaviour of the bulk structure of the paper. (Fig 10 & 11)

The opacity of both fine clay and GCC coated paper decreased as the consequent increase in gloss. The reduction in subsequent nip was more in case of fine clay/GCC blend coated sheets as compared to fine clay and GCC coated sheets alone (Fig. 12).

Calenderability and gloss of paper

It is obvious from above that gloss is an important optical property of coated paper that affects end use performance. The effect of calendering on gloss is quantified using delta (Δ) gloss i.e. difference between calendered gloss G_i and uncalendered gloss G_o. However, when comparing the gloss improvement for different coatings Δ gloss is meaningful, if the initial uncalendered gloss values of coating are identical. The potential increase in gloss is larger for coatings with low uncalendered gloss as compared to coatings with high uncalendered gloss. With different coating processes, applying coating formulations of different pigments and binders quantification using Δ gloss makes the problem still more complicated, so the effectiveness of calendering is described by a quantity given by $(G_i - G_0)/ (100 - G_0)$, where $(G_i - G_0)$ is Δ gloss and $(100 - G_0)$ is the potential for gloss improvement. The quantity $(G_i - G_0) / (100 - G_0)$, ranges from zero where calendering has no effect on gloss to 1, when G_i has increased to 100.

In this study calenderability κ is defined as

$\kappa = (100 - G_0) / (100 - G_i)$ where G_0 = uncalendered gloss & G_i = calendered gloss after i nips

The value of κ ranges from 1 where calendering has no effect on gloss to infinity. The quantity $(G_i - G_0)/ (100 - G_0)$ is related to κ by $1 - (1/ \kappa)$

i.e. $(G_i - G_0)/ (100 - G_0) = 1 - (1/ \kappa)$

Thus the effect of blade coating and film application coating process on gloss and print gloss, gloss increment due to calendering temperature with respect to calenderability was studied. The results are given in figures 13 to 18. The effect of different coating pigments on gloss and print gloss of coated papers was also studied with respect to calenderability. The results are given in Table II. The values of calenderability shows increasing trend with an increase in the gloss values due to increased number of nips.

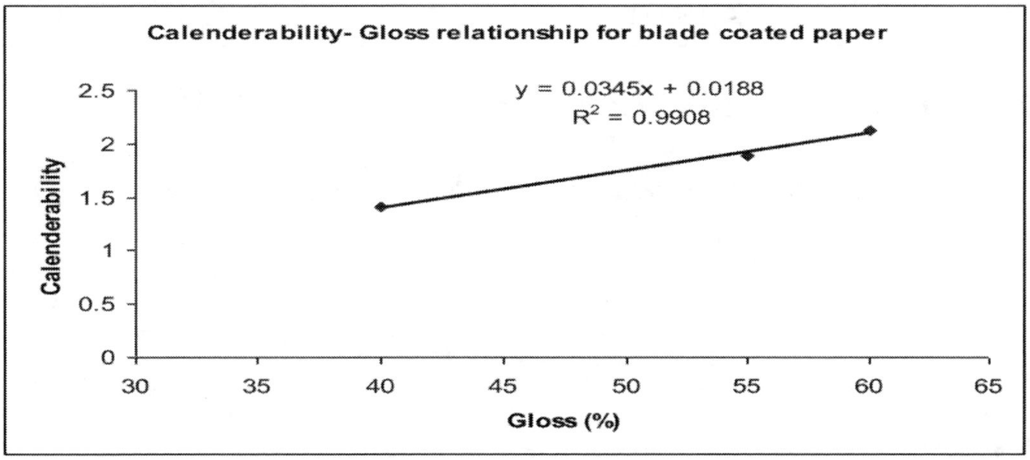

13. Calenderability- gloss relationship for blade coated paper

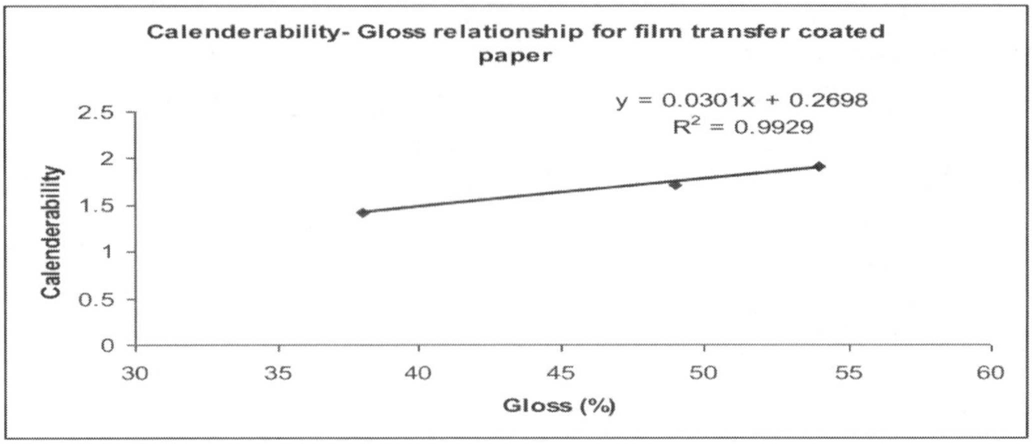

14. Calenderability- gloss relationship for film transfer coated paper

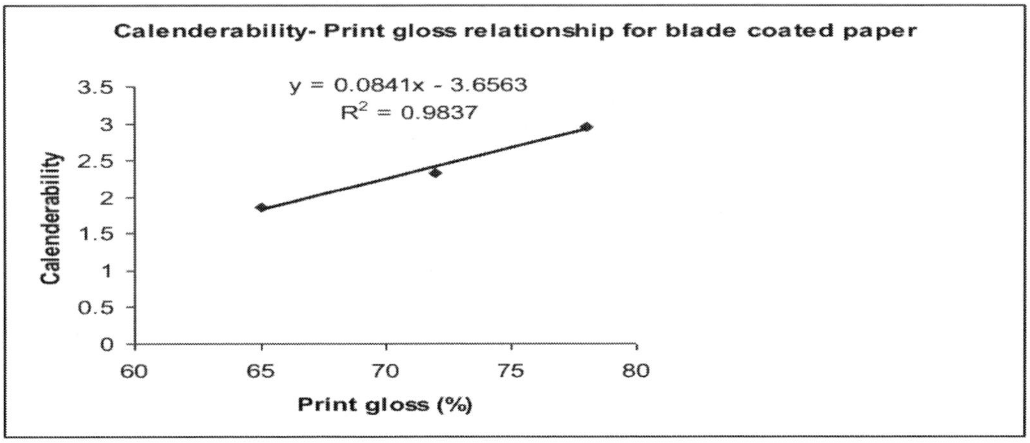

15. Calenderability- print gloss relationship for blade coated papers

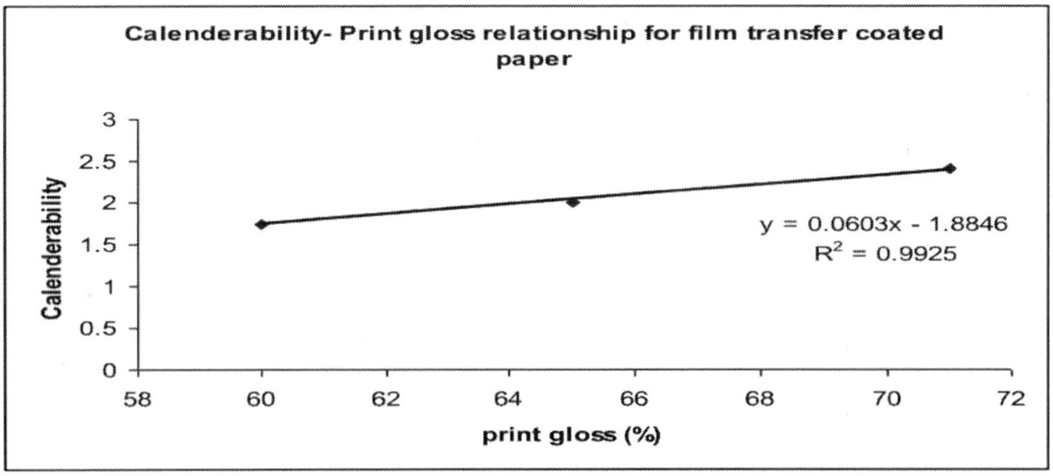

16. Calenderability- print gloss relationship for film transfer coated papers

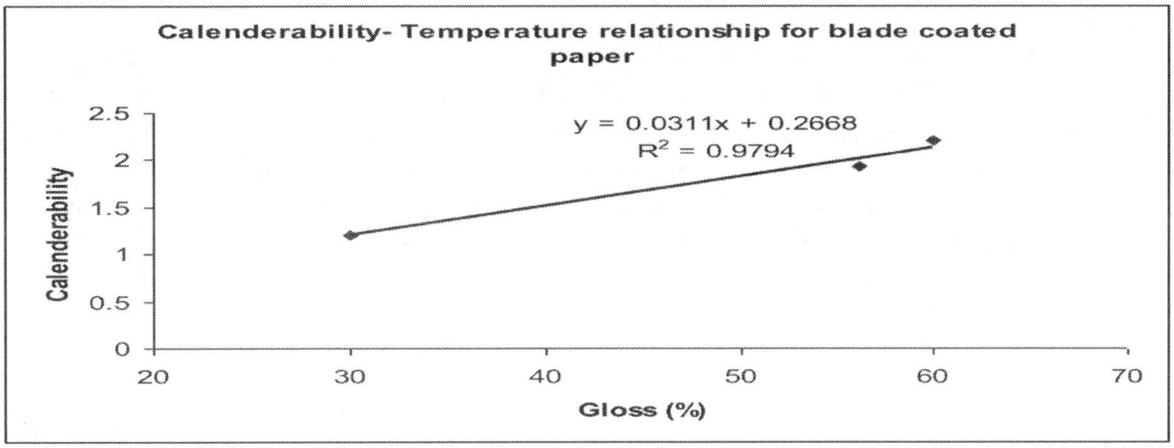

17. Calenderability- Temperature relationship for gloss development for blade coated papers

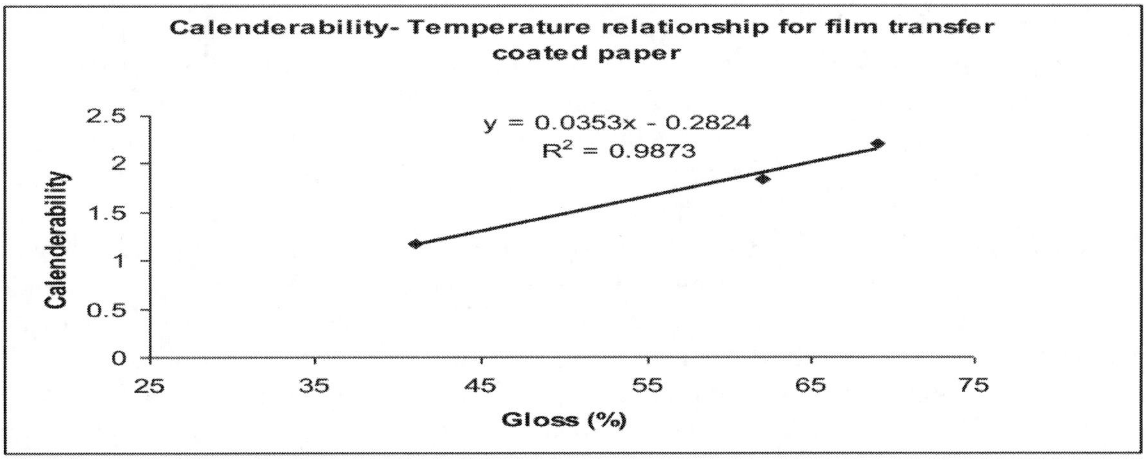

18. Calenderability- Temperature relationship for gloss development film transfer coated papers

II: Calenderability of paper coating – effect of coating pigments

	Sheet Gloss									
	Uncal.	1 pass	3 pass	5 nip 50 °C	Δ 1 pass	Δ 3 pass	Δ 5 pass	κ after 1 pass	K after 3 pass	K after 5 pass
Fine clay	20	45	59	64	25	39	44	1.45	1.6	2.22
GCC	12	37	48	52	25	36	40	1.39	1.39	1.83
Fine china clay/GCC	15	40	55	60	25	40	45	1.41	1.89	2.12
Print gloss (%)										
Fine clay	42	70	78	85	28	36	43	1.93	2.63	3.86
GCC	32	62	63	73	30	31	41	1.79	1.84	2.5
Fine china clay/GCC	35	65	72	78	30	37	43	1.85	2.32	2.95

Δ = difference between final value and initial value

The figures 13 to 18 and data given in Table II reveal that the value of calenderability increases as the value of gloss increases with increment in number of passes or nips. The same is true for print gloss also. This is well in accordance of results shown earlier in this work where both gloss and print gloss increases with number of calendar nips and calendar temperature. Also the vale of calenderability increases with an increase in calendering temperature increases from 50°C to 150°C as expected. The behaviour of calenderability on different coating pigments reveals that the value of calenderability is more in case of paper coated with china clay in comparison to GCC. For the blend of GCC/Clay the value lies in between two. It also in well accordance earlier findings in this chapter showing higher gloss for fine clay coated paper as compared to GCC. This is in excellent agreement with the results put forward by Kan et. al. [4]

CONCLUSIONS

The blade coating process reduces the fiber and flock scale unevenness more efficiently than film transfer coating process. Blade coating process produces better smooth sheets than film transfer coating process. The improvement in paper gloss and print gloss was found to be more for blade coated paper and that too after the first pass only. This indicates that deformation of blade coated paper occurs more easily in calendering. The decrease in sheet opacity in more gradual in film transfer coating and occurs throughout the calendering nip. The blade coated paper more readily showed improvement in sheet and print gloss as compared to film transfer coated paper with increasing calendar temperature. PPS roughness of blade coated paper shows only a minor change with increasing calendering temperature when different pigments are used in coating formulation. The roughness of GCC coated paper reduces more during subsequent calendering trials. At high temperature difference in roughness reduction between clay and carbonate coated paper becomes small. The reduction in roughness in subsequent calendering nip is more in case of fine clay as compared to GCC and fine clay/GCC blend coated sheets. The concept of calenderability founds to be in excellent correlation with gloss and print gloss values of coated paper and its value increases like gloss value with increasing number of nips. The value of calenderability also founds to be in good correlation with calendaring temperature. The results of calenderability for gloss value of paper coated with formulations of different coating pigments also correlates well and are in well accordance with results found earlier.

References:

1. Singh, S., De, A., Gupta, R., Krishnan, S. V. R and Umashankar, R., "Effect of Blade on Blade Coating on The print Results of High Quality Packaging Boards", *IPPTA convention Issue,* 61-64, (2003).
2. Linnonmaa J., and Trefz M., "*Pigment coating techniques*", Chapter 24 in *Pigment Coating and Surface Sizing of Paper.* Lehtinen E (editor), Fapet Oy, Helsinki , Finland, 415– 486, (2000)
3. Trefz, M. – "Film coated offset grades- theoretical aspects and practical experience", *Tappi Coating conference proceedings,* TAPPI PRESS, Atlanta, p. 435 (1995).
4. Kan, Charles S., Kim Linda H, Lee, Do I. & Gilder, Ronald L. Van, "Viscoelastic properties of paper coatings: Relationship between coating structure –Viscoelasticity and end use performance", *Tappi Journal*, 80 (5): 191 – 201,(May 1997).

(714a) The Role of Cellulose Accessibility On Enzymatic Saccharification of Lignocelluloses

Junyong Zhu, USDA Forest Products Laboratory, Madison, WI, Zhiguang Zhu, Biological Systems Engineering, Virginia Polytechnic Institute and State University (Virginia Tech), Blacksburg, VA, Y.-H. Percival Zhang, Biological Systems Engineering Department, Virginia Polytechnic Institute and State University, Blacksburg, VA, Yonghao Ni, 3Limerick Pulp and paper Center, University of New Brunswick, New Brunswick, Canada, ZiBin He, Limerick Pulp and Paper Centre, Virginia Polytechnic Institute and State University (Virginia Tech), Blacksburg, VA and Xiaolin Luo, South China University of Technology, Guangzhou, China

So much research has been dedicated to understand the effects of various factors on enzymatic hydrolysis of lignocelluloses. It has been understood that removal of hemicelluloses is very important to improve lignocellulosic substrate digestibility. It has also been identified that improving enzyme accessibility to cellulose is more important than delignification to enzymatic saccharification. This study is attempted to develop a comprehensive understanding of various factors on enzymatic hydrolysis of lignocelluloses by taking a progressive approach. In the first phase, a set of lignocellulosic substrates were produced by drying to different degrees from the same never dried sample. Therefore, these dried substrates have exactly the same chemical composition and structure. However, drying induced fiber hornification produces irreversible change of cell wall structure when rewetting, i.e., fibrils are tightly packed together through hydrogen bonding to produce permanent fiber pore shrinkage and loss of pores in cell wall. As a result, the enzyme accessibilities of these samples were significantly reduced comparing with the never- dried sample and varied with the degree of drying. This set of samples is an excellent candidate to verify the effect of enzyme accessibility on enzymatic hydrolysis. In the second phase, two sets of samples, one produced from fractionation of the same sample, and another from disk milling of the same chemically pretreated biomass to different degrees, were used to determine the effect of substrate size/specific surface or degree of size reduction on enzymatic hydrolysis. Finally, different pretreatment methods were used to explain the effect of chemical composition on enzymatic hydrolysis. Both enzymatic hydrolyses and measurements of cellulase accessiblity to cellulose (CAC) were carried out to quantify the contributions of various factors to enzymatic hydrolysis of lignocelluloses.

See more of this Session: Recalcitrance of Woody Biomass
See more of this Group/Topical: International Congress on Energy 2011

(714b) Structural Studies On Recrystallized Ionic-Liquid (IL) Treated Lignocellulosic Biomass

Indira Samayam[1], B. Leif Hanson[2], Paul Langan[3], Christopher J. Barr[1] and **Constance Schall**[1], (1) Chemical and Environmental Engineering, University of Toledo, Toledo, OH, (2)Instrumentation Center, The University of Toledo, Toledo, OH, (3)Biosciences Division - Los Alamos National Laboratory, Los Alamos, NM

Terrestrial lignocellulosic biomass is composed of three major components; cellulose, a linear polymer of glucose (a hexose sugar); hemicellulose, a complex branched polymer of xylose (a pentose sugar) and other sugar derivatives; and lignin, a polyphenyl propanoid macromolecular assembly that is covalently cross-linked to hemicellulose. These components can serve as a source of carbon based feedstock for fuel and chemical production in much the same way that crude oil serves as the carbon feedstock in petrochemical refineries.

The deconstruction of lignocellulosic biomass into simple sugars constitutes a core barrier for producing value added products from the sugar platform. The cellulosic component has a crystalline and oriented fibrillar structure in biomass, whereas the hemicellulosic and lignin components have less ordered structures that encrust the cellulose fibers. The crystalline structure of native cellulose fibers, called cellulose I, can be a major impediment to its hydrolysis to monomeric sugars. Pretreatment can disorder the crystalline cellulosic component and result in structural changes in the cellulose fibers. These structural changes in cellulose fiber can be achieved by modifying the pretreatment conditions. Depending upon the type of pretreatment, the conversion of cellulose fibers to other crystal forms, such as cellulose II, cellulose III_{II} and cellulose III_{I}, or amorphous forms, can greatly improve their susceptibility to hydrolysis. To investigate the effects of varying the parameters of IL pretreatment on the structure of their cellulosic components, X-ray powder diffraction (XRD) and X-ray fiber diffraction from biomass samples was used with unoriented and partially (fiber) orientated cellulosic components . A comparative study on enzymatic hydrolysis and sugar analysis related to the observed changes in cellulose structure to biomass digestibility was also performed. The improvement in hydrolysis is hypothesized to be primarily due to structural changes in the cellulose constituent.

See more of this Session: Recalcitrance of Woody Biomass
See more of this Group/Topical: International Congress on Energy 2011

(714c) Kinetics and Mechanism of Cellulase Hydrolysis of Nanocrystalline Cellulose

Xiao Zhang, Chemical Engineering and Bioengineering, Washington State University, Richland, WA and Xiaohui Ju, Washington State University, Richland, WA

Kinetics and Mechanism of Cellulase Hydrolysis of Nanocrystalline Cellulose

Xiaohui Ju, Xiao Zhang

School of Chemical Engineering and Bioengineering, Centre for Bioproducts and Bioenergy, Washington State University,

Abstract

Decrystallization of cellulose is a major bottleneck in enzymatic hydrolysis of lignocellulosic biomass. The kinetic and mechanism of enzyme degradation of nanocrystalline cellulose prepared from bleached kraft pulp were investigated. The efficiency of enzymatic degradation of nanocrystalline cellulose by three commercial enzymes was determined. All these cellulase preparations demonstrated the ability to degrade nanocrystalline cellulose. The mechanism of enzyme decrystallization of cellulose was investigated by XRD and NMR. This presentation will discuss some of the new findings in cellulose crystal structural changes during enzymatic hydrolysis.

See more of this Session: Recalcitrance of Woody Biomass
See more of this Group/Topical: International Congress on Energy 2011

(714d) Incremental Disassembling of Woody Biomass and Production of Biochemicals

Shijie Liu, Paper and Bioprocess Engineering, SUNY College of Environmental Science and Forestry, Syracuse, NY

Woody biomass is a reliable renewable source of raw materials for energy and chemicals. Woody biomass is structurally composed of cellulose, lignin, hemicelluloses and other minor components. The compositions vary from species to species. The complexity / heterogeneity of woody biomass presents challenges to conversion technologies, which is perceived as recalcitrance. Pretreatment and detoxification have been two unfavorable but widely discussed processes todate. At SUNY ESF, a strategy of incremental disassembling of woody biomass is adopted in which hemicelluloses and extractives are separated first through hot-water extraction, cellulose and lignin are next. Hot-water extraction uses water as solvent without any other added chemicals. Acetyl groups in the woody biomass dissolve into the liquor and serves as catalyst. In each step of the separation, high value applications are examined. For example, hemicelluloses are hydrolyzed and fractionated via membrane for xylo-oligomers, xylose, acetic acid, methanol, and aromatics, etc. The residual solid mass containing mostly cellulose and lignin has wide applications: fiber board, wood fuel pellet, pulp and paper, and sugars / aromatics. Incremental disassembling is emphasized for value preservation of various components. A clever integration of the processes makes the conventional pretreatment and detoxification processes obsolete. The sugar stream produced at SUNY ESF has been successfully fermented to ethanol, butanol, lactic acid, and PHA. There is no deliberate detoxification that generates waste byproducts.

See more of this Session: Recalcitrance of Woody Biomass
See more of this Group/Topical: International Congress on Energy 2011

(714e) Effect of Steam Explosion On Degradability and Accessiblity of Loblolly Pine In Bioethanol Applications

Yuzhi Kang, Sang Beom Kim, Prabuddha Bansal, Dr Matthew Realff and Andreas S. Bommarius, School of Chemical and Biomolecular Engineering, Georgia Institute of Technology, Atlanta, GA

Abstract

Among the various biomass pretreatment methods, steam explosion is one of the most widely used due to its cost-effective disruption of the plant structure and hence enhanced enzymatic accessibility. The operational parameters including temperature, residence time and acid concentration have pronounced effect on crystallinity (CrI) of the feedstock loblolly pine. CrI was determined from X-ray diffraction pattern with the method of least squares and it was found to decrease as pretreatment severity increase.

Three different substrates, Avicel, phosphoric acid swollen cellulose (PASC), and SO_2 steam exploded Loblolly pine (SELP), were subjected to enzymatic hydrolysis and subsequent adsorption study. Both SELP samples (2.9-419-5 and 5-417-5) were found more readily degraded by the CelluclastÒ enzyme system than Avicel. Despite the complexity of both substrates and enzyme mixtures, the SELP adsorption isotherm can be fitted to a Langmuir isotherm with $R^2 > 0.99$. The saturation point was reached at ~6.24 mg/ml total enzyme loading. Adsorption data was correlated with initial rates and a linear relationship between initial rate and adsorption amount was confirmed.

See more of this Session: Recalcitrance of Woody Biomass
See more of this Group/Topical: International Congress on Energy 2011

CHANGE OF STRUCTURE OF CORNCOB WITH DILUTE ACID AND LIME PRETREATMENT AND A POTENTIAL PROCESS FOR XYLOSE CO-PRODUCTION OF CELLULOSIC ETHANOL

Ji-hong Li[1], *Shi-Zhong Li*[1], *Zhi-pei Yan*[1], *Qing Li*[2], *Liang-cai Peng*[3]

[1]*Institute of New Energy Technology, Tsinghua University, Beijing, China*

[2]*Colledge of Science, Huazhong Agricultural University, Wuhan, China*

[3]*Colledge of Plant Science & Technology, Huazhong Agricultural University, Wuhan, China*

Abstract

Degree of polymerization (DP) of cellulose is considered an important factor to effect on the conversion of Lignocellulose. The higher DP of cellulose, the more difficult conversion of lignocellulose is. In this study, corncobs were pretreated with dilute acid process for production of xylose, and then the residues were pretreated with lime. The enzymatic hydrolysis and the structures involved the composition, DP and crystallinity index of these two types of feedstock were determined. Comparison of structures resulted from these two pretreatment revealed that DP was not the crucial factor to determined efficiency of enzymatic hydrolysis. Lime pretreatment could improve the conversion of cellulose although the residues with lime pretreatment had higher DP and percentage of crystallized cellulose. Moreover, the results also confirmed lignin was not the key factors for enzymatic hydrolysis of lignocellulose because the residues with dilute acid pretreatment had the similar percentages of lignin to that with lime pretreatment. The crucial effect of lime pretreatment was to remove the inhibitor for cellulase and increase the percentages of cellulose. With obtained xylose, the residues with further lime pretreatment had expected efficient enzymatic hydrolysis of cellulose. So this process could be a potential process for xylose co-production of cellulosic ethanol to decrease the cost of cellulosic ethanol.

(714g) Green and High Efficient Enzyme Hydrolysis Process for Sweet Sorghum Bagasse

Xinshu Zhuang, Wen Wang, Qiang Yu, Wei Qi, Qiong Wang, Xuesong Tan, Yu Zhang and Zhenhong Yuan, Guangzhou Institute of Energy Conversion,CAS, Guangzhou, China

In order to improve enzymatic hydrolysis efficiency, several approaches including liquid hot water (LHW) pretreatment, adding surfactant, magnetic stirring and fed-batch hydrolysis were investigated, in which Sweet sorghum bagasse (SSB) was chosen as the feedstock. The experiment results showed that with the LHW pretreatment, the SSB conversion ratio reached 35.91% , nearly 75% xylan and 33% lignin were removed. Using the LHW-pretreated SSB, glycan conversion ratio raised from 16.6% to 37.9% by adding surfactant Tween80 with the optimum dosage of 0.175ml/g DM. The virulence test showed that Tween80 nearly had no influence to the growing of ethanol fermentation yeast-*Saccharomyces cerevisiae*. Comparing stirring and shaking mode for the enzyme hydrolysis, the glycan conversion rate was 57.3% in stirring mode, higher than 44.4% obtained from shaking mode. The image of Scanning Electron Microscope (SEM) showed that LHW pretreatment could destroy the tight structure of SSB and the comminution of pretreated material could smaller the particles to enhance the hydrolysis process. Finally the LHW-pretreated SSB, Tween80 and stirring mode were applied to fed-batch process, the concentration of glucose, cellobiose and xylose reached 88.8g/l, 30.7g/l and 18.9g/l respectively, and the glycan conversion ratio was 89.3% under the optimum condition. These research works lay the foundation for the construction of a green and high efficient sugar platform to facilitate the cellulosic ethanol development.

These research work are financially supported by the National Key Technology R&D Program (2011BAD22B01), National High Technology Research and Development Program of China (2010AA101606) and the Knowledge Innovation Project of the Chinese Academy of Sciences (KSCX1-YW-11-A3, KSCX2-YW-G-063)

[1]Corresponding author. yuanzh@ms.giec.ac.cn (Z. Yuan)

See more of this Session: Recalcitrance of Woody Biomass
See more of this Group/Topical: International Congress on Energy 2011

AUTHOR INDEX

AUTHOR INDEX

AUTHOR INDEX

AUTHOR INDEX